ADVANCES IN METAL AND SEMICONDUCTOR CLUSTERS

Volume 4 • 1998

CLUSTER MATERIALS

ADVANCES IN METAL AND SEMICONDUCTOR CLUSTERS

CLUSTER MATERIALS

Editor: MICHAEL A. DUNCAN
Department of Chemistry
University of Georgia

VOLUME 4 • 1998

Stamford, Connecticut *London, England*

100 Prospect Street
Stamford, Connecticut 06904-0811

JAI PRESS LTD.
38 Tavistock Street
Covent Garden
London WC2E 7PB
England

ISBN: 0-7623-0058-2

ISSN: 1075-1629

Transferred to digital printing 2006
Printed and bound by Antony Rowe Ltd, Eastbourne

CONTENTS

LIST OF CONTRIBUTORS

A. P. Alivisatos	Department of Chemistry University of California Berkeley, California
U. Banin	Department of Chemistry University of California Berkeley, California
R. N. Compton	Departments of Chemistry and Physics The University of Tennessee Knoxville, Tennessee
J. M. Cowley	Department of Physics and Astronomy Arizona State University Tempe, Arizona
M. Samy El-Shall	Department of Chemistry Virginia Commonwealth University Richmond, Virginia
Y. Feldman	Department of Materials and Interfaces Weizmann Institute Rehovot, Israel
M. Gartz	Physikalisches Institut der RWTH Technische Universität Aachen Aachen, Germany
A. A. Guzelian	Department of Chemistry University of California Berkeley, California
R. E. Haufler	Comstock, Inc. Oak Ridge, Tennessee

R. L. Hettich Oak Ridge National Laboratory
Oak Ridge, Tennessee

A. Hilger Physikalisches Institut der RWTH
Technische Universität Aachen
Aachen, Germany

M. Homyonfer Department of Materials and Interfaces
Weizmann Institute
Rehovot, Israel

H. Hövel Experimentelle Physik
Universität Dortmund
Dortmund, Germany

Rajiv K. Kalia Department of Physics and Computer Science
Louisiana State University
Baton Rouge, Louisiana

U. Kreibig Physikalisches Institut der RWTH
Technische Universität Aachen
Aachen, Germany

Amer Lahamer Department of Physics
Berea College
Berea, Kentucky

J. C. Lee Department of Chemistry
University of California
Berkeley, California

Shoutian Li Department of Chemistry
Virginia Commonwealth University
Richmond, Virginia

Robert G. A. R. Maclagen Department of Chemistry
University of Canterbury
Christchurch, New Zealand

Sara Majetich Department of Physics
Carnegie Mellon University
Pittsburgh, Pennsylvania

Aiichiro Nakano	Department of Physics and Computer Science Louisiana State University Baton Rouge, Louisiana
Andrey Omeltchenko	Department of Physics and Computer Science Louisiana State University Baton Rouge, Louisiana
Gustavo E. Scuseria	Department of Chemistry and Rice Quantum Institute Rice University Houston, Texas
Hisanori Shinohara	Department of Chemistry Nagoya University Nagoya, Japan
R. Tenne	Department of Materials and Interfaces Weizmann Institute Rehovot, Israel
Kenji Tsuruta	Department of Physics and Computer Science Louisiana State University Baton Rouge, Louisiana
Priya Vashishta	Department of Physics and Computer Science Louisiana State University Baton Rouge, Louisiana
Lai-Sheng Wang	Department of Physics Washington State University Richland, Washington
Hongbin Wu	Department of Physics Washington State University Richland, Washington
Z. C. Ying	Department of Physics New Mexico State University Las Cruces, New Mexico

PREFACE

In Volume 4 of this series, we focus on *Cluster Materials* and the properties of clusters which determine their potential applications as new materials. Metal and semiconductor clusters have been proposed as precursors for materials or as actual materials since the earliest days of cluster research. In the last few years, a variety of techniques have made it possible to produce clusters in sizes varying from a few atoms up to several thousand atoms. While some measurements are performed in the gas phase on non-isolated clusters, many cluster materials can now be isolated in macroscopic quantities and more convenient studies of their properties become possible.

In this issue, we focus on the measurement of optical, electronic, magnetic, chemical, and mechanical properties of clusters or of cluster assemblies. All of these properties must fall into acceptable ranges of behavior before useful materials composed of clusters can be put into practical applications. However, as evidenced by the various work described here, the realization of practical products based on cluster materials seems to be approaching rapidly.

Michael A. Duncan
Editor

PREPARATION AND PROPERTIES OF INAS AND INP NANOCRYSTALS

A. A. Guzelian, U. Banin, J. C. Lee, and A. P. Alivisatos

Advances in Metal and Semiconductor Clusters
Volume 4, pages 1–34

ISBN: 0-7623-0058-2

ABSTRACT

Recently the selection of semiconductor nanocrystals available as high quality samples appropriate for optical and other size dependent studies has been extended to include the III-V compounds, InAs and InP. The particles are synthesized through a solution-phase chemical route that is in some ways similar to the methods developed for the more widely studied CdSe nanocrystals, but has significant differences due to the different chemistries of the systems. This chapter details some of these synthetic issues as well as the specific syntheses of both InAs and InP nanocrystals and then examines some of the structural and optical studies that have begun to be carried out on these systems. In addition, the quantum confinement effects of these systems are briefly compared with the II-VI systems through the shifts in their optical absorption spectra.

I. INTRODUCTION

Semiconductor nanocrystals consist of an inorganic crystalline core surrounded by coordinating organic ligands which serve to terminate and passivate the surface of the particle. The organic functionalities on these ligands impart solubility to the clusters and also allow for the further chemical manipulation of the particles. A variety of semiconductors can be produced as nanocrystals with clusters in the range of 20 to 200 Å in diameter. Particles of this size fall within the strong confinement regime for many semiconductors where the actual size of the particle determines the allowed electronic energy levels of the structure and thus the optical and electronic properties of the material. Examples of these quantum confinement effects include the shift of the band gap to higher energy with smaller particle size and concentration of oscillator strength into just a few transitions.[1]

Traditional semiconductor processing techniques such as lithography have been limited in their ability to fabricate structures with dimensions below a few hundred angstroms. Over the past decade, solution-phase colloidal syntheses and vapor-phase deposition into template structures such as glasses have been successfully applied to access these smaller sizes. Such chemical methods in general have the advantage of taking a "bottom up" approach where the nanocrystals are formed from molecular precursors. There is therefore no fundamental lower limit to the size of the particle as the transition from the molecular to solid state regime is entirely encompassed. In this chapter we will discuss colloidal solution phase syntheses.

Colloidal chemical techniques have several advantages over other chemical methods. First, when the timescales of nucleation and growth can be precisely controlled, more monodisperse samples may be obtained. This is crucial to the study of size-dependent properties and will be discussed in more detail below. Second, colloidally prepared systems remain chemically active structures after the synthesis is complete, allowing for a high degree of control over the particles through a variety of manipulations. Examples include the formation of optically clear solutions or polymer films, attachment to various surfaces[2] or other nanocrystals,[3] fabrication into LEDs,[4,5] integration into electrical circuits,[6] and the affixing of the nanocrystals to template structures.[7] This activity results primarily from the organic molecules attached to the particle surface. Through the substitution of different molecules with varying functionalities, properties such as solubility can be controlled and the nanocrystals may be manipulated. These molecules may also influence the fundamental properties of the nanocrystals through their effect on the surface of the particles. For example, luminescence from the nanocrystal is strongly dependent on the electronic passivation of the nanocrystal surface, and in some cases the nature of the derivatization of the surface with passivating ligands strongly affects this luminescence. This will be discussed further in a later section.

CdS and CdSe nanocrystals have been the most thoroughly studied systems initially due to the relative synthetic ease of preparing such systems and now due to the extremely high quality of the nanocrystal samples.[8,9] At present, CdSe is the prototypical nanocrystal system with a number of excellent sample characteristics. CdSe nanocrystals can be prepared in sizes ranging from 15 to about 120 Å in diameter and they are highly crystalline with each particle consisting of a single crystalline domain. The surface of the particles can be derivatized with a wide range of organic molecules and recently also with other semiconductor materials to give an inorganic passivating layer.[10–12] This ability to grow layers of different semiconductors provides the opportunity to develop more complex nanocrystal structures.[13–15] An important characteristic of the CdSe nanocrystals is the ability to prepare monodisperse samples. Size distributions with variations of only a few per cent in diameter corresponding to about one atomic layer are possible. Narrow size distributions increase the ability to tune size-dependent properties as inhomogeneous effects are reduced resulting in, for example, narrower and more resolved spectral features or more well-defined phase transitions.

Other compound semiconductor materials which are generally thought of as more technologically important, such as the group III-V materials, have lagged behind CdSe in their development in nanocrystal systems due to synthetic limitations. This chemistry as relating to nanocrystal formation is further discussed in a later section, but most generally, the covalent nature of these materials leads to more complex reaction mechanisms with a resulting loss of control over the timing and sequence of nanocrystal formation. This chapter will concentrate on recent progress in developing high-quality nanocrystals of III-V materials, in particular InP[16–18] and InAs.[19]

There are three major reasons for wanting to expand nanocrystal systems to include semiconductors besides II-VI materials. Most generally, nanocrystals of materials with a range of bulk properties will allow for a more thorough understanding of quantum confinement properties by identifying those properties that are controlled by size and those that are controlled by material characteristics. By distinguishing these effects, conclusions can be made regarding the validity of applying bulk properties such as dielectric constants or carrier mobilities to materials in this size regime. For instance, to what extent are carriers confined to the same band structure as in the corresponding bulk material? These questions are important to the second major reason for wanting to develop new nanocrystal materials, in this case particularly III-V materials. Based on bulk material characteristics, III-V materials should exhibit more pronounced quantum confinement effects compared to II-VI or other more ionic systems. The reasoning behind this expectation is based primarily on the effective mass approximation[20] which describes the extent to which carriers are delocalized over a lattice and the effect of confining those carriers to an unusually small volume. To get a sense of the variation in the delocalization of carriers in the bulk, consider that for InAs an exciton delocalizes over 680 Å, while in InP the bulk exciton diameter is 150 Å and in CdSe it is only 70 Å. Thus, one would expect that for comparably sized nanocrystals of materials with such different properties, the III-V nanocrystals would exhibit more pronounced quantum confinement effects with the carriers being more confined compared with the corresponding bulk system. Lastly, the development of III-V nanocrystals will allow many interesting comparisons with other types of quantum confined structures including quantum wells, wires, and larger dots as these systems are most highly developed in III-V materials.

This chapter will discuss recent progress in the development of InAs and InP nanocrystals, mentioning important contributions by other researchers and then concentrating on work in our laboratory. First synthetic issues will be discussed and then the characterization of the materials will be described. Structurally the nanocrystals are characterized using powder X-ray diffraction (XRD), transmission electron microscopy (TEM), and X-ray photoelectron spectroscopy (XPS). Optical studies also will be described including absorption and luminescence spectroscopy as well as transient holeburning experiments.

II. PREPARATION OF III-V SEMICONDUCTOR NANOCRYSTALS

Since some initial success in synthesizing GaAs nanocrystals in 1990 by Olshavsky et al.,[21] several groups have been attempting to prepare III-V nanocrystals, but only very recently has there been significant progress. The work of Olshavsky et al. used a dehalosilylation reaction of $GaCl_3$ and $As(Si(CH_3)_3)_3$ developed by Wells,[22] where quinoline was used as both the solvent and as the surface passivating ligand. These reactions produced nanocrystals with sizes of about 30 to 40 Å in diameter, but the particles suffered from interferences in their optical absorption behavior due to suspected Ga–quinoline complexes or quinoline oligomeric species.[23] These same quinoline species are formed in blank reactions of $GaCl_3$ and quinoline. In the nanocrystal samples much of the interfering species can be removed by washing in appropriate solvents, but due to the fact that quinoline is actually part of the nanocrystal as the surface capping group, it is not clear that the interfering species could ever be completely removed while maintaining the solubility of the nanocrystals. Most subsequent work on the preparation of III-V nanocrystalline materials has been based on this dehalosilylation reaction with variations on the reactants, reaction conditions, and solvents.

Looking first at work with GaAs, Nozik et al.[24] used $Ga(acac)_3$ and $As(Si(CH_3)_3)_3$ with triglyme as the solvent to produce GaAs nanocrystals. However, the absorption spectra do not show a well-defined feature that shifts with particle diameter, indicating a poor size distribution or other problems with nanocrystal quality. Wells et al.[25,26] have also prepared nanocrystalline GaAs as well as GaP using the metathetical reaction of $GaCl_3$ and $(Na/K)_3E$ (E = P, As) in glymes. These reactions produced nanocrystalline material, but there is no separation of sizes, and the optical spectra do not exhibit the characteristic well-defined absorption feature.

Within the last few years, Mićić and Nozik[16,17] have made an important contribution to the synthesis of III-V nanocrystals with the preparation of InP nanocrystals using the dehalosilylation reaction of an uncharacterized $InCl_3$–oxalate complex and $P(Si(CH_3)_3)_3$ with trioctylphosphine oxide (TOPO) serving as the solvent and surface passivating group. These nanocrystals are of high quality and a size- dependent feature is observed in the absorption spectra. They also report the synthesis of nanocrystals of the indirect gap semiconductor GaP and the ternary material $GaInP_2$ using analogous reactions.

In our laboratory we have prepared InP nanocrystals using adaptations of Mićić and Nozik's work and also expanded the synthesis to InAs nanocrystals. In both systems we can isolate discrete, well-passivated, crystalline nanocrystals for a range of sizes which are soluble in a variety of organic solvents.

A. Synthetic Strategy

The synthesis of nanocrystals can be broken down into three general steps for successful colloidal preparations.[27] The first step is the nucleation of the particles, meaning the initial formation of the appropriate semiconductor bond. This creates sites from which the particles proceed to grow. Nucleation is crucial for two reasons. First, successful nucleation is an indication of whether the reaction in question is appropriate for the formation of nanocrystals. The mechanism of the reaction must be such that the appropriate semiconductor bonds are formed and maintained in solution, thereby providing appropriate sites for the continued growth of the particles. If the initial reaction produces insoluble intermediate species, there will be no opportunity to grow discrete nanocrystals. The second crucial aspect is the timing of the nucleation event as this controls the monodispersity of the final product. For a monodisperse sample, the nucleation event must be well separated in time from the growth step which generally means that nucleation must occur on a short time scale, on the order of a fraction of a second. This type of nucleation event will be called "instantaneous". In this situation, all nucleation sites will undergo the same period of growth resulting in a narrow size distribution. In contrast, if nucleation is spread out in time such that nucleation and growth overlap, different nucleation sites will undergo growth of varying durations resulting in a large distribution of sizes. Practically speaking, an effective way to achieve "instantaneous" nucleation is to inject the reagents into the solvent at high temperature such that nucleation occurs immediately on contact with the hot solvent.[8,9] In contrast, reactions that are heated slowly from room temperature to high temperature result in nucleation that is poorly defined in time, thus resulting in a broad size distribution.

The second step in a nanocrystal synthesis is the growth of the particles, meaning the formation of a highly crystalline semiconductor core. As with nucleation, intermediate species and the growing particle must stay in solution throughout the process. Growth is completed when the reagents are consumed, although additional high-temperature annealing time may be needed to obtain good crystallinity. Also, with prolonged high-temperature annealing, Ostwald ripening-type processes are possible.[28]

The final element is the passivation of the nanocrystal surface. Surface passivation helps control the growth of the nanocrystal, prevents agglomeration and fusing of particles, and provides for the solubility of the nanocrystals in common solvents. This step cannot be separated in time since solvation is important not only for the nanocrystals, but for the reactants and intermediates as well. In general, surface passivation is achieved by using an organic molecule to coordinate or bond to the nanocrystal surface. Most commonly, this organic molecule is the solvent for the reaction. This strategy benefits from having the capping molecule in huge excess and from having a common solvation medium for the reactants, intermediates, and nanocrystals. The drawback of using the solvent as the capping group is that there is limited control over the passivation process since the concentration and amount of the cap are not variable; in all cases they are in huge excess.

Chemically, the capping group contains an electron-rich donating group, such as phosphine oxides, phosphines, amines, or thiols, and behaves as a Lewis base which coordinates to the electron-poor Lewis acid-like metal of the semiconductor such as cadmium or indium. This coordination passivates the dangling orbitals at the nanocrystal surface preventing further growth or agglomeration. The other end of the ligand imparts solubility to the nanocrystal by giving the particle a hydrophilic or hydrophobic surface. For example, the alkyl groups of trioctylphosphine oxide or trioctylphosphine result in nanocrystals that are soluble in relatively nonpolar solvents, while using ligands such as mercaptobenzoic acid results in particles that are soluble in polar solvents like methanol.

Surface passivation is not only a step in the initial nanocrystal synthesis as it can also be subsequently used to further derivatize the surface of the particle. Many of the properties of nanocrystals, such as photoluminescence, are sensitive to the surface passivation and manipulation of the surface capping groups can be used to tune these properties.[18] Due to the relatively weak coordination bonds of most surface capping groups, these molecules can be easily exchanged after the initial synthesis is complete, allowing for great flexibility in what is on the nanocrystal surface.

B. Mechanistic Considerations

While the precise chemical mechanisms of nanocrystal syntheses have not been investigated, several general comments can be made regarding the reaction mechanism's effect on the nanocrystal product. To accomplish the ideal nucleation event discussed above, the reaction needs to be rapid and complete. These characteristics point towards simple mechanisms using reagents with low barriers to reaction. For example, at one extreme is the aqueous or micelle synthesis of CdS or CdSe where cadmium salts such as $CdCl_4$ are the cadmium source and H_2S or H_2Se is the group VI source.[14] In these types of reactions the metal ion is available for reaction with a very simple group VI source. These reactions are rapid and can even occur at room temperature. On the downside, the reactions are so facile that it is difficult to control and isolate nucleation.

The most successful CdSe nanocrystal synthesis,[8] $(CH_3)_2Cd$ + Se[TOP] (TOP = tricoctylphosphine) in TOPO, provides a more ideal combination of rapid and complete reaction along with the ability to control the nucleation event. The initial reactants can be combined at room temperature with minimal reaction, allowing for a homogeneous source of both precursors. Injection into TOPO at high temperature provides the necessary energy to initiate the reaction which then proceeds quickly and completely providing for well-controlled nucleation and subsequent growth.

The chemistry of group III-V materials has proven to be more difficult as far as designing a system with the attributes of both rapid and complete yet controlled nucleation. In contrast to cadmium, the group III metals are too reactive to be used as ionic sources since they would react too strongly with the solvent medium. Attempts to develop reactions involving group III alkyls analogous to $(CH_3)_2Cd$ also appear to be too reactive since they tend to proceed uncontrollably to the bulk phase regardless of the capping group.

A general strategy to reduce the reactivity of these materials would be to attach ligands to the group III and group V centers and then react these molecules, presumably in a more controlled fashion. Over the past several years, there has been much interest in novel routes to bulk III-V materials,[29–33] particularly GaAs, due to the difficult and hazardous nature of traditional reactions involving $Ga(CH_3)_3$ and AsH_3. Substituted gallium and arsenic compounds were developed in the hope of reducing reactant toxicity, eliminating gaseous reagents, and lowering reaction temperatures. As mentioned, Wells and coworkers[22] developed a dehalosilylation reaction using $GaCl_3 + As(Si(CH_3)_3)_3$ that produced GaAs with the elimination of CH_3SiCl. Barron et al.[34] used an analogous reaction to produce bulk InP. These reactions are driven by the formation of the CH_3SiCl bond in combination with the good lability of the $Si(CH_3)_3$ ligand.

However, in all examples the barriers created by the ligands and the mechanistic pathway required to eliminate them results in the need for high reaction temperatures and prevents nucleation of the nanocrystals from occurring at a well-defined moment in time. As a result, while with the proper surface passivation nanocrystals can be prepared, the failure to separate nucleation and growth processes results in the formation of a wide size distribution of particles. These issues are discussed further in the next section.

C. Synthesis of InAs and InP Nanocrystals

The synthesis of InP and InAs are analogous, so InP will be treated in detail and specific differences for InAs will be noted. All steps of the reaction and the manipulation of the reagents are carried out under argon or in a dry box. The nanocrystals produced are crystalline and highly soluble in a variety of organic solvents including hexanes, toluene, and pyridine. The size distribution of the particles as produced ranges from 20 to 60 Å in diameter for InP and up to 70 Å for InAs. This is too broad for the investigation of size-dependent phenomena; however, applying size-selective precipitation techniques[8,18] produces relatively narrow size distributions of particles over the entire size range produced by the reaction. In this way, over 30 size distributions, distinguishable by optical and structural characterization techniques, may be generated from a single reaction product. The surface of the nanocrystals can be further derivatized with a variety of ligands including amines, thiols, and other phosphines.

The reaction sequence described here is:

$$InCl_3 + TOPO \xrightarrow{100\ ^\circ C,\ 12\ h} InCl_3\text{-(TOPO or TOP) complex} \tag{1}$$

$$InCl_3\text{-TOPO complex} + P(Si(CH_3)_3)_3 \xrightarrow{265^\circ C,\ 1\text{–}3\ days} InP\ (TOPO/TOP\ capped) + 3Si(CH_3)_3Cl \tag{2}$$

$$InP\ (TOPO\ capped) + surfactant \xrightarrow[+\ RNH_2,\ RSH,\ or\ RPH_2]{100\ ^\circ C} InP\ (TOPO\ and\ surfactant\ capped) \tag{3}$$

In step (1), the $InCl_3$ forms a complex with TOPO or TOP. The majority of InP nanocrystals have been synthesized in TOPO, while for InAs using TOP narrower size distributions result. While this complex has not been isolated, the literature has several examples of $InCl_3$–phosphine oxide complexes[35] and an interaction between the Lewis acid-like $InCl_3$ and the strongly donating phosphine or phosphine oxide ligand would be expected. This complex is necessary for the success of the reaction since it helps maintain the solubility of the reaction intermediates and establishes the interaction that ultimately passivates the surface of the nanocrystals.

Upon addition of $P(Si(CH_3)_3)_3$ at 100 °C in step (2), the solution turns clear yellow immediately. Over the course of 1 h this color intensifies and evolves to orange and on, to some degree, to orange/red or red. In the case of InAs, the reaction proceeds noticeably faster with the same color evolution occurring in seconds and continuing to a dark brown. Typically the reaction is maintained for 1 to 3 h at 100 °C to allow the color evolution to fully develop and then the reaction temperature is increased smoothly to 265 °C over 15 to 30 min. The reaction is maintained at 265 °C for at least 1 day to allow annealing time to improve the crystallinity. Typical reactions are heated for 1 to 2 days. Reactions have been run for up to 6 days, but no substantial change in nanocrystal properties is observed. The final reaction temperature of 265 °C was obtained empirically by varying the temperature from 240 to over 300 °C. Generally, a higher reaction temperature favors quality nanocrystals in terms of both crystallinity and size distributions, but it was found that over 290 °C the nanocrystal quality decreased and there was excessive formation of material resulting from the breakdown of TOPO at high temperatures. This by-product may also form to some degree at 265 °C, but it can easily be removed from the nanocrystals using selective precipitation techniques.

At this point in the reaction, phosphine and/or phosphine oxide capped nanocrystals can be isolated. For further surface derivatization, as shown in step (3), a surfactant ligand can be added to the reaction. A wide variety of ligands can be used with the general requirement of substantial electron donating ability to coordinate with the Lewis acid-like indium sites on the surface of the nanocrystal. Amines, particularly dodecyl and decyl amine have been used with substantial success and thiols and other phosphines will also coordinate to the surface. This surface exchange can also take place after the nanocrystals have been isolated by simply adding the surfactant ligand to nanocrystals dissolved in common solvents.

It is valuable to examine this nanocrystal synthesis using the methodology of nucleation, growth, and termination described earlier. Nucleation is most likely a complex process which is poorly defined in time as it occurs over several hours beginning with the addition of $P(Si(CH_3)_3)_3$ and possibly continuing until the reaction reaches high temperature. It is therefore not surprising that this reaction produces a wide range of nanocrystal sizes as nucleation is not an instantaneous event. While the details of the reaction's initial stages have not been worked out, we can look to the work of Barron[34] and more recently Wells[36] to provide insight on the structure of these initial reaction intermediates. Both groups have used the $InCl_3 + P(Si(CH_3)_3)_3$ reaction to synthesize bulk InP at temperatures above 400 °C while investigating intermediate products at lower temperatures. They report a stepwise elimination of the three equivalents of $(CH_3)_3SiCl$ beginning with the formation of $[Cl_2InP(SiCH_3)_2]_x$ at room temperature or below. Our exact intermediate product probably has a slightly different structure due to the action of the coordinating solvent and evidenced by the increased solubility of our product, but the important element is the formation of the In–P bond. Wells further establishes this In–P interaction by providing a crystal structure for the related 1:1 adduct $I_3In{\cdot}P(Si(CH_3)_3)_3$.

The growth of the particles occurs during the higher temperature portion of the reaction beginning with the further elimination of CH_3SiCl. Barron sees evidence of the elimination of additional CH_3SiCl as low as 150 °C along with the appearance of peaks in the XPS spectrum characteristic of indium in an InP environment. However, to eliminate the final equivalent of CH_3SiCl and obtain crystalline material, higher temperatures are required with our reactions producing InP nanocrystals at a minimum of about 240 °C.

The termination of the particle is accomplished by the coordinating solvent medium. Both TOPO and TOP are effective in passivating the particle surface. TOPO has been used with great success in passivating CdSe nanocrystals[8] and is known as a strong donor ligand with both high polarity and polarizability contributing to its donor strength.[35,37,38] Based on analogy to CdSe and standard donor–acceptor analysis, the TOPO would coordinate to acceptor indium surface sites providing a passivating shell to terminate growth, prevent agglomeration among particles, and through its alkyl groups provide for excellent solubility in organic solvents such as toluene and hexanes.

While the above synthesis gives crystalline, discrete, and highly soluble nanocrystals, it produces a wide range of nanocrystal sizes. In order to study size-dependent phenomena a variety of different sizes, each with a narrow size distribution, needs to be isolated. By taking advantage of the differential solubility of various sizes of nanocrystals narrow size distributions have been obtained using size-selective precipitation techniques.[8] The solubility of a specific nanocrystal is determined by several factors, most notably the size of the particle, its shape, and the nature of the particle surface. Surface considerations include the type of ligand on the nanocrystal surface and the degree of coverage of these ligands on the surface. To examine the influence of these factors, it is easiest to consider a model system which we will choose to be qualitatively similar to the InP nanocrystals. Take a sample of nanocrystals which are capped with TOPO. The sample has a large distribution in particle size and a slight distribution in particle shape with nanocrystals ranging from spherical to slightly ellipsoidal. There are some random variations in ligand coverage except there is a general trend of higher percent ligand coverage as one goes to smaller particle size.[9,18] Due to the alkyl groups of the TOPO cap, the particles are soluble in relatively nonpolar solvents. If a more polar solvent is added to a solution of nanocrystals, the ability of the combined solvent system to solvate the nanocrystals will be reduced. The mechanism of agglomeration is dominated by attractive van Der Waals forces among particles,[8,39,40–42] and at some composition of solvents, the solvation ability of the system will no longer overcome these attractive forces and the particles will begin to agglomerate and precipitate. The key to size-selective precipitation is that different sizes, shapes, and surface coverages will have different solubilities and thus may be separated. Assuming that the dominant difference in the nanocrystals is that of particle size, the ensemble of nanocrystals will be separated by size, while the degree of resolution among sizes will be determined by the smaller differences in the sample such as shape or surface structure. The largest particles will precipitate first due to their stronger attraction[8,40,42] and their lower percent ligand surface coverage.

The primary solvent system employed is toluene/methanol. The nanocrystals are soluble in toluene while in highly polar methanol they are insoluble. Starting with the initial reaction mixture diluted in toluene, the incremental addition of methanol results in the size-selective precipitation of the nanocrystals and the isolation of individual distributions ranging from 20 to 60 Å in diameter. This change in size is qualitatively observed through the evolution in the color of each of the size-selected nanocrystal samples dissolved in toluene. The first fractions result in dark brown solutions which slowly evolve through brown to deep red to finally bright red for the smallest sizes. This is an indication of the shifting band gap as the degree of quantum confinement is increased with decreasing size. As mentioned above, the limit to size-selective techniques is due to other factors besides size which affect particle solvation and is not an inherent limit to the ability to distinguish two particle sizes. If the initial distribution were more homogeneous, the reprecipitation techniques would be more selective.

While the nature of the reaction mechanism prevents an idealized "instantaneous" nucleation event from occurring and results in a wide distribution in initial particle size, the quality of the nanocrystals within the sample does not suffer. The use of size-selective precipitation overcomes the problem of the initial range of sizes to give greatly narrowed distributions. This allows for the study of the size-dependent properties of these systems using a wide variety of techniques, some of which are detailed in the remainder of the chapter.

III. STRUCTURAL CHARACTERIZATION

The evaluation of nanocrystal structure encompasses several important areas including the identity and crystallinity of the core, the morphology of the particles, and the composition of the nanocrystal surface. This section will discuss the crystallinity of the core using both powder XRD and electron diffraction and the morphology of the particles, including their shape and size distributions, using transmission electron microscopy (TEM). X-ray photoelectron spectroscopy (XPS) will be used to examine the nanocrystal composition, with particular attention to the surface capping groups. InAs and InP nanocrystals will be treated together, with representative data shown from one or both systems.

A. Powder X-Ray Diffraction

Powder X-ray diffraction of a series of InP nanocrystal sizes is shown in Figure 1. The peak positions index well with the bulk InP lattice reflections as shown by

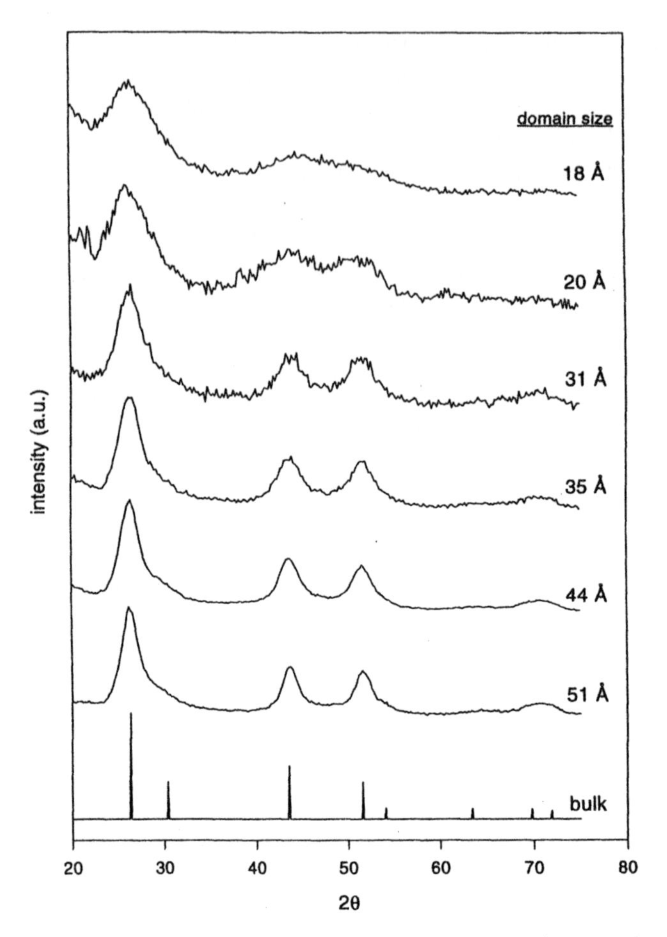

Figure 1. Powder XRD spectra of InP nanocrystals ranging in diameter from 18 to 51 Å along with the pattern for bulk InP.

the standard bulk InP pattern[43] in the figure. Information on the domain size of the sample can be obtained from the width of the lattice reflections. This relationship is quantified in the well-known Debye–Scherrer formula[44] and for spherical particles the appropriate formula is:

$$D = \frac{1.2\lambda}{[(\Delta 2\theta)(\cos\theta)]} \tag{1}$$

In this equation, λ is 1.5418 Å for CuKα radiation, $\Delta 2\theta$ is the measured linewidth in radians, θ is 1/2 of the measured diffraction angle, and D is the estimated particle size in angstroms. An average domain size for a given sample was obtained by averaging D for the (111), (220), and (311) peaks. The relatively weak (200) and (222) diffraction peaks were not included in the size determination. InAs nanocrystals also index to the bulk lattice reflections and the domain sizes ranging from 28 to 100 Å have been obtained, although the 100 Å domain size has not been isolated as a narrow size distribution.

B. Transmission Electron Microscopy

TEM is crucial to evaluating the structure of the nanocrystals as it allows a direct evaluation of both the size and shape of the particles. A typical image of InP nanocrystals is shown in Figure 2. Lattice fringes are clearly observed which index

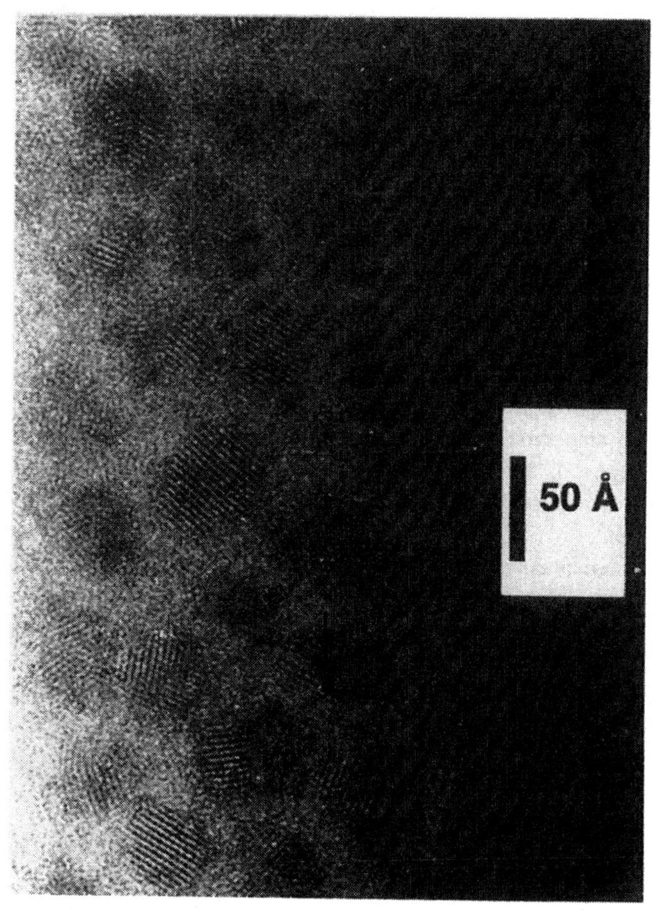

Figure 2. Transmission electron micrograph of a field of InP nanocrystals. The particles are discrete, roughly spherical, and are primarily made up of single crystalline domains. The visible lattice spacings correspond to the (111) zincblende lattice planes.

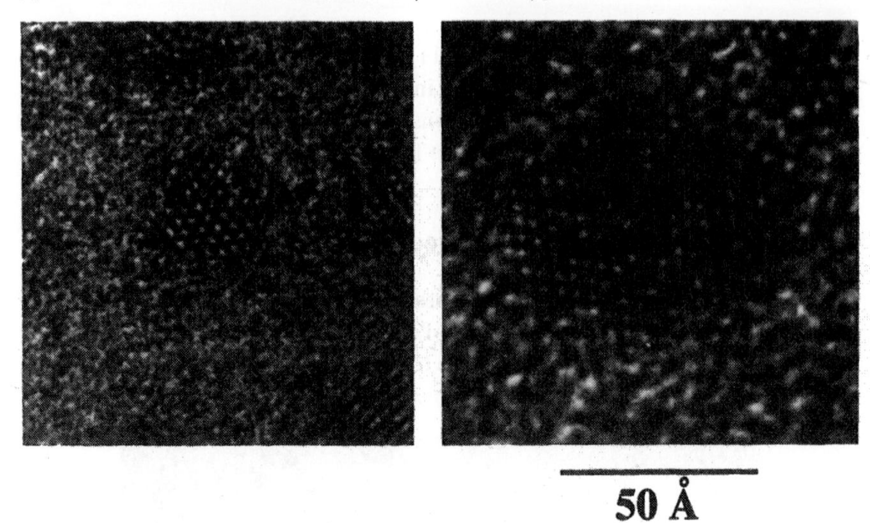

Figure 3. High-resolution TEM image of two InAs nanocrystals 20 and 60 Å in diameter.

to the InP (111) lattice spacing. The particles are dispersed, separate entities showing little agglomeration and no fusing. They are generally quite crystalline, with many unfaulted, single-domain particles although faults and multiple domains are certainly evident. The nanocrystals are roughly spherical to slightly oblong with aspect ratios on a variety of samples ranging from 1 to 1.15. Figure 3 shows a TEM image of two InAs nanocrystals 20 and 60 Å in diameter. Cross fringes are observed indicating a high degree of crystallinity. Faceting of the nanocrystal surface is also apparent. Figure 4 shows a lower resolution TEM image of a field of InAs nanocrystals. The relatively high degree of monodispersity in size and shape leads to spontaneous ordering of the nanocrystals on the TEM grid. The particles are generally spherical and have similar aspect ratios to InP.

TEM is also essential for calculating the size and size distribution of a nanocrystal sample. For the wide variety of size-dependent properties exhibited by the nanocrystals, this is clearly a crucial measurement. Size distributions were obtained by the direct measurement of nanocrystal images for a large number of particles, typically between fifty and a few hundred. The results of these measurements for InP show size distributions with standard deviations of about 20% in diameter, while for InAs the distributions are substantially narrower at 10%.

C. Comparison between TEM and Powder XRD

The data from powder XRD and TEM convey slightly different size information and a comparison between the two can lead to a more complete understanding of nanocrystal crystallinity. With TEM the measured size is a direct observation of the

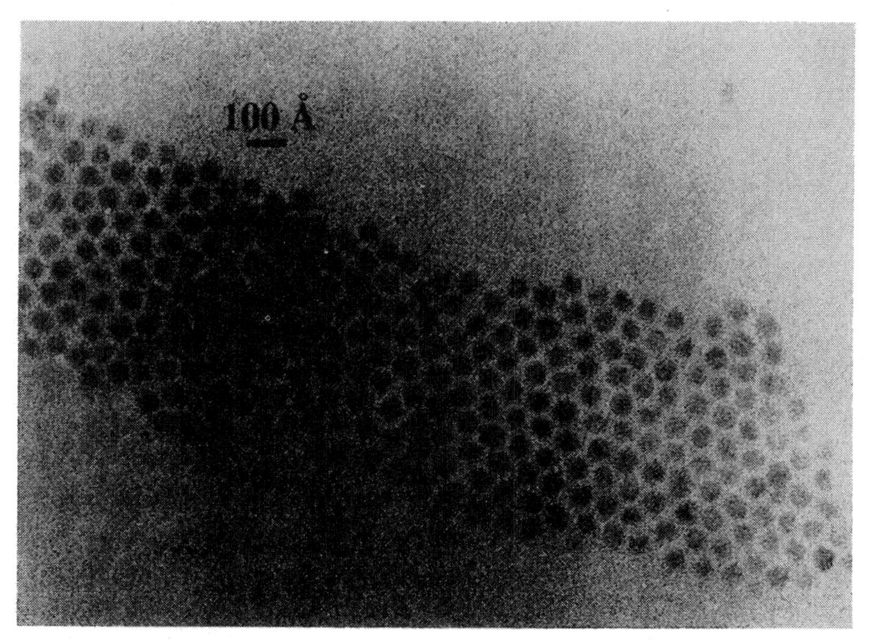

Figure 4. Lower resolution TEM image of a field of InAs nanocrystals, each about 50 Å in diameter.

diameter of the particle, regardless of the detailed nature of its crystallinity including such characteristics as faults or dislocations. In contrast, the powder XRD data provides little information about the actual size of the particles, but instead gives information on the crystalline domain size of the sample. Taken together these techniques can help evaluate the degree of crystallinity of the particles where a domain size close to the measured TEM size would indicate a relatively unfaulted nanocrystal sample. In contrast, a TEM size substantially larger than the X-ray domain size would indicate faults resulting in multiple crystalline domains within a single crystallite.

Figure 5 shows the powder XRD domain size plotted against the TEM size for several InP nanocrystal samples. The straight line has slope equal to one and corresponds to XRD domain sizes and TEM sizes in exact agreement as for perfectly unfaulted nanocrystals. The data is clustered around this line with scatter both above and below the idealized value. Within the error associated with these measurements, the size distributions from TEM compare closely with the domain sizes calculated from powder XRD, indicating the nanocrystals are not highly faulted. This is consistent with observations of many TEM images which show the majority of particles to be single domains although occasional faults are observed.

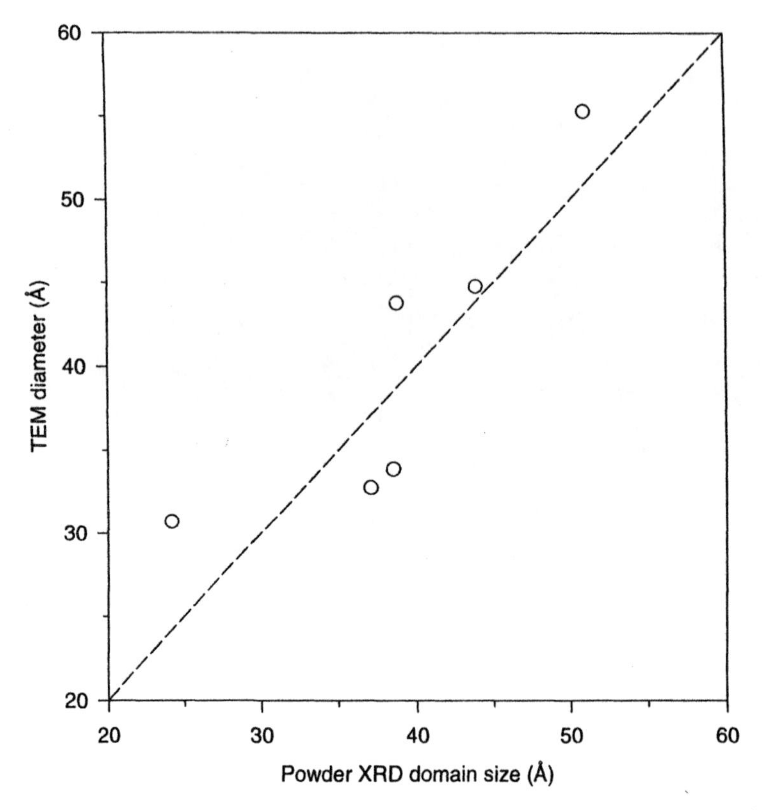

Figure 5. Powder XRD domain size versus TEM diameter for InP nanocrystals.

D. X-Ray Photoelectron Spectroscopy

X-ray photoelectron spectroscopy (XPS) is a powerful tool for the characterization of nanocrystals as it provides quantitative data on elemental composition and bonding environments which allows for the analysis of the nanocrystal surface.[9] XPS was performed on a series of InP nanocrystals of varying size capped with TOPO. Elemental compositions of indium and phosphorous for InP were obtained and also surface coverages of TOPO on the nanocrystals were calculated. Initial measurements have also been performed for InAs nanocrystals.

Due to the nature of XPS, it is important that the sample be electrically grounded. Otherwise, as electrons are ejected, the sample may charge resulting in shifts to higher binding energy and distorted, asymmetric peaks. To avoid this problem the nanocrystals were bound to gold surfaces via hexane dithiol linkages using techniques similar to those described elsewhere.[2] Briefly, hexane dithiol is allowed to self-assemble on ion-etched gold-coated aluminum plates. The gold substrates are then placed in a solution of nanocrystals where the nanocrystals bind to the thiols attached to the gold surface. Based on previous work with CdS and CdSe nanocrystals,[9] this method results in the formation of less than a single monolayer of nanocrystals covalently bound to the gold film. This is relevant to the calculation of surface coverages since it is important that there be no stacking of nanocrystals. The extent of oxidation of the InP nanocrystals affects their ability to bind to the surface, with unoxidized samples more readily binding.

A typical survey spectra is shown in Figure 6. The survey indicates the presence of In and P as well as Au from the substrate and C and O from the nanocrystal capping groups and adsorbed gaseous molecules. In higher resolution spectra, neither chlorine nor silicon was observed which is consistent with bulk elemental

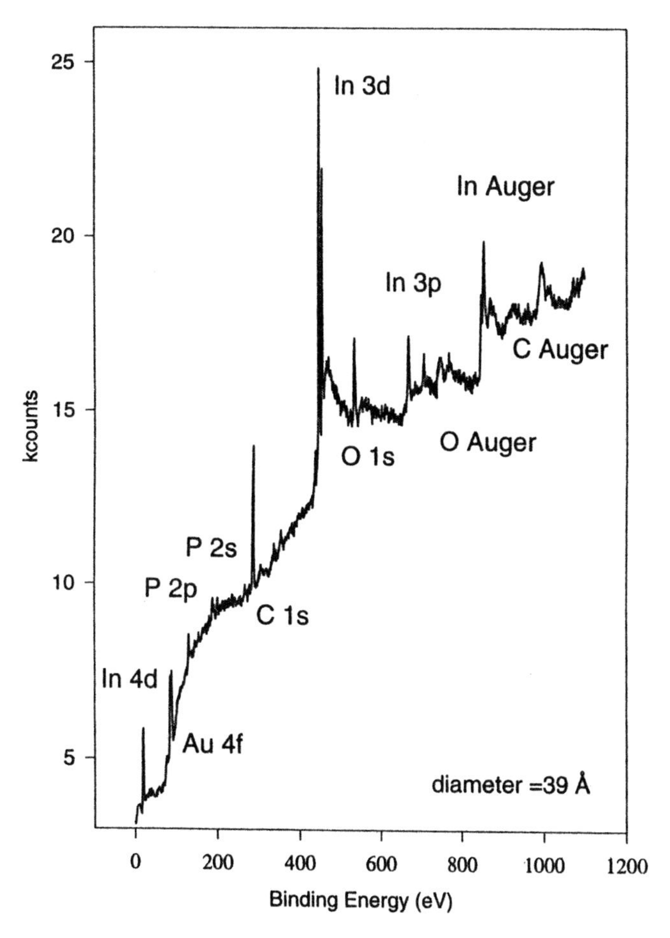

Figure 6. XPS survey scan of InP nanocrystals 39 Å in diameter.

analysis indicating less than 1% Cl or Si remaining from the initial reactants. Figure 7 plots higher resolution spectra of the In 3d region and P 2p regions for a series of sizes of unoxidized nanocrystals. The indium core is spin-orbit split to the $3d_{5/2}$ and $3d_{3/2}$ states. The P 2p core shows two peaks, one at about 130 eV corresponding to P from InP and the other at about 133 eV corresponding to oxidized P species, in this case TOPO.

Peak areas of these high-resolution scans were measured and used to calculate In to P ratios for the nanocrystals. For the In:P ratios, only the P 2p peak corresponding to InP was used. The raw In:P ratios are corrected for the relative cross sections of In and P,[45] as well as for the radius of the nanocrystals.[9] In addition, the spectra of a bulk sample of InP was run as part of each data set and all spectra

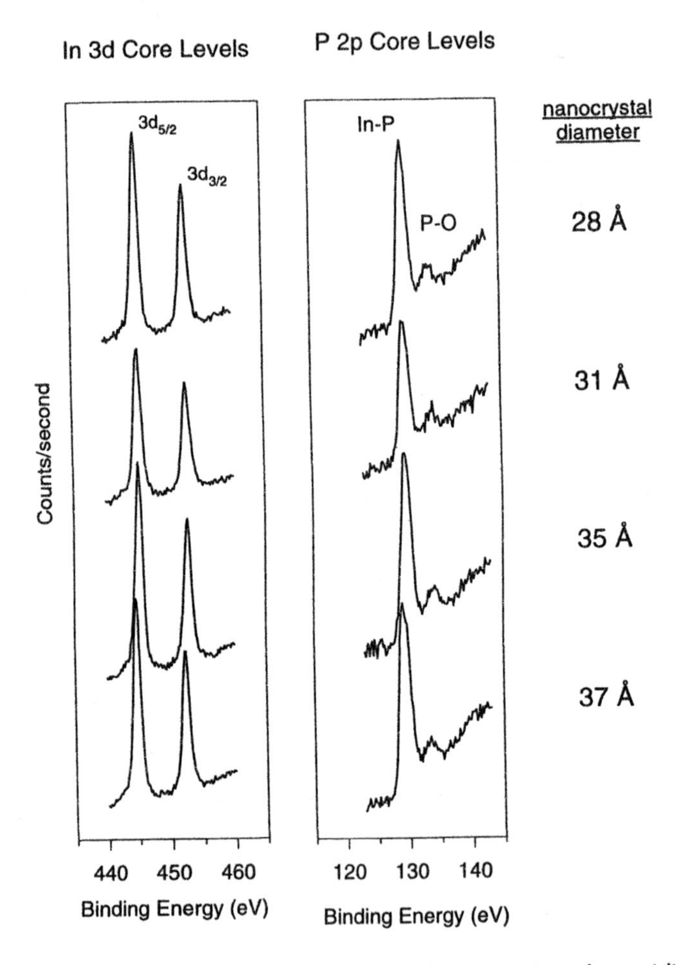

Figure 7. XPS spectra of In 3d and P 2p cores from a series of unoxidized InP nanocrystals. The nanocrystal diameter is given on the right.

are calibrated such that the In:P ratio for the bulk material is 1:1. Based on this assumption, the In:P ratio for the nanocrystals varied from 0.74 to 1.00 with an average of 0.88 ± 0.05. While previous studies on bulk InP indicate that the surface oxide has an In:P ratio of 1:1,[46,47] justifying our calibration, the formation of oxides resulting in higher than 1:1 ratios have also been observed.[48,49] Since the XPS technique samples less than the first 100 Å of the sample, the oxide would be a major contributor to the signal leading to a systematic error in our results with the true In:P ratio for the nanocrystals having a higher value. The signal from unoxidized nanocrystals would not have a contribution from the greater than 1:1 oxide and thus would give a low ratio if the bulk was set to 1:1. The magnitude of this error would not be large, leading to a maximum possible In:P ratio of 1.1:1. Thus, we may conclude that the nanocrystals are close to stoichiometric InP.

As mentioned above, the region of the P 2p core level exhibits two peaks, one due to P from InP (130 eV) and one at higher binding energy (133 eV) due to oxidized P. For oxidized samples the 133 eV peak will contain contributions from both TOPO and surface-oxidized P, but in unoxidized samples the contribution should come only from TOPO. Using a series of unoxidized nanocrystals we have examined TOPO surface coverages for a series of nanocrystal sizes using a method described previously.[9] This data is summarized in Figure 8 and shows a trend of higher coverages for smaller nanocrystals. The calculated total coverage of TOPO on the surface of unoxidized nanocrystals ranged from 0.29 for a 51 Å particle to 1.05 for a 25 Å particle. A coverage of 0.50 corresponds to TOPO binding to half of all surface atoms. The In and P surface atoms should exhibit very different coordination properties with the In acting more as a Lewis acid and the P more as a Lewis base. The TOPO, with its excellent donor ability, would then coordinate preferentially to the In sites. This situation is analogous to the coordination of TOPO to Cd surface sites in CdSe nanocrystals.[9] However, this analysis does not explain the coverages greater than 0.5 which would seem to indicate that P sites are also passivated. This result is most likely due to errors in the diameter assigned to the particle. The calculation is quite sensitive to the size of the nanocrystal, particularly for the smaller sizes where it is most difficult to obtain quality size distributions. Note that the only coverage that is much greater than 0.5 is for the smallest nanocrystal with a diameter of only 25 Å. Another possible explanation for coverages greater than 0.5 is the existence of an indium rich surface on the particles. This would allow greater than half of the surface atoms to be available for coordination by TOPO.

The trend of higher coverages for smaller particles can be explained in terms of steric hindrance where the bulky TOPO ligand is better able to access the smaller particles surface due to the higher curvature of the smaller particle surface. For larger nanocrystals the TOPO molecules, while possibly occupying a larger absolute number of sites, are able to access a smaller percentage of available surface sites. A similar phenomenon has been observed in CdSe nanocrystals.[9]

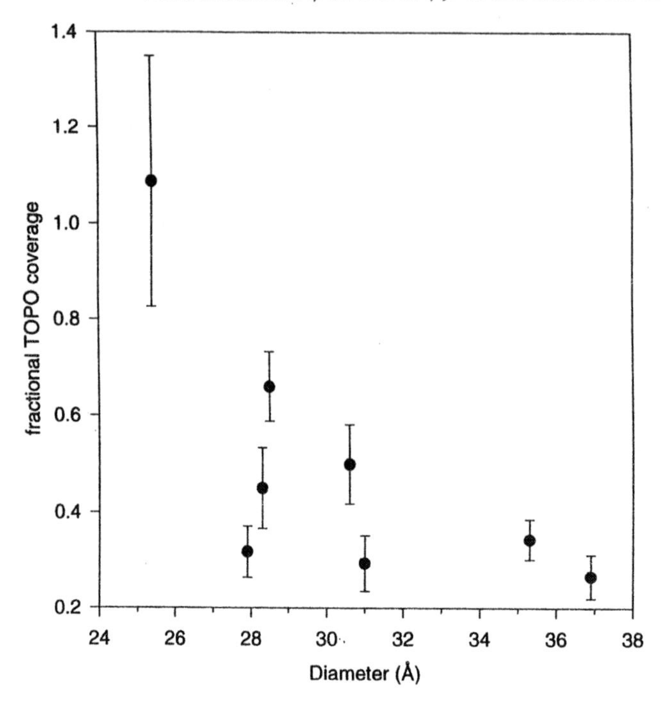

Figure 8. Fractional TOPO coverage on the surface of InP nanocrystals as a function of nanocrystal diameter. Coverage of 1.0 corresponds to TOPO binding to all surface atoms.

Upon exposure to air the nanocrystals form a surface oxide layer. XPS reveals that both the In and P surface atoms are oxidized as shown in Figure 9. The binding energy of oxidized indium is very close to that of indium in InP, making it difficult to quantify the amount of indium that is oxidized. The presence of the oxide is indicated by a slight broadening of the In cores in the oxidized samples. As mentioned previously, the oxidized P binding energy is well separated from the P in a InP environment. We may therefore quantify the amount of oxidized P present by integrating the peaks; however, the binding energies of P from TOPO and from a surface oxide are close in energy so we cannot independently determine the coverage of TOPO and surface phosphorous oxide. For oxidized samples we determine a surface coverage based on total oxidized P which ranges from 0.8 to 2.3 with an average of 1.4 ± 0.4. There is no correlation between the oxide coverage and nanocrystal size. In many samples the total oxide coverage is above 1, indicating more than a surface layer of oxide. In bulk InP, oxide layers are found to extend for a few tens of angstroms, so oxidation of more than a monolayer is not surprising. In the oxidized samples the In surface atoms are passivated with either

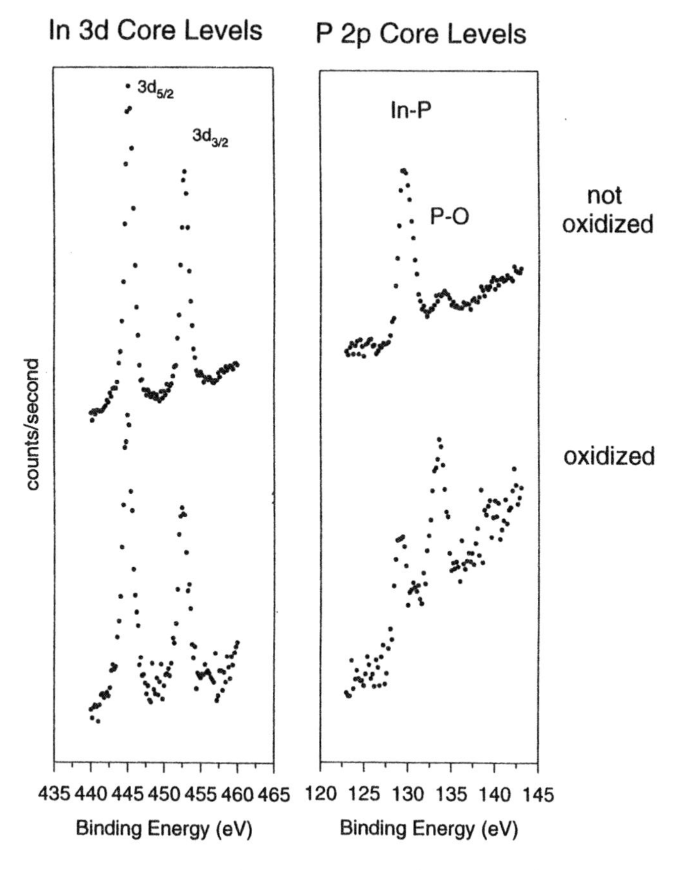

Figure 9. XPS spectra of In 3d and P 2p cores from unoxidized and oxidized InP nanocrystals.

TOPO, the additional capping group such as dodecylamine, or oxide while the P atoms are passivated with oxide.

IV. OPTICAL PROPERTIES

As discussed in the introduction, many of the quantum confinement effects of nanocrystals are evidenced in the optical properties of the system as the electronic energy levels of the clusters become a function of size. In this section, the absorption and emission properties of InAs and InP nanocrystals will be examined and transient holeburning spectroscopy will be utilized to study the homogeneous absorption properties of InP nanocrystals.

A. Absorption Spectroscopy

Room temperature absorption spectra for a series of InAs nanocrystal sizes are shown in Figure 10 along with the photoluminescence spectra. The quantum confinement effects are clearly evident from the size-dependent nature of the spectra. The smallest particles shown are 28 Å in diameter and the onset of absorption shifts smoothly across a number of samples to the largest size of 58 Å. The band gap in all samples is shifted substantially from the bulk InAs gap of 0.35 eV. In all samples the absorption onset is characterized by a distinct feature at the absorption edge corresponding to the first excitonic excited state. Additional features are resolved at higher energy corresponding to higher excited states. All states shift with nanocrystal size. The width of the features is primarily due to

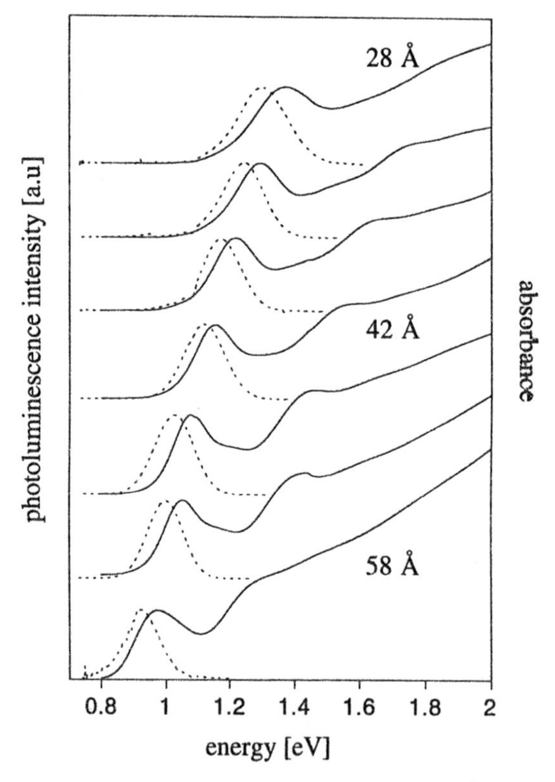

Figure 10. Absorbance and photoluminescence spectra of InAs nanocrystals with diameters ranging from 28 to 58 Å. The absorption onset shifts to the blue with decreasing size due to quantum confinement effects and all onsets are well shifted from the bulk InAs band gap of 0.35 eV. The emission was obtained using an excitation wavelength of 514.5 nm. All spectra have been normalized so that the initial feature is of comparable intensity.

inhomogeneous broadening resulting from size variations of the particles within a specific sample. As mentioned, for InAs the size distribution has a standard deviation of about 10% in diameter.

The absorption of InP nanocrystals show similar effects with an absorption onset shifting to the blue with decreasing cluster diameter as shown in Figure 11. The 37 Å particles show a shift of 0.5 eV from the bulk value of 1.35 eV, while for 20 Å clusters there is a shift of 1.1 eV from the bulk band gap. In InP samples the size distribution is broader than for InAs, so higher excited states are washed out and usually only the initial absorption feature is resolved.

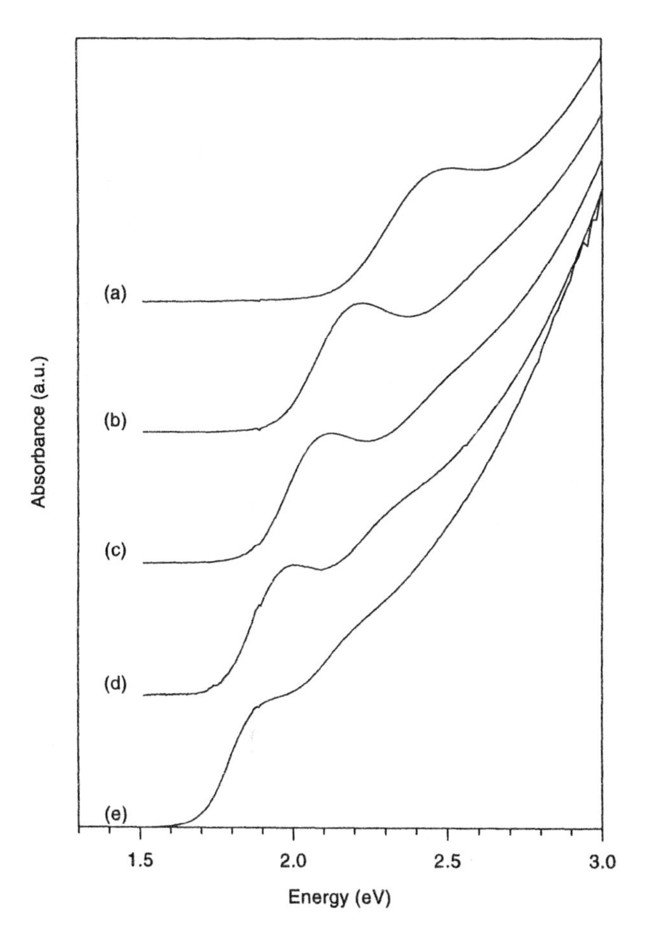

Figure 11. Absorption spectra of InP nanocrystals with diameters of: (**a**) <20 Å, (**b**) 23 Å, (**c**) 26 Å, (**d**) 33 Å, (**e**) 37 Å.

B. Transient Optical Holeburning Spectroscopy

The distribution in nanocrystal size within a sample results in substantial inhomogeneous broadening of the absorption spectrum as a sample with a range of particle sizes will absorb at a range of energies. These inhomogeneously broadened spectra can be thought of as a sum of homogeneously broadened "single-particle" spectra for a range of sizes. Transient optical holeburning spectroscopy can be used to examine the "single-particle" spectrum revealing information on the degree of inhomogeneous broadening in the sample as well as the homogeneous linewidth.[50,51]

More specifically, a narrow bandwidth nanosecond pump pulse is used to excite a subset of the sample at a particular wavelength near the band edge. After a delay of a few nanoseconds, a spectrally broad probe pulse follows to give absorption information on the sample. However, the particular subset of nanocrystals that absorbed the initial pulse will be bleached from the final spectrum leaving a "hole." The hole, plotted as the negative change in optical density ($-\Delta OD$) allows the extraction of the homogeneous absorption spectrum of a "single" nanocrystal size. Such a plot is presented in Figure 12 for four different sizes of InP nanocrystals. In panel (a), the inhomogeneous absorption spectrum for each size is plotted with the arrow indicating the wavelength of the pump pulse. Panel (b) shows the resulting holeburning spectrum for each size. The holeburning spectra are characterized by a narrow transition centered at the pump energy with a size-dependent width. In addition, there is a broader feature to higher energy.

To determine the homogeneous linewidth, the hole spectra must be examined as a function of excitation wavelength. This data is shown in Figure 13 where a 34 Å sample is excited at three different wavelengths. Figure 13a shows the inhomogeneously broadened absorption spectra at 6 K with the arrows indicating the excitation wavelength. Panel (b) shows the resulting holeburning spectrum for each wavelength. In Figure 13a, pump wavelength a1 is at the red edge of the sample's absorption onset, so only the largest nanocrystals in the distribution are excited resulting in a prominent narrow feature at the pump wavelength. This hole most closely resembles the "single-particle" absorption spectrum. A weak phonon sideband appears as a shoulder at a spacing of 310 cm^{-1} due to bleached absorption at high vibrational levels. There is also a broad feature shifted to higher energy by about 0.1 eV which is due to bleaching of the absorption of a second electronic transition. As the pump wavelength is shifted to the blue, to a2, and a3 in Figure 13, smaller particles in the sample are excited with the narrow transition in b2 and b3 corresponding to band gap absorption for these smaller sizes. However, while these higher pump energies are at the band gap of the smaller particles, they are also at the absorption energies of higher excited states for larger particles whose first excited state was accessed with a1. This results in additional absorption from the pump pulse, bleaching in the subsequent absorption spectrum, and a broad background in the holeburning spectra. This broad background increases as the

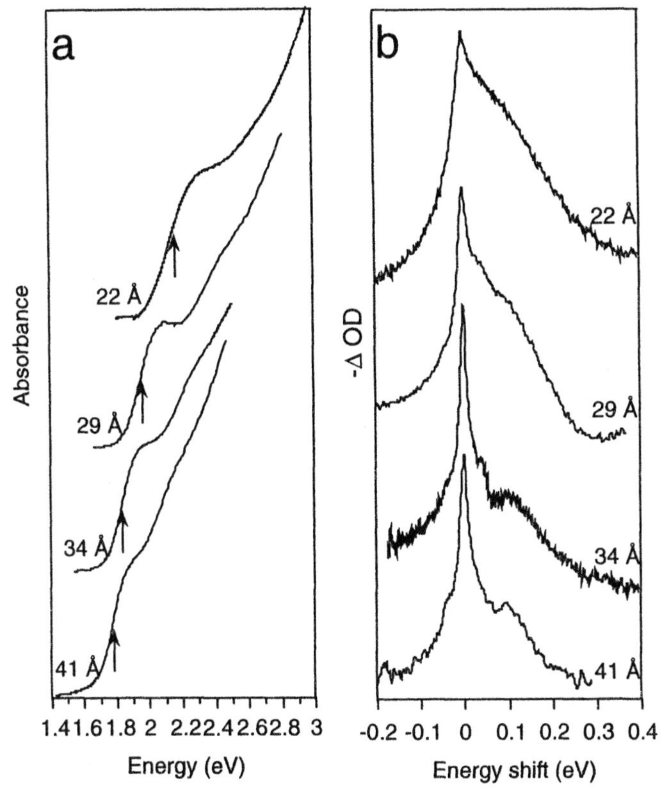

Figure 12. (**a**) Absorbance spectra for four sizes of InP nanocrystals. The arrow indicates the energy of the excitation pulse used for the corresponding holeburning spectra. (**b**) Holeburning spectra for four sizes of InP nanocrystals each of whose inhomogeneously broadened absorption spectrum is shown in panel (**a**).

pump wavelength moves further blue, exciting higher transitions of a wider range of sizes. At the highest pump wavelength, the width of the spectral hole is indicative of the inhomogeneous absorption spectrum.

To obtain quantitative information about the homogeneous linewidth, the holeburning spectra were simulated and applied to the series of wavelength-dependent spectra. The model includes the narrow band gap transition, a phonon sideband resembling the frequency of bulk InP optical phonons,[52] and a substantially broadened second excited state separated by 0.11 eV. The results of such a simulation for the 34 Å nanocrystals are shown in Figure 14 for b1, b2, and b3, the same spectra shown in Figure 13. The homogeneous linewidth of the InP nanocrystals is about 10 meV for the largest size and increases to about 20 meV for the

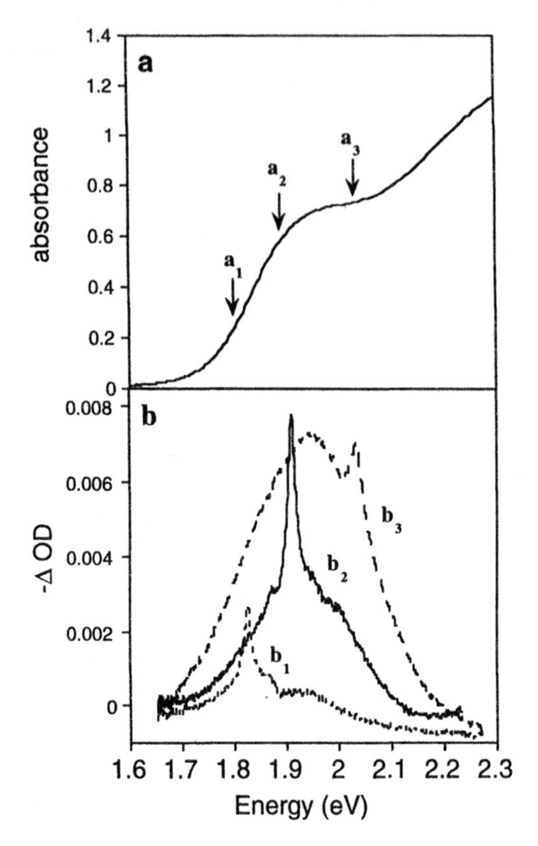

Figure 13. Holeburning spectra as a function of excitation energy. (**a**) Absorbance spectra for 34 Å sample of InP nanocrystals. The arrows indicate the position of the holeburning excitation. (**b**) Holeburning spectra for the same sample as in (**a**), each excited at a different energy, as indicated by the arrows in (**a**).

smallest particles.[50,51] These values are comparable for similar sized CdSe nanocrystals.[53,54]

C. Photoluminescence

While the absorption behavior of semiconductor nanocrystals is relatively insensitive to the nature of the particle surface, the photoluminescence may be strongly affected by how the surface of the particle is passivated. In the case of InAs nanocrystals, emission occurs in a relatively narrow band shifted slightly to the red of the absorption band edge, as seen in Figure 10 for nanocrystals ranging in size

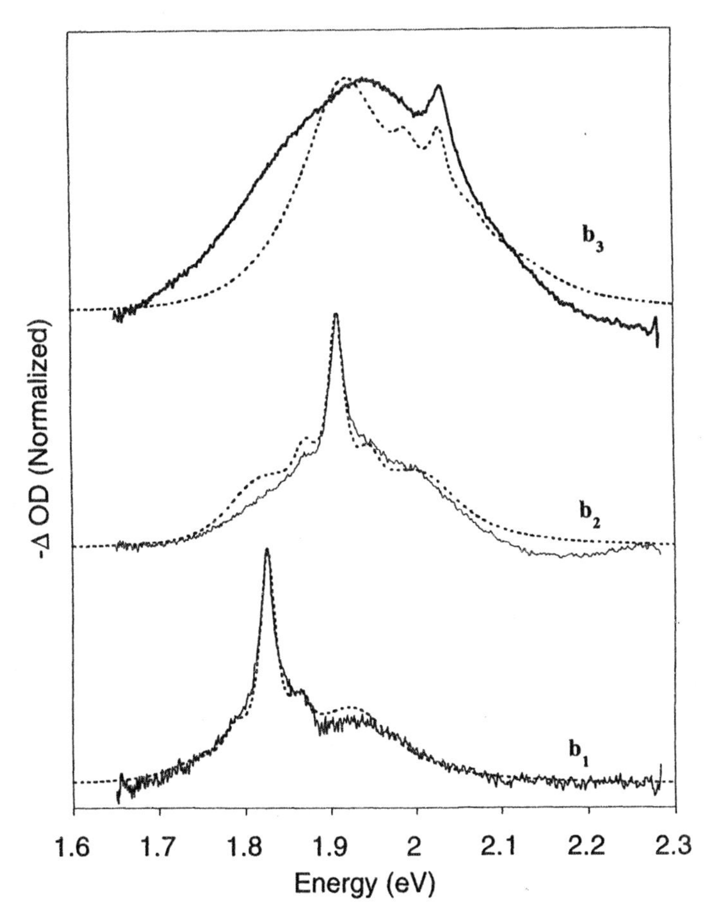

Figure 14. Holeburning data (—) along with the simulated spectra (···) for one sample excited at three different energies, as in Figure 13.

from 28 to 58 Å in diameter. The proximity of the emission to the absorption feature and its well-defined width indicate this luminescence is due to electron-hole recombination of the initial excitation near the band gap. As expected, this band gap feature is size-dependent and shifts along with the absorption onset as the size of the nanocrystal changes. The structure in the luminescence appearing between 1.05 and 1.15 eV is due to reabsorption of the dot luminescence by solvent modes. Absolute quantum yields at room temperature of a few percent with no substantial size dependence have been calculated relative to the laser dye LDS821.

InP nanocrystals also show the characteristic shifting of emission with nanocrystal diameter, but the emission exhibits two bands, one near the band edge and another broader band further to the red. Analogously to InAs, the higher energy band is assigned to electron-hole recombination near the band gap. The luminescence quantum yield for this band gap emission is substantially lower than for InAs with yields on the order of only 0.01%. The lower energy feature is an indication of trapped emission where at least one carrier has relaxed into a lower state before emission. Incomplete surface passivation provides a source for a variety of trap sites resulting in a broad emission feature as carriers fall into traps of varying depth. Not unexpectedly, this surface trapped emission is generally more pronounced in smaller particles where the surface is a more significant portion of the total number of atoms. The trapped emission feature is less sensitive to particle size as it is due to recombination from an assortment of energy levels which are only indirectly tunable with size. The initial absorption transition is obviously the same for both band edge and surface trapped emission, but during relaxation to the surface trapped states, the carriers lose this direct connection to the band gap energy and become spread out in energy as they relax to a variety of trapped states before emitting.

The relative intensities of these two bands depends on the particular type of surface passivation. TOPO capped particles show emission that is dominated by surface traps, while InP nanocrystals capped with a combination of TOPO and decylamine show emission due primarily to band edge luminescence. In both cases the samples are oxidized while in corresponding unoxidized samples, no luminescence is observed. In InP nanocrystals oxidation is a necessary condition for luminescence; however, the decylamine cap results in a more complete or different oxidation process than TOPO and thus provides improved electronic passivation.

At this point it is important to more clearly define surface passivation by distinguishing between chemical passivation and electronic passivation. The surface ligand serves to chemically passivate the nanocrystals if it maintains the dispersity and provides for the solubility of the particles. The nanocrystal is electronically passivated if the surface states which lead to trapped emission are eliminated. Electronic passivation can lead to increased band edge emission with decreased trap emission if the carriers no longer relax to the surface, or to a decrease in all forms of luminescence if surface states are created which quench luminescence by providing nonradiative pathways. The two forms of passivation do not necessarily go together nor are they mutually exclusive. Therefore, when the effect of a surface ligand on luminescence is considered, other factors including passivating processes such as oxidation must be included.

Comparing InAs and InP, the luminescence of InAs nanocrystals is dominated by band edge emission with substantial yields while InP luminescence shows both surface trapped and band edge emission with very poor yields. In InAs nanocrystals, the coordinating ligands are able to provide substantial electronic passivation to give a particle surface that is relatively insensitive to oxidation or further surface derivatization using organic ligands. Meanwhile, the InP luminescence is sensitive to the surface derivatization of the particle, yet in no case has good electronic passivation been obtained using only derivatization with organic ligands and oxidation. It should be emphasized that the luminescence behavior is to a large extent a function of the particle surface and that the superior characteristics of InAs nanocrystals are likely due to more effective surface passivation rather than any inherent property of the material. For instance, other means of surface passivation of nanocrystals have begun to be explored which can show pronounced effects on luminescence behavior including growing another semiconductor layer on the surface of the particle or etching of the particle surface. In the case of InP nanocrystals, Mićić has observed substantially enhanced luminescence yields by etching with HF.[55]

V. CONCLUSIONS

The range of semiconductor materials that can be produced as high-quality nanocrystals has been extended to include the group III-V materials InAs and InP. The development and characterization of these new nanocrystal materials will allow for a more general and complete understanding of quantum confinement effects in two important ways. First, the available range of materials that can be configured as nanocrystals now includes semiconductors with a much broader range of bulk properties. For instance, when compared with the two prototypical systems of CdSe and CdS, the addition of InAs and InP extends the range of nanocrystal band gaps by almost 1.5 eV to the red. This is shown in Figure 15 which plots band gap versus particle diameter for InAs, InP, CdSe,[8] and CdS.[56] The band gap may be finely tuned by the choice of both semiconductor material and nanocrystal size. The addition of InAs and InP also reduces the effective electron masses by a factor of 5 and the hole masses by 2, doubles the highest value of the dielectric constant, and increases the range of carrier mobilities by 1 order of magnitude. By studying materials with such different bulk properties, it is possible to distinguish which quantum confinement effects are more or less sensitive to the bulk properties of the material and the degree of that sensitivity. More generally, there is the important question of the validity of applying bulk material parameters to systems of this size. How large does a lattice have to be for these "constants" to assume the accepted bulk value? By examining size effects in materials with a wide range of bulk properties, properties that are sensitive to size versus those that are sensitive to material may be distinguished.

The most obvious comparison to make is the degree of quantum confinement in nanocrystals of different materials. To examine this quantity it is convenient to look at the change in energy gap from the respective bulk material versus particle diameter. Such a plot is shown in Figure 16 for InAs, InP, CdSe, and CdS. In general, the confinement effects behave consistently across the different materials with shifts of comparable magnitude, particularly for InP, CdSe, and CdS. InAs shows slightly larger shifts from the bulk band gap for a given size of nanocrystal. According to the effective mass approximation,[20] the III-V nanocrystals, with their lower effective carrier masses and larger dielectric constants, should show larger

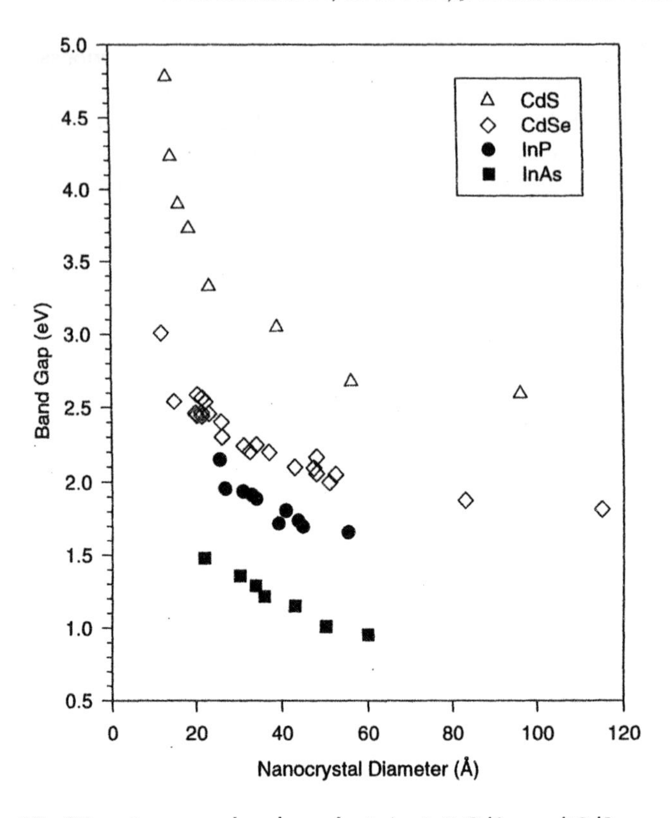

Figure 15. Diameter versus band gap for InAs, InP, CdSe, and CdS nanocrystals.

shifts than the more ionic II-VI materials. While this approximation is known to fail quantitatively due to simplifications including ideally parabolic bands and an infinite potential at the nanocrystal surface, it is interesting that, even qualitatively, substantially enhanced confinement is not observed in either InP or InAs nanocrystals. InAs, whose exciton bohr diameter is nearly 10 times that of CdSe, shows only slightly increased quantum confinement. Clearly more detailed theoretical examinations are necessary and several more complex theoretical models[57–60] have been applied to quantum confinement effects in II-VI nanocrystals. Hopefully the development of these III-V nanocrystal materials will expand the theoretical studies to include these new materials.

The second way group III-V nanocrystals will provide for an increased understanding of quantum confinement is through comparisons to other quantum confined structures prepared by molecular beam epitaxy (MBE) and various lithographic techniques. Examples include quantum wells and wires where the confinement is respectively in one or two dimensions, and quantum dots which,

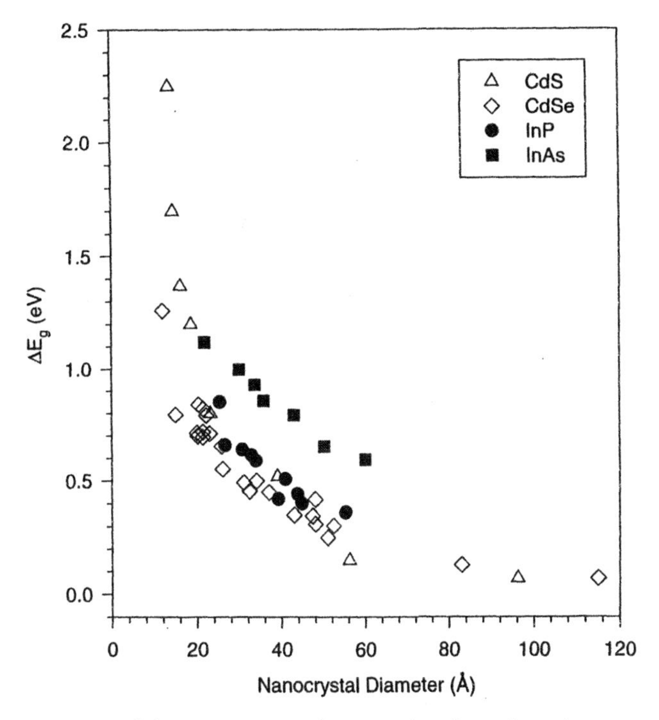

Figure 16. Nanocrystal diameter versus change in band gap from the respective bulk material for InAs, InP, CdSe, and CdS.

like nanocrystals, are confined in three dimensions but are several times to an order of magnitude larger. These structures are most highly developed in III-V semiconductors, so III-V nanocrystals are needed for direct comparisons to these structures. An exciting recent development is the fabrication of InAs quantum dots in the same size regime as the InAs nanocrystals.[61–66] In this case, direct comparisons of similarly sized MBE structures and nanocrystals will be possible for the first time.

Fabrication of InAs and InP nanocrystals have proven to be a difficult synthetic problem due to the nature of their chemistry. The lack of control over the nucleation of the particles during synthesis results in a broad initial size distribution for a given reaction. However, this broad distribution does not necessarily reduce the quality of the nanocrystals within the sample. Through the use of size-selective precipitation techniques, greatly narrowed size distributions of high-quality nanocrystals may be obtained that are appropriate for a host of size-dependent studies and as building blocks for more complex quantum confined structures. A major synthetic challenge that remains is the improved electronic surface passivation of the nanocrystals. The most promising technique for improved passivation is the recently developed method of inorganic surface passivation through the growth of another semiconductor material on the surface of the particle. This has been accomplished in the case of HgS on CdS,[13–15] ZnS on CdSe,[10] and CdS on CdSe.[12] In the latter two cases dramatic increases in luminescence yields were observed as evidence of the reduction of surface traps. Also, the growth of an outer layer gives an entirely new variable to control as the relative band gaps of the two materials may be tuned to achieve desired properties.

Regardless of possible future improvements, the current preparations of InAs and InP are suitable for a wide variety of important studies. Experiments that have been carried out or are in progress include the investigation of homogeneous linewidths using photon echo and holeburning measurements,[50,51] the study of the evolution of electronic level structure with size using fluorescence line narrowing (FLN) and photoluminescence excitation (PLE) studies, examination of surface structure with ^{31}P NMR, and high-pressure diffraction studies of solid–solid phase transitions.[67] The field of semiconductor nanocrystals has grown tremendously over the last 10 years and continues to do so at present, but the vast majority of studies and conclusions have been based on a rather narrow segment of the wide range of semiconductors available. Through the development of new materials in nanocrystal configurations, a broader range of properties is now accessible, thereby expanding the possibilities for viable applications. Also, the examination of these properties may now be extended to another class of semiconductors which will add to the fundamental understanding of quantum confined systems.

ACKNOWLEDGMENTS

We would like to thank Andreas Kadavanich for the TEM photographs, Janet Katari for contributions to the XPS studies and Professor Jim Heath of UCLA for contributions to many of the InP studies. U.B. thanks the Rothschild and Fulbright Foundations for fellowships.

REFERENCES

1. Alivisatos, A. P. *Science* **1996**, *271*, 933.
2. Colvin, V. L.; Goldstein, A. N.; Alivisatos, A. P. *J. Am. Chem. Soc.* **1992**, *114*, 5221.
3. Peng, X. G.; Wilson, T. E.; Alivisatos, A. P.; Schultz, P. G. *Angew. Chem. Int. Ed. Engl.* **1997**, *36*, 145.
4. Colvin, V. L.; Schlamp, M. C.; Alivisatos, A. P. *Nature* **1994**, *370*, 354.
5. Dabbousi, B. O.; Bawendi, M. G.; Onitsuka, O.; Rubner, M. F. *Appl. Phys. Lett.* **1995**, *66*, 1316.
6. Klein, D. L.; Mceuen, P. L.; Katari, J. E. B.; Roth, R.; Alivisatos, A. P. *Appl. Phys. Lett.* **1996**, *68*, 2574.
7. Alivisatos, A. P.; Johnsson, K. P.; Peng, X.; Wilson, T. E.; Loweth, C. J.; Bruchez, Jr., M. P.; Schultz, P. G. *Nature* **1996**, *382*, 609.
8. Murray, C. B.; Norris, D. J.; Bawendi, M. G. *J. Am. Chem. Soc.* **1993**, *115*, 8706.
9. Katari, J. E. B.; Colvin, V. L.; Alivisatos, A. P. *J. Phys. Chem.* **1994**, *98*, 4109.
10. Hines, M. A.; Gunyotsionnest, P. *J. Phys. Chem.* **1996**, *100*, 468.
11. Danek, M.; Jensen, K. F.; Murray, C. B.; Bawendi, M. G. *Chem. Mater.* **1996**, *8*, 173.

12. Peng, X. G.; Schlamp, M. C.; Kadavanich, A. V.; Alivisatos, A. P. *J. Am. Chem. Soc.* **1997**, *119*, 7019.
13. Eychmuller, A.; Mews, A.; Weller, H. *Chem. Phys. Lett.* **1993**, *208*, 59.
14. Mews, A.; Eychmuller, M.; Giersig, M.; Schooss, D.; Weller, H. *J. Phys. Chem.* **1994**, *98*, 934.
15. Mews, A.; Kadavanich, A. V.; Banin, U.; Alivisatos, A. P. *Phys. Rev. B* **1996**, *53*, 13242.
16. Mićić, O. I.; Curtis, C. J.; Jones, K. M.; Sprague, J. R.; Nozik, A. J. *J. Phys. Chem.* **1994**, *98*, 4966.
17. Mićić, O. I.; Sprague, J. R.; Curtis, C. J.; Jones, K. M.; Machol, J. L.; Nozik, A. J.; Giessen, B.; Fluegel, B.; Mohs, G.; Peyghambarian, N. *J. Phys. Chem.* **1995**, *99*, 7754.
18. Guzelian, A. A.; Katari, J. E. B.; Kadavanich, A. V.; Banin, U.; Hamad, K.; Juban, E.; Alivisatos, A. P.; Wolters, R. H.; Arnold, C. C.; Heath, J. R. *J. Phys. Chem.* **1996**, *100*, 7212.
19. Guzelian, A. A.; Banin, U.; Kadavanich, A. V.; Peng, X.; Alivisatos, A. P. *Appl. Phys. Lett.* **1996**, *69*, 1432.
20. Brus, L. E. *J. Chem. Phys.* **1984**, *80*, 4403.
21. Olshavsky, M. A.; Goldstein, A. N.; Alivisatos, A. P. *J. Am. Chem. Soc.* **1990**, *112*, 9438.
22. Wells, R. L.; Pitt, C. G.; McPhail, A. T.; Purdy, A. P.; Shafieezad, S.; Hallock, R. B.; *Chem. Mater.* **1989**, *1*, 4.
23. Uchida, H.; Curtis, C. J.; Nozik, A. J. *J. Phys. Chem.* **1991**, *95*, 5382.
24. Uchida, H.; Curtis, C. J.; Kamat, P. V.; Jones, K. M.; Nozik, A. J. *J. Phys. Chem.* **1992**, *96*, 1156.
25. Kher, S. S.; Wells, R. L. *Chem. Mater.* **1994**, *6*, 2056–2062.
26. Kher, S. S.; Wells, R. L. *Nanostructured Materials* **1996**, *7*, 591.
27. LaMer, V. K.; Dinegar, R. H. *J. Am. Chem. Soc.* **1950**, *72*, 4847.
28. Smith, A. L. *Particle Growth in Suspensions*; Academic Press: London, 1983.
29. Cowley, A. H.; Jones, R. A. *Angew. Chem. Int. Ed. Engl.* **1989**, *28*, 1208.
30. Wells, R. L. *Coord. Chem. Rev.* **1992**, *112*, 273.
31. Cowley, A. H.; Benac, B. L.; Ekerdt, J. G.; Jones, R. A.; Kidd, K. B.; Lee, J. Y.; Miller, J. E. *J. Am. Chem. Soc.* **1988**, *110*, 6248.
32. Cowley, A. H.; Harris, P. R.; Jones, R. A.; Nunn, C. M. *Organometallics* **1991**, 652.
33. Miller, J. E.; Kidd, K. B.; Cowley, A. H.; Jones, R. A.; Ekerdt, J. G.; Gysling, H. J.; Wernberg, A. A.; Blanton, T. N. *Chem. Mater.* **1990**, 2, 589.
34. Healy, M. D.; Laibinis, P. E.; Stupik, P. D.; Barron, A. R. *J. Chem. Soc., Chem. Commun.* **1989**, 359.
35. Robinson, W. T.; Wilkins, C. J.; Zeing, Z. *J. Chem. Soc., Dalton Trans.* **1990**, 219.
36. Wells, R. L.; Aubuchon, S. R.; Kher, S. S.; Lube, M. S.; White, P. S. *Chem. Mater.* **1995**, *7*, 793.
37. Jin, S.; McKee, V.; Nieuwenhuyzen, M.; Robinson, W. T.; Wilkens, C. J. *J. Chem. Soc., Dalton Trans.* **1993**, 3111.
38. Inoue, K.; Yoshizuka, K.; Yamaguchi, S. *J. Chem. Eng. Japan* **1994**, *27*, 737.
39. Sato, T.; Ruch, R. *Stabilization of Colloidal Despersions by Polymer Adsorption*; Marcel Dekker: New York, 1980, pp. 46–51.
40. Kimura, K. *J. Phys. Chem.* **1994**, *98*, 11997.
41. Hamaker, H. C. *Recl. Trav. Chim.* **1937**, *56*, 3.
42. Verwey, E. J.; Overbeek, J. Th. G. *Theory of Stability of Lyophobic Colloid*; Elsevier: Amsterdam, 1948, p. 160.
43. JCPDS-ICDD PDF card 32-452. Gong, P., Polytechnic Institute of New York, Brooklyn, NY, 1981.
44. Guinier, A. *X-Ray Diffraction*; Freeman: San Francisco, 1963.
45. Wagner, C. D.; Riggs, W. M.; Davis, L. E.; Moulder, J. F.; Muilenberg, G. E. *Handbook of X-ray Photoelectron Spectroscopy*; Perkin-Elmer: Eden Prairie, MN, 1978.
46. Schwartz, G. P.; Sunder, W. A.; Griffiths, J. E. *J. Electrochem. Soc.* **1982**, *129*, 1361.
47. Bergignat, E.; Hollinger, G.; Robach, Y. *Surf. Sci.* **1987**, *189/190*, 353.
48. Besland, M.-P.; Louis, P.; Robach, Y.; Joseph, J.; Hollinger, G.; Gallet, D.; Viktorovitch, P. *Applied Surface Sci.* **1992**, *56–58*, 846.
49. Hollinger, G.; Gallet, D.; Gendry, M.; Besland, M. P.; Joseph, J. *Appl. Phys. Lett.* **1991**, *59*, 1617.

50. Banin, U.; Mews, A.; Kadavanich, A. V.; Guzelian, A. A.; Alivisatos, A. P. *Mol. Cryst. Liq. Cryst.* **1996**, *283*, 1.
51. Banin, U.; Cerullo, G.; Guzelian, A. A.; Bardeen, C. J.; Alivisatos, A. P.; Shank, C. V. *Phys. Rev. B.* **1997**, *55*, 7059.
52. Hellwege, K.-H., Ed. *Landolt-Bornstein, Group III*; Springer-Verlag: Berlin, 1982, Vol. 17a.
53. Alivisatos, A. P.; Harris, A. L.; Levinos, N. J.; Steigerwald, M. L.; Brus, L. E. *J. Chem. Phys.* **1988**, *89*, 4001.
54. Norris, D. J.; Sacra, A.; Murray, C. B.; Bawendi, M. G. *Phys. Rev. Lett.* **1994**, *72*, 2612.
55. Mićić, O. I.; Sprague, J.; Lu, Z. H.; Nozik, A. J. *Appl. Phys. Lett.* **1996**, *68*, 3150.
56. Vossmeyer, T.; Katsikas, L.; Giersig, M.; Popovic, I. G.; Weller, H. *J. Phys. Chem.* **1994**, *98*, 7665.
57. Efros, Al.L.; Rosen, M.; Kuno, M.; Nirmal, M.; Norris, D. J.; Bawendi, M. *Phys. Rev. B* **1996**, *54*, 4843.
58. Hill, N. A.; Whaley, K. B. *J. Chem. Phys.* **1994**, *100*, 2831.
59. Zorman, B.; Ramakrishna, M. V.; Friesner, R. A. *J. Phys. Chem.* **1995**, *99*, 7649.
60. Lippens, P. E.; Lannoo, M. *Phys. Rev. B.* **1989**, *39*, 10935.
61. Grundmann, M.; Christen, J.; Ledentsov, N. N.; Bohrer, J.; Bimberg, D.; Ruvimov, S. S.; Werner, P.; Richter, U.; Gosele, U.; Heydenreich, J.; Ustinov, V. M.; Egorov, A. Yu.; Zhukoov, A. E.; Kop'ev, P. S.; Alferov, Zh. I. *Phys. Rev. Lett.* **1995**, *74*, 4043.
62. Leonard, D.; Krishnamurthy, M.; Reaves, C. M.; Denbaars, S. P.; Petroff, P. M. *Appl. Phys. Lett.* **1993**, *63*, 3203.
63. Leonard, D.; Pond, K.; Petroff, P. M. *Phys. Rev. B* **1994**, *50*, 11687.
64. Mui, D. S. L.; Leonard, D.; Coldren, L. A.; Petroff, P. M. *Appl. Phys. Lett.* **1995**, *66*, 1620.
65. Grundmann, M.; Stier, O.; Bimberg, D. *Phys. Rev. B* **1995**, *52*, 11969.
66. Alonso, M. I.; Ilg, M.; Ploog, K. *J. Appl. Phys.* **1995**, *78*, 1980.
67. Herhold, A. B.; Tolbert, S. H.; Guzelian, A. A.; Alivisatos, A. P. In *Fine Particles Science and Technology*; Kluwer Academic Publishers, Dordrecht, 1996, p. 331.

MAGNETIZATION REVERSAL IN NANOPARTICLES

Sara Majetich

Advances in Metal and Semiconductor Clusters
Volume 4, pages 35–65

ISBN: 0-7623-0058-2

I. INTRODUCTION

Nanoparticles have generated great interest because of their size-dependent properties. There are many types of magnetic behavior, but not all are size dependent within the nanoparticle size range (2–1000 nm). For example, in a paramagnetic material individual spins respond independently to an external field. Without collective behavior, the magnetization of paramagnetic nanoparticles is identical to that of the bulk.

Size-dependent phenomena are observed in nanoparticles formed from ferromagnetic materials. The ferromagnetic interaction leads to parallel spins on adjacent atoms. However, the exchange forces responsible are short range, and magnetostatic forces dominate at greater distances. Magnetic domains arise spontaneously in bulk ferromagnets to minimize their overall energy. The size of a magnetic domain depends on the material, but typical sizes are in the nanoparticle size range. Below a certain size, it is energetically favorable for a particle to be monodomain.

Here we will concentrate on the physics of monodomain particles, with particular emphasis on how they reverse their magnetization directions. The magnetization per unit volume has not been found to be size-dependent except in clusters of a few hundred atoms or less,[1] but magnetic metastability or hysteresis and the value of the switching field are strongly size-dependent in the nanoparticle size range.

Following a review of the relevant background concerning bulk ferromagnets, the criteria for monodomain particles will be presented. Superparamagnets, which have no hysteresis, will be discussed, and the conditions for superparamagnetism and hysteretic behavior will be explained in terms of an energy barrier model. The remainder of this article focuses on understanding the magnetic behavior of monodomain $SmCo_5$ particles predicted to have very high energy barriers. While experimental results indicate substantial barriers to magnetization reversal, they are a factor of 50 lower than expected. Mechanisms providing a physical basis for this reduction are investigated, and the energy barrier model is refined.

II. BULK FERROMAGNETISM BACKGROUND

Magnetism arises from the spin angular momentum of electrons. For some atoms or ions, the contributions from different electrons couple such that there is a net magnetic moment, but moments on different atoms or ions act independently of each other, leading to paramagnetism. Statistical mechanics can be used to show that the shape of the magnetization versus fields curve is an S-shaped Brillouin function,[2] which depends on the ratio of the magnetic field, H, to the temperature. When $H = 0$, paramagnets have zero magnetization.

If the net moments of different atoms or ions couple to each other so that the moments of nearest neighbors are in the same direction, there is ferromagnetic response, and the magnetization of the sample in high fields is generally orders of magnitude larger than in a paramagnet. Ferromagnetic ordering occurs due to the quantum mechanical exchange interaction, which competes with magnetic dipole interactions that favor antiparallel ordering perpendicular to the spin direction. The exchange interaction falls off exponentially with separation, while the dipolar energy drops off with the cube of the distance. Dipole interactions therefore dominate at large distances, and magnetic domains are formed spontaneously in bulk ferromagnets (Figure 1). Real ferromagnets have a distinctively different magnetization curve than paramagnets, with hysteretic behavior (Figure 2).

Before turning to fine particle magnetism, we will define a few technical parameters used to describe ferromagnets. A ferromagnetic sample, such as a magnetic recording disk, does not normally act as a magnet by itself. However, one of its bits placed in a strong magnetic field, such as that from the magnetic recording heads, will remain magnetized even when the external field is removed. If a material starts out unmagnetized at zero field ($M = 0$, $H = 0$), the moments due to different domains cancel. When a field is applied, the domain walls move through the crystal such that favorably oriented domains grow at the expense of those unfavorably oriented. When a domain wall passes through a region with a grain boundary, a structural imperfection, or an impurity, it can be pinned temporarily. Additional energy (supplied by the external field) is required to overcome this barrier and continue its motion. If the magnetic field direction is reversed, it will also take extra energy for the domain wall to pass through the defect in the opposite direction. This leads to hysteretic behavior. The magnetization curve in general has both irreversible sections where domain wall motion occurs, and reversible regions where the magnetization changes due to rotation of the overall moment to orient with the field direction. These irreversible sections are metastable; given a long enough time at finite temperature no hysteresis would be observed. Fortunately for the many

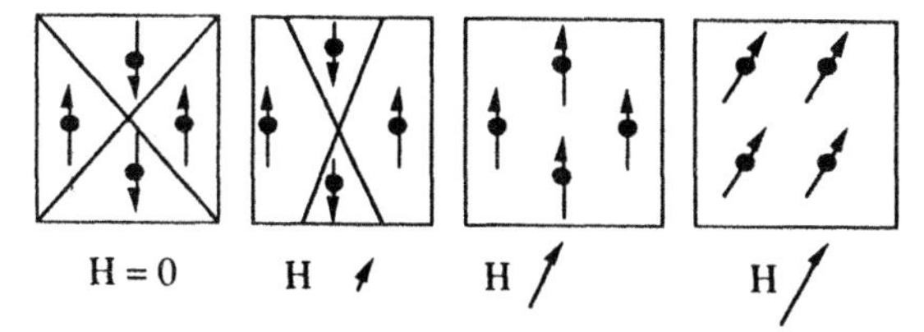

Figure 1. Magnetic domains in a ferromagnet, as a function of the applied magnetic field. The magnetization increases by domain wall motion, which can be irreversible, until there is a single domain in the sample. Further increases in the magnetization arise from rotation of the spins to align with the applied field.

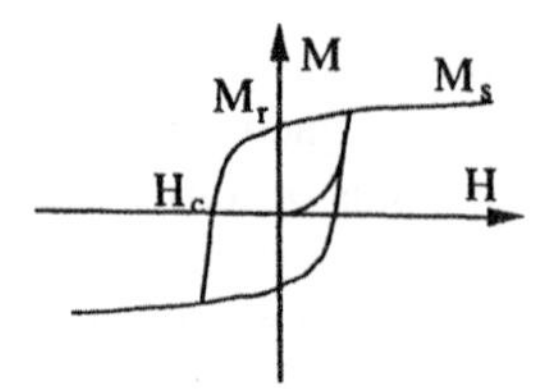

Figure 2. Magnetization curve of a ferromagnet. Domain walls move in the double-valued region, and reversible rotation occurs at high fields.

applications involving ferromagnets the equilibration time can be many years. However, a steel paperclip can be temporarily magnetized by a permanent magnet, but loses its magnetization within a few minutes. At high fields the magnetization reaches its saturation value, M_s. If the field is reduced to zero following saturation, it does not immediately return to zero, but to its remanent magnetization, M_r. Only with a field in the opposite direction equal to the coercivity, H_c, does the magnetization return to zero.

Some materials reach their M_s at lower fields than others. The saturation field generally depends on the crystallographic direction, due to spin-orbit coupling. The electron density associated with orbital angular momentum, and therefore chemical bonds, does not freely rotate with an arbitrary applied field direction, so M_s is achieved at lower fields along certain crystallographic directions. The direction with the lowest field required for saturation is called the easy axis. The preference of a material in magnetizing along a certain crystallographic direction is described quantitatively by the magnetocrystalline anisotropy, K. Shape and stress can also cause magnetic anisotropy, but these effects are negligible in the roughly spherical particles described here.

On an atomic scale, it requires less energy to have magnetic moments aligned in certain directions. The crystallographic direction with the minimum energy for alignment is called the "easy axis" (E. A.). In highly symmetric crystals there can be multiple equivalent easy axes, but in uniaxial crystals there is a unique direction. The degree to which a particular direction is favored is quantified in terms of the magnetocrystalline anisotropy energy density, K. In general the anisotropy energy can be expressed in terms of direction cosines between the magnetic moment and the crystal axes.[3] For a uniaxial crystal this expression simplifies, since only the angle between the c axis (the E. A.) and the magnetic moment, ϕ, is important. While in general, there may be several terms in the anisotropy energy,

$$E = K_o V + K_1 V \sin^2\phi + K_2 V \sin^4\phi + \ldots \tag{1}$$

the first term is merely a constant offset, and frequently $K_2 << K_1$, so that including only the term in $\sin^2\phi$ is a reasonable approximation. We will make this assumption here to simplify the mathematics, and let $K_1 = K$, but in general the constants K_1 and K_2 for each material should be checked to see if this is appropriate.

Magnetocrystalline anisotropy arises from spin-orbit coupling. Exchange-coupled spins rotate with the magnetic field, regardless of the crystallographic direction. However, orbital angular momentum is strongly associated with the crystal lattice, and not free to rotate with the applied field. Unfortunately, it is currently impossible to calculate the magnetocrystalline anisotropy constants from first principles.[3] In general K_1 and K_2 are determined from torque magnetometry of large single crystals.

There are two qualitative classes of ferromagnetic materials, hard and soft. They are differentiated by the width of the hysteresis loop, which is related to the magnetocrystalline anisotropy of these materials. Compounds with large values of K have a strong preference for magnetizing in a particular crystallographic direction, and are known as hard magnets. They generally have anisotropic crystal structures such as hexagonal or rhombohedral. Hard magnetic materials have large coercivities and large values of M_r/M_s. They are used in applications such as permanent magnets and magnetic recording media. The permanent magnets have the widest loops, and a figure of merit for permanent magnets is the energy product, $(BH)_{max}$, where the magnetic induction, $B = H + 4\pi M_s$.

Compounds with small magnetocrystalline anistropy are called soft magnets. They usually possess cubic crystal structures, and therefore have a number of equivalent easy axes. Soft magnetic materials have very narrow hysteresis loops, and are used in applications where they are magnetized and demagnetized frequently, such as magnetic recording heads, transformer cores, or the stators and rotors of motors and generators.

Because the coercivity is strongly size-dependent in the nanoparticle range, a ferromagnetic material can be made magnetically harder or softer by changing the size of the particle, or equivalently by changing the grain size in a solid. The experimental results presented here will focus on magnetically hard materials, but there are many exciting possibilities and interesting questions about the behavior of magnetically soft materials in the nanoparticle size range.

III. CRITERIA FOR MONODOMAIN PARTICLES

The coercivity as a function of grain size has been shown to be strongly size-dependent both in fine particles[4] and in recrystallized amorphous magnets.[5] While the magnetic moment of a monodomain particle depends on the number of parallel spins, the magnetization per unit volume is not. M_s and M_r are usually expressed in emu/cm^3, so they do not vary with size, though the moment of a single monodomain particle is clearly size-dependent. There have been reports of size dependence in the magnetocrystalline anisotropy, K,[6] and the Curie temperature, T_c,[7] but we have not observed any conclusive deviations from bulk behavior.

The size dependence of the coercivity is intimately connected with the domain structure and the grain size. In bulk samples, magnetization reversal occurs due to domain wall motion. Favorably oriented domains grow at the expense of those not aligned with the applied field. As domain walls move through a sample, they can become pinned at grain boundaries, and additional energy is needed for them to continue moving. Pinning is one of the two main sources of the coercivity, the other being the ease of nucleating a new domain.[8] Reducing the grain size creates more pinning sites and increases H_c. This is true for larger grain sizes, but below a certain value, H_c decreases rapidly. In monodomain particles with $d < d_{cr}$, such as the nanoparticles discussed here, the energy needed to rotate the spins depends on the number of spins. Therefore, H_c decreases as the particle size is reduced for monodomain particles. Below a second size threshold, d_{sp}, the coercivity is zero.

The critical size d_{cr}, below which particles are monodomain is estimated from the relation,[9]

$$d_{cr} = \frac{18\gamma}{M_s^2} \tag{2}$$

where γ is the domain wall energy per unit area and M_s is the saturation magnetization. This size can be quite large, as in $SmCo_5$, where $d_{cr} = 2$ μm, but is typically much smaller. The monodomain threshold is about 45 nm for BCC Fe and 50 nm for FCC Co. The particles described here are all within the monodomain range.

IV. SUPERPARAMAGNETS

Below a certain size, d_{sp}, the coercivity drops to zero, and the particles are said to be superparamagnetic. This critical size is temperature-dependent, so a particle can be superparamagnetic at one temperature and have hysteresis at another. Thermal effects are very important in determining many types of magnetic behavior. For example, above the Curie temperature of a ferromagnet, there is sufficient thermal energy to overcome the magnetic forces which tend to align the spins, and the sample becomes paramagnetic, but thermal forces are still active well below this temperature. A soft material like an iron nail can be magnetized in a strong field. For a short time, the nail itself attracts other ferromagnetic objects, but this ability decays rapidly in time because thermal fluctuations cause the formation and growth of domains aligned in different directions. Magnetic recording media are made from harder magnetic materials so that thermal fluctuations are unlikely to erase information. While in a magnetized nail, the time decay of the remanent magnetization, M_r, can be measured, in a superparamagnet, the decay time is by definition shorter than the measurement time. While above the Curie temperature individual spins have sufficient thermal energy to decouple from their neighbors, in a superparamagnet the entire particle moment flips back and forth.

Superparamagnetic particles have single magnetic domains, but they are small enough that thermal fluctuations can easily reverse the direction of their magnetic moment. Examples of superparamagnetic particles are found in ferrofluids[10] and in some granular giant magnetoresistance materials.[11] The experimental signature of superparamagnetism is $H_c = 0$ and a magnetization proportional to a Langevin function, L(x),

$$\frac{M}{M_s(T=0)} = \mathrm{L}(x) = \coth(x) - \frac{1}{x} \tag{3}$$

where T is the temperature, and $x = \mu H/kT$, and $\mu = M_s V_{avg}$ is the particle magnetic moment (Figure 3).[12] From fitting experimental data at different temperatures and fields to a Langevin function, the average magnetic moment per particle, μ, can be determined. Typical numbers of coupled spins in superparamagnets can range from hundreds to tens of thousands. The Langevin function is a classical analog of the Brillouin function which describes the behavior of paramagnets, with the particle moment substituting for the atomic moment. Whether or not the particle moments have been reversed depends both on the timeframe of the magnetization measurement, and the time required for equilibration.

However, when the sample of Figure 3 is cooled to cryogenic temperatures, hysteresis reappears and the sample is no longer superparamagnetic (Figure 4). Evidently the behavior of monodomain particles is more complex. The current

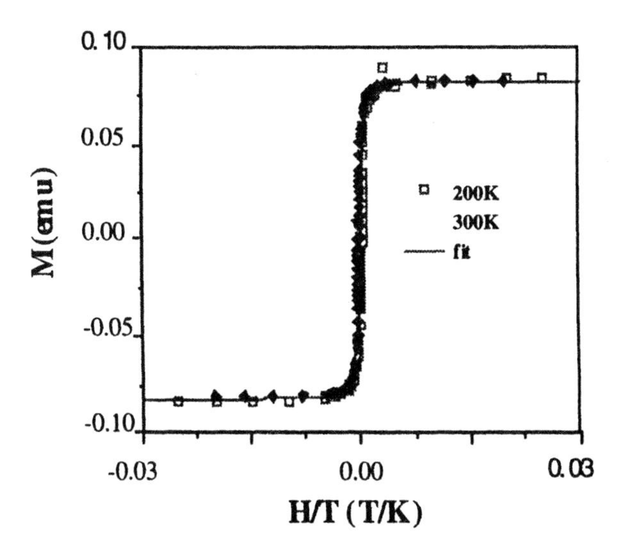

Figure 3. Superparamagnetic magnetization curve at high temperatures for FCC Co nanoparticles made in a carbon arc. Data taken from ref. 10.

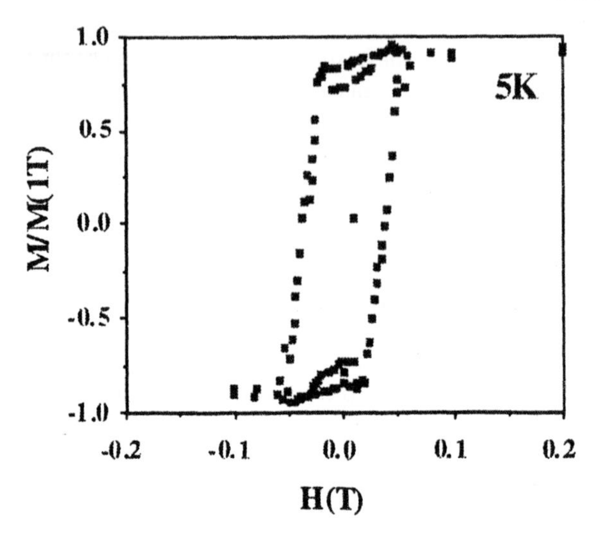

Figure 4. Hysteresis at low temperatures in FCC Co nanoparticles. Here the average particle size is 5 nm. Data taken from ref. 10.

understanding of this phenomenon involves no pinning of defects, as in bulk ferromagnets, but an energy barrier to coherent rotation of the magnetic moment.

V. THE COHERENT ROTATION MODEL

A. Energy Considerations

While the behavior of bulk ferromagnets is extremely complex and not yet understood quantitatively, significant progress has been made in understanding magnetization reversal in ellipsoidal monodomain particles. In Stoner–Wohlfarth theory for a single domain particle,[9] all spins within the particle are assumed to be aligned, and magnetization reversal occurs by coherent rotation of these spins. To see this, we examine the energy of such a particle. There is a potential energy, M_sVH, associated with a magnetic moment M_sV aligned with an external magnetic field H. Here M_s is the saturation magnetization per unit volume.

The total energy of a uniaxial single domain particle can be written as the sum of the anisotropy and potential energy terms,[13]

$$E = KV \sin^2(\phi) - VHM_s \cos(\theta - \phi) \tag{4}$$

where the relationships between the angles, θ and ϕ, and the directions of the external field, the easy axis, and the particle's magnetization vector are shown in Figure 5.

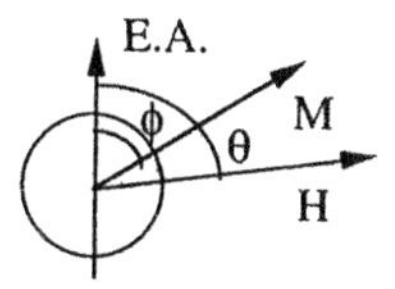

Figure 5. A spherical single domain particle in an external field H, with crystallographic easy axis (E.A.) and magnetic moment $M_S V$. θ is the angle between H and the easy axis, and ϕ is the angle between M and the easy axis.

If the crystallographic easy axis is aligned with the H, $\theta = 0$, and there is a simple energy barrier model for magnetization reversal. The maximum and minimum energies are found by setting the $dE/d\phi = 0$, yielding the conditions:

$$\cos\phi = -\frac{H}{H_K} \tag{5}$$

and,

$$\sin\phi = 0 \Rightarrow \phi = 0{,}180° \tag{6}$$

Here the anisotropy field, $H_K = 2K/M_s$. From the sign of $d^2E/d\phi^2$, the condition of Eq. 5 represents a maximum, while that for Eq. 6, is a minimum. In the latter case $\phi = 180°$ is a local minimum, and $\phi = 0°$ is a global minimum. By subtracting the local energy minimum from the energy maximum, the energy barrier to magnetization reversal is found to be:

$$E_B = E_{\max} - E_{\min} = KV\left(1 - \frac{H}{H_K}\right)^2. \tag{7}$$

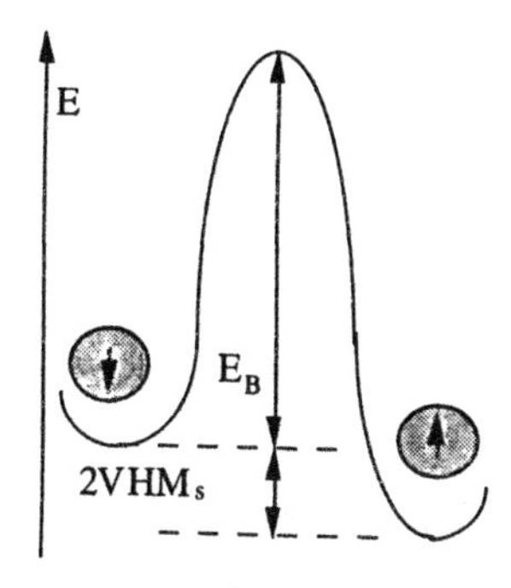

Figure 6. The energy barrier model for a spherical monodomain particle of a uniaxial ferromagnet, showing the local and global minima separated by an energy barrier, E_B.

Thus, for a system with a large magnetocrystalline anisotropy, the energy barrier will be large except when $H \approx H_K$. The difference between the local and global minima is due to the potential energy supplied by the external field:

$$\Delta E = E_{\text{min(local)}} - E_{\text{min(global)}} = 2M_s VH \tag{8}$$

The energy barrier magnitude is shown for small fields in Figure 6. It varies with the applied field, and will be at a minimum when $H/H_K = 1$. Equations 5–8 and Figure 5 assume that the external field direction is parallel to the particle's easy axis. If θ is nonzero, then Eq. 4 has no analytic solution, but the energy barrier picture is still useful for visualizing magnetization reversal.

B. Time Dependence of Equilibration

Moving back to the case of perfect alignment, the transitions rates can be calculated. The transition probability for a particle to go from its local to its global minimum in Figure 6, reversing its magnetization direction, is given by,

$$\omega_f = \tau_o^{-1} \exp\left[-\frac{(E_{\text{max}} - E_{\text{min(local)}})}{kT}\right] = \exp\left[\frac{E_B}{kT}\right] \tag{9}$$

for $T > 0$. Here ω_f is the forward transition rate, τ_o^{-1} is the attempt frequency which is on the order of the Larmor precession frequency (10^9 Hz), k is Boltzmann's constant, and T is the temperature. The equilibration time, τ, is the time it takes for a system to reach 1/e of its equilibrium magnetization value. Its inverse, the equilibration rate, is a sum of the forward and reverse transition rates over the energy barrier,

$$\tau^{-1} = \omega_f + \omega_r. \tag{10}$$

In Figure 6, the forward and backward transition probabilities may be comparable for small fields, but for larger fields, the probability of a reverse transition is negligible. For 25 nm diameter nanoparticles of $SmCo_5$ in zero field at room temperature, E_B is roughly 25,000 times kT, and so magnetization reversal almost never occurs at small fields. The equilibration rate can therefore be approximated by the equation,[14]

$$\tau^{-1} = \tau_o^{-1} \exp(-E_B/kT) \tag{11}$$

with minimal error. Equilibration times are short when E_B is small compared to the thermal energy. High anisotropy materials have large energy barriers, and therefore long equilibration times.

C. Conditions for Superparamagnetism

Different magnetization curves are observed, depending on the ratio of the equilibration time to the magnetization measurement time, τ/τ_{exp}. When $\tau/\tau_{exp} << 1$, the sample equilibrates before the magnetization is measured, and no hysteresis is observed even though the particles are made from a ferromagnetic material. This is one definition of a superparamagnet. The equilibration time for a given sample can be changed by varying the temperature, and a sample which is superparamagnetic at high temperature may therefore have hysteresis at lower temperatures. The temperature of the onset of hysteresis is called the Blocking temperature, T_B. An estimate of the Blocking temperature for a given sample can be made by combining Eq. 7 (with $H = 0$) and Eq. 11 to get:

$$T_B = \frac{KV}{k\ln(\tau/\tau_o)} \approx \frac{KV}{25k} \tag{12}$$

The typical measurement time τ for a given apparatus and experimental conditions can vary slightly, but can easily be determined. For the $SmCo_5$ data presented here, the value was found to be 29.3. Alternatively we can define a critical diameter, d_{sp}, such that particles that size or smaller are superparamagnetic:

$$d_{sp} = \left[\frac{6kT\ln(\tau/\tau_o)}{\pi K}\right]^{1/3} \tag{13}$$

For $SmCo_5$ nanoparticles, d_{sp} is on the order of 2 nm.

VI. MAGNETIZATION REVERSAL IN MONODOMAIN PARTICLES WITH HYSTERESIS

A. Temperature Dependence of the Coercivity

At the Blocking temperature there is barely enough thermal energy to equilibrate via thermal fluctuations within the measurement time, and the coercivity, H_c, equals zero. Below T_B, the coercivity increases. For many systems, such as the carbon-coated Co nanoparticles in Figure 7, the coercivity obeys the relation,

$$H_c(T) = H_{co}[1 - (T/T_B)^{1/2}] \tag{14}$$

where H_{co} is the zero temperature value of the coercivity. The origin of this equation can be seen by combining Eq. 7 for the energy barrier in a magnetic field with Eq. 11 for the equilibration rate, and simplifying the expression using Eq. 12 for the Blocking temperature to get:

$$H = H^*(T) = H_K[1 - (T/T_B)^{1/2}]. \tag{15}$$

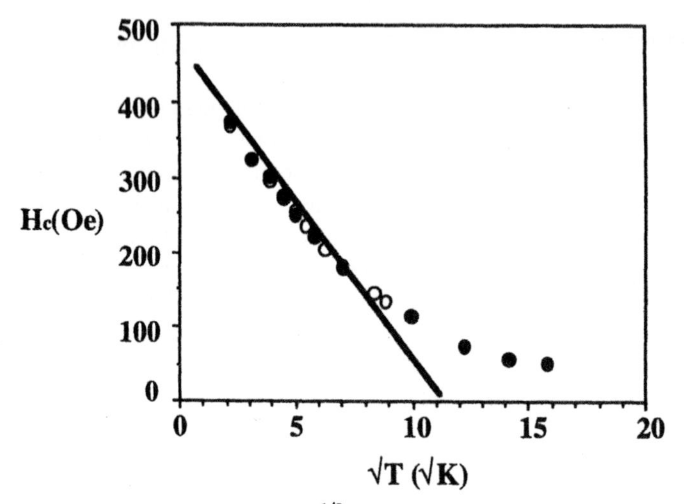

Figure 7. Coercivity as a function of $T^{1/2}$ for carbon-coated Co nanoparticles with $\bar{d} = 5$ nm. Extrapolation to $H_c = 0$ yields a Blocking temperature of 160 K. Open circles from magnetically aligned samples; filled circles from randomly aligned samples. Data taken from ref. 10.

In the energy barrier model, the sample will not equilibrate within the measurement time for $H = 0$, but will if a sufficient external field, $H^*(T)$, is applied. Unfortunately there is not in general exact agreement between the theoretical value for the coercive field and H_K and the experimental H_c and H_{co}, even for fine particle magnets. In permanent magnet materials, H_c is often an order of magnitude smaller than H_K.[15]

Equation 12 can be exploited to design fine particles with room temperature hysteresis ($T_B > 300$ K), either by increasing the particle size, or by making particles of a magnetic material with a large K value,[16] or both. Bulk $SmCo_5$ has a large magnetocrystalline anisotropy constant, 1.77×10^8 erg/cm^3 at 300 K,[17] which gives rise to large coercivities and the use of samarium cobalt alloys as permanent magnet materials. There are materials with greater K and H_c, but the enhanced values are associated with the presence of multiple phases, plus interactions between grains. Here we focus on the properties of single-phase particles, which are more straightforward to interpret, but complicated nonetheless.

The $SmCo_5$ nanoparticle samples showed large coercivities at all temperatures studied, ranging from a maximum of 2.17 T at 25 K to room temperature value of 1.11 T (Figure 8). Fine particle magnetism theory predicts a $T^{1/2}$ temperature dependence of the coercivity for non-interacting particles.[18] However, in bulk $SmCo_5$, a nearly linear temperature dependence has been reported.[19] However, the bulk coercivity drops off by over a factor of 6 between 4 and 300 K, whereas our H_c falls by only a factor of 2. In addition, the bulk temperature dependence is attributed to domain wall pinning mechanisms believed to be absent in our fine

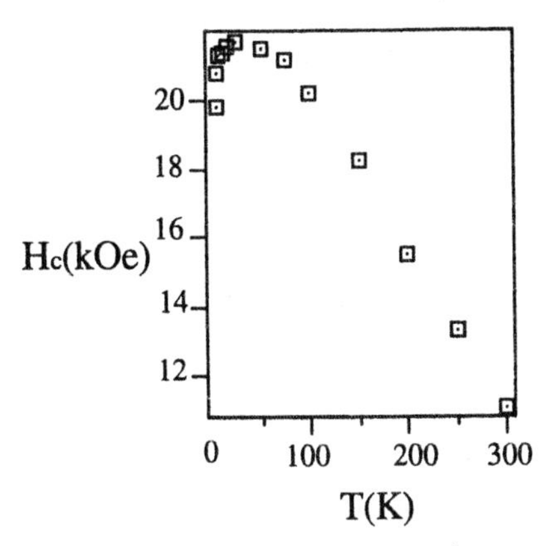

Figure 8. Coercivity as a function of temperature for $SmCo_5$ nanoparticles with $\bar{d}$ = 25 nm.

particle samples, and to the magnetic Peierls effect, which drops off much more rapidly with increasing temperature.[20]

There are standard methods for extracting the value of the magnetocrystalline anisotropy for either perfectly aligned or randomly oriented samples,[20–22] but they could not be used reliably since our samples were highly but not completely aligned. The bulk $SmCo_5$ magnetocrystalline anisotropy is known to decrease linearly with increasing temperature.[23,24] An empirical fit shows that our data could be understood in terms of an effective anisotropy with the same temperature dependence as the bulk but only 0.0265 times its magnitude. Including the temperature dependence of H_K and T_B in Eq. 15, and assuming that the effective anisotropy has the same temperature dependence as the bulk value, we were able to fit the experimental values for H_c quantitatively as well as qualitatively.

B. Time Dependence of the Magnetization

Suppose we have an assembly of monodomain particles with a single E_B, which are aligned and at their saturation magnetization. If this field is turned off, the magnetization decays as,

$$M(t) = M_o \exp(-t/\tau) \tag{16}$$

where M_o is the magnetization at $t = 0$ when the field direction changes. If a field is suddenly applied in the opposite direction, the magnetization decays in time, so that,

$$M(t) = (M_o - M_{eq}) \exp(-t/\tau) + M_{eq} \tag{17}$$

where t is the time elapsed after the reverse field is applied, and M_{eq} is the equilibrium magnetization of the sample at $t \gg \tau$. If the reverse field is zero, M_{eq}

will be zero. The magnetization decay rate can be found by taking the derivative of Eq. 16 and substituting for τ using Eq. 11, yielding:

$$\frac{dM}{dt} = \frac{-(M_o - M_{eq})}{\tau} \exp(-t/\tau) = \frac{-(M_o - M_{eq})}{\tau_o} \exp\left(\frac{-E_B}{kT}\right) \exp(-t/\tau) \tag{18}$$

All real fine particle systems have a distribution of particle volumes,[25] and from Eq. 7 therefore a distribution of energy barriers. Let the number of particles with energy barriers between E_B and $E_B + dE_B$ be $f(E_B)dE_B$. dM/dt for a multiple energy barrier system is then:

$$\frac{dM}{dt} = -\frac{(M_o - M_{eq})}{\tau_o} \int_0^\infty f(E_B) \exp(-t/\tau) \exp(-E_B/kT) dE_B \tag{19}$$

Using Eq. 11 to change the integration variable from dE_B to $d[1/\tau]$, we obtain:

$$\frac{dM}{dt} = M_o kT \int_{\tau_o^{-1}}^{0} f(E_B) \exp(-t/\tau) d\left[\frac{1}{\tau}\right] \tag{20}$$

If $f(E_B)$ is slowly varying compared to $\exp(-t/\tau)$ over the range of integration, it can be taken out of the integral, and the integral can be evaluated analytically.[26]

For the case that $t/\tau_o \gg 1$ the solution to Eq. 20 simplifies even further and one can obtain an expression for the magnetic viscosity, S:

$$\frac{dM}{dt} = -\frac{(M_o - M_{eq})kTf(E_B)}{t} [1 - \exp(-t/\tau_o)]$$

$$\approx -\frac{(M_o - M_{eq})kTf(E_B)}{t} = -\frac{S}{t}. \tag{21}$$

($\frac{dM}{dt}$ has the same H dependence as $f(E_B)$.) Since $\tau_o^{-1} \sim 10^9$ Hz, and measurement times are at least 10 s which is orders of magnitude greater, this is an excellent approximation. Integrating Eq. 21 and applying the boundary condition that $M(t = 0) = M_o$ yields:

$$M(t) = M_0 - S \ln t \tag{22}$$

In order for the natural log to have a dimensionless argument, it is convention to change variables from t to (t/t_o), where t_o is the first measurement time. The slope remains the same, and since most experiments focus on the measurement of S and not M_o, the change has no effect.

In magnetic viscosity experiments, after initial saturation, a field in the reverse direction H was applied, and the time dependence of the magnetization at the field H was measured. The magnetic viscosity, $S = dM/d\ln(t)$, was determined from plots

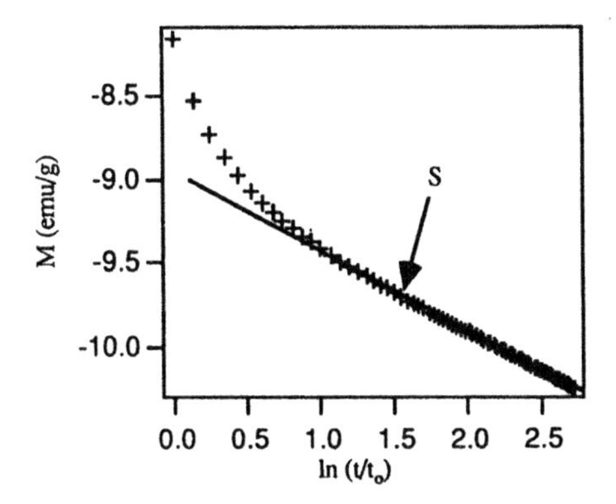

Figure 9. Time dependence of the magnetization at 10.5 kOe and 300 K.[27]

of M as a function of normalized time, t/t_0, where t_0 was the time of the first measurement after the reverse field was applied. Its precise value did not affect the magnitude of S, which was found by fitting the linear portion of the plot, corresponding to a typical time range of 1400–8000 s. This process was repeated for different fields and temperatures.

Figure 9 illustrates the time decay of the sample magnetization at a temperature of 300 K in a reverse field of 10.5 kOe. The magnetization follows the $\ln(t)$ behavior of Eq. 22 at long times. At short times there is an additional component which decreases exponentially with time, just as in a single barrier model. Here we focus on the long time data and the magnetic viscosity.

C. Field Dependence of Magnetization Reversal Rate

By combining Eqs. 22 and 7, we see that S depends on the applied field, H. For an ideal sample with a single energy barrier, the magnetization decay rate is proportional to the equilibration rate, τ^{-1}, which is given by Eq. 11. Substitution for the energy barrier height using Eq. 7, we can show that,

$$\tau^{-1} = \tau_0^{-1} \exp[E_B/kT] = \tau_0^{-1} \exp\left[\left(\frac{VM_s^2}{4\,KkT}\right)(H - H_K)^2\right] \tag{23}$$

which has a Gaussian dependence on the magnetic field and is peaked at $H = H_K$.

An alternative method of studying magnetization reversal and the distribution of energy barriers is through DC demagnetization (DCD) experiments.[28] Here $M(H)$ refers to the remanent magnetization measured after a field H was applied to the sample for a short time, on the order of 30 s. Though the exposure time to the field H seems short, it is sufficient for $SmCo_5$ because of its large anisotropy energy. For a particular field, a certain portion of the particles will have energy barriers of a few kT or less, and they will switch their magnetization direction within the exposure

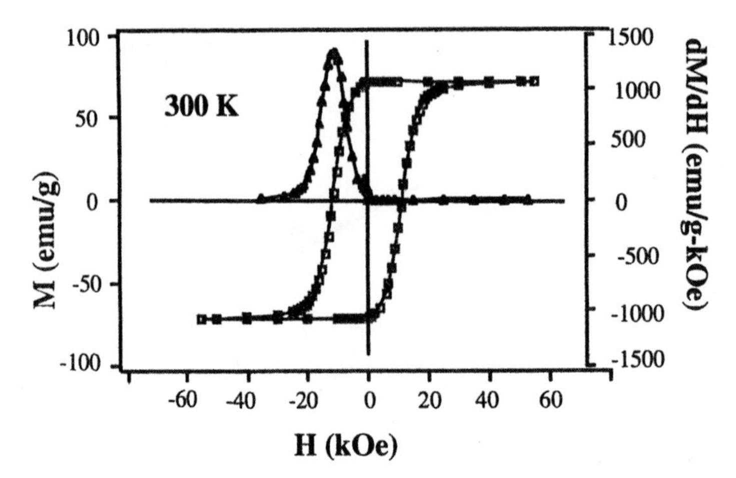

Figure 10. DC demagnetization hysteresis loop for $SmCo_5$ nanoparticles with $\overline{d} = 25$ nm, along with the irreversible susceptibility, *dM/dH*.

time, while most others would remain unaffected even for much longer times at the field *H*. Because the field at the time of measurement equals zero, the particle moments are restricted to lie parallel or antiparallel to the easy axis (Figure 6). The magnetization rotation contribution to the energy at high magnetic fields (Figure 1) are removed, as are superparamagnetic contributions, and the coherent rotation model can be used to calculate the change in magnetization.

The derivative of the DCD loop is *dM/dH* or the irreversible susceptibility, χ_{irr}, as shown in Figure 10. From this one can obtain the height, $\chi_{irr,max}$, and the 1/e width of the distribution, 2σ. The relationship between *S* and χ_{irr} is given by,[29]

$$S = \chi_{irr} H_f \tag{24}$$

where H_f is the fluctuation field. The form of H_f for fine particles is,[29]

$$H_f = \frac{kT}{vM_s} \tag{25}$$

where v is the nucleation volume and for monodomain particles has been postulated to correspond to either the particle volume,[30] or to *dE/dH*, which in the case of the coherent rotation model and Eq. 7 is given by:

$$v = \frac{dE}{dH} = M_s V \left(1 - \frac{H}{H_K}\right) \qquad (26)^{31}$$

Figure 11 illustrates $\chi_{irr}(H)$ in comparison with $S(H)$, for 300 K. Both functions had maxima at the coercivity and were Gaussian in shape, with comparable

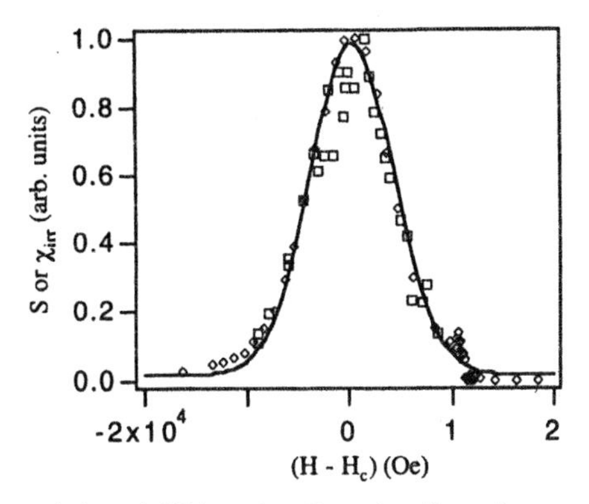

Figure 11. Scaled χ_{irr} (H) and $S(H)$ and a Gaussian fit to the experimental results, at 300 K. From ref. 27.

linewidths. However, the magnetic viscosity data had considerably lower signal-to-noise ratios, leading us to concentrate on quantitative analysis of the χ_{irr} data.

Because the coercivity of the sample changed with temperature, S and dM/dH were plotted as a function of a normalized magnetic field, H/H_c. As the temperature was increased, the peak magnitude of the irreversible susceptibility, dM/dH_{max} increased, as shown in Figure 12. At the same time, the 1/e linewidth 2σ of the peak in dM/dH decreased, as shown in Figure 13.

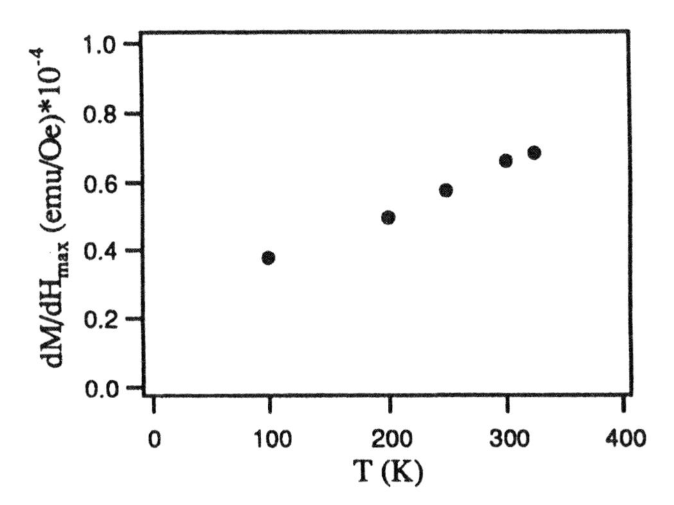

Figure 12. The maximum value of dM/dH (H), as a function of temperature. From ref. 27.

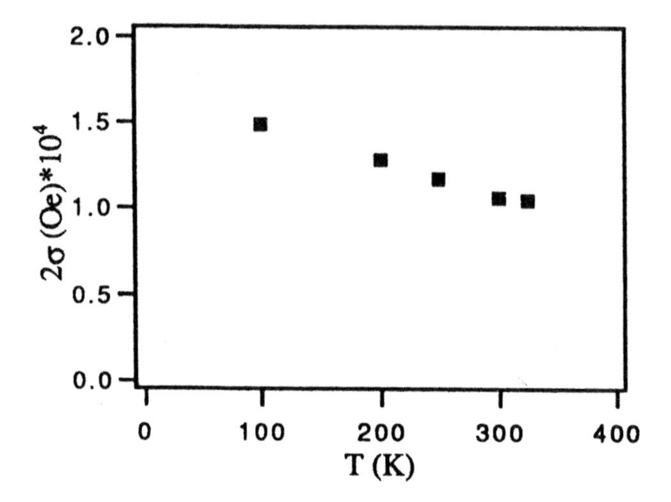

Figure 13. The linewidth σ of the peak in *dM/dH*, as a function of temperature. From ref. 27.

To interpret the functional dependence of these parameters, the raw data were scaled, first by plotting as a function of a reduced field, H/H_c. With this step data for all temperatures peaks at the same value of the reduced field, and curves for different temperatures have the same linewidths. The second step was to plot $dM/d(H/H_c)$ as a function of the reduced field, which we define as $\chi_{irr}(H/H_c)$, such that:

$$\chi_{irr}\left(\frac{H}{H_c}\right) = \frac{dM}{d\left(\frac{H}{H_c}\right)} = H_c\frac{dM}{dH} = H_c\chi_{irr}(H) \tag{27}$$

With this adjustment all the amplitudes scale to a universal curve as well. The formula for this curve is given by:

$$\chi_{irr}(H/H_c) = \frac{H_c(A(T)/H_c)}{\sqrt{2\pi}(\sigma/H_c)}\exp\left\{-\frac{(1-H/H_c)^2}{2(\sigma/H_c)^2}\right\}. \tag{28}$$

Here $A(T)$ is the temperature-independent amplitude and is the same for both the raw and the scaled data, and σ refers to the linewidth of the raw data, while $\sigma_{rel} = \sigma/H_c$ is the scaled and temperature-independent linewidth. Figure 14 illustrates how the scales data fit a universal curve given by Eq. 27. The area under the curve now equals twice the DC demagnetization M_s, as it should for particles switching from magnetization $+M_s$ to $-M_s$.

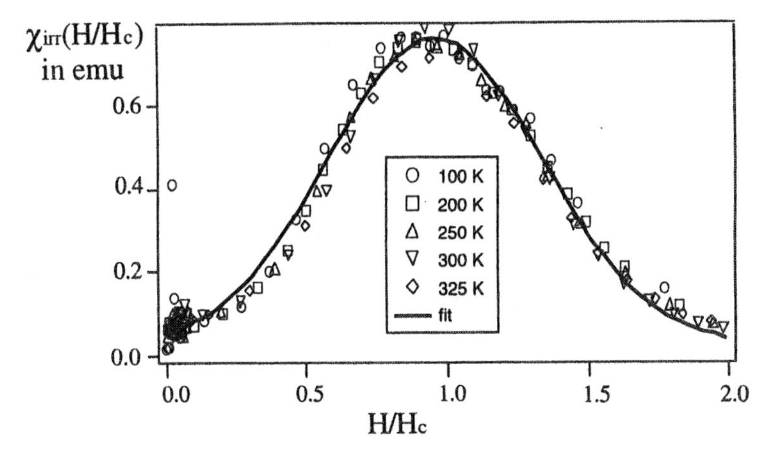

Figure 14. Scaled irreversible susceptibility as a function of temperature. From ref. 27.

Because data from a range of temperatures can be scaled to a universal function of the applied field relative to the coercivity or average switching field, H/H_c rather than H/H_K is the critical variable. Just as the experimental coercivities are substantially less than the theoretical values, there is a discrepancy between the theoretical model with the bulk magnetocrystalline anisotropy. The same adjustments which would empirically make the experimental coercivity data in agreement with theory would also lead to agreement with the irreversible susceptibility results.

Regardless of the values of K used to calculate the anisotropy field,[23,24] H/H_K is small and therefore the factor $[1 - H/H_K]$ is roughly constant for all fields where the values of S and χ_{irr} are significant. Because of this, we cannot distinguish the two possibilities experimentally. The ratios of $S(H)/\chi_{irr}(H)$ near the coercivity fall within a factor of 2 of the fluctuation field values calculated from Eqs. 24 and 25, but with a stronger temperature dependence than that predicted.

VII. MODELING OF THE SWITCHING FIELD DISTRIBUTION

A. Switching Field Distribution Due to Particle Size

The experimental results for $\chi_{irr}(H)$ and $S(H)$ are both related to the distribution of switching fields, and therefore the distribution of energy barriers in the samples. From the dependence of the energy barrier on the particle volume, it seems logical to expect variations in particle size dominate the distribution of energy barriers, $f(E)$.

SEM of the ball milled nanoparticles (Figure 15) confirmed earlier transmission electron microscopy results, which indicated nominally spherical particles which

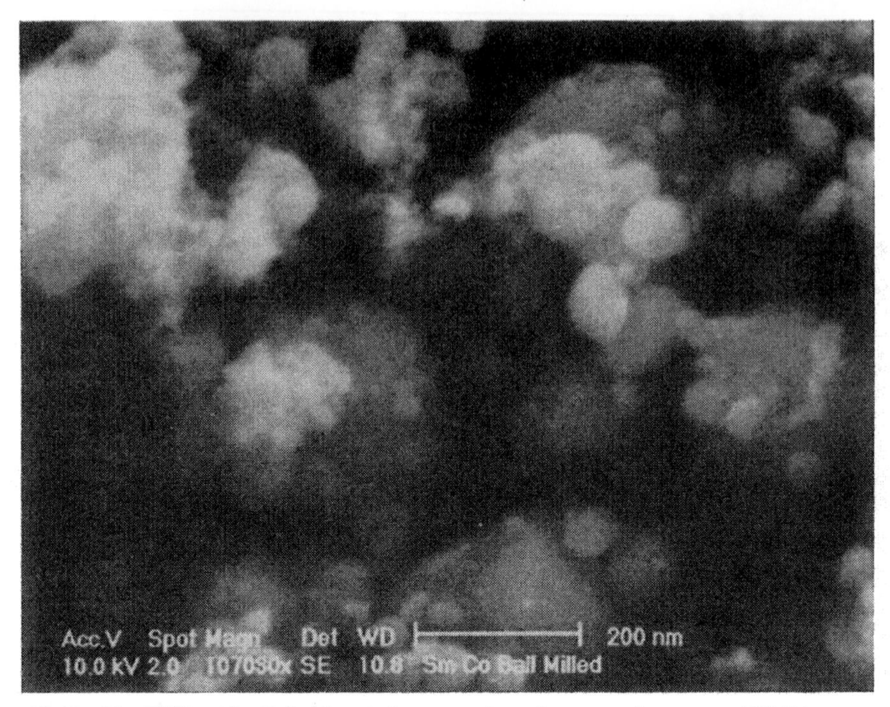

Figure 15. Philips XL-40 field emission scanning electron microscope (SEM) image of ball milled $SmCo_5$ nanoparticles.

were clearly within the monodomain size range. From the micrographs, the size distribution of the nanoparticles was determined. The diameters were fit to a log normal function, $f(d)$, given by,

$$f(d) = \frac{1}{\sqrt{2\pi}\ln\sigma}\exp\left\{-\frac{(\ln d - \ln \bar{d})^2}{2(\ln\sigma)^2}\right\} \tag{29}$$

where the most probable size $\bar{d} = 16.4$ nm, and the 1/e linewidth, $2\sigma = 4.36$, as shown in Figure 16. The most probable particle size was close to the average size of 25 nm obtained from Scherrer analysis of the X-ray diffraction peak widths.[27] Note that cursory examination of the micrograph or measurements on only a few particles could lead to a significant overestimate of the average particle size.

Computer modeling was performed assuming that all of the particles were aligned with their easy axes parallel to the applied field. They were assumed to be initially at saturation, so that all of the moments pointed along the original field direction. Larger monodomain particles have larger magnetic moments, and the

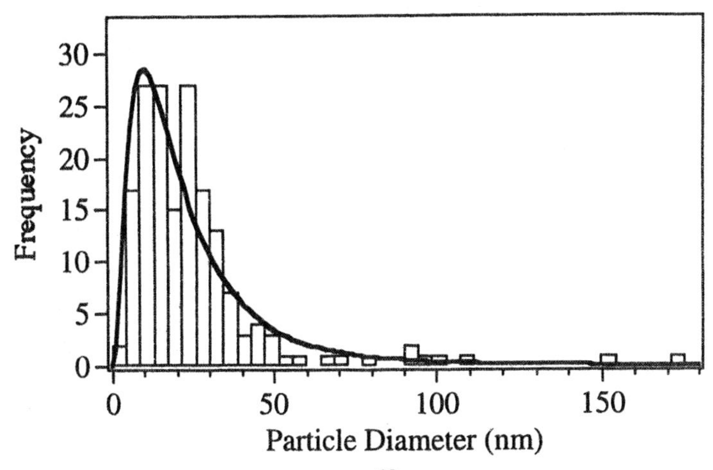

Figure 16. Log normal distribution function[25] fit to particle sizes obtained from ball milled $SmCo_5$ nanoparticles shown in Figure 15.

total magnetization in emu is given by the weighted sum of moments form individual particles, and is given at saturation by:

$$M = \int M_s \frac{\pi}{6} D^3 f(D) dD \tag{30}$$

After a reverse field H is applied, the moments may reverse their direction:

$$M_s V_i \rightarrow -M_s V_i, \text{ if } |H| > |H_c(V_i)|. \tag{31}$$

From Eq. 7, the barrier height is proportional to the particle volume. From Figure 16, it is clear that the particle volume distribution in this sample has a sharp peak

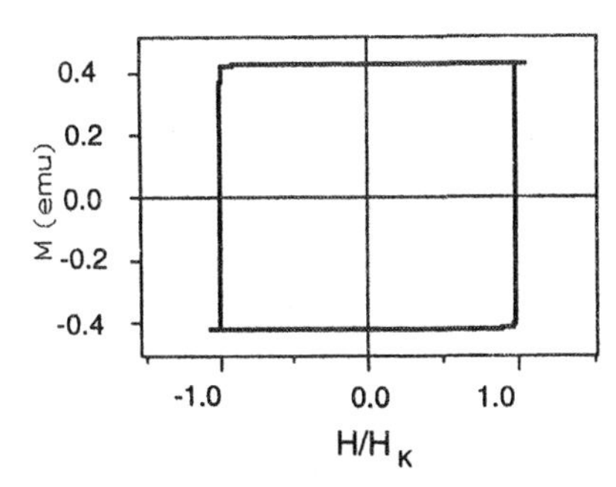

Figure 17. *M*(*H*) calculated from the experimental size distribution of Figure 16 for $SmCo_5$ nanoparticles.

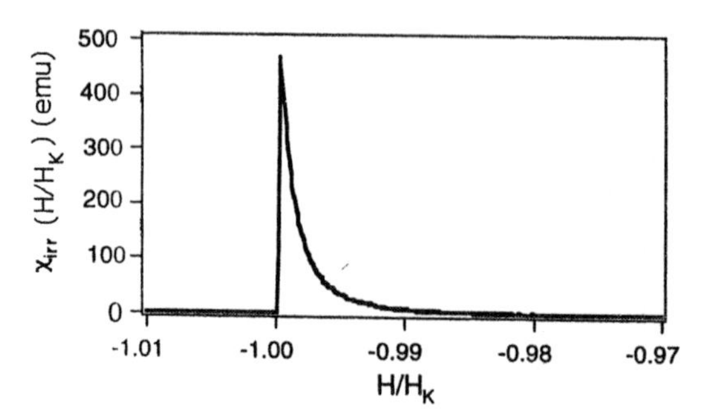

Figure 18. Calculated $\chi_{irr}(H)$ based on the experimental size distribution.

in the neighborhood of $(\pi/6)(16 \text{ nm})^3$, and an extremely long tail. While the single particle energy barrier heights vary dramatically, this does not necessarily mean that the individual particles have different coercivities. By combining Eqs. 15 and 13, the volume dependence of the coercivity can be shown to be,

$$H_c(V) \approx H_K\left[1-\left(\frac{V_{sp}}{V}\right)^{1/2}\right] = H_K\left[1-\left(\frac{d_{sp}}{d}\right)^{3/2}\right] \tag{32}$$

where V_{sp} is the volume of the largest superparamagnetic particle and V is the particle volume. From Eq. 13, d_{sp} is approximately 2 nm for $SmCo_5$, and so for a large fraction of the particles in the sample of Figure 15, $H_c \approx H_K$.

This procedure yielded a very square hysteresis loop (Figure 17). There was therefore an extremely sharp peak in $dM/dH = \chi_{irr}(H)$ at $H = H_K$, much sharper than the experimental width and asymmetric (Figure 18). Though $f(d)$ decreases with increasing particle size, it always falls off more slowly than $1/D^3$, so the switching fields for the largest sizes dominate the switching field distribution. Note that with a narrower size distribution this is not necessarily true. While this model neglects thermally activated switching of particles within a few kT of $H_c(V)$, this is not expected to make a substantial difference for a material with high anisotropy such as $SmCo_5$.

The switching field distribution obtained from the coherent rotation model and the particle size distribution for the $SmCo_5$ nanoparticles of Figure 15 is inconsistent both with the peak position of $\chi_{irr} = dM/dH$ and with the experimental linewidth, σ. $H_K = 61.8$ T at 300 K, while the peak in the irreversible susceptibility occurs at the coercivity, 1.14 T. This leaves two related mysteries: first why the coercivity is so much less than the anisotropy field, and second, if this is so, why the temperature dependence of the coercivity can be explained by Eq. 15, with an effective anisotropy much less than the bulk value.

B. Experimental Distribution of Energy Barriers, $f(E)$

To learn more about the dominant magnetization reversal mechanism, we sought to determine the distribution of energy barriers from the experimental data. For any mechanism, the energy barrier height varies with the field, and magnetization reversal occurs at fields for which the energy barriers are zero or within a few kT of zero. We want to learn about the distribution of barriers at zero or very small values of the applied field. Here the distribution, $f(E)$, is a function of energy barrier heights, E, analagous to the anisotropy energy, KV, in Eq. 7. Earlier we considered what happened if K were constant and V varied according to the measured size distribution. Now we take the opposite approach, using the average particle volume and assuming all variations are due to fluctuations in K.

The experimentally determined function, $\chi_{irr}(H)$, indicated the distribution of switching fields, H_{Ki}, for the i particles in this sample. If the switching volumes are all equal, this function is equivalent to $N(H_{Ki})$, the number of particles with a reversal field H_{Ki}. This function can be related to the distribution of energy barriers, given a relation between the energy and the switching field. In the variable anisotropy model:

$$E = K\bar{V}\left[1 - \left(\frac{HM_s}{2K}\right)\right]^2 = \left(\frac{H_{Ki}M_s\bar{V}}{2}\right)\left[1 - \left(\frac{H}{H_{Ki}}\right)\right]^2 \tag{33}$$

With a small applied field, $H << H_{Ki}$, and $E \approx (H_{Ki}M_s\bar{V}/2)$, or $H_{Ki} = 2E/M_s\bar{V}$. Substituting into the distribution of switching fields,

$$N(H_{Ki}) = \frac{A(T)}{\sqrt{2\pi}\sigma}\exp\left\{-\frac{(H_{Ki} - H_c)^2}{2\sigma^2}\right\} \tag{34}$$

we obtain an expression for the distribution of energy barriers at $H = 0$,

$$f(E) = \frac{(M_s\bar{V}/2)A(T)}{\sqrt{2\pi}\sigma_E}\exp\left\{-\frac{(E - \bar{E})^2}{2\sigma_E^2}\right\} \tag{35}$$

where the average barrier height and linewidth are given by,

$$\bar{E} = \frac{M_s\bar{V}H_c}{2} \tag{36}$$

and,

$$\sigma_E = \frac{M_s\bar{V}\sigma}{2} \tag{37}$$

respectively.

With the analysis of the experimental data, the average barrier height at 300 K is approximately 8.1×10^{-19} *J*, or almost 200 *kT*, and half the 1/e linewidth, $\sigma \approx 2.9 \times 10^{-19}$ *J*. The barrier height $K\bar{V}$ calculated using the bulk value of the anisotropy is almost 10,600 *kT*. This is consistent with the reduced anisotropy found to fit the temperature dependence of the coercivity. Any magnetization reversal mechanism should therefore explain the large reduction in the energy barrier height, or equivalently, the greatly reduced anisotropy. Two such possibilities are considered next: misalignment of the particle easy axes, and interactions between particles.

C. Misalignment of Easy Axes

The value of M_s from the DC demagnetization experiment at 300 K is found to be approximately 30% smaller than that for an ordinary hysteresis loop, suggesting that a significant fraction of the particles do not have their easy axes perfectly aligned with the applied field. The energy barrier shown in Figure 6 is only for a particle with the field applied along its easy axis. In other cases there is no simple analytical solution, but the energy as a function of θ and ϕ can still be calculated. Equation 7 can be rewritten in terms of a function, $j(\theta, \phi, H/H_K)$, that depends only on the angles and the field strength relative to the anisotropy field, H/H_K:

$$j\left(\theta, \phi, \frac{H}{H_K}\right) = \frac{E_B}{KV} = \sin^2\phi - 2\left(\frac{H}{H_K}\right)\cos(\theta - \phi) \tag{38}$$

Figure 19 illustrates this function, $j(\theta, \phi, H/H_K)$, for $\theta = 0°$ and three different field strengths. In zero field, there are two equivalent energy minima, but as soon as a field is applied, one becomes lower than the other. As the field strength increases, so does the energetic preference for one of the states. At the same time, the angle

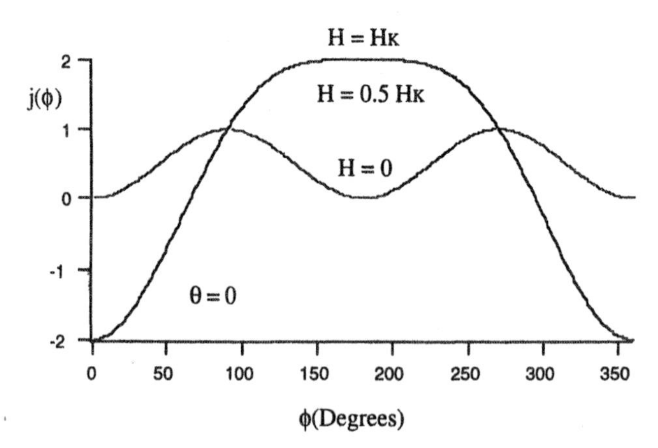

Figure 19. $j(\phi)$ for perfect alignment of the applied field along the easy axis ($\theta = 0°$), for $H = 0$, $0.5\ H_K$, and H_K.

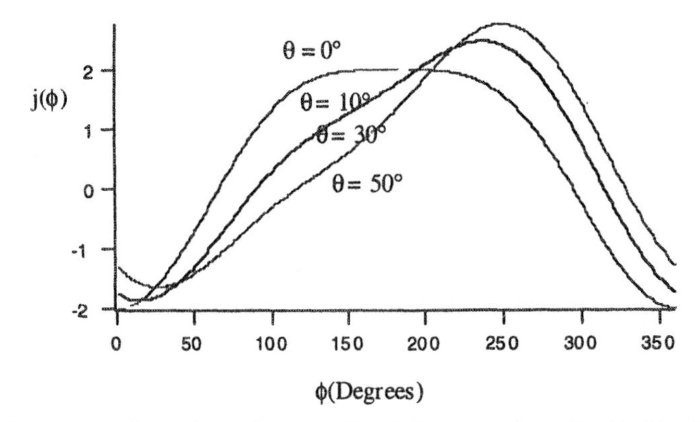

Figure 20. Effect of misalignment for different angles, all with $H = H_K$.

of maximum energy shifts and there is a flattening of the energy in the region of the local minimum. At the anisotropy field, what was once the local minimum has become the overall maximum energy angle and is unstable to reversal of the magnetization. Because the anisotropy energy, KV, for $SmCo_5$ is roughly 10,600 times the thermal energy at room temperature, perfectly aligned particles are not expected to reverse their magnetization until the barrier height becomes comparable to the thermal energy, or very close to $H = H_K$.

Now suppose the particles are slightly misaligned. Figure 20 illustrates the normalized energy as a function of ϕ for various misalignment angles θ, all at $H = H_K$, where a perfectly aligned particle has zero energy barrier. With misalignment, instead of zero energy barrier at this field, the energy is highly unstable with respect

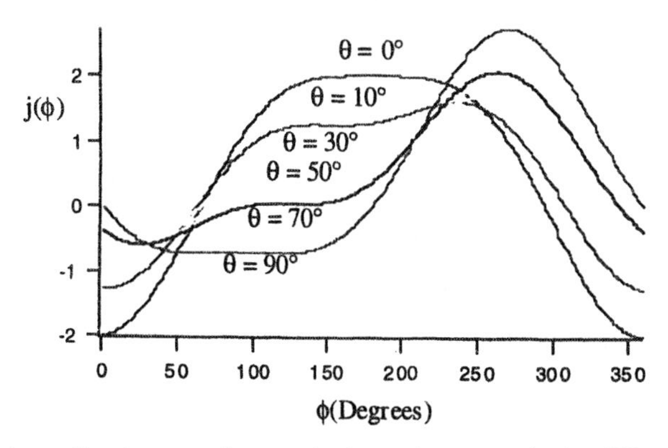

Figure 21. Normalized energy $j(\phi,\theta, H/H_K)$ as a function of ϕ for different misalignment angles θ, plotted for fields H such that f has no energy barrier between the initial local minimum and the final equilibrium state.

Table 1. Zero Barrier Fields as a Function of the Misalignment Angle, θ

θ *(Degrees)*	*Field* (H_K)
0	1.0
10	0.65
30	0.50
50	0.50
70	0.55
90	0.75

to rotation of the magnetization vector. Misalignment causes particles to reach the zero barrier configuration at lower values of the applied field, and therefore permits switching at reduced fields.

Figure 21 shows that the field required for zero energy barrier between the zero applied field local minimum state and the global minimum state at the switching field; these are the fields where the particles reverse their magnetization. Table 1 lists the zero barrier or switching fields for the misalignment angles of Figure 21.

D. Interacting Nanoparticles

So far we have considered the magnetic behavior of isolated nanoparticles, prepared by diluting and dispersing the particles in a nonmagnetic matrix. Magnetic particles can have both magnetostatic and exchange interactions. The exchange energy decays exponentially with distance, and the exchange length, L_{ex}, is given by,

$$L_{\mathrm{ex}} = \left(\frac{A}{K}\right)^{1/2} \tag{39}$$

where A is the exchange stiffness, and is roughly equal to kT_c divided by the spacing between coupled spins.[9] For $SmCo_5$, the exchange length is slightly over one nanometer, and so exchange interactions between particles, even when they are touching, are likely to be much smaller than the magnetic dipole interactions.

Between point dipoles, like those shown in Figure 22, the dipole–dipole magnetostatic energy, E_{dd}, can be calculated exactly:

$$E_{\mathrm{dd}} = \left(\frac{m_1 m_2}{r^3}\right)\left[\cos\left(\theta_1' - \theta_2'\right) - 3\cos\theta_1'\cos\theta_2'\right] \tag{40}$$

For simplicity we approximate the magnetic particles as point dipoles. If the two particles have the same volume and are touching, the dipole energy becomes:

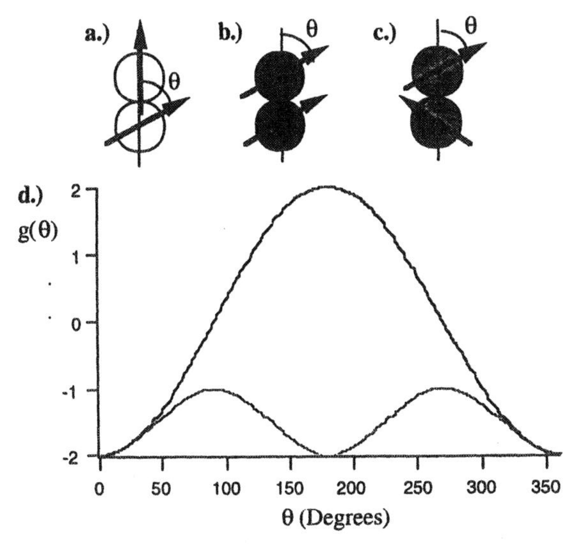

Figure 22. (**a.**) $\theta_1' = 0°$, $\theta_2' = \theta$; (**b.**) $\theta_1' = \theta_2' = \theta$; (**c.**) $\theta_1' = -\theta_2' = \theta$; (**d.**) Energy as a function of θ for cases (**a.**), (**b.**), and (**c.**).

$$E_{dd} = \left(\frac{M_s^2 \pi V}{6}\right)\left[\cos\left(\theta_1' - \theta_2'\right) - 3\cos\theta_1' \cos\theta_2'\right] = \left(\frac{M_s^2 \pi V}{6}\right) g\left(\theta_1', \theta_2'\right) \tag{41}$$

Consider a few special cases of Eq. 41, where the function $g\left(\theta_1', \theta_2'\right)$ is plotted as a function of angle in Figure 22. Here the particles are the same size, they are touching, and magnetocrystalline anisotropy energy is ignored. When θ_1' is fixed and $\theta_2' = \theta$, there is a large energy barrier at 180°, relative to the minimum energy. When $\theta_1' = \theta_2' = \theta$, the particle moments rotate together, but there are still considerable energy barriers at 90° and 270°. The lowest energy barriers occur when $\theta_1' = -\theta_2' = \theta$, and the particle moments rotate in opposite directions, a mechanism known as fanning. Note that in this case for all angles the magnetostatic interaction lowers the overall energy of the system.

For the case of side-by-side particles, the minimum energy configuration has the particle moments antiparallel (Figure 23). Once again, magnetization reversal via fanning has a lower energy barrier than coherent rotation of the particle moments.

The overall sample energy is affected by both magnetocrystalline anisotropy and anisotropy due to dipole–dipole interactions. The dipole–dipole interaction strength is independent of the applied field, unlike the barrier height due to magnetocrystalline anisotropy. For touching particles with average size,

$$\frac{M_s^2 \pi \bar{V}}{6} \approx 0.013 K \bar{V} \tag{42}$$

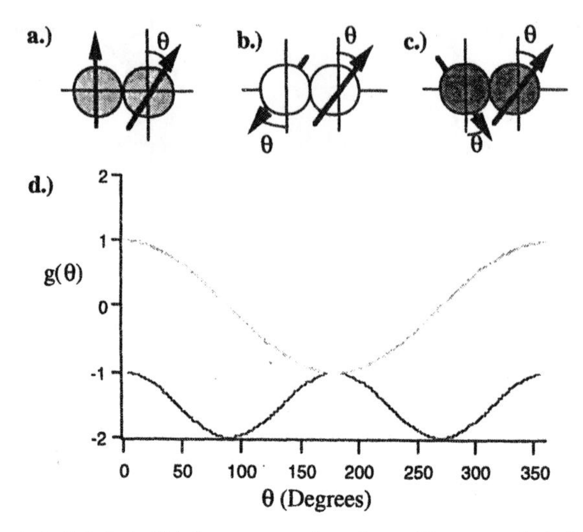

Figure 23. (**a.**); (**b.**); (**c.**); (**d.**) Energy as a function of for cases (**a.**),-(**b.**), and (**c.**).

and so the orientation of the magnetic moments at zero field will not be shifted away from the easy axes by magnetic dipole interactions. However, these forces can be substantial in large $SmCo_5$ nanoparticles, due to the large particle moments, and agglomeration or clumping of particles frequently occurs. This is observed experimentally through transmission electron microscopy (TEM), where even under intense sonication of the powder in methanol, many of the particles are found to be clumped, unlike identically prepared Fe or Co particles.

As the applied field increases, the energy barrier due to magnetocrystalline anisotropy decreases. When the energy destabilization from magnetocrystalline anisotropy becomes comparable to the stabilization energy from magnetostatic interactions, there is zero energy barrier and the particles reverse their magnetization.

Magnetostatic interactions, like misalignment of the easy axis relative to the applied field direction, can reduce the magnitude of the switching field. The fanning interaction reduces the energies of the most stable configurations more than that of the energy barrier. In the case of two identical particles touching, the overall barrier height including magnetocrystalline anisotropy is reduced by roughly 12% at zero applied field. The field for which there is zero barrier with respect to magnetization reversal is also reduced by approximately 12%.

In the actual sample, there are magnetostatic interactions among many particles, and there are variations in particle size and spacing. While no analytic results for such complicated system are possible, clusters of magnetostatically interacting magnetic particles have been investigated by a Monte Carlo method using simulated annealing.[33] In superparamagnetic particles, more rapid relaxation times have been

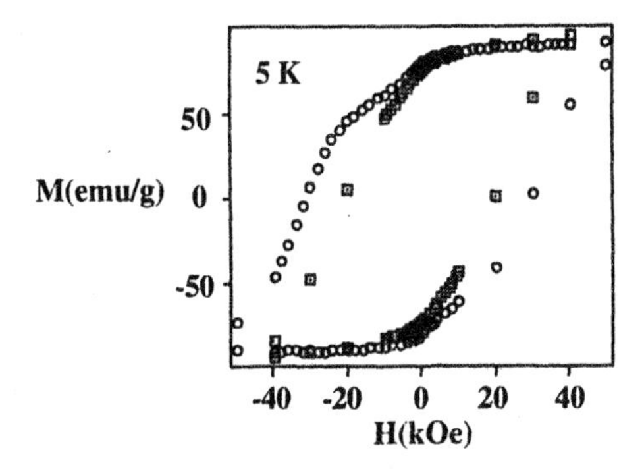

Figure 24. *M*(*H*) at 5 K for dilute (*circles*) and concentrated (*squares*) ball milled $SmCo_5$ nanoparticle samples with Scherrer $\overline{d}$ = 25 nm. Data from ref. 34.

observed by Mössbauer spectroscopy for clumps of particles than for those which are spatially separated.[33] Unlike in many systems of superparamagnetic particles, we observed no evidence of linear or circular chain formation which are energetically most favored. However, because the anisotropy energy barrier is so much higher in the $SmCo_5$ nanoparticles, this result is unsurprising.

Experimental evidence for the importance of magnetostatic interactions in the $SmCo_5$ nanoparticles was obtained by comparison of concentrated and diluted samples. For magnetic measurements, the degree of agglomeration and therefore the amount of particle interactions was minimized by a 10^4:1 by weight dilution of the powder in epoxy, and by dispersing the particles using ultrasound as the epoxy cured. Though many clumped particles remained, this increased the coercivity substantially (Figure 24), and contributions to the magnetization from more isolated particles become apparent. The smallest particles which are superparamagnetic or switch at very low fields cause a shoulder in the magnetization curve near $H = 0$. Isolated larger particles experience only the external field, and so they reverse their magnetization direction at greater values of H. This sample does not completely saturate below fields of 8 T, so it must include a small fraction of particles with $H_c > 3.75\ T$, which is the largest previously reported coercivity for $SmCo_5$.[9] The ability to further separate the particles from each other would enable H_c to exceed this value.

VIII. SUMMARY

Monodomain nanoparticles of ferromagnetic materials have unique and controllable magnetic properties. Though the saturation magnetization per unit volume is unchanged at the nanometer size range, the coercivity is dramatically different from that of the bulk material. It can be zero for small superparamagnetic particles, where thermal fluctuations overcome metastability due to ferromagnetic coupling. Alternatively, H_c can theoretically be much larger than in the bulk material. Unfortunately this prospect is not realized for high magnetocrystalline anisotropy nanoparticles of $SmCo_5$ or $Nd_2Fe_{14}B$.[35]

Even for a well characterized sample, such as the $SmCo_5$ nanoparticles described here, the microscopic basis for the experimental distribution of switching fields could not be explained quantitatively. The size distribution yielded an asymmetric and much narrower distribution of energy barriers. Misalignment of the particle easy axes and cooperative magnetization reversal through magnetostatic interactions were both demonstrated in simplified systems to be capable of substantial reductions in the switching field magnitude. It is plausible in either case that random orientations would lead to a Gaussian distribution of switching fields, as was observed experimentally. There is experimental support for the presence of misaligned particles from the 30% deviation of the ordinary and DC demagnetization measurements of M_s, and evidence that magnetostatic interactions are important from the changes in the hysteresis loop found when the sample is substantially diluted.

Clearly there are many remaining questions concerning magnetization reversal mechanisms, even in monodomain particles, which are the simplest of all ferromagnetic systems. For example, defects could easily reduce the barrier to magnetization reversal, and their random occurrence would lead to a Gaussian switching field distribution. By probing the fundamental physics of magnetization reversal in these particles, it is hoped that more complex phenomena in bulk magnets will be understood.

ACKNOWLEDGMENTS

S. A. M. thanks the National Science Foundation for support through grants DMR-9283508 and DMR-9500313. This material is also based in part on work supported by the NSF under grant ECD-8907068. E. M. Kirkpatrick in particular is gratefully acknowledged for magnetic measurements. S. A. M. would also like to thank N. T. Nuhfer for scanning electron microscopy, J. H. Scott for transmission electron microscopy, and M. E. McHenry and the CMU Buckyball Project for work on carbon-coated magnetic nanoparticles.

REFERENCES

1. Billas, I. M. L.; Chatelain, A.; de Heer, W. A. *Science* **1994**, *265*, 1682.
2. *Introduction to Solid State Physics*; Kittel, C., Ed.; John Wiley and Sons: New York, 1996, p. 422.
3. *Introduction to Magnetic Materials*; Cullity, B. D., Ed.; Addison-Wesley: Reading, MA, 1972, pp. 211–214.
4. Luborsky, F. E. *J. Appl. Phys.* **1961**, *32*, 171S.
5. Herzer, G. *J. Magn. Magn. Mater* **1992**, *112*, 258; Herzer, G. *IEEE Trans. Magn.* **1990**, *26*, 1397.
6. Vassiliou, J. K.; Mehrotra, V.; Russel, M. W.; Giannelis, E. P.; McMichael, R. D.; Shull, R. D.; Ziolo, R. F. *J. Appl. Phys.* **1993**, *73*, 5109.
7. Tang, Z. X.; Sorensen, C. M.; Klabunde, K. J.; Hadjipanayis, G. C. *J. Appl. Phys.* **1991**, *69*, 5279.

8. Kronmüller, H. In *Science and Technology of Nanostructured Magnetic Materials*; Hadjipanayis, G. C.; Prinz, G. A., Eds.; Plenum Press: New York, 1991, p. 657.
9. *Ferromagnetic Materials: Structure and Properties*; McCurrie, R. A., Ed.; Academic Press: New York, 1994, p. 251.
10. McHenry, M. E.; Majetich, S. A.; Artman, J. O.; DeGraef, M.; Staley, S. W. *Phys. Rev. B* **1994**, *49*, 11358.
11. Scott, J. H.; Majetich, S. A. *IEEE Trans. Mag.* **1996**, *32*, 4701.
12. Néel, L. *Ann. Geophys.* **1949**, *5*, 99; *Compt. Rend. Acad. Sci.* **1949**, *228*, 664.
13. Stoner, E. C.; Wohlfarth, E. P. *Phil. Trans. Roy. Soc. London* **1948**, *A240*, 599; Brown, W. F., Jr.; *J. Appl. Phys.* **1958**, *29*, 470; *J. Appl. Phys.* **1959**, *30*, Suppl. 130S.
14. Jacobs, I. S.; Bean, C. P. In *Magnetism*; Rado, G. T.; Suhl, H., Eds.; Academic Press: New York, 1963, Vol. 3.
15. Durst, K. D.; Kronmüller, H. *J. Magn. Magn. Mat.* **1987**, *68*, 63.
16. Kirkpatrick, S.; McHenry, M. E.; De Graef, M.; Smith, P. A.; Nakamura, Y.; Laughlin, D. E.; Brunsman, E. M.; Scott, J. H.; Majetich, S. A. *Scripta Metallurgica et Materialia* **1995**, *33*, 1703.
17. Zhao, T. S.; Jin, H.-M.; Grössinger, R.; Kou, X.-C.; Kirchmayr, H. R. *J. Appl. Phys.* **1991**, *70*, 6134.
18. Jacobs, I. S.; Bean, C. P. In *Magnetism*; Rado, G. P.; Suhl, H., Eds.; Academic Press: New York, 1963, Vol. 3.
19. Kütterer, R.; Hilzinger, H.-R.; Kronmüller, H. *J. Magn. Magn. Mater.* **1977**, *4*, 1.
20. Sucksmith, W.; Thompson, J. E. *Proc. Roy. Soc. London, Ser. A* **1954**, *225*, 362.
21. Hadjipanayis, G.; Sellmyer, D. J.; Brandt, B. *Phys. Rev. B* **1981**, *23*, 3349.
22. Ram, U. S.; Gaunt, P. *J. Appl. Phys.* **1983**, *54*, 2872.
23. Sankar, S. G.; Rao, V. U. S.; Segal, E.; Wallace, W. E.; Frederick, A. G. D.; Garrett, H. J. *Phys. Rev. B* **1975**, *11*, 435.
24. Benz, M. G.; Martin, D. L. *J. Appl. Phys.* **1972**, *11*, 4733.
25. Granqvist, A.; Buhrman, R. *J. Appl. Phys.* **1976**, *47*, 2200.
26. Street, R.; Woolley, J. C. *Proc. Roy. Phys. Soc.* **1949**, *A62*, 562.
27. Majetich, S. A.; Kirkpatrick, E. M. *IEEE Trans. Mag.* **1997**, *33*, 3721.
28. Gangopadhyay, S.; Hadjipanayis, G. C.; Sorensen, C. M.; Klabunde, K. J. *IEEE Trans. Mag.* **1993**, *29*, 2619.
29. El-Hilo, M.; O'Grady, K.; Pfeiffer, H.; Chantrell, R. W.; Veitch, R. J. *IEEE Trans. Mag.* **1992**, *28*, 2689.
30. Gaunt, P. *J. Appl. Phys.* **1986**, *59*, 12.
31. Chantrell, R. W.; Fearon, M.; Wohlfarth, E. P. *Phys. Stat. Sol. (a)* **1986**, *97*, 213.
32. Hendriksen, P. V.; Morup, S.; Christiansen, G.; Jacobsen, K. W. In *Science and Technology of Nanostructured Magnetic Materials*; Hadjipanayis, G. C.; Prinz, G. A., Eds.; Plenum Press: New York, 1991, p. 573.
33. Tronc, E.; Jolivet, J. P.; Livage, J. *J. Chem. Res.* **1987**, *S*, B6.
34. Majetich, S. A.; Scott, J. H.; Kirkpatrick, E. M.; Chowdary, K.; Gallagher, K.; McHenry, M. E. *Nanostructured Materials* **1997**, *9*, 291.
35. Brunsman, E. M.; Scott, J. H.; Majetich, S. A.; Huang, M. Q.; McHenry, M. E. *J. Appl. Phys.* **1996**, *79*, 5293.

ELECTRON NANODIFFRACTION AND STEM IMAGING OF NANOPARTICLES AND NANOTUBES

J. M. Cowley

I. INTRODUCTION

Particles of solid materials having dimensions of the order of 1 nm, or "nanoparticles," may contain as few as 50, or as many as 10,000 atoms. The arrangement of the atoms within the particle determines the physical and chemical properties of

Advances in Metal and Semiconductor Clusters
Volume 4, pages 67–113

ISBN: 0-7623-0058-2

the material in this form, which have been the topics of intense interest in recent years. Electron diffraction using electron beams of diameter 1 nm or less, i.e. nanodiffraction, together with related high-resolution imaging techniques, offers a relatively new approach to the determination of the atomic arrangements of nanoparticles which has considerable potential value but is practiced by relatively few people. Various other physical techniques have been pushed to their limits in attempts to derive information on these arrangements, but with limited success.

X-ray diffraction, the most powerful tool for structure analysis of larger crystals, cannot be applied to study individual particles less than about 1 μm in diameter even when the most intense synchrotron radiation sources are used. For smaller crystals, powder patterns can be obtained for the analysis of the average structures, sizes and, to some extent, shapes of assemblies of crystals, but such averaging can often obscure important details. Also, the derivation of useful information becomes much more difficult as the particle diameters approach the nanometer scale. Spectroscopic methods using electromagnetic radiation necessarily involve an averaging over even larger numbers of particles and give even less information on crystal shapes and sizes.

The much stronger interactions of electrons with matter suggest that the most effective way to determine the structures of nanoparticles is the use of electron beams for diffraction and imaging experiments. The possibility of focusing electron beams by using strong electromagnetic lenses leads to the imaging of assemblies of atoms with a resolution of the order of 0.2 nm, i.e. "atomic resolution" in electron microscopes. Also the same lenses may be used to form electron beams of diameter as small as 0.2 nm so that diffraction patterns can be obtained from regions of such a diameter. The combination of the imaging and diffraction experiments of this sort should, in principle, provide all the information needed to specify the structures of individual nanoparticles. The objective of this article is to report the progress that has been made towards this end.

The scattering of electrons by atoms is so strong that a single heavy atom may give sufficient contrast in an electron microscope image to allow its detection with any current high-resolution instrument. Individual heavy atoms were first imaged in an electron microscope by Crewe[1] with dark-field imaging in a scanning transmission electron microscopy (STEM) instrument of his own design. Soon after, individual atoms were imaged in standard, conventional transmission electron microscopy (TEM) instruments, for example by Hashimoto et al.[2] in dark-field images and in bright-field images by Iijima.[3] The arrangements of atoms in inorganic crystals were first imaged clearly by Iijima.[4] Since then, the resolution limit for commercial electron microscopes has been improved from about 0.25 to 0.15 nm (or better, approaching 0.1 nm for ultrahigh voltage microscopes) with a corresponding improvement in the ability to produce clear images of the arrangements of the atoms in thin crystals.

The electron microscope images of thin specimens give what is essentially a projection of the structure of the specimen.[5,6] The electron wavelengths are very short, of the order of 2–3 pm, so the angles of scattering by atoms are of the order of 10^{-3} to 10^{-2} radians and, for thin specimens, the electrons may be considered to pass through the specimen without deviation. The structures of crystals are most easily seen when the incident beam is parallel to a principal axis of the crystal so that the rows of atoms passing through the crystal are well separated in projection and are seen as individual spots in the image. For example, Figure 1 shows a high-resolution TEM image of nanocrystals of Pt on a support of CeO_2.[7] For these nanocrystals and the crystals of the support that are in principal orientations relative to the incident beam, the regular arrangement of the atom rows is clearly visible. For other orientations, the periodicity is not resolved.

Figure 1 shows atoms or rows of atoms as dark spots. However the contrast in such images is highly dependent on the operating parameters of the microscope. With a different setting of the defocus of the objective lens of the microscope, the atom rows can appear as bright spots. The contrast also depends strongly on the specimen parameters; particularly the crystal thickness and orientation. The available theoretical treatments of the scattering of electrons by crystals are well developed. The effects of the various parameters on image contrast are adequately understood so that the prediction and interpretation of image contrast is possible with a high level of accuracy.[5,6] The theory will not be discussed here except for the case of very thin specimens, for which some simplifying approximations may be made.

The case of Figure 1 is a very favorable one in that some of the nanocrystals are seen almost in isolation with no interference from the support. In the more common cases of nanocrystals, or imperfectly ordered nanoparticles, imbedded in, or supported on thin films of the same or different composition, complications arise. The

Figure 1. High-resolution TEM image of Pt nanocrystals on CeO_2. The Pt particles are epitaxed on well-facetted CeO_2 surfaces. (From Yao et al.[7])

contrast from the support may seriously interfere with the imaging of the nanoparticles.

For thin amorphous films, the contrast of the image under the optimum defocus conditions depends on the projection of the electrostatic potential of the film which modifies the phase of the incident electron wave. Because the film is made of an almost random arrangement of atoms, each of which gives a peak of projected potential, the image consists of the superposition of a random array of unresolved dark dots, giving the "phase noise" image contrast as illustrated in Figure 2. Such a background tends to obscure the images of nanoparticles. It has been shown by image simulation, for example, that a crystal of silicon of diameter 2 nm cannot be distinguished if embedded in film of amorphous silicon of thickness 10 nm.[8] Furthermore, the randomness of the contrast features given by a thin amorphous film may be modified by the imaging process in such a way that patches of almost parallel fringes, suggesting the presence of nanocrystals, may appear for even a completely random arrangement of atoms[9] (see Figure 2) Such appearances have led in several instances to questionable deductions regarding the presence of nanocrystals.

To a lesser extent, as we shall see, the adverse effects of a supporting film apply also in the case of diffraction from nanometer-size regions ("nanodiffraction"). The scattering from the atoms of the support can give rise to a mottled background in

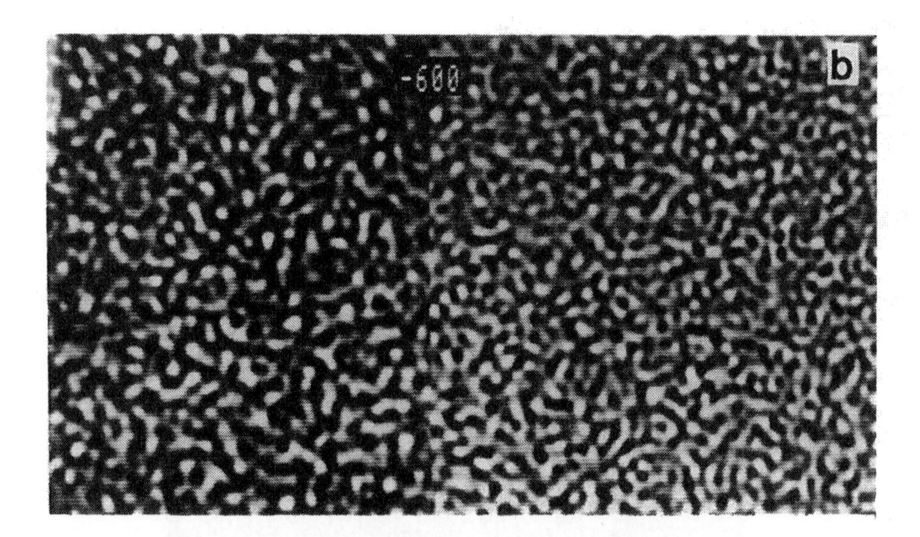

Figure 2. High resolution TEM images of a film of amorphous silicon about 2 nm thick. (**a**) Experimental image obtained with 200 keV, Cs = 1.2 mm and defocus –60 nm. (**b**) Simulated image for same experimental conditions but assuming a completely random arrangement of Si atoms. (From Fan and Cowley[9]).

the diffraction pattern. However there are factors which can reduce the adverse effects of this background in the case of nanodiffraction.

Electron diffraction patterns are commonly obtained from small areas of the specimen in TEM instruments by use of the selected-area electron diffraction (SAED) technique. A small aperture is placed in the image plane of the objective lens where the image magnification is a few hundred times. When the magnifying lenses of the microscope are refocussed to project the diffraction pattern, formed in the back-focal plane of the objective lens, on the final viewing screen, only those electrons coming through the part of the image selected by the SAED aperture can contribute to the diffraction pattern. The smallness of the region of the specimen that can be selected in this way is, however, limited by the spherical aberration of the objective lens. For electron microscopes operating with energies in the normal range of 100 to 400 keV, diffraction patterns can be obtained from regions of diameter about 50 nm or more. While these selected regions are considerably smaller than those from which X-ray diffraction patterns can be obtained, they are usually much too big for the study of diffraction from individual nanoparticles. Hence other approaches, involving the illumination of only a very small area of the specimen, must be used, and these will be the subject of the remainder of this article.

II. NANODIFFRACTION AND STEM IMAGING

In a scanning transmission electron microscopy (STEM) instrument, the strong electromagnetic objective lens is used to form an electron probe of very small diameter by demagnifying a small bright electron source. The small probe is scanned over the specimen in a two-dimensional raster. Some part of the electron beam transmitted through the specimen is detected to form the image signal which is displayed on a cathode ray tube with a raster scan synchronized with that at the specimen level. The magnification of the image is given simply by the ratio of the scan dimensions on the display tube and at the specimen. A simplified diagram of a STEM instrument is given in Figure 3. More detailed descriptions have been given elsewhere.[10,11]

The electron source is normally a cold field-emission gun, with accelerating voltages of up to 100 kV or sometimes 300 kV. The gun must be operated under ultrahigh vacuum conditions in order to obtain relatively stable electron emission. The effective source diameter for such a gun is about 4 nm. Hence if the strong objective lens gives a demagnification of the source by a factor of 100 or more, the probe size at the specimen is determined almost entirely by the aberrations of the objective lens which limit the diameter of the probe to about 0.2 nm or more. Usually one or two condenser lenses, and sometimes also a "gun lens" (a weak magnetic lens incorporated in the gun assembly), are added before the objective lens in order to provide a range of probe characteristics: probes of diameter 1–2 nm and high intensity are needed for analytical purposes and some nanodiffraction

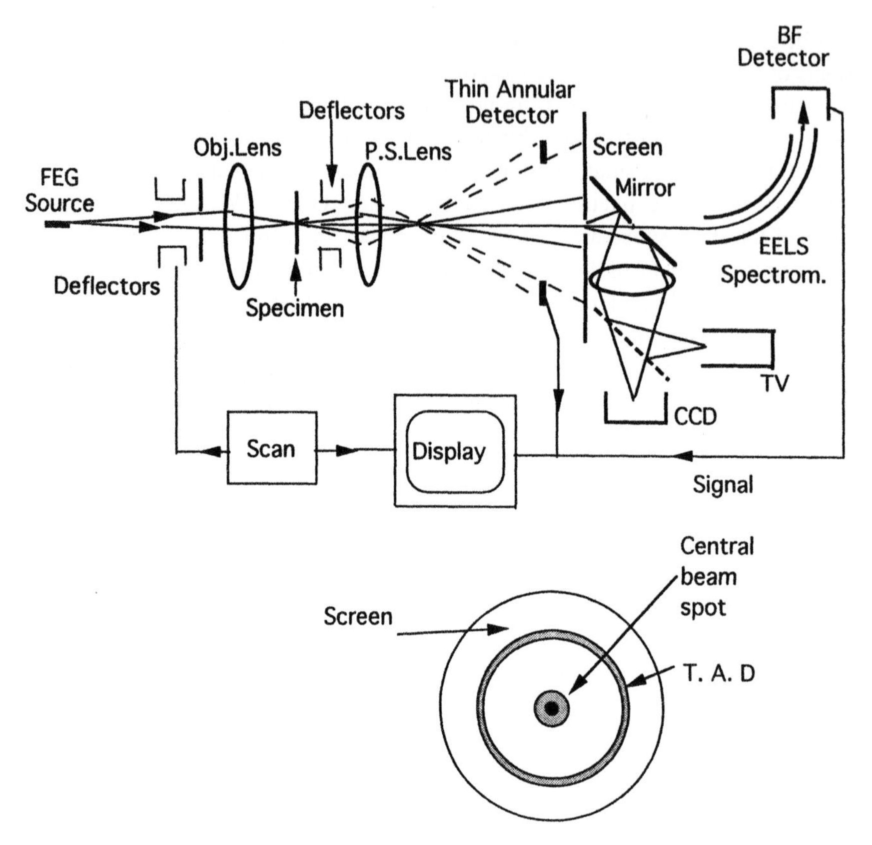

Figure 3. Diagram of a STEM instrument with one post-specimen (P.S.) lens and diffraction patterns formed on a transmission phosphor screen which is viewed by means of a mirror, a large-aperture optical lens and either a TV camera or a CCD camera, or by both if a half-silvered mirror is used.

applications, and probes of the smallest possible diameter and, inevitably, low intensity, are required for high-resolution STEM imaging.

The objective lens is a strong lens of the type used in TEM instruments, with the specimen normally immersed in the magnetic field. Often an asymmetric lens, with one pole-piece of relatively large bore, is preferred so that there is adequate space for the insertion of an energy-dispersive X-ray detector, or else a secondary-electron detector to record the emission of low-voltage secondary electrons or Auger electrons. The detection of the characteristic X-rays given from the area of the specimen bombarded by the incident beam allows the use of spectroscopic methods for the chemical analysis of specimen areas of diameter as small as 1–2 nm,[12] or a mapping of the distribution of particular elements within the specimen when particular X-ray emission lines are detected. Similarly, microanalysis of nm-size regions or a mapping of the distributions of particular elements may be made by using electron energy-loss spectroscopy (EELS) to analyze and distinguish the characteristic energy losses of electrons transmitted through the specimen.[12]

The detection of low-energy secondary electrons, as in secondary-electron microscopy (SEM), allows the imaging of the surface morphology of the specimen with spatial resolutions which may be as small as 0.6 nm.[13] The detection of Auger electrons having energies characteristic of particular elements allows the chemical composition of the surface layers of the specimen to be determined, or images may be obtained showing the distribution of particular elements in thin surface layers. If special arrangements are made for the highly efficient collection and analysis of Auger electrons, the resolutions of such images may be in the nanometer range.[14]

Specimens are prepared as for normal TEM. Small particles may be held on a thin film of amorphous carbon or on a holey carbon grid so that the particles extend over the vacuum. Self-supporting films of specimen material, 1 to 20 nm thick, may be prepared in some cases, and thick specimens may be thinned down to similar thicknesses by ion bombardment or chemical etching. Specimens in STEM are commonly held in a vacuum of 10^{-8} torr or better, but in some special instruments, ultrahigh vacuum conditions (better than 10^{-10} torr) may be approached.[15]

The electron beam incident on the specimen is necessarily a convergent cone produced by the objective lens. On any plane of observation beyond the specimen, the distribution of electrons at any time represents a convergent-beam electron diffraction (CBED) pattern of the small specimen area illuminated by the beam. The transmitted incident beam forms a bright central disk of diameter proportional to the incident cone angle. Around this, extending out to about 10^{-1} radians, is the distribution of elastically scattered electrons, the Fraunhofer diffraction pattern of the illuminated area of the specimen. A background of inelastically scattered electrons is also present. This is relatively weak for thin specimens but may become important for thick specimens.

For a crystalline sample, individual diffraction spots appear in the pattern as disks of the same diameter as the central spot. The diffraction pattern may be observed and recorded by use of a fluorescent screen. Often one or more post-specimen lenses are inserted so that the size of the diffraction pattern on the screen can be adjusted for specific purposes.

The fluorescent screen is usually of the transmission type and is viewed from behind. An arrangement of mirrors and lenses serves to convey the optical image to the detectors outside the vacuum. In the arrangement of Figure 3, a YAG phosphor screen is used and the diffraction pattern, or shadow image, is viewed and recorded either by use of a low-light-level TV camera and a video cassette recorder (VCR), or else with a CCD camera and an associated digital recording system.

If there are holes in the middle of the fluorescent screen and mirror, or if the screen and mirror are removed, the central beam or some selected part of the diffraction pattern is transmitted to the electron energy analysis spectrometer

(EELS) to allow the EELS spectrum to be displayed or images of the specimen with electrons of selected energy-loss values may be recorded. When the zero-loss electrons are selected, the image formed with the central beam gives the standard, energy-filtered, bright-field STEM image. Dark-field STEM images formed by deflecting the diffraction pattern so that a particular diffraction spot, or other feature of the diffraction pattern, passes through the EELS spectrometer, are analogous with the dark-field images formed in TEM instruments by suitably tilting the incident beam.

Following the initial scheme of Crewe et al.,[1] dark-field STEM images may also be obtained by using an annular detector to collect all the electrons scattered outside the central beam of the diffraction pattern. This annular-dark-field (ADF) imaging mode is highly efficient in that it gives the maximum signal possible for a dark-field mode. It is appropriate when there is no interest in the form of the diffraction pattern, as when the specimen is amorphous or consists of a well-resolved distribution of single heavy atoms. For micro-crystalline materials, some contrast variations appear in the ADF image as a result of the appearance of diffraction spots in the diffraction patterns and the strong variations of their intensities with crystal orientation and thickness. It was suggested by Howie[16] that this complication could be avoided and an image contrast, depending only on the number and nature of the atoms present, could be obtained if an annular detector of large inner radius is used to collect only those electrons scattered to high angles, beyond the angular range of the diffraction spots. This high-angle annular dark-field (HAADF) mode has been used very effectively for the detection of very small heavy-atom particles on light-atom supports (such as occur in supported metal catalysts)[17] and has also been used by Pennycook and associates[18] for obtaining clear atomic-resolution images of interfaces between crystalline phases. Another possibility, recently explored,[19] is that of using a thin annular detector for which the ratio of outer and inner radii of the detector is about 1.1. This mode is discussed in more detail below.

The resolution and contrast of STEM images may be related to those of TEM images by application of the Reciprocity Principle.[20] In general, for any optical system (not including vectorial components) and for an elastic scattering process (not involving energy changes), it may be said that the complex wave amplitude at any point B due to a point source at point A is the same as the wave amplitude at A due to a point source at B. If the direction of propagation of an electron beam in a STEM instrument, such as in Figure 3, is reversed, the instrument becomes a TEM instrument. For point sources and point detectors, reciprocity suggests that the image signal is the same in the two cases. The argument can be extended to include finite incoherent sources and finite detectors. For each TEM configuration, there is an equivalent STEM configuration and vice versa. The differences in imaging modes and areas of application come from practical considerations of instrument construction and experimental convenience.

Considerable advantages of STEM, relative to TEM, result from the serial nature of the image recording and the possibility of using several detectors simultaneously.

Thus, in principle, an energy-filtered bright-field image, or a dark-field, energy-loss image, may be recorded at the same time as a dark-field image from a diffraction spot and also a HAADF image. Since each image signal is a time-varying voltage or current, on-line manipulation of each signal is possible so that its intensity and contrast may be modified or it may be differentiated. Then the various image signals may be added, subtracted or multiplied at will.

As a STEM image is being observed, a particular feature of the image may be selected and indicated by a marker spot on the screen. The scan may then be stopped with the beam at the position of the marker, and the nanodiffraction pattern from the selected spot on the specimen may be recorded. Alternatively, a linear scan across a specimen area may be chosen, or a two-dimensional scan over a rectangular specimen area may be made, and nanodiffraction patterns may be recorded during the scan. With a TV camera and VCR, for example, 300 diffraction patterns may be recorded during a 10 second scan across a distance of 10 nm so that a pattern is recorded for each movement of the beam by 0.03 nm.[21]

Figure 4 shows a bright-field STEM image of some small gold particles, 1 to 5 nm in diameter, embedded in a thin film of plastic.[22] The bright marker spot may be placed in turn over each of the dark images of the gold particles and the nanodiffraction patterns may be recorded. The diffraction patterns are not as sharp nor as extensive as for the SAED patterns obtained in TEM instruments, but are sufficient to provide the essential information required. The main point of interest for such particles is the determination of whether the particles are single crystals or whether they are twinned or multiply twinned. Theoretical studies and TEM observations on larger particles suggest that, with decreasing size, gold particles tend to be multiply-twinned, with icosahedral or decahedra form.[23] The analysis of many nanodiffraction patterns such as those of Figure 4 showed clearly that, at least for gold particles encased in plastic films, the percentage of multiply-twinned particles tends to decrease as the particle size is reduced beyond about 5 nm.

In some TEM instruments special pre-specimen lenses have been added so that the beam incident on the specimen may be made convergent and the illuminated area of the specimen may be reduced to a diameter as small as 1 nm. Then the pre-specimen lenses can imitate those of a STEM instrument and the microscopes are described as TEM/STEM. With a thermionic emission gun, the intensity that can be concentrated in a 1 nm diameter spot is usually too small to give useful nanodiffraction patterns, but with the increasing use of a field-emission gun (FEG) the possibilities for nanodiffraction and STEM imaging in a TEM/STEM instrument have been greatly enhanced. The capabilities of a FEG-TEM instrument for nanodiffraction have been demonstrated, for example, by Matsushita et al.[24] in studies of local ordering in thin films of amorphous Pd–Si alloys. The dedicated STEM instrument, however, retains important advantages for many purposes, largely because of the much greater flexibility possible for the detector system and the positioning of the strong objective lens before the specimen.

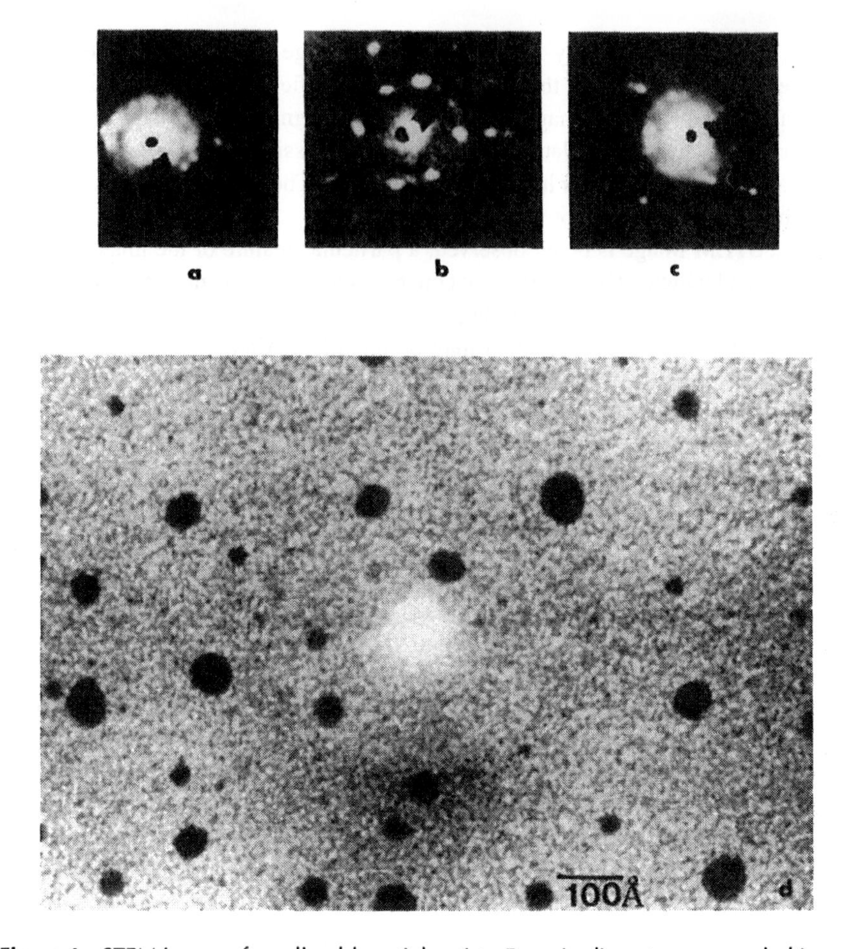

Figure 4. STEM image of small gold particles, 1 to 5 nm in diameter, suspended in a plastic film. The overexposed position marker appears as a blurred white dot. Nanodiffraction patterns, such as (**a**,**b**,**c**) showing multiple twinning of the gold crystals, are obtained with the beam stopped at the marker, placed over the image of a small particle.[22]

The TEM instruments may be used very effectively for obtaining CBED patterns from near-perfect regions of crystalline specimens for beam diameters of the order of 10 to 20 nm (or smaller if a field-emission gun is used) and for crystal thicknesses up to about 100 nm. With the convergent beam, the individual diffraction spots have the form of circular disks which do not overlap. The intensities within the disks vary as the diffraction intensities vary with the angle of incidence and, in the absence of overlapping spot effects, it may be assumed that there are no interference effects due to incident-beam coherence, whether a field-emission gun is used or not.

Studies of the intensity distributions within the spots of these CBED patterns have been highly effective for the determination of crystal symmetries, for the determination of crystal thicknesses and strains, and for the derivation of electron density, or electrostatic potential, distributions in crystals with very high accuracy.[25,26] However this technique is not usually applicable to nanoparticles and so will not be described further here.

The nanodiffraction patterns, such as in Figure 4, obtained in a STEM instrument with a very fine beam, may be interpreted in much the same way as for the SAED patterns obtained in TEM. The convergence of the incident beam in this case is sufficient to broaden the spots, but is not so great that the spots cannot be individually located and measured. The incident-beam convergence angle in this case is in the order of 10^{-3} radians, corresponding to a beam diameter at the specimen, and also a STEM image resolution, of about 1 nm, whereas the separation of the diffraction spots is about 10^{-2} radians. For other cases of smaller spot spacings, imperfect crystals, larger angles of beam convergence, or for more detailed interpretations of pattern intensities, it is necessary to consider in more detail the process of diffraction in nanocrystals, taking into account the coherence of the incident beam.

III. PRINCIPLES OF NANODIFFRACTION

When a parallel electron beam is incident on a thin specimen, it may be considered that the specimen modifies the amplitude and phase of the plane electron wave to produce the wavefunction on the exit surface of the specimen. Then the wavefunction at some large distance from the specimen is given by Fraunhofer diffraction, expressed as the Fourier transform of the exit wave.

For high-voltage electrons, which are scattered through only small angles in the order of 10^{-2} radians, the simplest approximation valid for very thin specimens assumes that the electrons travel in straight lines through the sample and the electron wave is given a phase change, relative to the vacuum wave, depending on the refractive index of the sample for electrons. From the wave equation for electrons it is readily deduced that the refractive index is given by,

$$n = (1 + \phi(\mathbf{r})/E)^{1/2} \approx 1 + \phi(\mathbf{r})/2\,E \tag{1}$$

where $\phi(\mathbf{r})$ is the electrostatic potential at position r and E is the accelerating voltage for the electron beam. The approximation of the latter part of Eq. 1 is possible when, as is typical, ϕ is of the order of 10 volts and E is of the order of 10^5 volts. If the integral of $\phi(\mathbf{r})$ along the beam direction, assumed to be the z-axis, is $\phi(x,y)$, the phase change of the electron wave is $(2\pi/\lambda)\cdot\phi(x,y)/2E$, and, putting $\sigma = \pi/E\lambda$, as the "interaction constant," the wave at the exit plane of the specimen may be written as,

$$\psi(x,y) = \exp\{-i\sigma\phi(x,y)\} \tag{2}$$

where the use of the minus sign is consistent with the definition of a plane wave as $\exp\{i(\omega t - \mathbf{k}\cdot\mathbf{r})\}$. Equation 2 expresses the "phase-object approximation" (POA) which neglects any change of amplitude of the wave due to effective absorption of electrons resulting from inelastic scattering, and also neglects any spread of the wave in the specimen due to elastic scattering or Fresnel diffraction effects. For 100 keV electrons, it can usually be considered as valid for thicknesses up to 2–5 nm and so can be used for most discussions of nanodiffraction of nanoparticles. For thicker specimens, the complications of three-dimensional diffraction must be introduced and, except when only very light atoms are present, dynamical diffraction effects (the coherent interactions of all diffracted beams) must be considered.[27]

The amplitude distribution in the diffraction pattern in the POA, is given by Fourier transform of Eq. 2. The Fourier transform is defined as,

$$\Psi\,(\mathrm{u,v}) = \mathrm{F}\cdot\psi(x\cdot y) \equiv \int \psi(x,y)\cdot\exp\{2\pi i(ux + vy)\}\cdot dx\cdot dy \tag{3}$$

and, for Fraunhofer diffraction, u,v are the angular variables, $u = (\sin\theta_x)/2\lambda$ and $v = (\sin\theta_y)/2\lambda$, where θ_x and θ_y are the components of half the scattering angle. Then the intensity distribution in the plane of the diffraction pattern is:

$$I(u,v) = |\,\mathrm{F}\,\{\cos(\sigma\phi(x,y)\} - i\,\mathrm{F}\{\sin(\sigma\phi(x,y))\}\,|^2 \tag{4}$$

When only light atoms are present in the sample, it may be possible to assume that the product $\sigma\phi(x,y)$ is very much less than unity. Then Eq. 4 simplifies to,

$$\begin{aligned} I(u,v) &= |\,\delta(u,v) - i\,\mathrm{F}\,\{\,\sigma\phi(x,y)\,\}\,|^2 \\ &= \delta(u,v) + \sigma^2\,|\,\Phi(u,v)\,|^2 \end{aligned} \tag{5}$$

where $\Phi(u,v)$ is the Fourier transform of $\phi(x,y)$ and so represents the amplitude of the diffraction pattern for the simple kinematical diffraction theory as used for X-ray diffraction. The δ function represents the undeflected central beam of the pattern.

For a convergent incident beam on the specimen, one can assume in the POA that, for each incident beam direction, represented by u_0,v_0, the diffraction pattern is shifted sideways to give $\Phi(u-u_0,v-v_0)$, and then all such patterns are added together. If the incident beams from all the different directions are incoherent, it is the intensity distributions of the patterns that are added. If all incident beams are coherent, the amplitude distributions are added and then the modulus squared of the sum gives the intensity distribution. Hence,

$$\text{Incoherent: } I_{\text{Obs.}} = \int |\Psi(u-u_0,v-v_0)|^2\cdot|T(u_0,v_0)|^2\cdot du_0\cdot dv_0$$

$$\text{Coherent: } I_{\text{Obs.}} = \left|\int \Psi(u-u_0,v-v_0)\cdot T(u_0,v_0)du_0\cdot dv_0\right|^2 \tag{6}$$

where $T(u_0,v_0)$ is the complex amplitude of the corresponding incident wave, affected by the aberrations of the probe-forming lens.

The coherence of the waves coming in different directions depends on the lateral coherence of the electron waves at the position of the aperture which defines the incident beam cone angle. This coherence depends on the angle subtended by the source at the aperture position.[5,27] It may generally be assumed that when a hot-filament electron gun is used in an electron microscope and the effective source diameter is of the order of 1–2 μm, the incoherent approximation is valid. When a field-emission gun is used, however, the effective source size is as small as 4 nm, and it is possible to assume complete coherence and the second form of Eq. 6 is used.

The integrals of Eq. 6 have the form of convolution integrals. The convolution of two functions is defined, in one dimension, as:

$$f(x) * g(x) \equiv \int f(X) \cdot g(x - X)\, dX \tag{7}$$

For the incoherent case, $I(u,v) = |\Phi(u,v)|^2 * |T(u,v)|^2$. The 'transfer function, $T(u,v)$, includes the limitation of the amplitude by the aperture function, $A(u,v)$, which is zero except within the aperture of radius $u = u_1$, and a phase function $\exp\{i\chi(u,v)\}$ given by the lens aberrations. In electron microscopy it is usually sufficient to include in $\chi(u,v)$ only the second-order term due to the defocus, Δ, and the fourth-order term due to spherical aberration, with coefficient C_s, so that we assume:

$$\chi(u,v) = \pi\Delta\lambda(u^2 + v^2) + \pi C_s\lambda^3(u^2 + v^2)^2/2 \tag{8}$$

Hence $|t(u,v)|^2 = A(u,v)$, and, for the incoherent case, the diffraction pattern intensity distribution is smeared out by the convolution with the aperture function: each diffraction spot from a crystal becomes an image of the aperture, a disk of uniform intensity.

From the last part of Eq. 6, the intensity for the coherent case becomes the square of the convolution of the two complex functions,

$$\Psi(u,v) = \Phi(u,v) * T(u,v) \tag{9}$$

and so is less easily interpreted. The Convolution Theorem of Fourier transforms states that the Fourier transform of the convolution of two functions is the product of their Fourier tranforms. Hence, as may be expected, this diffraction pattern amplitude comes from the Fourier transform of the exit wave at the specimen level, the product $q(x,y) \cdot t(x,y)$, where $q(x,y)$ is the transmission function of the object and $t(x,y)$ is the Fourier transform of $T(u,v)$ and is generally called the "spread function" because it represents the spreading or "smearing out" of the wave at the specimen due to the aperture and aberrations of the lens.

If the specimen is a perfect thin crystal, the parallel-beam diffraction pattern consists of an array of sharp spots[5] which can be represented by δ functions. From Eq. 9, for a convergent beam each spot, with indices h,k, is spread out by the function $T(u,v)$ and has a weighting $F_{h,k}$, the structure amplitude for the reflection. If the extended diffraction spots do not overlap, the diffraction pattern intensity distribution becomes just a set of broadened spots of the form $|F_{h,k}|^2 \cdot |T(u,v)|^2$, which is exactly the same as for the incoherent case.

If, however, the aperture size is made larger so that the diffraction spots overlap, as suggested in Figure 5, then, for the coherent case, the wave functions for two or more extended spots are added together in the regions of spot overlap and interference effects occur, giving interference fringes.[27,28] For the case of the overlap of the h,0 and $h + 1$,0 spots, for example, it is readily shown that the phase difference between the waves of the h and $h + 1$ spot is given by,

$$2\pi\{(\Delta\lambda/d) + (C_s\lambda^3/4d^3)\}\cdot\varepsilon \qquad (10),$$

where ε is the distance from the midpoint of the overlap region in nm^{-1}, and d is the lattice plane spacing. Hence fringes appear in the region of overlap of the spots with a periodicity equal to $d/\Delta\lambda$ if the defocus is sufficiently large. Excellent examples of such fringe patterns, have been shown recently in CBED patterns taken from much larger specimen areas of single crystals by use of TEM instruments equipped with field-emission guns.[29,30]

With a FEG STEM instrument, spectacular patterns of fringes, arising in the same way, may be observed if a very large objective aperture, or no aperture at all, is

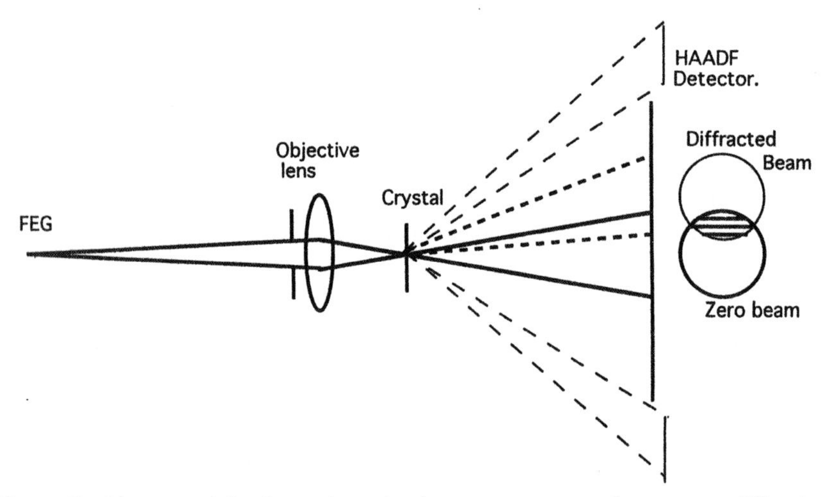

Figure 5. Diagram of the formation of coherent convergent beam nanodiffraction pattern for which adjacent diffracted beam disks overlap and interference fringes form in the area of overlap.

used, as in Figure 6. For these under-focus cases, the characteristic patterns of curved fringes beyond the region of straight fringes, are clearly seen.[31] These patterns are referred to as "electron Ronchi fringes" because of their similarity with the Ronchi fringes of light optics[32] used for testing the aberrations of large telescope mirrors.

For the cases of Figures 5 and 6, the positions of the fringes depend on the relative phases of the reflections giving the overlapping spots, and the relative phases depend on the both the relative phases of the structure amplitudes, $F_{h,0}$ and $F_{h+1,0}$ and on the position of the incident beam relative to the origin of the unit cell of the crystal. As the incident beam is moved from one side of the unit cell to the other, the fringes are translated by one periodicity. Hence, if a small detector is placed in the region of overlap and the beam is scanned over the crystal, the intensity of the STEM image formed oscillates with the crystal periodicity, i.e. the image shows lattice fringes.[33] Application of the reciprocity relationship shows that this configuration for obtaining lattice fringe images in STEM is exactly equivalent to the tilted-beam method for obtaining lattice fringe images of crystals in TEM.

For the incoherent case of Figure 5, no interference effects occur in the nanodiffraction pattern. When diffraction spots overlap, the intensities are added in the overlap region. The pattern can then be interpreted just as for SAED patterns, except that the greater size of the spots sometimes makes it more difficult to measure interplanar distances.

If the specimen is not a perfect thin crystal and if it has boundaries, thickness variations, or disorder of the atomic sites or their occupancies, the plane-wave diffraction pattern is no longer an array of δ-function spots. There are streaks, diffuse spots, or diffuse patches between the spots, or else the spots are extended and/or distorted. Then the coherent and incoherent nanodiffraction patterns are different, even for the cases where there is no overlap of the main diffraction spots.

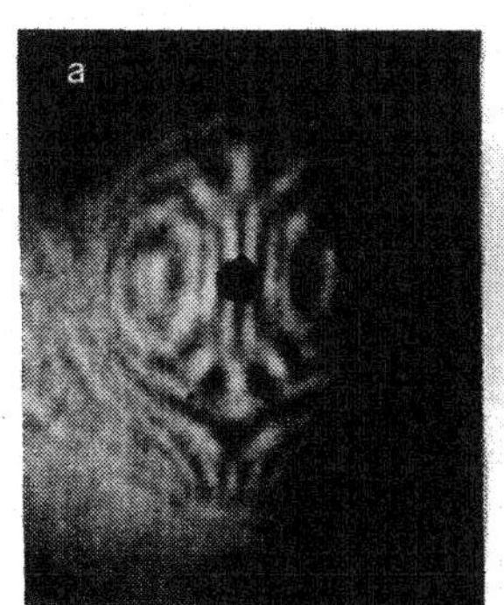

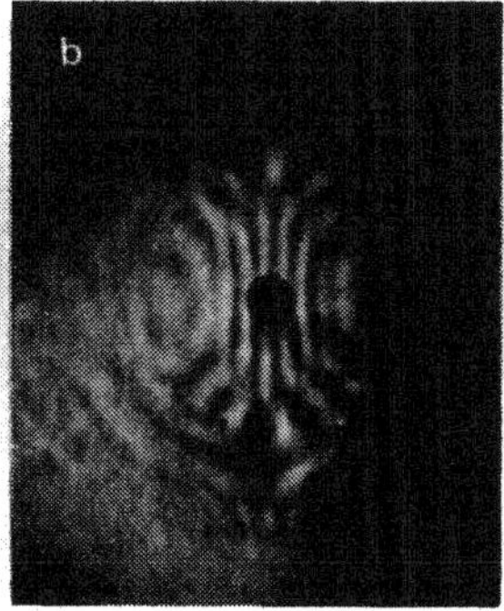

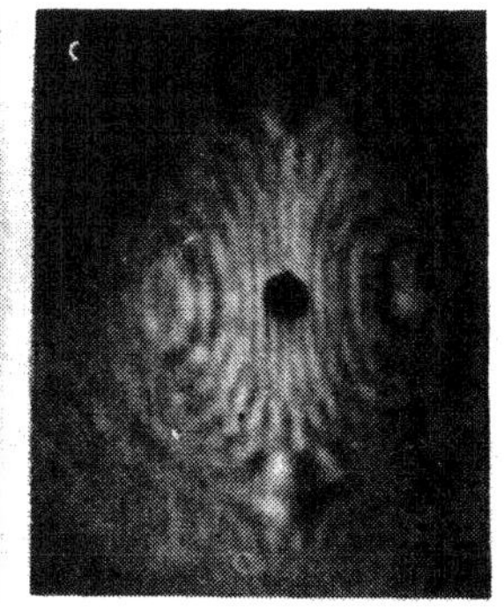

Figure 6. Interference fringes (electron Ronchi fringes) formed in the shadow image of a thin crystal (beryl, with a periodicity of 0.8 nm) by interference between adjacent diffracted beams.[27]

The incoherent pattern is given just by smearing out the intensity distribution of the plane-wave pattern. The coherent nanodiffraction pattern contains interference effects due to the overlapping of complex amplitude distributions.

The case that the coherent incident beam illuminates the linear edge of a crystal has been studied in some detail. In the plane-wave diffraction pattern, the discontinuity at the crystal edge gives rise to streaks running through the diffraction spots perpendicular to the edge. In coherent nanodiffraction patterns, the diffraction spots, instead of being disks of uniform intensity, become rings, or hollow disks, with high intensity around the edges of the disks, particularly along the edges that are nearly parallel to the crystal edge.[34,35] For example, Figure 7 shows circular spots generated when an incident beam illuminates the edge of a crystal of MgO. The form of the spots varies with the position of the beam relative to the edge. The circular form is most clearly seen when the crystal edge is at the middle of the beam probe (Figure 7c). Such observations were extended by Pan,[36] who considered the case of the edges of crystals so small that the crystal edge may be of limited length, comparable with the beam diameter. It was found by calculation and experimentally that the diffraction spots could become either almost complete circles or short arcs, depending on the length of the crystal edge. The details of the intensity distributions within the spots could, in principle, be related to the form of the edge, whether the crystallite is terminated in a plane parallel to the beam or by two planes tilted with respect to the beam to form a wedge. Thus a means is provided for determining the detailed three-dimensional morphology of crystals which are in the nanometer size range.

The spots of nanodiffraction patterns may also be turned into hollow rings if there is an internal discontinuity in the projection of the crystal structure in the beam direction, but in this case not all diffraction spots may be affected. For example, for crystals of ordered binary alloys there are out-of-phase domain boundaries at which the origin point of the unit cell can shift sideways, from one lattice site to another, on a boundary plane.[37] In the ordered Cu_3Au structure, for example, the Au atom

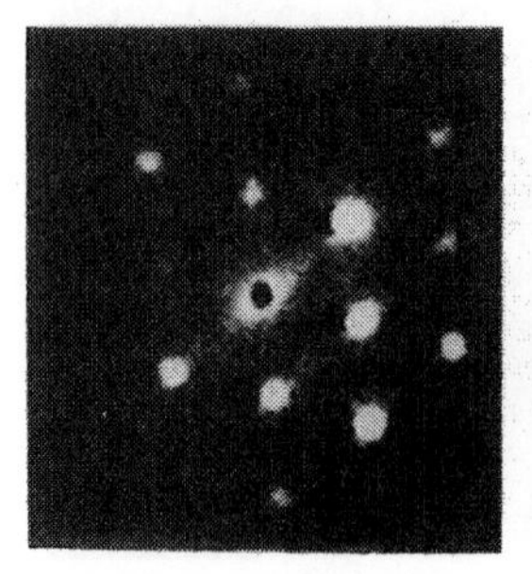

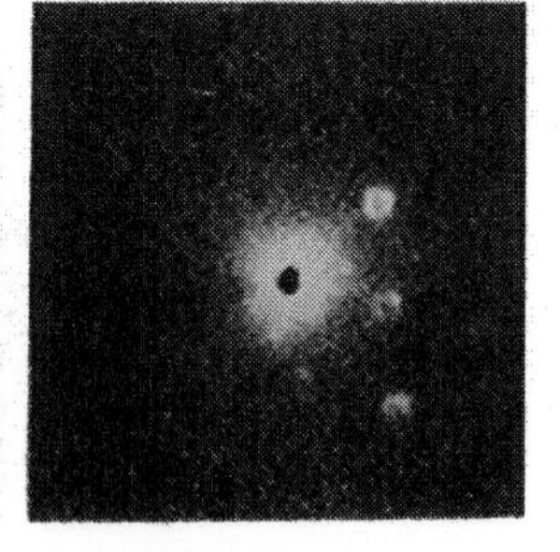

Figure 7. Nanodiffraction patterns obtained as the incident beam, parallel to a (001) face, is translated across the edge of a MgO crystal in [110] orientation. For (**a**) the beam passes through the crystal. For (**c**) the beam is centered on the crystal edge.

occupies one of the four atomic sites in a face-centered cubic unit cell and copper atoms occupy the other three sites so that, within a domain, the structure is simple-cubic. At an out-of-phase domain boundary, the gold atom shifts from one of the four sites to another. For the lattice planes corresponding to the average, face-centered-cubic structure (giving the F.C.C. 200, 020, 220, etc. reflections) there is no discontinuity at the boundary, but for the other simple-cubic planes (giving the 100, 110, 210 etc. reflections), there may be a discontinuous change of phase of the reflection and the corresponding coherent-nanodiffraction spots are split into circular arcs.[37] There are, in fact, several ways in which the Au atoms may be translated at the boundary and these may be recognized by observing which of the simple-cubic diffraction spots are split.

Similarly, faults may occur in the stacking of successive planes of atoms in a crystal lattice. Such stacking faults occur frequently on [111] planes of a F.C.C. metal structure. When the incident electron beam is parallel to such a fault, the fault represents a linear discontinuity in the projection of the structure, affecting some reflections but not others. Nanodiffraction patterns obtained with the beam illuminating such a fault then show a splitting of a set of diffraction spots which is characteristic of the nature of the fault.[38]

In general, however, for nanocrystals of varying shapes, sizes and states of disorder, it is not feasible to carry out the detailed examination of the intensity distributions, splitting, and distortions of the coherent diffraction spots that would be necessary to determine the form of each crystallite. While some valuable indications on particular features may be available, the patterns are often interpreted as for the incoherent case, taking into consideration any distortions of the spots that may result from interference effects.

On the other hand, when very small coherent incident beams are used and the diffraction spots are of such large diameter that many diffraction spots may overlap and interfere at any point of the diffraction pattern, important structural information may sometimes be derived. The overlap of the diffraction spots is an indication that the diameter of the incident-beam probe is smaller than the periodicity of the crystal structure seen in projection in the direction of the incident beam. Thus the region of the specimen illuminated by the beam is smaller than the "unit cell" of the projected structure. The periodicity of the projected crystal structure is no longer reflected in the diffraction pattern. As the beam is moved about within the area of the unit cell, the diffraction pattern intensity distribution changes because different groupings of atoms are illuminated. The symmetry of the pattern reflects the local symmetry of the atom arrangement relative to the center of the incident beam.[39]

An example of an application of such patterns is given by the study of the structure of planar faults in diamond crystals.[21] Such faults have been known to occur and have been studied for many years by X-ray diffraction, electron microscopy, and other techniques with no clear indication of their structure or nature, except that evidence seemed to show that they are associated with the presence of nitrogen in the diamond lattice. In a STEM instrument, nanodiffraction patterns were obtained as a beam of diameter 0.3 nm was scanned slowly across one of these defects lying in a plane parallel to the incident beam. The patterns were recorded with a TV-VCR system at intervals of 0.02 nm along the scan. Figure 8 shows a selection of the patterns from a series taken from every tenth frame of the recording, i.e. for separations of 0.2 nm along the scan. The individual diffraction patterns do not show the individual diffraction spots for the [110] beam direction, but only diffuse intensity distributions. They do, however, show the local symmetry of the illuminated areas of the structure, and the change of symmetry as the beam traverses the defect is clearly seen. The diffraction patterns could not be directly interpreted in terms of the local structure at the defect. Instead it was necessary to make calculations, using many-beam dynamical diffraction algorithms for the many

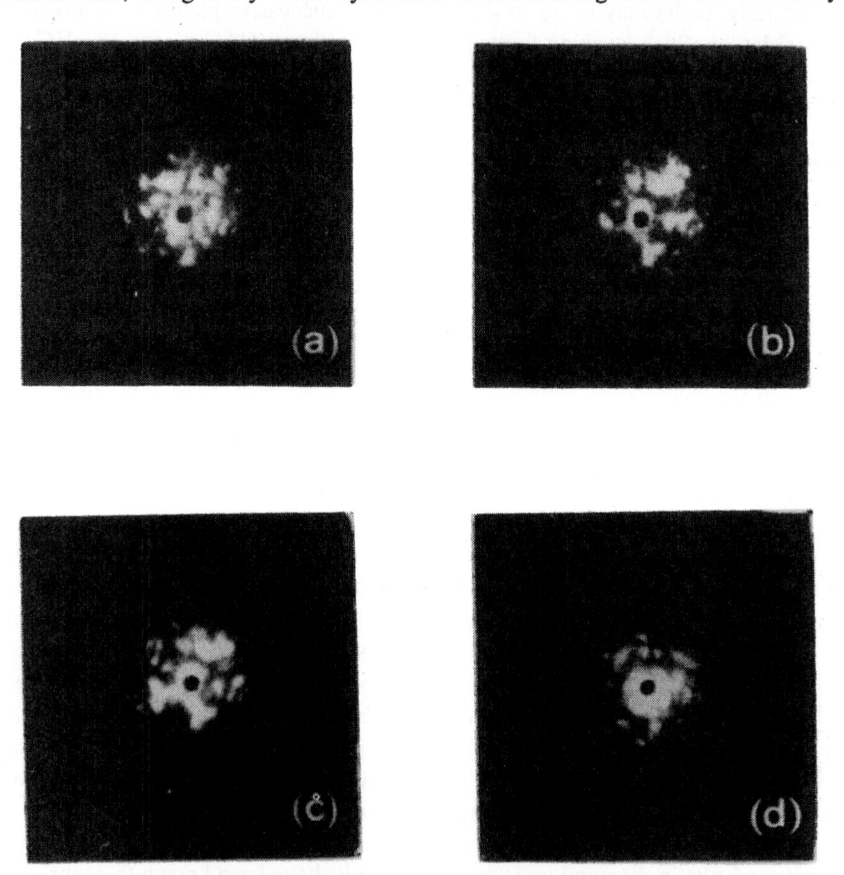

Figure 8. Nanodiffraction patterns from a series recorded at intervals of 0.2 nm as a beam of diameter 0.3 nm was scanned over a planar defect in a diamond crystal in approximately [110] orientation. (**a**) and (**d**) are from the perfect crystal on the two sides of the defect and (**b**) and (**c**) are obtained with the beam on the defect.[21]

different models that have been proposed for the structure of the defect and compare the experimental results with the calculated intensities. The results favored one particular model not involving nitrogen atoms.[40]

For an amorphous material the diffraction pattern from a relatively large area, greater than 100 nm in diameter, even if only a few nm thick, consists of a set of very diffuse halos which arise from the presence of preferred interatomic distances in the sample, occurring at random in all possible orientations. In the case of nanodiffraction with a beam of diameter about 1 nm and for an amorphous film less than about 5 nm in thickness, the number of atoms illuminated by the beam is not sufficient to give the same averaging over atomic configurations and their orientations. The pattern does not show smooth halos, but consists of irregular maxima and minima of intensity given by scattering from the particular configurations of atoms illuminated (see Figure 9a). The pattern of maxima and minima changes if the incident beam is moved by a fraction of 1 nm so that a different configuration of atoms is illuminated. It should, in principle, be possible to make deductions from a large number of such diffraction patterns concerning the statistics of local configurations, and of the short-range ordering of amorphous materials, but this has not yet been attempted.[41] Similar considerations apply to the possibility of determining the statistics of the local ordering of atoms in disordered binary alloys or related materials.[42]

Because amorphous materials give nanodiffraction patterns consisting of more-or-less random distributions of diffuse maxima and minima, it is often difficult to distinguish their diffraction patterns from those of more ordered regions, such as those containing very small crystallites which tend to be very diffuse although somewhat more regular. In general it may be stated that for well-ordered regions of diameter greater than 1 nm the characteristically regular array of diffraction spots, or the occurrence of spots on rings of particular diameters, may be distinguished even in the presence of considerable scattering from surrounding amorphous material (see, for example, Figure 9b), but for ordering on this or a smaller scale some preknowledge of the type of ordering to be expected is an advantage.

One important characteristic of nanodiffraction patterns, and particularly for those of near-amorphous materials, is that the intensity distribution does not often have a center of symmetry at the origin. For the simple kinematical theory, the center of symmetry at the origin is a requirement for diffraction patterns from thin samples when the diffraction pattern intensities are calculated with the usual assumption of a plane-wave incident or an incoherent convergent beam. For the coherent incident beams used for nanodiffraction, however, there is a center of symmetry in the pattern only when both the incident beam and the illuminated part of the specimen have a center of symmetry at the same point. Even for a very thin single crystal which has a center of symmetry at some point in the unit cell, the pattern does not show a center of symmetry unless the incident beam is centered on that point. The prediction of a lack of a center of symmetry has been discussed and demonstrated for the case of amorphous thin films[43] and for carbon nanotubes.[44]

a

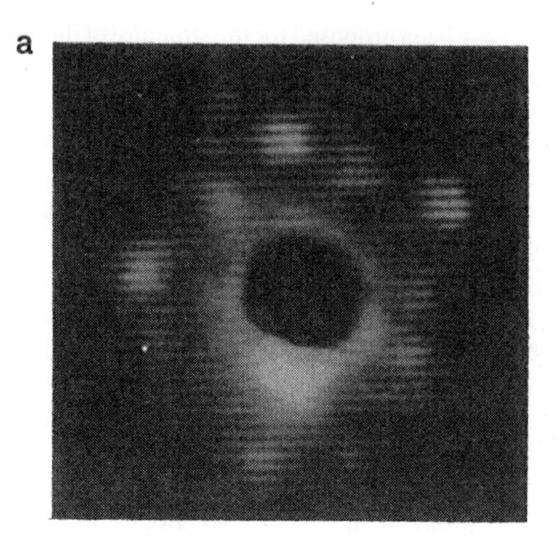

b

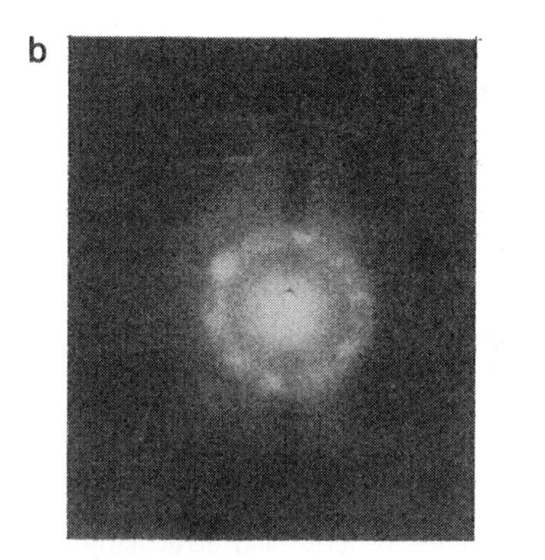

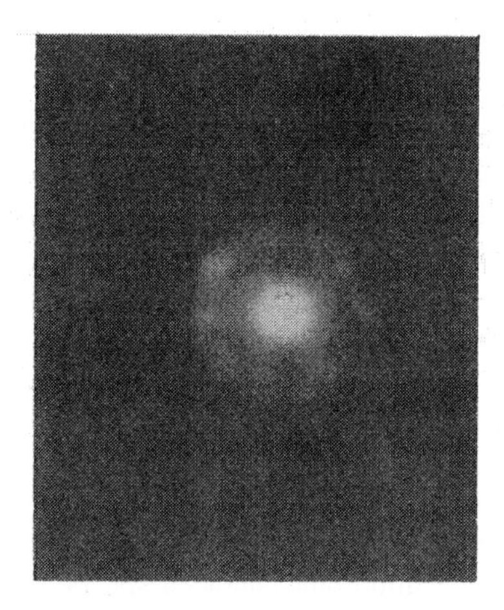

Figure 9. Nanodiffractions patterns from amorphous and nanocrystalline specimens. (**a**) a Na–Al silicate glass (Courtesy of Dr. J. Konnert); (**b**) a quenched Mn–Al thin film.

IV. PRINCIPLES OF STEM

As the incident beam is scanned over the specimen in a STEM instrument, such as is sketched in Figure 3, the signal collected to form the image depends on the form of the detector. The reciprocity principle suggests that in order to get the same image as in bright-field TEM one should collect the electrons transmitted through the specimen and propagated along the axis to the center of the central spot of the diffraction pattern. Although it is often possible to derive the image intensity and contrast by using the reciprocity principle to relate the imaging conditions to those of the TEM, for which the theory of imaging has been very well explored and exploited, it is preferable for many cases to derive the STEM image intensities directly from appropriate theory, especially for cases for which no equivalent TEM experiment or theory has been developed.

From Eqs. 6 and 9 it can be seen that the intensity in the diffraction pattern for coherent illumination is $|Q(u,v) * T(u,v)|^2$, which is the square of the Fourier transform of wave function at the exit face of the specimen, $q(x,y)\cdot t(x,y)$, where $q(x,y)$ is the transmission function of the specimen, $Q(u,v)$ is its Fourier transform, and $t(x,y)$ is the "spread function", giving the amplitude distribution of the focused incident beam, spread by the effects of the lens aberrations, and given by Fourier transform of the transfer function, $T(u,v)$ of the lens. We wish to find the intensity of the signal as the incident beam is moved across the specimen to positions designated X,Y. Thus we want to deal with the exit wave function,

$$\psi(x,y) = q(x,y)\cdot t(x - \mathrm{X}, y - \mathrm{Y}) \tag{11}$$

and the corresponding diffraction pattern intensity distribution:

$$I(u,v) = |\, Q(u,v) * T(u,v)\cdot \exp\{2\pi i(u\mathrm{X} + v\mathrm{Y})\}\,|^2 \tag{12}$$

Then the signal produced by a detector for which the sensitivity to incoming radiation is defined by the "detector function" $D(u,v)$, is

$$J(\mathrm{X,Y}) = \int D(u,v)\cdot |\, Q(u,v) * T(u,v)\cdot \exp\{2\pi i(u\mathrm{X} + v\mathrm{Y})\}\,|^2\, du\cdot dv. \tag{13}$$

For a small detector, on axis, the detector function may be approximated by a δ function, $\delta(u,v)$. It is readily shown that expression 13 then reduces to,

$$J(\mathrm{X,Y}) = |\, q(\mathrm{X,Y}) * t(\mathrm{X,Y})\,|^2 \tag{14}$$

and this is exactly the expression for the intensity of a bright-field image for a parallel incident beam in TEM. For a sufficiently thin object, neglecting absorption, the transmission function of the object can be written, as in Eq. 2: $q(x,y) = \exp\{-i\sigma\phi(x,y)\}$. And for a weakly scattering object, it is possible to make the "weak phase object approximation" (WPOA), $\exp\{-i\sigma\phi(x,y)] = 1 - \sigma\phi(x,y)$, so that Eq. 14, becomes

$$J(\mathrm{X,Y}) = 1 + 2\sigma\phi(\mathrm{X,Y}) * s(\mathrm{X,Y}) \tag{15}$$

and higher order terms in $\sigma\phi$ have been neglected. Here $s(x,y)$ is the imaginary part of $t(x,y) = c(x,y) + is(x,y)$, and $c(x,y)$ and $s(x,y)$ are the Fourier transforms of $A(u,v)\cdot\cos\{\chi(u,v)\}$ and $A(u,v)\cdot\sin\{\chi(u,v)\}$ where $A(u,v)$ is the aperture function and $\chi(u,v)$ is the phase factor given by Eq. 8. As in the standard theory for TEM imaging of weak phase objects,[6] the best resolution is given from Eq. 15, when

$s(x,y)$ is a single sharp peak of minimum width, and this is the case for the so-called, "Scherzer optimum defocus", given by $\Delta = -(4C_s\lambda/3)^{1/2}$. The "image resolution", or, more correctly, the least resolvable distance, is then taken to be:

$$\varepsilon = 0.62.\ C_s^{1/4} \cdot \lambda^{3/4} \tag{16}$$

Various efforts to improve the resolution of electron microscopes have therefore been concentrated on the decrease of the electron wavelength by increasing the accelerating voltage, or else the reduction of the spherical aberration constant, C_s. For C_s of the order of 1 mm, Eq. 16 gives resolutions of about 0.2 nm for 200 keV electrons and about 0.12 nm for 1 MeV electrons.

The results of Eqs. 15 and 16 apply only for the case that the STEM detector has zero diameter, and hence the signal has zero magnitude. It is customary to make the detector as large as is reasonably possible. It has been shown[45] that as the diameter of a circular axial detector is increased from zero, the resolution in the WPOA is slightly improved, but the contrast of the image decreases slowly at first and then more rapidly. When the detector size becomes equal to the size of the central beam spot of the diffraction pattern, the WPOA image contrast becomes zero. Some contrast may then be given by the second- and higher-order terms in the expansion of the POA Eq. 2, and then the resolution is better than for the WPOA case by a factor of about 1.4.[45]

The dark-field mode of STEM, as originally used by Crewe and associates[1] to see individual heavy atoms, is said to give resolutions better than Eq. 16 by a factor of about 1.4. In this mode an annular detector is used to collect all those electrons scattered outside the incident-beam cone, i.e. all of the diffraction pattern outside the central-beam spot. In order to distinguish between the central beam and the remainder of the diffraction pattern, we write the specimen transmission function as $q(x,y) = 1 - p(x,y)$, and its Fourier transform as $Q(u,v) = \delta(u,v) - P(u,v)$, where the δ function represents the sharp central spot for a plane-wave diffraction pattern. Then Eq. 13 becomes,

$$\begin{aligned} J(X,Y) &= \int D(u,v)\cdot A(u,v)\, du\cdot dv, \\ &\quad + \int D(u,v)|\, P(u,v) * T(u,v)\cdot \exp\{2\pi i(uX + vY)\}|^2 \cdot du\cdot dv \\ &\quad - \int D(u,v)\cdot T^*(u,v)\cdot \exp\{-2\pi i(uX + vY)\}\cdot [P(u,v) * \exp\{2\pi i(uX + vY)\}]\cdot du\cdot dv \\ &\quad - c\cdot c \end{aligned} \tag{17}$$

where $c\cdot c$ represents the complex conjugate of the previous term. The first term of this expression comes from the incident beam spot and contributes a background to the bright-field image. The last two terms represent the effects of interference between the incident and diffracted beams, and so are important for considerations of crystal lattice imaging, holography, and related effects.[46] These terms, and the first term of Eq. 17, are zero if the detector does not overlap the central beam so that $D(u,v)\cdot A(u,v) = 0$.

The remaining, second term of Eq. 17 is the integral over the diffraction pattern which occurs both inside and outside the central beam. A dark-field image must be given by a detector which does not overlap the central beam. If an annular detector, collecting everything except the central beam is used, the second term of Eq. 17 is complicated. A relatively simple result can be obtained if it is assumed that the intensity of scattering within the central beam is proportional to that outside the central beam. This assumption is good, for example, if the specimen consists of isolated atoms for which the plane-wave diffraction pattern falls smoothly away from the central spot. It is a poor assumption if the specimen consists of closely spaced atoms so that the diffraction pattern intensity oscillates strongly on a scale comparable with the dimensions of the central spot, i.e. near the resolution limit for imaging assemblies of atoms.

If this assumption is made, the image intensity may be derived from the second term of Eq. 17 by assuming that $D(u,v) = 1$, so that all the diffracted electrons are detected equally. Then the second term of Eq. 17 gives, by application of Parseval's Theorem[5]:

$$J(X,Y) = |p(X,Y)|^2 * |t(X,Y)|^2 \tag{18}$$

In the WPOA, $|p(X,Y)|^2 = \sigma^2\phi^2(X,Y)$ and, since $|t(X,Y)|^2$ represents the intensity distribution of the incident beam on the specimen, this approximation suggests that the image contrast is given as for incoherent imaging of the square of the projected potential. If the spread function, $t(X,Y)$, could be approximated by a gaussian, its square would have a half-width $2^{1/2}$ times smaller. Hence the conclusion that the dark-field mode has a resolution better than the small-detector bright-field mode by a factor of about 1.4.

The atomic scattering factors, f, for electrons are approximately proportional to the atomic number, Z, for high angles of scattering, and the total scattered intensity is roughly proportional to $Z^{3/2}$ for isolated atoms. The resulting Z contrast of ADF images has proved to be very useful for the imaging of individual heavy atoms, or small clumps of heavy atoms supported on or in thin films of amorphous light-atom material. There have been many applications to the imaging of the metal particles of supported metal catalysts. However, for most such catalyst specimens, both the light-atom support and the metal particles are nanocrystalline rather than amorphous and contrast in the image can arise from the occurrence of relatively strong diffraction peaks from crystallites in particular orientations. The high-angle annular dark-field (HAADF) mode, proposed by Howie,[16] avoids this difficulty, while retaining the strong Z-contrast effect. In this case, the inner radius of the annular detector for 100 keV electrons is about 50 mrad and the outer radius is 100 mrad or more. The electrons collected to form the signal are mostly those scattered to high angles by thermal diffuse scattering[18,47,48] arising from the vibrations of the atoms around their mean positions, but the approximation that this scattering is similar to the high-angle Rutherford scattering from isolated atoms, proportional to Z^2, is sometimes useful.

The use of a thin annular detector for which the width of the annulus may be only about 10% of the inner radius, offers several interesting possibilities.[19] A thin-annular-detector bright-field (TADBF) image is given when post-specimen lenses in the STEM instrument are used to magnify the diffraction pattern so greatly that the outer edge of the central beam spot overlaps the inner edge of the detector. Then, in the WPOA, the third and fourth terms of Eq. 17 dominate the image contrast, and these terms can be rearranged to show that the effective transfer function is given by the convolution of the detector aperture function with the normal phase-contrast WPOA transfer function, so that the spread function for the image is given by the product of the $s(x,y)$ function of Eq. 15 and the Fourier transform of the detector function.[49] The result is to reduce the width of the spread function, and hence reduce the resolution limit by a factor of about 1.6 relative to the small-detector bright-field image or the TEM BF image.

A different type of contrast is given in the so called "marginal imaging mode" when the magnification of the diffraction pattern is such that the central spot comes just within the inner radius of the thin annular detector.[50] Some image contrast is given by the inner portion of the diffraction pattern which strikes the detector, but this is relatively small for weak-phase objects. Strong signals are given if the central beam is deflected so that it overlaps the detector. Deflection of the beam may result from any magnetic or electric field in the specimen, or if some change in the specimen thickness or scattering field gives a variation of the projected potential which acts as a prism for the electrons. Thus the marginal-mode image has a contrast depending on the square of the differential of the projected potential and serves to emphasize edges or discontinuities in the specimen.

When the magnification of the diffraction pattern by the post-specimen lenses is increased further so that the thin annular detector inner radius is greater than the central spot radius, a dark-field image is produced by an annular region of the diffraction pattern of variable radius. For crystalline or amorphous materials, the image contrast for a given TAD radius is given by the scattering to the corresponding range of diffraction angles, and since the diffraction pattern is characteristic of the material, various components of different structure may often be imaged separately.[50] Thus, perfect metal crystallites and twinned regions of crystallites may be imaged separately and, as we shall see below, various forms of crystalline or amorphous carbon may be distinguished.

Dark-field images from particular diffraction spots from crystalline specimens, similar to those formed in TEM instruments by tilting the incident beam on the specimen, are readily obtained by deflecting the diffraction pattern so that the diffracted beam, rather than the incident beam, passes into the entrance aperture of the EELS spectrometer (see Figure 3).

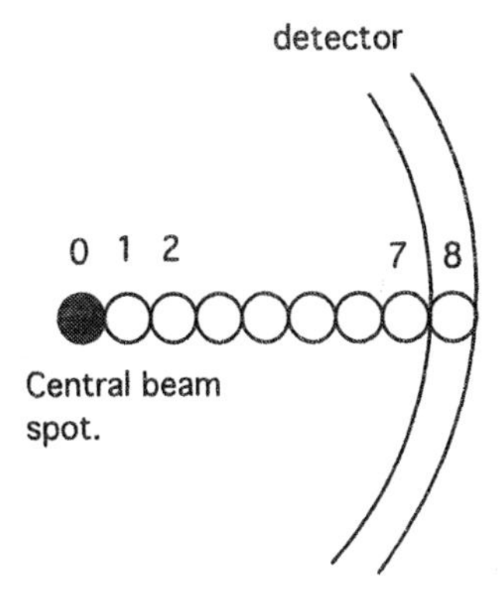

Figure 10. Diagram of the incident beam spot deflection positions for various "quick and easy" applications of a thin annular detector. Position 0 gives HAADF; Position 8 gives bright-field: Position 7 gives dark-field with inner diffracted beams.

For many purposes it is more convenient to obtain a dark-field STEM image, quickly and easily, by deflecting the unmagnified incident beam spot to the neighborhood of a thin annular detector which then collects, preferentially, all the diffraction spots in one sector of the diffraction pattern for a poorly defined range of scattering angles, as suggested by the diagram of Figure 10.[51] For the incident beam at the position "1" in this figure, the detector collects only high-angle scattering so that something like the usual HAADF image is formed. For the detector at position "8" a bright-field image is given. For position "7", the strongest signal comes from inner diffraction spots corresponding to d-spacings of about 0.3 to 1 nm. For position "6", the image shows regions giving reflections with d in the range of about 0.2 to 0.3 nm, and so on.

V. NANOCRYSTALS

The possibility of obtaining nanodiffraction patterns from very small crystals included in, or supported on, thin films of near-amorphous material, as illustrated in Figure 4, has been of importance for a number of studies of supported catalysts. The important information required for such samples is usually the size distribution, the crystal structure, and the faulting of the crystallites. The size distribution is best seen from dark-field STEM images in which the lack of complications due to the Fresnel fringes at the edges and the phase-contrast noise of bright-field images makes the measurement of particle dimensions more reliable. By the use of the Z-contrast ADF or HAADF STEM modes, particles of high atomic number can readily be distinguished even when the light-atom support material is relatively thick. For example, small Pt particles included in thick films of near-amorphous alumina are readily detected with HAADF STEM imaging even when they are invisible in bright-field images.[17]

In any transmission mode of imaging with high-voltage electrons, some ambiguity remains in the interpretation of images of nanocrystals in amorphous supports. The image represents a projection of the structure and it is not possible to derive three-dimensional data to determine whether the nanocrystals are on the surfaces or in the interior of the film. This difficulty may be overcome if images are obtained from the low-energy secondary electrons emitted when the incident high-energy beam strikes the specimen.

In the UHV MIDAS STEM instrument,[15] the low-energy secondary electrons emitted from the specimen are first accelerated by biasing the specimen with a few hundred volts, negative. These electrons initially follow close helical paths in the strong magnetic field of the objective lens, but as they drift out of the magnetic field they are "parallelized". Their helical paths increase in pitch until, when they enter a field-free space outside the lens casings, they are confined to a narrow beam having a divergence of only about 5°. They may then be deflected into an energy analyzer and detector and used to form a secondary electron microscopy (SEM) image of the specimen surface. In the MIDAS instrument, secondary electrons may be collected in this way from either the entrance or the exit side of the specimen.[52] Hence SEM images with resolutions of 1 nm or better may be obtained and compared with DF transmission images, obtained simultaneously, to determine the positions of particles on either the surface or in the bulk of the supporting film. The additional information concerning the structure of an individual particle may then be obtained by stopping the incident beam at the individual particle position in any of the images and recording the nanodiffraction pattern. Also for those particles on the surfaces, an analysis of the composition may be made by observing the spectra of the Auger electrons emitted, collected by the parallelizer system, and energy-analyzed with a suitable hemispherical electron-energy analyzer.[53]

In the case of light-atom nanocrystals or films on light-atom supports, the Z-contrast DF STEM methods are ineffective, but sufficient image contrast may often be obtained by use of a thin annular detector in the TADDF mode. One example is that of thin films, about 1 nm thick, of amorphous carbon deposited on supporting films of amorphous silica, 6 nm thick. In bright-field TEM or STEM imaging, the randomly mottled phase-contrast images given by the two types of film are not distinguishable. With a thin annular detector it is possible to take advantage of the fact that the diffraction patterns from the two types of film are different. In each case the diffraction pattern consists of diffuse haloes. However for amorphous carbon there is a strong amorphous halo corresponding to a spacing of about 0.12 nm, whereas for the amorphous silica there is no strong scattering in this region: the only strong peak occurs at a radius corresponding to a spacing of about 0.42 nm. Hence a TADDF image, taken with the post-specimen lenses set so that the annular detector collects electrons scattered at angles corresponding to the 0.12 nm spacing, shows preferentially the amorphous carbon. Such an image may be compared with images taken with slightly higher or lower currents in the post-specimen lenses in order to distinguish more clearly the contributions of the amorphous carbon.[50] It is clearly shown in this way that the amorphous carbon film has a distribution different from that of the amorphous silica, with a concentration in patches having relatively sharp edges.

When such films are heated to about 700 °C, the amorphous carbon component crystallizes to form nanocrystals. The TADDF images obtained with the spacing of 0.12 nm then show very little indication of contrast except that from the amorphous silica. Instead, TADDF images may be obtained with the post-specimen lenses set so that the reflections corresponding to a spacing of $d = 0.34$ nm, the interlayer spacing for a graphitic structure, are collected. Such images show good contrast of spots corresponding to the graphitic nanocrystals, as shown in Figure 11.[51] From such TADDF images, and from images obtained by deflecting the incident beam of the unmagnified diffraction pattern to the position "7" of Figure 10, the nanocrystal sizes could be derived as 1.0 ± 0.2 nm. In the images obtained with the beam deflected to "7", a few of the smaller bright spots in the image could be attributed to the amorphous silica since it was found that films of silica without the carbon layers give an occasional bright spot of diameter less than 1 nm.

Nanodiffraction patterns from individual nanocrystals, imaged in Figure 11, showed only the strong diffraction spots with spacing $d = 0.34$ nm, and higher orders

Figure 11. Dark-field STEM image with a thin annular detector set to collect diffracted beams for $d = 0.34$ nm from nanocrystals of carbon supported on 6 nm of amorphous silica. Marker = 10 nm. The inset is a nanodiffraction pattern showing the 0.34 nm carbon reflections.

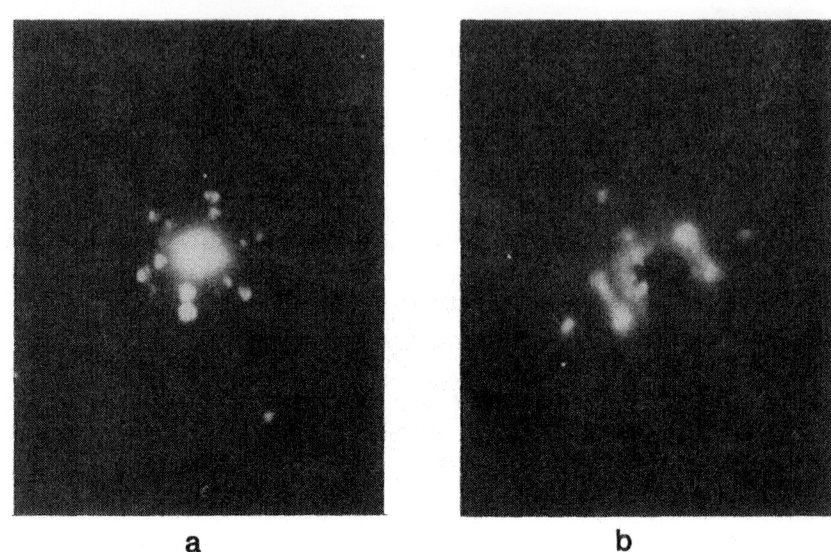

Figure 12. Nanodiffraction patterns from Au–Ru catalyst particles on MgO. (**a**) The new phase, B.C.C. Ru, shows a hexagonal set of spots inside the hexagonal MgO [111] set. (**b**) Pattern from the new phase in [110] orientation showing strong diffuse streaks between the inner spots.[56]

of these, as seen in the insert of Figure 11. The absence of any reflections with indices *hkl* with $l \neq 0$, lends support to the suggestion that the graphitic nanocrystals have disordered "turbostratic" structure, with arbitrary rotations of the planar graphitic sheets about the normal to the sheets.

The application of nanodiffraction to the study of catalyst particles may be illustrated by the relatively straightforward cases of Pt particles on γ-alumina,[54] Rh on ceria,[55] and the somewhat more complicated case of Au–Ru particles on MgO and silica supports.[56] For the first two of these cases, the metal particles are single crystals of diameter about 2 nm. The substrate material, alumina or ceria, is microcrystalline rather than amorphous and gives characteristic nanodiffraction patterns. Then the epitaxial relationship of the metal particles and the substrate crystals can be established. The surprising result for the platinum-on-alumina sample is the presence of the oxide, α-PtO_2, in both the calcined and reduced forms of the catalyst and the evidence that this oxide occurs with unit cell dimensions about 6% smaller than previously reported for the bulk material. Such a dimensional anomaly may arise either as a result of the small particle size or possibly because of nonstoichiometry.

In the case of the Ru–Au catalysts,[56] an explanation was sought for the rather surprising result that, on a MgO support, the catalytic activity of the Ru appears to increase with the addition of the inert Au. For particle sizes greater than about 5 nm, conventional techniques of microscopy and microanalysis could show that, as in bulk samples, no mixing of the Au and Ru occurs, thus it was envisaged that the enhancement of the catalytic activity might result from the properties or interactions of the metals for particle sizes of 1–3 nm. Nanodiffraction patterns from this material included those of cubic MgO and Au and of Ru in its normal hexagonal form, but there were also patterns from some other phase, as shown in Figure 12. These patterns could not be attributed to a mixed Au–Ru phase, but appeared to come from a heavily faulted, body-centered cubic form of Ru. Hexagonal patterns of spots, as in Figure 12a are consistent with such a B.C.C. Ru structure seen in [111] orientation. The pattern of Figure 12b is consistent with this structure seen in a [110] orientation, but the strong continuous lines of scattering between the [110]-type diffraction spots suggest disorder in the stacking of [110]-type planes of atoms. This result is consistent with other observations that, in very small crystals, metals may form with structures other than their structures in bulk, but it does not resolve the question of the influence of Au on the catalytic activity since the occurrence of the B.C.C type of patterns shows no clear correlation with the presence of Au.

VI. SINGLE-WALLED CARBON NANOTUBES

The discovery by Iijima and Ichihashi[57,58] of the existence of carbon nanotubes, each consisting of a single atomic layer of graphitic carbon wrapped into a cylinder of diameter about 2 nm, excited a great deal of attention, particularly because of the very interesting mechanical and electrical properties that may be expected if such tubes could be assembled to form macroscopic fibers.[59] Important progress has been made towards the production of ropes of these single-walled nanotubes (SWnTs), for example by Thess et al.[60] These authors reported the production of ropes 4 to 30 nm in diameter and up to 100 μm long using a process of double laser irradiation. They showed, using X-ray diffraction and electron microscopy, that the SWnT within the ropes tend to all have the same diameter of 1.37 nm and are stacked in hexagonal array with a periodicity of 1.69 nm. A theory, which has been proposed for the mechanism of formation of the SWnT and the ropes,[61] suggests that all the SWnTs in a rope should have the same diameter and helix angle, with the structure designated as [10,10] which has the symmetry C_{5V} and should give the ropes metallic properties. The nanodiffraction and STEM techniques have been applied to the experimental confirmation of these predictions.[62]

The diffraction patterns to be expected from a SWnT, according to the kinematical approximation, have been described, for example, by Lucas et al.[63] using the theory developed earlier for X-ray diffraction by helical structures.[64] For our immediate purposes, a simpler construction is sufficient.

For a single, planar sheet of hexagonal graphitic carbon structure, such as illustrated in Figure 13, the distribution of scattering power in reciprocal space is represented in Figure 14a. In the *hk*0 plane there is a set of maxima of scattering power at the hexagonal set of *hk*0 reciprocal lattice points. Through each of these

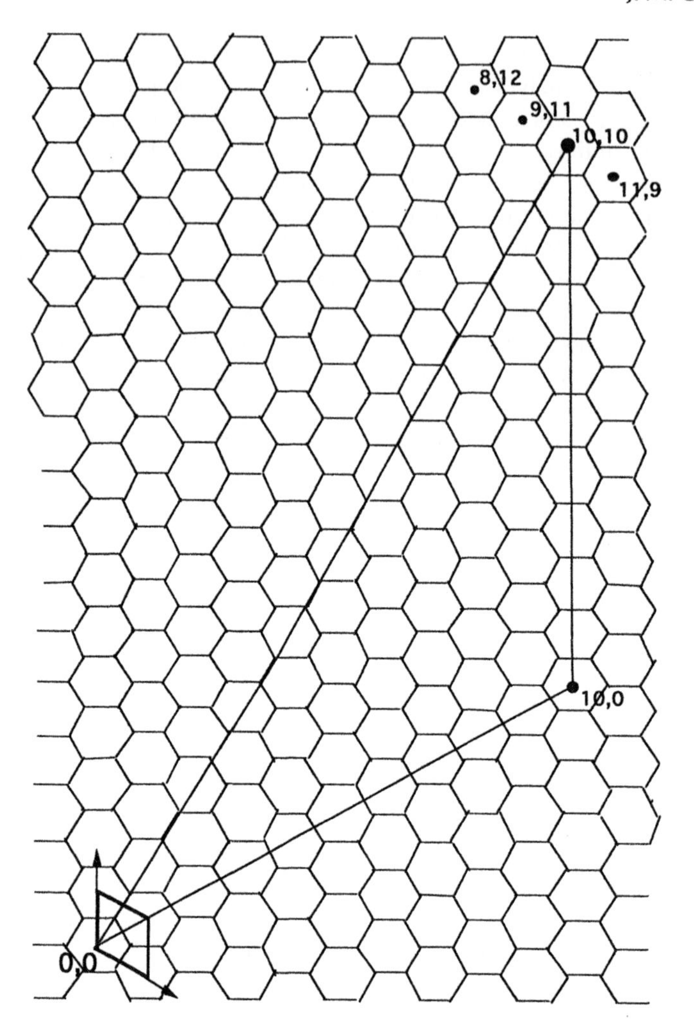

Figure 13. Diagram of a graphitic C sheet, with designations of the hexagons (10,10), (11,9), etc. The (10,10) single-walled nanotube is formed by rolling the sheet into a cylinder so that the origin (0,0) point coincides with the (10.10), and so on.

points a continuous line of scattering power extends perpendicular to the plane of the carbon sheet. For high-energy electrons, the intensities in the diffraction pattern are given by the intersection of the Ewald sphere, approximated by a plane perpendicular to the incident beam, with this distribution of scattering power. Thus, for the incident beam perpendicular to the carbon sheet, the diffraction pattern consists of a hexagonal array of spots, and it may readily be shown that the inner-most set of six [1,0] spots are stronger than the next group of [1,1] spots, in

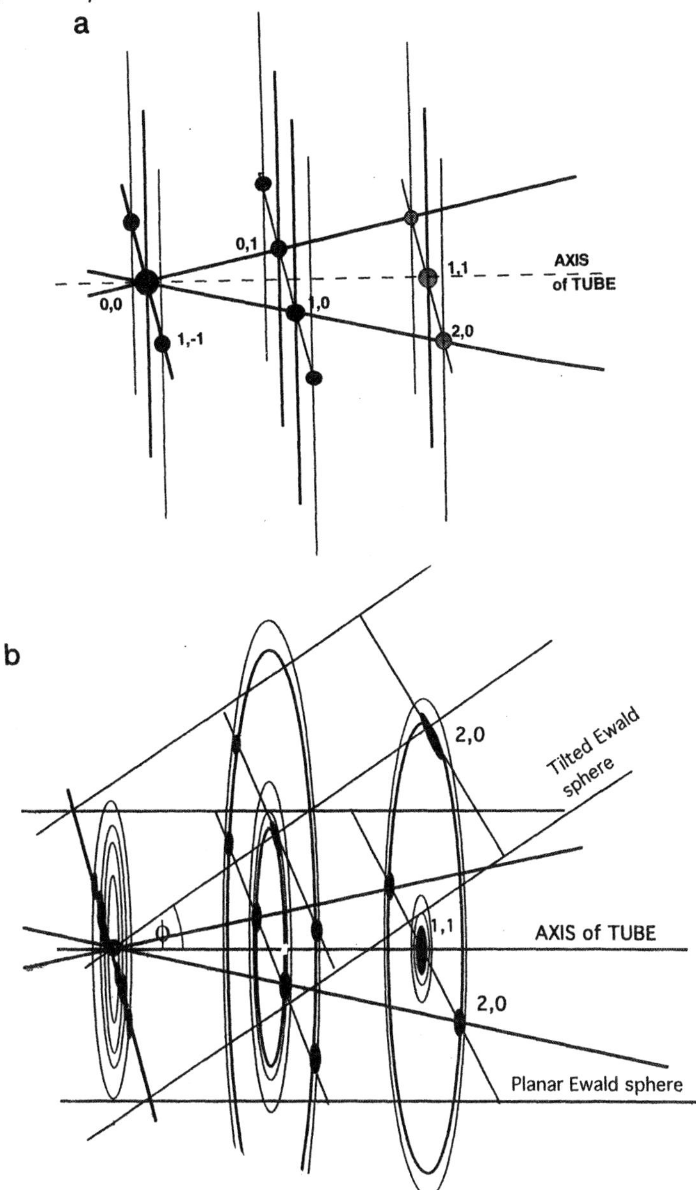

Figure 14. Reciprocal space construction for a C-sheet and a (10,10) SWnT. (**a**) For a single flat graphitic sheet, there are continuous rods of scattering power, perpendicular to the plane of the sheet, through each *h,k* point of the hexagonal array of reciprocal lattice points. For a SWnT, the distribution of (**b**) is rotated about the axis of the tube, assumed to pass through the 1,1 point, to create planes of scattering. The intersection of the planar Ewald sphere with this distribution gives the observed intensity distribution.

contrast to the case for the ordered graphite structure for which the [1,1] spots are stronger. If the carbon sheet is tilted away from the position normal to the beam, the planar Ewald sphere cuts some of the continuous lines of scattering power at points away from the *hk*0 plane. Hence the hexagonal array of diffraction spots becomes increasingly distorted, with increased dimensions in directions perpendicular to the axis of tilt.

A SWnT is formed by wrapping the planar graphitic carbon sheet of Figure 13 into a cylinder so that two of the hexagons of carbon, separated by a suitable distance, coincide.[65] For example, if the sheet is bent around so that the center of the hexagon, labeled [10,10] in Figure 13 is made to coincide with the origin, the [10,10] SWnT formed has an axis parallel to the [1,1] direction of the carbon sheet, the circumference of the tube is equal to $10 \times \sqrt{3.a_0} = 4.25$ nm, its diameter is 13.5 nm, and there are C–C bonds perpendicular to the tube axis. If the center of the hexagon labeled [11,9], for example, is made to coincide with the origin, the SWnT formed has a helical structure. The lines of C–C hexagons are continuous around the tube but are inclined to the direction perpendicular to the axis by about 5°. The circumference of the tube is slightly larger than for the [10,10] tube. Similarly other SWnTs with a wide range of diameters and helix angles are possible.

The reciprocal space distribution of scattering power can be approximated with sufficient accuracy for nanodiffraction considerations by rotating the distribution of scattering power for the single sheet (Figure 14a) around the direction of the axis of the cylinder, as suggested by Figure 14b for the case of the [10,10] SWnTs. Each continuous line of scattering of Figure 14a gives rise to a distribution of scattering in a plane, with scattering power rising sharply to, and terminated on, an inner circle. For a reciprocal lattice point which lies on the axis of the cylinder, the rotation forms a plane in which the scattering power falls off rapidly from the reciprocal lattice point as center.

The diffraction pattern is then given by considering the intersection of the planar Ewald sphere with this distribution of scattering power, located in a set of planes, as suggested in Figure 15. If the incident beam is perpendicular to the axis of the tube, the diffraction pattern contains a hexagonal array of spots, similar to those for a planar sheet except that the individual spots have linear tails, or streaks, directed away from the tube axis. If the tube axis is tilted with respect to this orientation, the hexagonal spot pattern is distorted and the individual spots become increasingly streaked before disappearing. Figures 15a and b show diagrams of patterns for tilts of 0° and 30°. Nanodiffraction patterns of approximately these forms are reproduced in Figures 16a and c. The form of the patterns given for other tilts can be derived from Figure 14b.

If the helix angle for the [10,10] tube is taken as zero, any deviation from this zero angle causes a doubling of the diffraction spots. For example, for a [11,9] tube, with the incident beam perpendicular to the axis, the beam sees the top and bottom layers of the graphitic sheet rotated by plus or minus about 5°, respectively, and so the individual [1,0]-type spots are split into pairs of spots separated by about 10°,

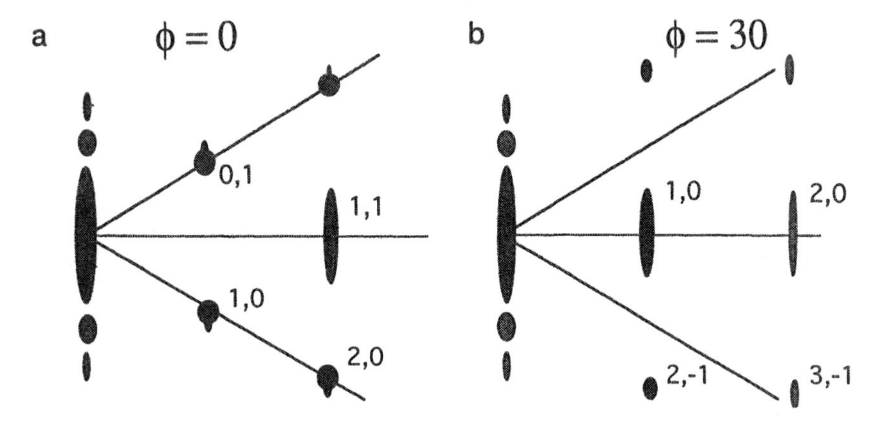

Figure 15. Diffraction patterns from a (10,10) SWnT, derived from Figure 14b, for the cases that the incident beam is perpendicular to the axis of the tube and tilted by 30°.

as is evident in Figure 16d. Hence the helix angles for the SWnTs can be deduced readily from the nanodiffraction patterns.

The nanodiffraction patterns of Figure 16 however, were not obtained from a single SWnT, but from ropes of SWnTs, 10 nm or more in diameter, with a stationary beam illuminating a region about 1 nm in diameter and extending through the rope.[66] In these patterns, the strong line of spots through the origin gives information about the stacking of the individual tubes within the ropes. The observation that the other diffraction spots in the patterns (Figure 16a–c) are similar to those expected from a single SWnT is a clear indication that all the SWnTs illuminated by the beam have the same orientation and helix angle.

Patterns were also recorded while the incident beam was scanned very fast over a rectangular area, about 10 × 15 nm or twice that size, so that several scans were made while the diffraction pattern was being recorded. Such patterns could then give the average of the patterns for all parts of a segment of a rope. In a pattern, such as Figure 16e, the spots are not just split into two or more spots but are smeared out over an arc subtending 20° or more at the origin. This suggests that a range of helix angles from 0° up to about 10° is present in the rope. The arcs are sometimes of nonuniform intensity, suggesting that one or more helix angles occurs more often than others in the rope.

Observations of patterns from particular spots in the rope image and of the averaged patterns, obtained from a large number of ropes, have confirmed that the ropes tend to have the 0° helix angle of the [10,10] structure, predicted by the theory.[61] It was concluded that 44% of SWnTs have this zero helix angle, 30% have a 5° helix angle corresponding to an [11,9] structure, and 20% have a helix angle close to 10° which could be the [12,8] structure.[66] The theory suggests that the [11,9] and [12,8] are the configurations which, after the [10,10], are the ones most

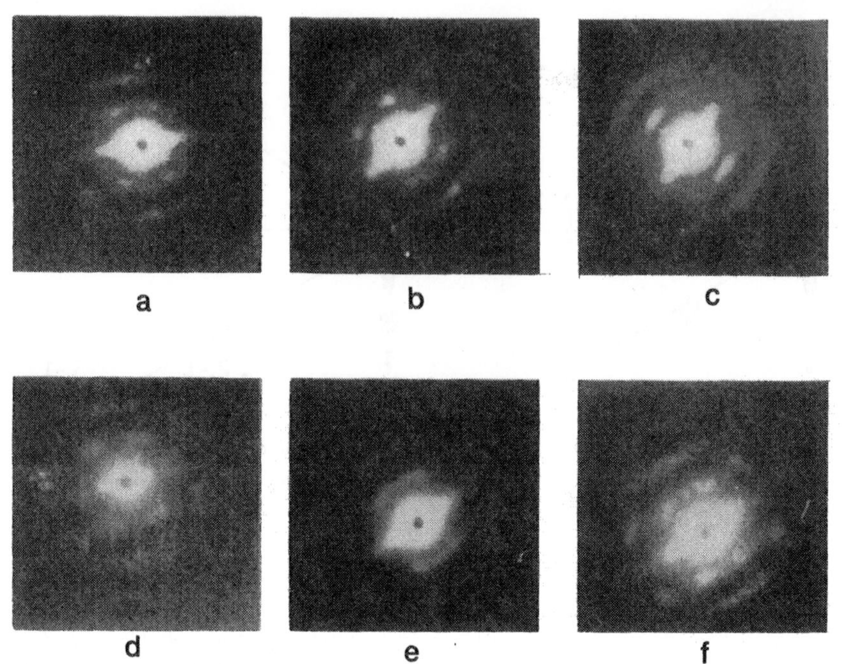

Figure 16. Nanodiffraction patterns from SWnT ropes. (**a**) Averaged over a (10,10) rope; helix angle 0, tilt angle $\phi = 0°$. (**b**) Pattern from one spot on rope of (**a**). (**c**) Pattern from a similar rope with $\phi = 30°$. (**d**) Pattern from one spot on a rope with helix angle 5°. (**e**) Pattern averaged over a rope with a spread of helix angles up to 10°. (**f**) Pattern from one spot on rope of (**e**).

energetically favored. The appearance of patterns such as Figure 16b, d and f from small parts of the ropes indicates that large parts of the ropes, 10 nm or more in diameter, may be made up of a perfect hexagonal stacking of equal SWnTs, all having the same helix angle. This observation is significant for the modeling of the process by which the ropes are formed.

The information from the strong equatorial line of spots in the nanodiffraction patterns of Figures 15 and 16 confirms that the SWnTs within a rope are, in general, of equal diameter and are stacked in a regular hexagonal two-dimensional lattice. However, considerable variations of the orientation of the hexagonal lattice axes are observed as the incident beam is moved over the image of a rope. The indication of these variations is best given by dark-field STEM images, such as Figure 17, obtained with the central beam deflected to a position such as "7" of Figure 10 so that the annular detector collects diffraction maxima corresponding to reflections from lattice spacings of about 0.3 to 0.7 nm. Bright streaks, almost parallel to the rope, appear in this image with a lateral spacing of 1.4 to 1.7 nm. These streaks appear along bright bands which cross the rope image diagonally, and their

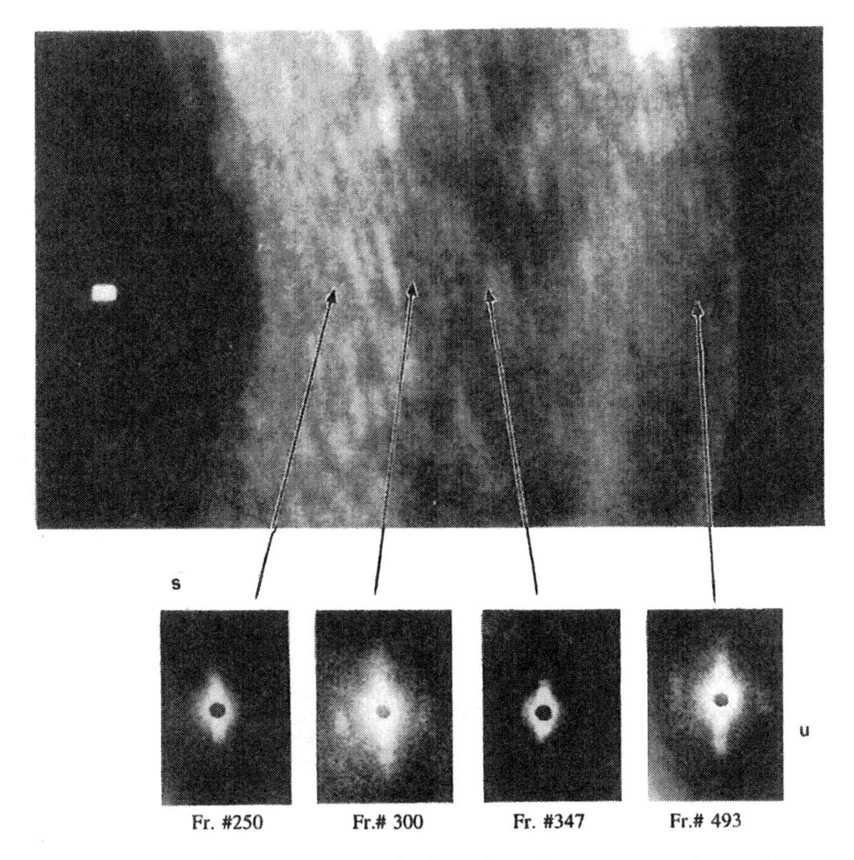

Figure 17. Dark-field STEM image of a large bent SWnT rope and nanodiffraction patterns obtained from the indicated points as the beam was scanned across the line through the marker. Patterns were recorded with a TV camera and played back, frame-by-frame from a VCR recorder and photographed. The frame numbers for the patterns from the beginning of the scan are indicated. Rope diameter, 45 nm.

intensities vary strongly and rapidly along these bands. The relative strengths of the streaks and the bands vary widely for images of different ropes.

The interpretation of the dark-field STEM images was assisted by recording series of nanodiffraction patterns with a TV camera as the incident beam was scanned over the rope image, either along the axis of the rope or across a diameter of the rope.[62] For example, for a 20 s scan across the image magnified two million times, 600 diffraction patterns are recorded for a scan of 12 cm on the display tube or 60 nm on the specimen. Hence a diffraction pattern is recorded with the VCR for each 0.1 nm movement of the beam, and these patterns may be played back and photographed, one at a time. Figure 17 shows the nanodiffraction patterns recorded at selected position of the DF STEM image for a scan across the rope along a line through the bright marker spot.

The variation of the spot positions and intensities along the equatorial line of the nanodiffraction patterns as the axes of the hexagonal packing of the SWnTs in the rope are rotated can be derived by consideration of the corresponding two-dimensional reciprocal space arrangement, shown in Figure 18. The hexagonal array of reciprocal lattice points have been enlarged into circular disks of area proportional to the relative intensities. The intersection of the planar Ewald sphere with the plane of the diagram gives a line through the zero point in the diagram. The intensities of the one-dimensional line of spots, produced by the intersection of this line with the pattern of enlarged points, is seen to vary as the array is rotated with respect to the Ewald sphere, or vice versa. The actual spots in the diffraction pattern are enlarged to the size of the circle drawn about the origin point by the convergence of the incident beam. Hence, it is seen that for the orientation giving the [1,0] reflection

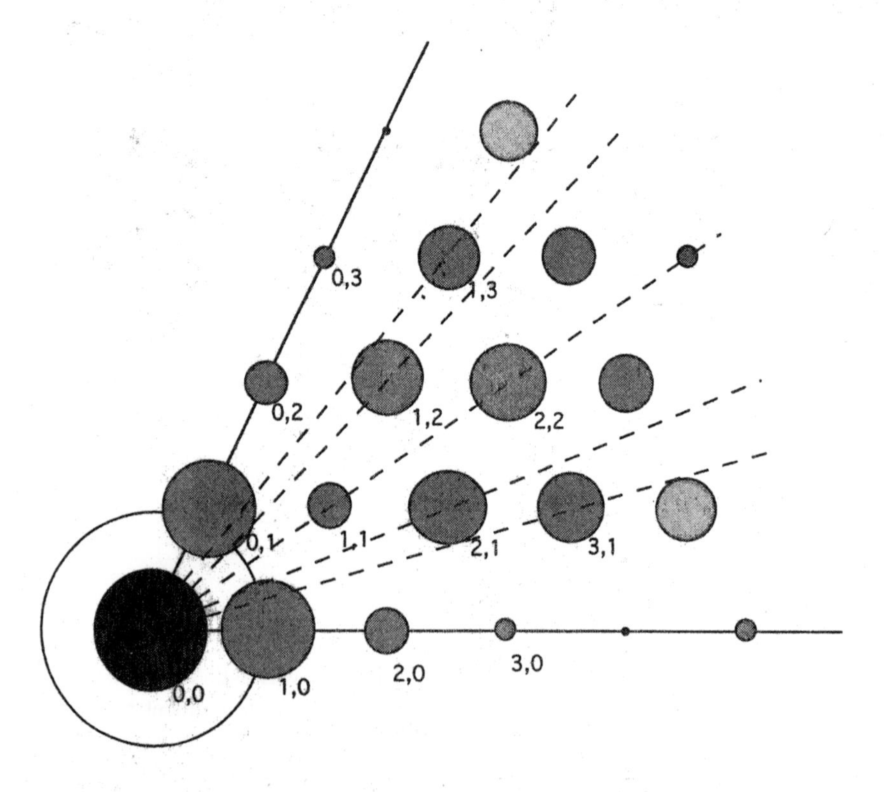

Figure 18. Reciprocal lattice construction for reflections in the plane perpendicular to a rope axis for a hexagonal packing of SWnTs. The spot areas are proportional to the intensities of the reflections. The actual diffraction spot diameters are indicated by the unfilled circle about the origin.

and higher orders, the diffraction spots overlap considerably, the individual spots are difficult to distinguish and the intensities are strongly influenced by coherent interference effects in the areas of overlap. Clear lines of intense spots are given when the [2,1] or [3,1] type of reflections, and, to some extent, the [1,1] and [2,2] reflections are produced. These are the reflections that give high intensities in the DF STEM images.

For a perfectly ordered, straight rope one would expect uniform diffraction conditions and the DF image would be of uniform intensity, except that some modulation of intensity with the periodicity of the stacking of the SWnTs is expected. For almost any orientation there are only 5 to 20 SWnTs in the thickness of the rope so that some remanent modulation of the projected potential distribution with this periodicity is to be expected.

The appearance of the ropes in low-resolution TEM images,[60,62] however, suggests that the ropes are usually bent, forming tangled arrays, and are often twisted. For a twisted straight rope, the DF image should show bright bands at right angles to the axis as the rotation of the lattice brings the various strong reciprocal lattice points of Figure 18 into diffracting condition. Series of nanodiffraction patterns recorded during sweeps of the incident beam parallel to the axes of several ropes have allowed the rotation of the lattice to be deduced.[62] Along one rope, for example, a twist angle of 70° was deduced for a translation along the rope of 100 nm.

The bright bands in the DF images of the ropes, however, are usually not perpendicular to the axis of the rope, but may be inclined to this direction by a considerable angle. This implies that, as well as being twisted, the ropes are bent. The bending of the rope in Figure 17, for example, is obvious. In this case, there is a change of orientation of the hexagonal stacking axes across the diameter of the rope. Sets of nanodiffraction patterns obtained with a scan of the incident beam across the rope diameter, as in Figure 17, have shown changes of orientation of about 40° across a rope of diameter 26 nm.

The observation that then remains to be explained is that the bright streaks, separated by the projected spacing of the individual SWnT, do not have uniform intensity along the inclined bright bands, but appear bright only in local patches. This appearance has been attributed to the occurrence of faults in the hexagonal stacking of the SWnT. Evidence from the nanodiffraction patterns, previously discussed, suggests that there are regions of perfect order of stacking of SWnTs, having uniform diameter and helix angle, that may extend over distances of 10 nm or so. But these regions may be surrounded by regions of SWnTs having different helix angles and hence slightly different diameters so that defects at the boundaries of these regions may arise. Other defects in the stacking may arise from the occurrence of occasional tubes of very different diameter or vacancies in the array, or from dislocation structures induced in the nonequilibrium conditions of formation of the ropes.

A further observation of interest is that, very rarely, ropes of considerable diameter are seen to end abruptly instead of showing a gradual decrease in diameter.[62] It has been suggested that such terminations may arise when a rope is broken during the specimen-preparation process or, alternatively, such terminations may represent the places where the SWnTs grew out from a relatively flat area of a metallic catalytic surface and then became detached from this base. Nanodiffraction patterns from near such rope terminations give results that are of interest in terms of either of these possible origins, namely that within about 100 nm of the abrupt end of the rope, the structure is almost completely disordered, with nanodiffraction patterns similar to those for completely amorphous material. Further away from the termination, the ordered stacking of the SWnTs develops slowly.

VII. MULTIWALLED CARBON NANOTUBES

The discovery of multiwalled nanotubes predated that of the single-walled tubes.[67] They are assemblies of concentric cylindrical graphitic carbon layers with successive radii increasing by approximately the graphitic interlayer spacing of 0.34 nm, as suggested in Figure 19. The inner tube may have a diameter in the range of about 1–5 nm, and there may be as few as 3 or 4, or as many as 20 or 30 concentric tubes, so that the outer diameter may be as great as 30 nm. Specimens containing these multiwalled tubes are commonly obtained from material produced in carbon arcs

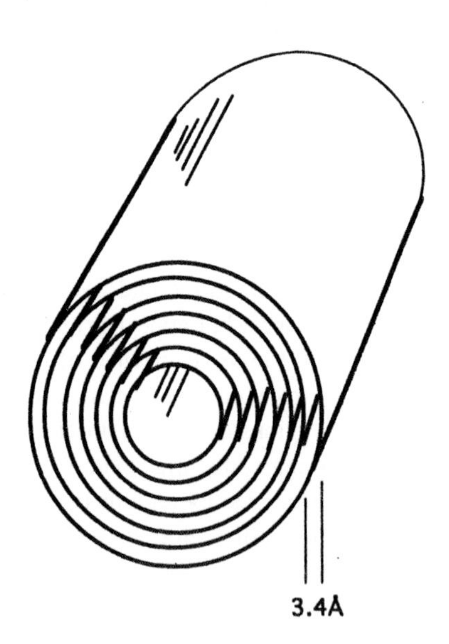

Figure 19. Diagram of a multiwalled carbon nanotube.

struck in appreciable pressures of inert gas,[68] but may also result from chemical or catalytic reactions.[69]

Within one multilayer tube, there may be one or more helix angles present. Evidence from high-resolution TEM and from selected-area diffraction patterns obtained in a TEM suggests that groups of three or four adjacent monotubes tend to have the same helix angle, and there may be slightly greater interlayer spacings between the groups where the helix angle changes.[70]

Because the multiwalled tubes are larger and scatter more strongly than SWnTs, it is much easier to obtain nanodiffraction patterns from them and, in practice, it is easy to observe and record the changes of the nanodiffraction patterns as the incident beam is scanned across the diameter of the multitube.[71] When the incident beam is approximately perpendicular to the tube axis and strikes one of the outer walls of the tube, where the beam is parallel to the graphitic planes, a strong line of [00*l*] diffraction spots appears, corresponding to the 0.34 nm interlayer spacing. The parallel lines of *hkl* diffraction spots do not appear because the lateral ordering of the graphitic layers that occurs in crystalline graphite cannot take place. Since the circumferences of successive tubes increase with the radius, the relative positions of atoms in neighboring tubes must change continuously around the circumference, and so the stacking of the layers must be disordered. The *hkl* lines of spots are replaced by continuous streaks.

If the incident beam passes through the middle of the tube image, the beam is perpendicular to the graphitic planes. If all the graphitic sheets have the same helix angle and this helix angle is zero or 30°, the pattern shows the simple hexagonal arrangement of diffraction spots, as in Figure 15a. Because the layers are disordered laterally, the intensity distribution is the same as for a single sheet, as in Figure 15a, and not as for the ordered graphite structure, for which the [1,1] spots are stronger than the inner [1,0] spots. If there is only one, non-zero helix angle, the *hk*0 spots are all doubled, as in Figure 16b. For a large-diameter tube with many helix angles, each helix angle gives a pair of spots in place of each *hk*0 spot, and these multiple spots may overlap to the extent that they appear to form continuous arcs (c.f. Figure 16c).

An electron beam striking the multiwalled tube in between the middle position and the walls sees the graphitic planes tilted at some angle. As a result, the *hk*0 spots or arcs and the *hkl* streaks combine to form elliptical rings, for which the ratio of the major and minor axes of the ellipse depends on the inverse of the cosine of the angle of tilt. Examples of these various forms of the nanodiffraction pattern are given in Figure 20.

It has been generally assumed that all nanotubes have a circular cross section, as illustrated in Figure 19. However, particularly for some arc-discharge preparations, there is clear evidence from the nanodiffraction patterns and the HREM images[71] that the cross sections of some tubes may be polygonal, as suggested in Figure 21. The HREM images, in general, show contrast only when the incident beam is parallel to the graphitic carbon layers, so that the images show only the sets of

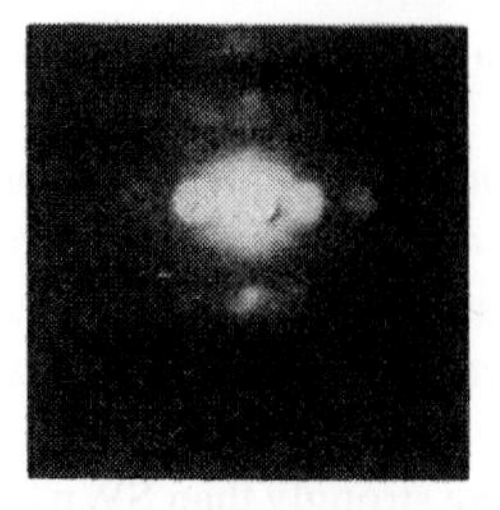
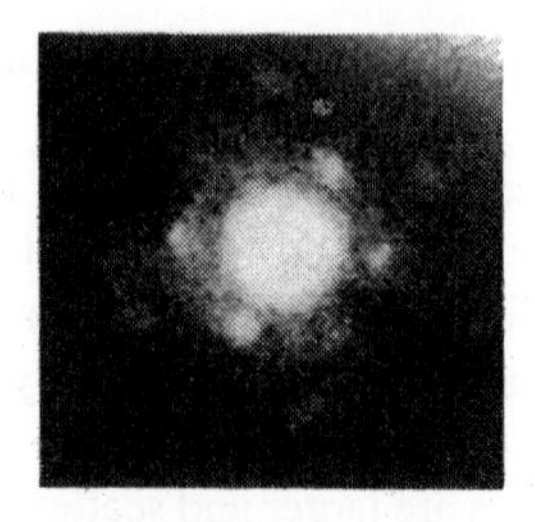
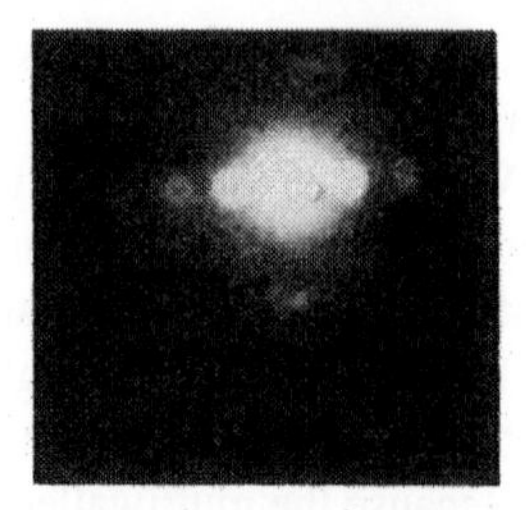

Figure 20. Nanodiffraction patterns from a tube of circular cross section, such as illustrated in Figure 19, with the incident beam on one wall of the tube, in the middle of the tube, and on the other wall.

fringes with a spacing of about 0.34 nm. For a tube oriented as in Figure 21, the left side of the image shows strong 0.34 nm fringes since the beam is parallel to extended planar layers with this spacing. On the right side, however, weaker fringes are seen with a considerably larger spacing, up to 0.4 nm or more, corresponding to the spacing of carbon planes sharply bent at the corner.

Correspondingly, the nanodiffraction patterns (Figure 22) are strong with the beam on the left side and relatively sharp *hkl* spots may appear indicating that for the planar parallel layers of carbon there may be some ordering of the layers into a crystalline graphite-type structure. With the beam on the right side, the line of $00l$ spots has the spacing corresponding to the 0.4 nm spacing. The spots are weaker and higher orders fade out more rapidly and there is no sign of *hkl* spots. For the

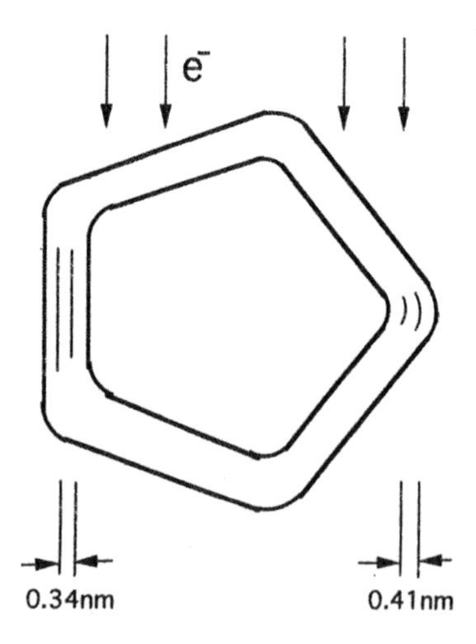

Figure 21. Diagram of a multiwalled carbon nanotube having a polygonal (pentagonal) cross section.

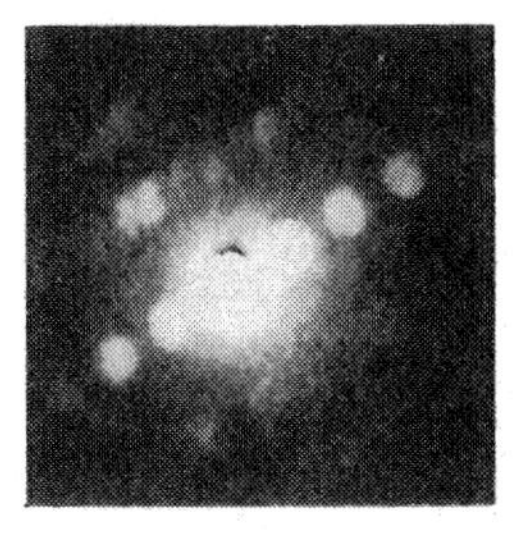
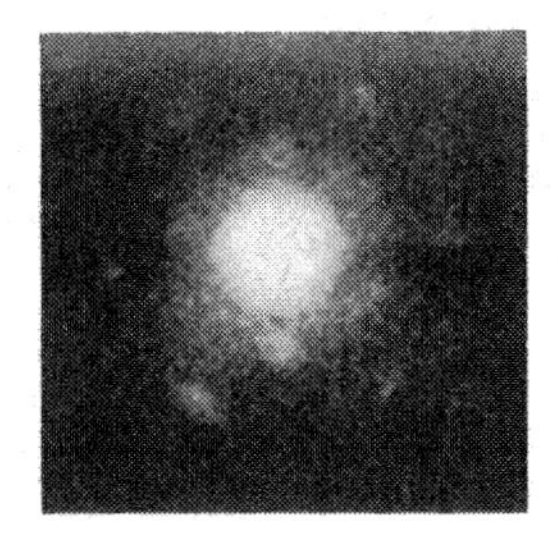
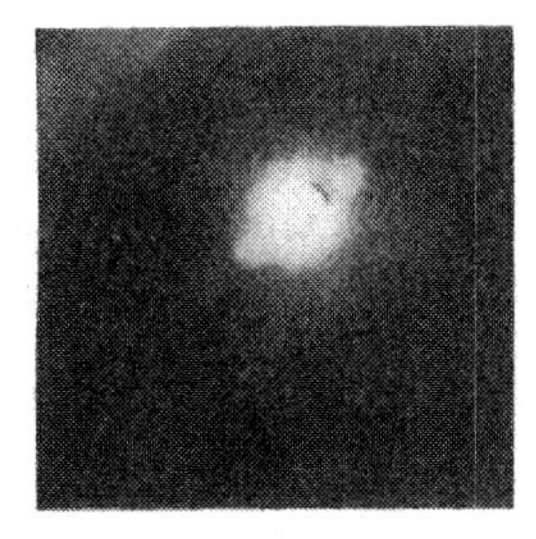

Figure 22. Nanodiffraction patterns from a tube of polygonal cross section, such as illustrated in Figure 21, with the beam running from top to bottom. Patterns are from the left side, the middle, and the right side (corner) of the tube.

beam in the middle of the tube image, the patterns of *hk*0 spots are more nearly consistent with those expected from ordered graphite structure.

Hence, all the evidence is consistent with the proposal of a polygonal cross section of the layers.[71] The dimensions of the fringes in the HREM images and the forms of the diffraction patterns are usually consistent with the model of a pentagonal cross section as sketched in Figure 21. It seems reasonable to suppose that the sharp bending of the carbon sheets at the corners of such a structure should introduce sufficient flexibility in the positioning of the sheets in the flat faces to allow the sheets to order in the energetically favored stacking sequence of the three-dimensional-ordered graphite structure. The occurrence of sharp bends through 108° between planar section of parallel sheets in the polygonal tubes is reminiscent of the observation of similar angles between planar sections of walls in the images of the ends of some tubes and in the images of the polyhedral shells of graphitic-layer carbon which often accompany the multiwall carbon nanotubes in specimens formed by carbon-arc preparation methods.[68,72]

The use of the CCD camera for the quantitative measurement of diffraction intensities in the scheme of Figure 3 introduces the possibility of testing one of the known theoretical results for thin-specimen electron microscopy, namely that the diffraction pattern of an asymmetric specimen region does not have a center of symmetry. The simple kinematical approximation used for most X-ray diffraction and for much electron diffraction predicts that for an object, thin enough to avoid three-dimensional diffraction effects, the diffraction pattern must always have a center of symmetry; but for a coherent convergent electron beam this will be so only if both the incident beam amplitude distribution and the specimen have centers of symmetry and these centers coincide. For a nm-size beam moving over a thin specimen, this coincidence is likely to occur, to the necessary precision of 0.1 nm or less for only a very small fraction of the beam positions. This prediction has been supported by some observations on the diffraction patterns from thin amorphous film,[43] but in this case the structures of the illuminated regions of the specimen cannot be known. The multilayer walls of carbon nanotubes, however, offer sufficiently small objects of known structure and so serve to test the theoretical prediction. For tubes of circular cross section, the projections of the structure in the beam direction are independent of the tube orientation.

When a coherent beam probe of diameter about 1 nm illuminates one of the side walls of a multiwalled tube, it sees an object which is essentially nonsymmetric because all-carbon layers are bent in one direction. Calculations predict that the inner diffraction spots, the [0,0,1] and [0,0,–1] reflections should differ in intensity by about 20%. Measurements made with the CCD camera show asymmetries very close to this value,[44] always in the correct sense. For tubes of polygonal cross section, such as illustrated in Figure 20, the beam on one side of the tube can be diffracted as from a relatively thick crystal for which the asymmetry can be very large and strongly dependent on the orientation of the tube. Hence the observation of the asymmetries of the diffraction patterns can be used as a tool to distinguish regions of the tubes which are smoothly curved, as in the case of tubes of circular cross section, from regions which have flat faces and sharp edges of high curvature. Thus a means is provided for determining three-dimensional information not available from HREM images, on the shapes of tubes, especially in the regions of the tube-ends, which often have complicated shapes.[68,72]

An interesting possibility for the multiwalled tubes is that they may be filled with metals, carbides, or other compounds[73–77] so that the properties of materials having well-defined small dimensions may be investigated. The materials may be inserted in the tubes by capillary action after the ends of the tubes have been opened by, for example, refluxing with concentrated acid.[73,76] Otherwise, some metal or compound may be inserted in one electrode of the carbon arc.[74,75] Then the usual result is that it is the carbide of the metal that fills the tubes. The carbides may also be seen to fill, or partially fill, the three-dimensional multilayer polyhedral balls that are formed with the nanotubes in the arc.

For many metals or carbides, the material within the multiwall tubes appears to form single crystals, filling the tube bore. Then the lattice fringes for the crystal may be seen in high-resolution TEM or STEM images, and the diffraction patterns from the crystals may be seen in selected-area electron diffraction (SAED) patterns. In other cases, the materials in the tubes consist of very small crystals or appear to be amorphous to these techniques. Then nanodiffraction is required for the determination of the structures of the materials and the determination of relationships of the crystal structures to the structures of the carbon tubes. For example, in the case of yttrium,[77] the carbide in the tubes did not show any recognizable structure in the HREM images or SAED patterns. Nanodiffraction, however, revealed the structure as being that of the carbide and it was clear that the carbide crystals grow in an epitaxial relationship to the inner graphitic surface layer of the tube. The fitting of the tetragonal crystal structure to the curved surface of the graphitic layer inevitably involves some distortions of the structure.

The report of the application of a TEM/STEM instrument, fitted with a field-emission gun for the study of several filled-tube specimens,[76] is of particular interest in that, in addition to the acquisition of nanodiffraction patterns from regions a few nm in diameter of the filling material, the method of energy-dispersive X-ray microanalysis (EDX) was applied with the same small probe sizes to determine the chemical compositions of the material in the tubes.

VIII. DISCUSSION AND CONCLUSIONS

For the studies of nanoparticles and nanotubes, reported in the previous sections, the nanodiffraction patterns have been interpreted as if given by incoherent convergent beams and this basis for interpretation is sufficient when there are clearly defined diffraction spots which do not overlap. For the occasional patterns with overlapping spots, or from regions of disorder or faulting, the complications of interference effects arising from the coherence of the incident beam are apparent and must be taken into account to the extent that they are not wrongly interpreted on the basis of incoherent beam effects. With this understanding, the use of nanodiffraction, in conjunction with the appropriate imaging modes, can be widely applied to systems having structural variations on a nanometer scale. Applications to studies of near-amorphous thin films and to disordered crystal structures have been cited above.[41–44]

A further area of application involves the study of the structures of the surfaces of solids. Reflection electron microscopy (REM) has been extensively developed in recent years as a means for imaging the surface layers, or particles or deposits on the flat surfaces of crystals, with resolutions of 1 nm or better.[78,79] The equivalent technique employing the STEM, scanning reflection electron microscopy (SREM) has been exploited to give comparable resolutions.[80] In these techniques, a high-energy beam of electrons (usually 100 keV, but ranging from about 20 to 1000 keV) strikes a flat surface at a grazing angle of a few degrees. In the forward-scattering direction, a reflection high-energy electron diffraction (RHEED) pattern is seen. One strong diffraction spot, usually the specularly reflected beam, is selected by the objective aperture and used to form an image of the surface. Because of the grazing angle of incidence, the image is fore-shortened by a factor of from 20 to 100 for a flat surface, but for the directions in the surface perpendicular to the beam the resolution approaches that of the dark-field images in TEM or STEM. Similarly, any small projection from the surface can be imaged with normal TEM or STEM resolution.

The particular advantage of the SREM mode of surface imaging is that, as in the case of STEM, the imaging may be combined with nanodiffraction from any part of the specimen image. For example, nanodiffraction patterns have been obtained from several nanocrystalline reaction products formed by very thin layers of gold or silver on crystals of MgO.[81,82] In the case of palladium evaporated on MgO crystal faces, the transformation of nanocrystals of Pd to PdO has been followed by nanodiffraction and SREM imaging.[83] Also observations have been made of the oxidation of copper surfaces.[84]

If large objective aperture sizes are used so that there is considerable overlapping of the spots in the nanodiffraction patterns, and if full advantage is taken of the coherence of the incident beam and the resulting interference effects, the nanodiffraction technique is, in principle, very much more powerful. In this case, each nanodiffraction pattern may be regarded as an in-line electron hologram, as proposed by Gabor when he introduced the concept of holography in 1948.[85] From such a hologram, the positions of the illuminated atoms may, theoretically, be derived with resolution better than 0.1 nm, although complications due to the presence of a conjugate image may create difficulties.[46] In an even more sophisticated approach, a nanodiffraction pattern may be recorded for each pixel of a STEM image, and then the correlation of the diffraction intensities for neighboring image pixels may be used to deduce an image with ultrahigh resolution. In this way an image of a thin crystal of Si in [110] orientation has been derived with an apparent resolution of better than 0.1 nm.[86] A thorough exploration of related techniques has been made by Rodenburg and associates.[87]

The practical applications of nanodiffraction, particularly in the forms making use of the full coherence of the incident beam, are currently limited by the experimental factors of specimen preparation, specimen stability, and instrumental stability. The specimens must be made in such a way that the structural feature to be investigated must be well separated from irrelevant scattering material. For the study of nanocrystals, some form of supporting film or crystal is usually necessary, and any support, even of a light-atom material of thickness greater than a few nm, is liable to give scattering which, added coherently in amplitude to the scattering from the sample, can confuse the detailed interpretation of the pattern. The irradiation of the specimen by the high-voltage electrons of the focused incident beam can in many cases cause radiation damage, modifying the specimen structure. Crystals of organic material and of some inorganics, such as alkali halides or clay minerals, may become amorphous after irradiation for a small fraction of 1 s. While the possibility of obtaining patterns at TV rates, i.e. in 1/30 s, may help in some cases, the correlation of diffraction patterns with image features is then often difficult.

Because nanodiffraction pattern intensities can change considerably for a movement of the incident beam by as little as 0.1 nm, the stability of the STEM instrument is an important factor. Any vibration, sudden shift, or steady drift of the specimen can destroy the clear correlation of the diffraction pattern with the imaged structure. To some extent the effects of such instabilities may be compensated by taking a series of nanodiffraction patterns from sets of neighboring points with a one-dimensional or two-dimensional scan of the beam, but this involves the collection and correlation of a very large amount of data. Reference points giving characteristic patterns, such as sharp edges of a crystal giving hollow-ring spots, may serve to locate particular features.

The experimental results described in this report may serve to indicate the extent to which these experimental difficulties may be overcome with the currently available equipment. They suggest that there are many structural problems for materials of current interest involving nanometer-scale structural variations for which the technique of nanodiffraction, together with the associated imaging techniques, may provide information not accessible by any other approach.

REFERENCES

1. Crewe, A. V.; Wall, J.; Langmore, J. *Science* **1970**, *168*, 1338.
2. Hashimoto, H.; Kumao, A.; Hind, K.; Yotsumoto, Y.; Ono, A. *Japan J. Appl. Phys.* **1971**, *10*, 1115.
3. Iijima, S. *Optik* **1977**, *48*, 193.
4. Iijima, S. *J. Appl. Phys.* **1971**, *42*, 5891.
5. Cowley, J. M. *Diffraction Physics*, 3rd. revised edn.; Elsevier: Amsterdam, 1995.
6. Cowley, J. M. In *High Resolution Transmission Electron Microscopy*; Buseck, P. R.; Cowley, J. M.; Eyring, L., Eds.; Oxford Univ. Press: New York, Oxford, 1988.
7. Yao, M.-H.; Smith, D. J.; Datye, A. K. *Ultramicroscopy* **1993**, *52*, 282.
8. Howie, A. In *High Resolution Transmission Electron Microscopy*; Buseck, P. R.; Cowley, J. M.; Eyring, L., Eds.; Oxford Univ. Press: New York, Oxford, 1988.
9. Fan, G.-Y.; Cowley, J. M. *Ultramicroscopy* **1987**, *21*, 125.
10. Crewe, A. V. *Rep. Prog. Phys.* **1980**, *43*, 621.
11. Cowley, J. M. In *Handbook of Microscopy, Methods II*; Amelinckx, S.; Van Dyck, D.; Van Landuyt, J. F.; Van Tenderloo, G., Eds.; V.C.H. Verlag: Weinheim, 1996, p. 563.
12. Joy, D. C.; Romig, A. D., Jr.; Goldstein, J. I., Eds. *Principles of Analytical Electron Microscopy*; Plenum: New York, 1986.
13. Cowley, J. M.; Liu, J. *Surface Sci.* **1993**, *298*, 456.
14. Hembree, G. G.; Drucker, J. S.; Luo, F. C. H.; Krishnamurthy, M.; Venables, J. A. *Appl. Phys. Lett.* **1991**, *58*, 1890.
15. Hembree, G. G.; Crozier, P. A.; Drucker, J. S.; Krishnamurthy, M.; Venables, J. A.; Cowley, J. M. *Ultramicroscopy* **1989**, *25*, 183.
16. Howie, A. *J. Microscopy* **1979**, *117*, 11.
17. Liu, J.; Cowley, J. M. *Ultramicroscopy* **1993**, *52*, 335.
18. Pennycook, S. J.; Jesson, D. E. *Ultramicroscopy* **1991**, *37*, 4.
19. Cowley, J. M. *Ultramicroscopy* **1993**, *49*, 4.
20. Cowley, J. M. *Appl. Phys. Lett.* **1969**, *15*, 58.
21. Cowley, J. M.; Osman, M.; Humble, P. *Ultramicroscopy* **1984**, *15*, 311.
22. Cowley, J. M.; Roy, R. A. In *Scanning Electron Microscopy, 1981*; Johari, O., Ed.; SEM: AMF O'Hare (Chicago), 1982, p. 142.
23. Allpress, J. G.; Sanders, J. V. *Surface Sci.* **1967**, *7*, 1.
24. Matsushita, M.; Hirotsu, Y.; Ohkubo, T.; Oikawa, T. *J. Electron Micros.* **1996**, *45*, 105.
25. Tanaka, M.; Terauchi, M. *Convergent Beam Electron Diffraction*; JEOL: Tokyo, 1985.
26. Spence, J. C. H.; Zuo, J. M. *Electron Microdiffraction*; Plenum Press: New York, London, 1992.
27. Cowley, J. M. In *Electron Diffraction Techniques*; Cowley, J. M., Ed.; Oxford University Press: Oxford, New York, 1992, Vol. 1, p. 439.
28. Cowley, J. M. *J. Electron Micros.* **1996**, *45*, 3.
29. Tanaka, M.; Terauchi, M.; Tsuda, K. *Convergent Beam Electron Diffraction, III*; JEOL: Tokyo, 1994.
30. Vine, W. J.; Vincent, R.; Spellward, P.; Steeds, J. W. *Ultramicroscopy* **1992**, *41*, 423.
31. Lin, J. A.; Cowley, J. M. *Ultramicroscopy* **1986**, *19*, 31.
32. Ronchi, V. *Appl. Optics* **1964**, *3*, 437.
33. Spence, J. C. H.; Cowley, J. M. *Optik* **1978**, *50*, 129.
34. Cowley, J. M. *Ultramicroscopy* **1979**, *4*, 435.
35. Cowley, J. M.; Spence, J. C. H. *Ultramicroscopy* **1981**, *6*, 367.

36. Pan, M.; Cowley, J. M.; Barry, J. C. *Ultramicroscopy* **1989**, *30*, 385.
37. Zhu, J.; Cowley, J. M. *Acta Cryst.* **1982**, *A38*, 718.
38. Zhu, J.; Cowley, J. M. *J. Appl. Cryst.* **1983**, *16*, 171.
39. Ou, H. J.; Cowley, J. M. In *Proc. 46th Annual Meeting, EMSA*; Bailey, G. W., Ed.; San Francisco Press: San Francisco, 1988.
40. Humble, P. *Proc. Roy. Soc. (London)* **1982**, *A381*, 65.
41. Cowley, J. M. In *Diffraction Studies of Non-Crystalline Substance*; Hargittai, I.; Orville-Thomas, W. J., Eds.; Akademia Kiado: Budapest, 1981.
42. Chan, I. Y. T.; Cowley, J. M. In *Proc. 39th Annual Meeting, EMSA*; Bailey, G. W., Ed.; Claitors Publ: Baton Rouge, 1981.
43. Hÿtch, M. J.; Chevalier, J. P. *Ultramicroscopy* **1995**, *58*, 114.
44. Cowley, J. M.; Packard, S. D. *Ultramicroscopy* **1996**, *63*, 39.
45. Cowley, J. M.; Au, A. V. In *Scanning Electron Microscopy 1978*; Johari, O., Ed.; SEM: AMF O'Hare (Chicago), 1978, p. 53.
46. Wang, S.-Y.; Cowley, J. M. *Micros. Res. Tech.* **1995**, *30*, 181.
47. Wang, Z. L.; Cowley, J. M. *Ultramicroscopy* **1990**, *32*, 275.
48. Hillyard, S.; Silcox, J. *Ultramicroscopy* **1993**, *52*, 325.
49. Cowley, J. M.; Hansen, M. S.; Wang, S.-Y. *Ultramicroscopy* **1995**, *58*, 18.
50. Liu, R.-J.; Cowley, J. M. *J. Micros. Soc. Amer.* **1995**, *2*, 9.
51. Cowley, J. M.; Merkulov, V. I.; Lannin, J. S. *Ultramicroscopy* **1996**, *65*, 61.
52. Liu, J.; Cowley, J. M. *Ultramicroscopy* **1991**, *37*, 50.
53. Liu, J.; Hembree, G. G.; Spinnler, G.; Venables, J. A. *Catalysis Lett.* **1992**, *15*, 133.
54. Pan, M.; Cowley, J. M.; Chan, I. Y. *J. Appl. Crystallogr.* **1987**, *20*, 300.
55. Pan, M.; Cowley, J. M.; Garcia, R. *Micron Microscopia Acta* **1987**, *18*, 165.
56. Cowley, J. M.; Plano, R. J. *J. Catalysis* **1987**, *108*, 199.
57. Iijima, S.; Ichihashi, T. *Nature* **1993**, *363*, 603.
58. Kiang, C.-H.; Goddard, W. H.; Beyers, R.; Bethune, D. S. *Carbon* **1995**, *33*, 903.
59. Chico, L.; Crespi, V. H.; Benedict, L. X.; Louie, S. G.; Cohen, M. L. *Phys. Rev. Lett.* **1996**, *76*, 971.
60. Thess, A.; Lee, R.; Nikolaev, P.; Dai, H.; Petit, P.; Robert, J.; Xu, C.; Lee, Y. H.; Kim, S. G.; Rinzler, A. G.; Colbert, D. T.; Scuseria, G.; Tománek, D.; Fischer, J. E.; Smalley, R. E. *Science* **1996**, *273*, 483.
61. Xu, C.; Colbert, D. T.; Smalley, R. E.; Scuseria, G. E. In preparation.
62. Cowley, J. M.; Sundell, F. A. *Ultramicroscopy* **1997**, *68*, 1.
63. Lambin, Ph.; Lucas, A. A. *Phys. Rev. B* **1997**, *56*, 3571.
64. Cochran, W.; Crick, F. H. C.; Vand, V. *Acta Cryst.* **1952**, *5*, 581.
65. Hamada, N.; Sawada, S.; Oshiyama, A. *Phys. Rev. Letts.* **1992**, *68*, 1579.
66. Cowley, J. M.; Nikolaev, P.; Thess, A.; Smalley, R. E. *Chem. Phys. Lett.* **1997**, *265*, 379.
67. Iijima, S. *Nature* **1991**, *354*, 56.
68. Ebbesen, T. W.; Ajayan, P. M. *Nature* **1992**, *358*, 220.
69. José-Yacaman, M.; Miki-Yoshida, M.; Rendon, L.; Stantiesteben, J. G. *Appl. Phys. Lett.* **1993**, *62*, 657.
70. Liu, M.; Cowley, J. M. *Carbon* **1994**, *32*, 393.
71. Liu, M.; Cowley, J. M. *Ultramicroscopy* **1994**, *53*, 333.
72. Liu, M.; Cowley, J. M. *Materials Sci. Eng.* **1994**, *A185*, 131.
73. Ajayan, P. M.; Iijima, S. *Nature* **1993**, *361*, 333.
74. Seraphin, S.; Zhuo, D.; Jiao, J.; Withers, J. C.; Loutfy, R. *Appl. Phys. Lett.* **1993**, *63*, 2073.
75. Liu, M.; Cowley, J. M. *Carbon* **1995**, *33*, 749.
76. Cook, J.; Sloan, J.; Chu, A.; Heesom, R.; Green, M. L. H.; Hutchison, J. L.; Kawasaki, M. *JEOL News* **1996**, *32E*, 2.
77. Cowley, J. M.; Liu, M. *Micron* **1994**, *25*, 53.

78. Yagi, K. In *Electron Diffraction Techniques*; Cowley, J. M., Ed.; Oxford University Press: Oxford, 1993, Vol. 2.
79. Cowley, J. M. In *Surface and Interface Characterization by Electron-Optical Methods*; Howie, A.; Valdre, U., Eds; Plenum Press, New York, 1987.
80. Liu, J.; Cowley, J. M. *Ultramicroscopy* **1993**, *48*, 381.
81. Kang, Z.-C.; Cowley, J. M. *Ultramicroscopy* **1983**, *11*, 131.
82. Lodge, E. A.; Cowley, J. M. *Ultramicroscopy* **1984**, *13*, 215.
83. Ou, H.-J.; Cowley, J. M. *Phys. Status Solidi* **1988**, *107*, 719.
84. Milne, R. H. In *Reflection High Energy Electron Diffraction and Relection Electron Imaging of Surfaces*; Larson, P. K.; Dobson, P. J., Eds.; Plenum Press: New York, 1988.
85. Gabor, D. *Nature* **1948**, *161*, 777.
86. Konnert, J.; D'Antonio, P.; Cowley, J. M.; Higgs, A.; Ou, H.-J. *Ultramicroscopy* **1989**, *30*, 371.
87. Rodenburg, J. M.; McCallum, B. C.; Nellist, P. D. *Ultramicroscopy* **1993**, *48*, 304.

SYNTHESIS AND CHARACTERIZATION OF METAL AND SEMICONDUCTOR NANOPARTICLES

M. Samy El-Shall and Shoutian Li

Advances in Metal and Semiconductor Clusters
Volume 4, pages 115–177

ISBN: 0-7623-0058-2

I. INTRODUCTION

In recent years, there has been an intense interest in the synthesis and characterization of nanoparticles.[1–11] Due to their finite small size, nanoparticles often exhibit novel properties which are different from bulk materials' properties.[1–11] The characterization of these properties can ultimately lead to identifying many potential uses, particularly in the field of catalysis. Research in this area is motivated by the possibility of designing nanostructured materials that possess novel electronic, optical, magnetic, photochemical, and catalytic properties. Such materials are essential for technological advances in photonics, quantum electronics, nonlinear optics, and information storage and processing.

A wide range of scientifically interesting and technologically important nanoparticles has been produced by both chemical and physical methods.[1–11] The most common physical methods involve gas condensation techniques where oven sources are usually used to produce metal vapors. In spite of the success of this method, there are, however, some problems and limitations, such as: possible reactions between metal vapors and oven materials, inhomogeneous heating which can limit the control of particle size distribution, limited success with refractory metals due to low vapor pressures, and difficulties in controlling the composition of the mixed metal particles due to the difference in composition between the alloys and the mixed vapors. Laser vaporization techniques provide several advantages over other heating methods such as the production of a high density vapor of any metal, the generation of a directional high-speed metal vapor from the solid target which can be useful for directional deposition of the particles, the control of the evaporation from specific spots on the target, as well as the simultaneous or sequential evaporation of several different targets.

Recently we described a novel technique to synthesize nanoparticles of controlled size and composition.[12–17] Our technique combines the advantages of pulsed laser vaporization with controlled condensation (LVCC) in a diffusion cloud chamber under well-defined conditions of temperature and pressure. It allows the synthesis of a wide variety of nanoparticles of metal oxides, carbides, and nitrides.

This chapter deals with the vapor phase synthesis of nanoscale particles by the LVCC method. The chapter consists of three major sections. The first is a brief review of the processes of nucleation and growth in supersaturated vapors for the formation of clusters and nanoparticles. The second section deals with the application of laser vaporization for the synthesis of nanoparticles in a diffusion cloud chamber. In the third section, we present some examples of nanoparticles synthesized using this approach and discuss some selected properties.

II. CLASSICAL NUCLEATION THEORY FOR THE FORMATION OF CLUSTERS AND NANOPARTICLES FROM SUPERSATURATED VAPORS

In discussing clusters' formation it is instructive to provide an abbreviated account of the nucleation process. Nucleation is the formation of new particles from a continuous phase and can occur heterogeneously and/or homogeneously. Heterogeneous nucleation from a vapor phase can occur on foreign nuclei or dust particles, ions, or surfaces. Homogeneous nucleation occurs in the absence of any foreign particles or ions when the vapor molecules condense to form embryonic droplets or nuclei. The theory of nucleation from the vapor has been reviewed on many occasions.[18–23] This theory (usually referred to as the classical nucleation theory) is based on the assumption (known as the capillarity approximation) that embryonic clusters of the new phase can be described as liquid drops having a well-defined radius with the bulk liquid density inside and the vapor density outside this radius. The free energy of these clusters, relative to the uniform vapor, is assumed to be the sum of two terms: a positive contribution from the surface free energy and a negative contribution from the bulk free energy difference between the supersaturated vapor and the liquid. The surface free energy results from the reversible work involved in forming the interface between the liquid drop and the surrounding vapor, and is proportional to the droplet surface area. For a cluster containing n atoms or molecules, the interface energy is $\sigma A(n)$ where σ is the interfacial tension, and $A(n)$ is the surface area of the cluster. Therefore one may write,

$$\sigma A(n) = 4\pi\sigma(3v/4\pi)^{2/3} n^{2/3} \tag{1}$$

where v is the volume per molecule in the bulk liquid. Since n molecules are transferred from the vapor to the cluster (droplet), the bulk contribution to the free energy of formation can be represented as $n(\mu_l - \mu_v)$ where μ_l is the chemical potential in the bulk liquid and μ_v is the chemical potential per molecule in the vapor. Assuming an "ideal" vapor it can be shown that,[18,22]

$$(\mu_l - \mu_v)\, n = -n\, k_B\, T \ln S \tag{2}$$

where k_B is the Boltzmann constant, T is the temperature, and S, the supersaturation, is

$$S = P/P_e \tag{3}$$

In Eq. 3, P is the vapor pressure and P_e is the equilibrium or "saturation" vapor pressure at the temperature of the vapor. The sum of the contributions in Eqs. 1 and 2 is the reversible work (free energy), $W(n)$, of formation of a cluster of size n and is given by:

$$W(n) = -n\, k_B T \ln S + 4\,\pi\,\sigma \left(\frac{3v}{4\pi}\right)^{2/3} n^{2/3} \tag{4}$$

The interface contribution in Eq. 1 represents the free energy barrier which impedes nucleation. The condensation nucleus, the smallest cluster which can grow with a decrease in free energy, lies at the top of the barrier where $n = n^*$. The quantity n^* is determined from the condition, $\partial W/\partial n = 0$. Thus:

$$n^* = 32\,\pi\,\sigma^3\, v^2/3\,(k_B T \ln S)^3 \tag{5}$$

It follows that the critical radius of the condensation nucleus, r^*, is:

$$r^* = 2\,\sigma\, \mathrm{v}/kT \ln S \tag{6}$$

Substituted n^* into Eq. 4 yields the barrier height $W(n^*)$:

$$W(n^*) = 16\,\pi\,\sigma^3\, v^2/3\,(k_B\, T \ln S)^2 \tag{7}$$

Equation 4 expresses the competition between "bulk" and "surface" behavior in determining cluster stability and, ultimately, cluster concentration in the supersaturated vapor. For a saturated vapor where $S = 1$, the bulk term vanishes and $W(n)$ is proportional to $n^{2/3}$. For $S > 1$, however, the first term provides a negative contribution to $W(n)$. Increasing S reduces the barrier height $W(n^*)$ and the critical size n^*, and therefore enhances the probability that fluctuation processes will allow some clusters to grow large enough to overcome the barrier and grow into stable clusters. Figure 1 exhibits curves for the dimensionless free energy of formation of different metal clusters as a function of size. The calculations were made using Eq. 4 with $S = 50$ and a partial pressure of the metal vapor of 0.5 torr. This implies that P_e is 0.01 torr. For the metals K, Al, and Fe, the temperatures at which the equilibrium vapor is 0.01 torr are 464 K, 1472 K, and 1678 K, respectively. Literature values of the bulk surface tension and density were used in the calculations.[24] It is important to emphasize that S can be increased either by increasing P or decreasing P_e. The pressure P can be increased by increasing the rate at which atoms are placed in the vapor or decreasing the rate at which they leave the region where the particle nucleation and growth is occurring. The equilibrium vapor pressure can be decreased by decreasing the temperature since P_e is approximately given by,

$$P_e \approx \exp\,(-L(0)/RT) \tag{8}$$

where $L(0)$ is the latent heat of evaporation per mole at zero temperature and R is the gas constant. It should be noted that a more exact expression for P_e would include the temperature dependence of the latent heat of evaporation. In some of the results discussed later, the lower temperature in a region of the apparatus separated from the source permits nanoparticles to nucleate and grow. This occurs

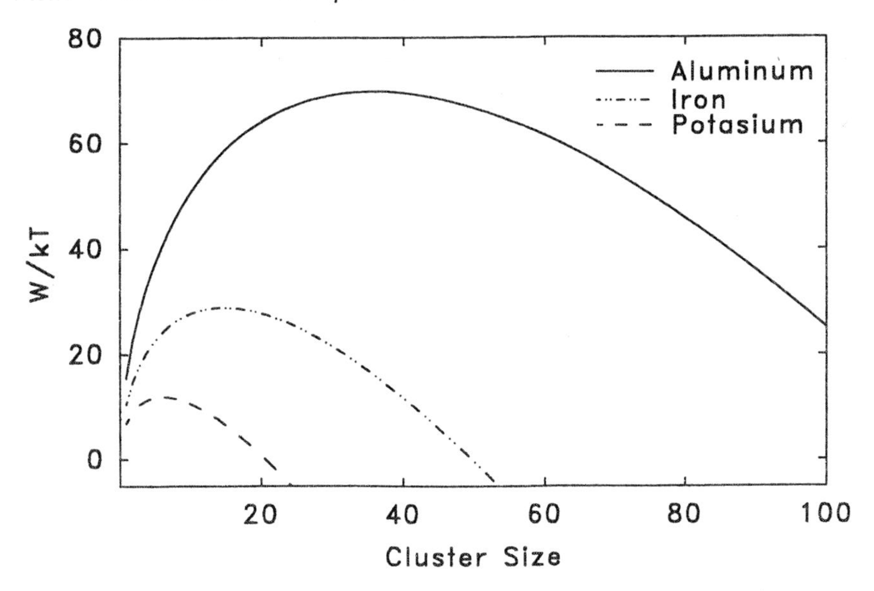

Figure 1. Free energy of cluster formation plotted vs. *n*, the number of molecules in the cluster for vapor supersaturation, $S = 50$ and temperatures = 1472 K, 1678 K, and 464 K for Al, Fe, and K, respectively.

because the lower temperature decreases P_e sufficiently that supersaturation is achieved.

The rate of homogeneous nucleation J in a supersaturated vapor, usually defined as the number of drops nucleated per cubic centimeter per second, is given by:

$$J = K \exp[-W(n^*)/k_B T] \tag{9}$$

The preexponential factor K incorporates both the effective collision rate of vapor molecules from a nucleus of size n^* as well as the departure of the cluster distribution from equilibrium. The factor K varies much more slowly with P and T than does the exponential term, and is usually expressed as,[22,25]

$$K = \alpha \left(\frac{S P_e}{k_B T}\right)^2 \left(\frac{2\sigma v^2}{\pi m}\right)^{1/2} \tag{10}$$

where m is the mass of a single vapor molecule and α is a "sticking" coefficient, normally set equal to unity. A critical supersaturation (S_c) is typically defined as the supersaturation at which $J = 1/\text{cm}^3/\text{s}$.

The rate of nucleation J, is clearly very sensitive to the height of the free energy barrier W^*. Thus, if the vapor becomes sufficiently supersaturated, the barrier is reduced to the point where nucleation occurs at a high rate. Preexisting surfaces (e.g. aerosol or dust particles), ions, or large polymer molecules greatly accelerate the rate of nucleation by lowering W^*, defined by Eq. 7. Such surfaces accomplish this by reducing the amount of work required to provide the interface in nucleation (since a surface already exists) while ions accomplish this (especially with polar molecules) by dielectric polarization so that the barrier W^* can be lowered to the point where a single ion can induce the formation of a macroscopic liquid drop. This represents almost the ultimate in amplification and detection.[26,27]

In case of binary nucleation involving a two-component system, the free energy of formation of a binary nucleus, $W(n_1,n_2)$, is given by,

$$W(n_1,n_2) = (\mu_{11} - \mu_{v1})_{n_1} + (\mu_{12} - \mu_{v2})_{n_2} + 4\pi r^2 \gamma(n_1,n_2) \tag{11}$$

where μ_{11} and μ_{12} are the chemical potentials in the bulk liquids 1 and 2, respectively and μ_{v1} and μ_{v2} are the chemical potentials in the vapors 1 and 2, respectively, and $\gamma(n_1,n_2)$ is the surface energy per unit area. For a binary system it may be shown that the free energy hill in Figure 1 becomes a free energy surface since now the free energy of formation of a cluster depends upon the number of molecules, n_1 and n_2, of both species in the droplet. The clusters, on the way to becoming droplets, flow over the surface and through a mountain pass. The saddle point marking the formation of this pass represents an energy barrier that embryos have to overcome in order to grow and become stable. The saddle point can be determined by solving the two equations:

$$(\partial W/\partial n_1)_{n_2} = 0, \quad (\partial W/\partial n_2)_{n_1} = 0 \tag{12}$$

Several review articles are available for more details on the nucleation in two-component systems.[21,23,28]

Aside from the approximations made and the subtleties not considered in the classical theory,[29–33] the theory is expected to fail at low temperatures in the case of high supersaturations. High supersaturations are usually involved when the nucleated clusters are strongly bound (as in metal clusters) because the vapor pressure of the condensed species is very low. In this case, the nucleus for condensation would be very small and the steady-state nucleation kinetics (used in the classical theory) is not likely to be achieved. Also, at very high supersaturations where the nucleus is less than 10 atoms, it is unreasonable to treat the nucleus as a macroscopic entity having macroscopic properties such as surface tension and density. When the supersaturation is achieved by a very rapid expansion as in nozzle devices commonly used in free jets and cluster beams, the transit time is so rapid that the steady-state nucleation kinetics may not be valid. This problem arises when the change of state of the gas is faster than the time required to establish a local metastable equilibrium for the supersaturated state.

III. SYNTHESIS OF NANOPARTICLES FROM THE VAPOR PHASE

Preparing nanoparticles directly from a supersaturated vapor is one of the earliest methods for producing nanoparticles. It has the advantages of being versatile, easy to perform and to analyze the particles, produces high purity particles, and naturally produces films and coatings. It has the disadvantage that, despite preparing larger quantities of material than by a free-jet expansion, the cost per gram of material is still very high. Further, it may be difficult to produce as large a variety of materials and microstructures as one can produce by chemical means.

To produce nanoparticles from the vapor it is necessary to produce a supersaturated vapor. This can be achieved by different techniques based on adiabatic expansions of the vapor as in free jets and molecular beams.[34–36] Other ways for achieving supersaturation are thermal evaporation,[37] sputtering,[38] laser ablation,[39] laser pyrolysis,[40] and photodissociation of molecular and organometallic species.[41–43] Thermal evaporation has the disadvantage that the operating temperature is limited by the choice of crucible material. Possible reactions with the crucible is also a concern. Sputtering has the advantage that it can be applied to almost any material, whereas, thermal evaporation is largely limited to metals.

The diffusion of the atoms once they leave the source must be inhibited by collisions with a gas. If this is not done, then supersaturation is not achieved and individual atoms or very small clusters of atoms are deposited on the collecting surface. Usually the gas that is used to limit the diffusion by shortening the mean free path is an inert gas, but one could also use a mixture that includes a component which will react with the sputtered atoms to produce molecular nanoparticles. The high pressure needed to limit the mean free path and thus confine the vapor to achieve supersaturation decreases the sputtering rate. This occurs because the ions lose energy by collisions with the inert gas and thus fewer of them have the necessary threshold energy to sputter atoms from the target. In most sputtering systems, there is a broad pressure range between the lowest pressure required to achieve supersaturation and the higher pressure at which the sputtering rate becomes unacceptably low.[44]

A. Laser Ablation/Vaporization of Metals and Cluster Beam Sources

In laser ablation, a high-energy pulsed laser with an intensity flux of about 10^6–10^7 W/cm^2 is focused on a metal target of interest. The resulting plasma causes highly efficient vaporization and the temperature at the focusing spot can exceed 10,000 K. This high temperature can vaporize all known substances so quickly that the surrounding vapor stays at the ambient temperature. Typical yields are 10^{14}–10^{15} atoms from a surface area of 0.01 cm^2 in a 10^{-8} s pulse. The local atomic vapor density can exceed 10^{18} atoms/cm^3 (equivalent to 100 torr pressure) in the microseconds following the laser pulse.

The coupling of laser vaporization with supersonic expansion beams has been introduced by Smalley and coworkers to overcome the limitations of the oven sources for the synthesis of metal clusters.[45–48] In this method, a high-energy pulsed laser (visible or ultraviolet radiation) is focused onto a target composed of the atoms or molecules to be made into clusters. The hot metal vapor is entrained in a pulsed flow of a carrier gas (typically He) and expanded through a pulsed nozzle into high vacuum. The cool, high density helium flowing over the target serves as a buffer gas in which clusters of the target material form, thermalize to near room temperature, and then cool to a few K in the subsequent supersonic expansion. Larger clusters (several hundred atoms) can be produced by adding a "growth channel" to the nozzle which serves to constrain the expansion. Discussions of different cluster sources based on supersonic beams can be found in ref. 35.

B. The LVCC Method

The LVCC method is based on the formation of nanoparticles by condensation from the vapor phase. Previous work was based on using thermal evaporation or sputtering to produce supersaturated metal vapors.[37,49–51] In the LVCC method, the process consists of pulsed laser vaporization of a metal target into a selected gas mixture in a modified diffusion cloud chamber. The laser vaporization produces a high-density vapor within a very short time, typically 10^{-8} s, in a directional jet that allows directed deposition. Desorption is possible from several targets simultaneously, yielding mixed particles.

An important feature is the use of an upward diffusion cloud chamber at well-defined temperatures and pressures. A temperature differential between the end plates produces a convection current into which the metal is evaporated. This chamber has been commonly used for the production of steady-state supersaturated vapors for the measurements of homogeneous and photoinduced nucleation rates of a variety of substances.[25,52,53] Detailed description of the chamber and its major components can be found in several references.[25,52,53] Here, we only offer a brief description of the modifications relevant to the synthesis of the nanoparticles. A sketch of the chamber with the relevant components for the production of nanoparticles is shown in Figure 2.

The chamber consists of two horizontal, circular stainless steel plates, separated by a glass ring. A metal target of interest is set on the lower plate, and the chamber is filled with a pure carrier gas such as helium or Ar (99.99% pure) or a mixture containing a known composition of a reactant gas (e.g. O_2 in case of oxides, N_2 or NH_3 for nitrides, CH_4 for carbides, etc.). The metal target and the lower plate are maintained at a temperature higher than that of the upper one (temperatures are controlled by circulating fluids). The top plate can be cooled to less than 150 K by circulating liquid nitrogen. The large temperature gradient between the bottom and top plates results in a steady convection current which can be enhanced by using a heavy carrier gas such as Ar under high pressure conditions (10^3 torr).

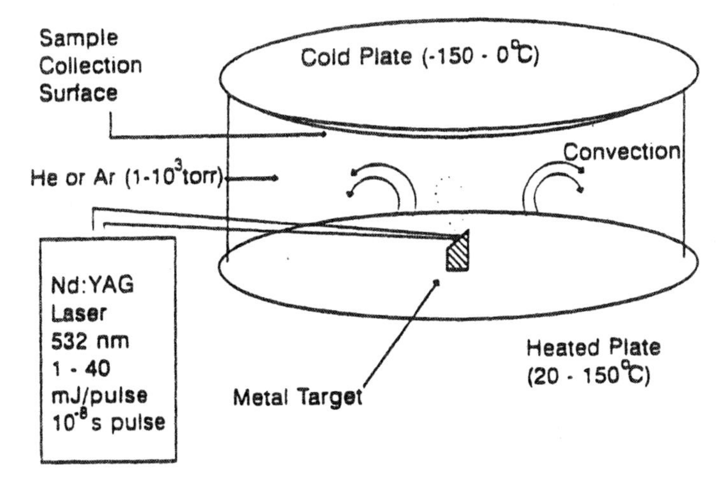

Figure 2. Experimental set-up for the synthesis of the nanoparticles using laser vaporization in a convective atmosphere.

The metal vapor is generated by pulsed laser vaporization using the second harmonic (532 nm) of a Nd-YAG laser (15–30 mJ/pulse, 10^{-8} s pulse). The laser beam is moved on the metal surface in order to expose new surface to the beam and assure good reproducibility of the amount of metal vapor produced. Following the laser pulse, the ejection of the metal atoms and their eventual interaction with the ambient atmosphere take place. Since the target surface where evaporation occurs is located near the middle of the chamber (about 0.5 reduced height) and the ambient temperature rapidly decreases near the top plate, it is likely that maximum supersaturation develops within the upper half of the chamber above the surface target (closer to the target than to the top plate). The heat and mass flux equations in a convective transport system, as in the pulsed laser vaporization in a diffusion cloud chamber, are not simple. These equations have been solved for one-dimensional plane parallel diffusion as in the conventional diffusion chamber.[25,52,53] In this case, the solution of the boundary value problem yields the supersaturation and temperature profiles as a function of elevation within the chamber. In the case of laser vaporization we assume that a supersaturation profile will develop within the chamber with a maximum value near the top plate. Since the equilibrium vapor pressure (P_e) is approximately an exponential function of temperature, the metal vapor can easily be supersaturated in the condensation zone near the top plate. This supersaturation can be made as large as desired by increasing the temperature gradient between the chamber plates. The higher the supersaturation, the smaller the size of the nucleus required for condensation. The role of convection in the experiments is to remove the small particles away from the nucleation zone (once condensed out of the vapor phase) before they can grow into larger particles. The rate of convection increases with the temperature gradient in the chamber. Therefore, by controlling the temperature gradient, the total pressure and the laser power (which determines the number density of the metal atoms released in the vapor phase), it is possible to control the size of the condensing particles.

Nichrome heater wires are wrapped around the glass ring and provide sufficient heat to prevent condensation on the ring and to maintain a constant temperature gradient between the metal target and the top plate. In a typical run, the laser operates at 20 Hz for about 1–2 h. Then the chamber is brought to room temperature and the particles are collected under atmospheric conditions. Glass slides or metal wafers can be attached to the top plate when it is desired to examine the morphology of as-deposited particles. No particles are found on any other place in the chamber except on the top plate; this supports the assumption that nucleation takes place in the upper half of the chamber and that convection carries the particles to the top plate where deposition occurs.

IV. PROPERTIES OF SOME SELECTED NANOPARTICLES

In the following sections, we present a brief summary of the significant results relating to different nanoparticles produced by the LVCC method.

A. Silica Nanoparticles

The SEM micrographs of the as-deposited silica particles on glass substrates reveal highly organized web-like structures characterized by micropores with pore diameters up to 1–2 μm and wall thicknesses of 10–20 nm as shown in Figure 3. The three dimensional arrays exhibit several remarkable features: (1) well-defined pore sizes and shapes, (2) fine control of particle size as indicated by the homogeneous particle diameter which is typically 10–20 nm as determined from the TEM micrograph as shown in Figure 4, and (3) a very high degree of pore ordering over micrometer length scales with regular interconnecting arrays (network). These structures are more pronounced when the temperature of the condensation plate (upper plate) is kept lower than –40 °C. At higher temperatures, a smoother, less porous deposit is obtained.[14]

The observed web-like structure is different from the general structure of conventional silica which consists of small, spherical particles of anhydrous SiO_2 linked together into a three-dimensional network through interparticle siloxane linkages.[54] Since many of the properties of silica are related to its aggregate structure, a great deal of effort has been made to understand and control the growth process of the primary particles, their linking to branch chains, and ultimately the formation of the aggregated network structure. The web-like aggregations with large uniform cavities could provide special catalytic activities for the silica nanoparticles.

Figure 3. SEM micrographs of the web-like agglomeration of the silica nanoparticles.

The X-ray diffraction analysis of silica nanoparticles prepared in 20% O_2 in He reveals a typical amorphous pattern very similar to that of the commercial fumed silica as shown in Figure 5.

The FTIR spectrum of the nanoparticles (displayed in Figure 6) shows strong absorption bands associated with the characteristic stretching, bending, and wagging vibrations of the SiO_2 group. In addition, bands common to moisture-induced surface species on highly active silica (large surface area) are also observed. In particular, the band at 957 cm^{-1}, which can only be seen if the specific surface area of the silica is high enough, is clearly observed in the sample. This band is directly related to the surface silanols and is assigned to the ν (Si–O) stretching vibration

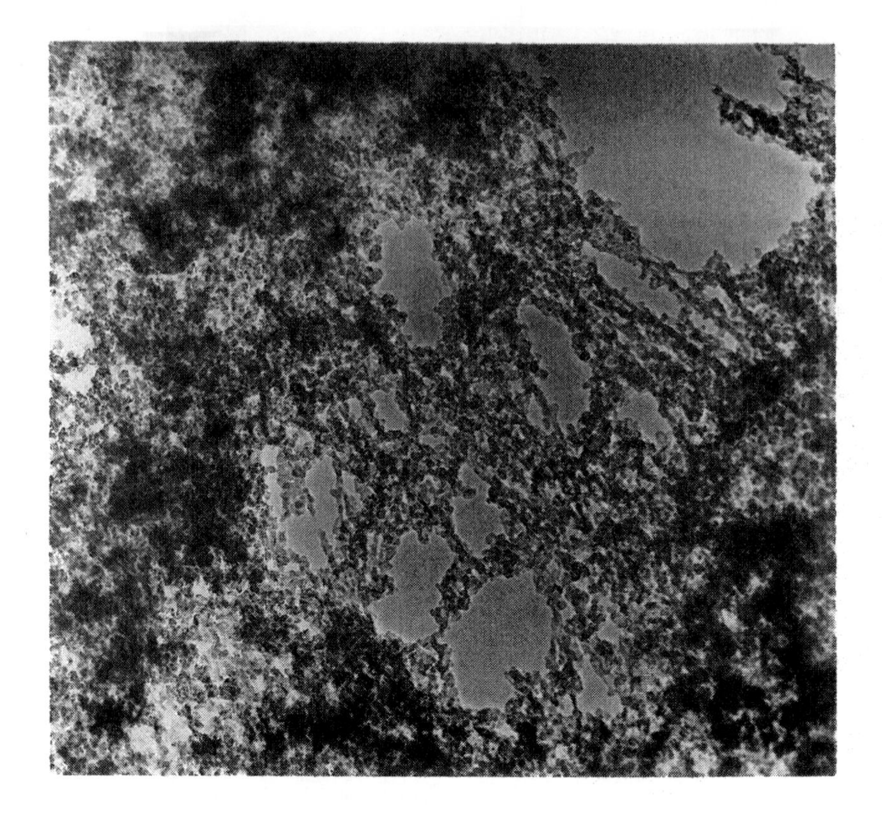

Figure 4. TEM micrograph of the silica nanoparticles.

in the Si–OH surface groups. Infrared surface analysis indicated that all the Si–OH groups are located on the surfaces of the nanoparticles.[14] The surface area of the particles depends on the experimental parameters used during the synthesis and values between 380 and 460 m^2/g were determined by the BET method.[14] In comparison, the largest surface area of a commercially available fumed silica has a surface area of 300 m^2/g; the nominal particle size of this material is 7 nm (Cabosil EH-5, fumed silica).

The silica nanoparticles exhibit bright blue photoluminescence (PL) upon irradiation with UV light.[14] The emission and excitation spectra are shown in Figure 7. The origin of this surprisingly strong blue photoluminescence appears to be related to defects in the SiO_2 structure. Several such defect models have been discussed in the literature.[55–58] The emission at 2.65 eV has been assigned to a new intrinsic defect in amorphous SiO_2 for which a twofold coordinated Si is proposed, i.e. a Si $(II)^0$ (neutral) center.[57] Chemically, this is equivalent to the quasi-molecule SiO_2^{2-} with a $^1A_1$ ground state, first excited $^1B_1$ singlet state, and a $^3B_1$ triplet state.

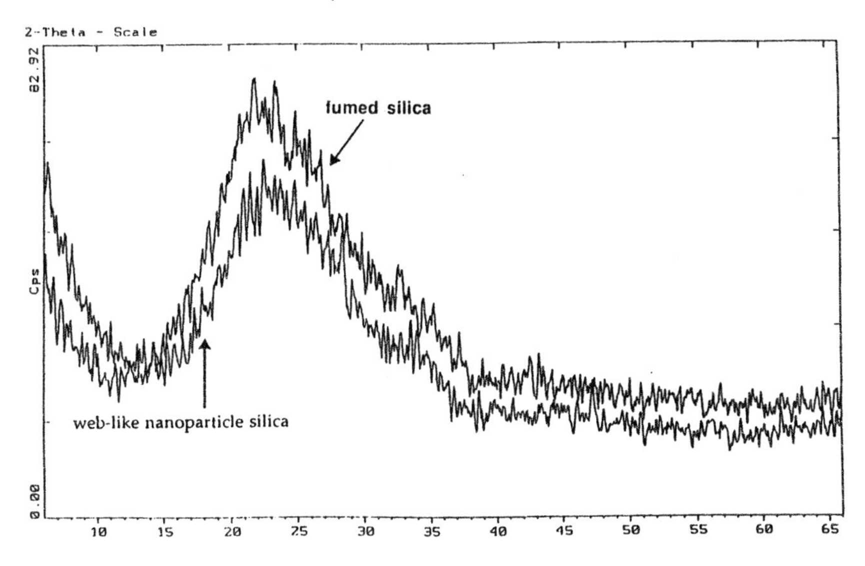

Figure 5. X-ray diffraction of the silica nanoparticles.

The transition from the latter to the singlet ground state, $S_0 \leftarrow T_1$, consists of a radiative part at 2.65 eV in conjunction with a nonradiative transition to S_0.[57]

It appears that this Si (II)^0 center is present in significant amounts in oxygen-deficient silica. This conclusion is also supported by data obtained with time-resolved PL in amorphous SiO_2. Therefore, the Si nanoparticles prepared by the LVCC

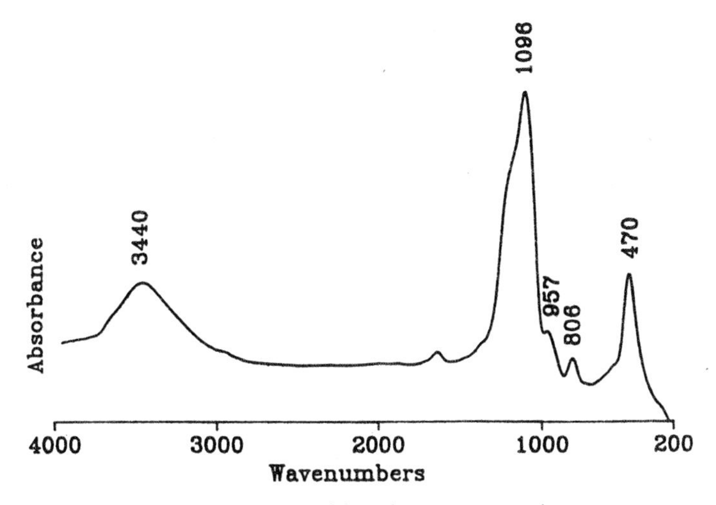

Figure 6. FTIR of the silica nanoparticles.

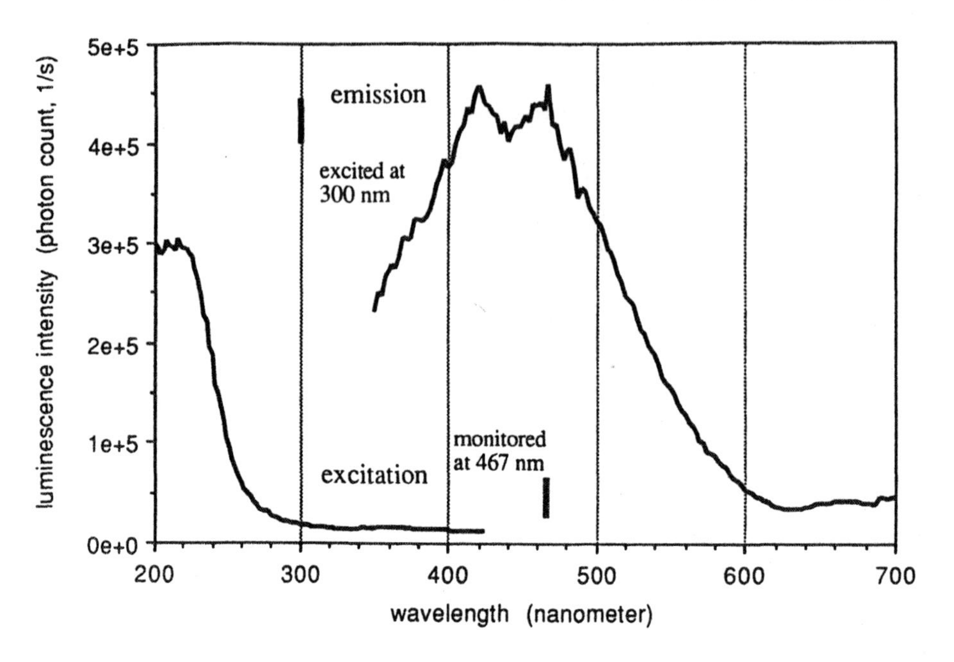

Figure 7. Emission and excitation photoluminescence spectra of the silica nanoparticles.

method contain a large concentration of the Si (II)° defect, especially at its internal surfaces, which in turn gives rise to the observed bright photoluminescence.

B. Silicon Nanocrystals

Among the many interesting nanoparticles, Si nanocrystals show important promise for applications in Si-based devices for optical communication. This exciting possibility has been hampered by the indirect band gap of bulk Si which prevents efficient electron–photon energy conversion. However, the discovery that porous and Si nanocrystals emit visible light with a high quantum yield has raised hopes for new photonic Si-based devices.[59–66] This discovery has also stimulated interest in the synthesis of Si nanocrystals which are believed to be the luminescent centers in porous silicon.[67–75]

Methods Used to Generate Si Nanocrystals

Various methods have been used to make Si nanocrystals. They have been generated by the slow combustion of silane,[76] reduction of $SiCl_4$ by Na,[77] separation from porous Si,[78] UV or IR laser photolysis of silane-type precursors,[79] thermal evaporation,[80] microwave discharge,[81] RF sputtering,[82] and high-temperature aerosol techniques.[68] In most cases, some control over particle size can be achieved by lowering the concentration of the nucleating particles. As compared to other methods, the LVCC method has the advantage of eliminating the need for high temperatures and for chemical precursors.

The Si nanoparticles prepared by the LVCC method appear as a yellow powder.[17] The SEM micrographs of the as-deposited particles on glass substrates reveal highly organized weblike structures characterized by micropores with wall thicknesses of 10–20 nm. The weblike morphology is similar to that of the silica nanoparticles prepared by the same method using O_2/Ar or O_2/He gas mixtures.[14] The web-like structure with strings of aggregated Si nanoparticles is shown in Figure 8. The

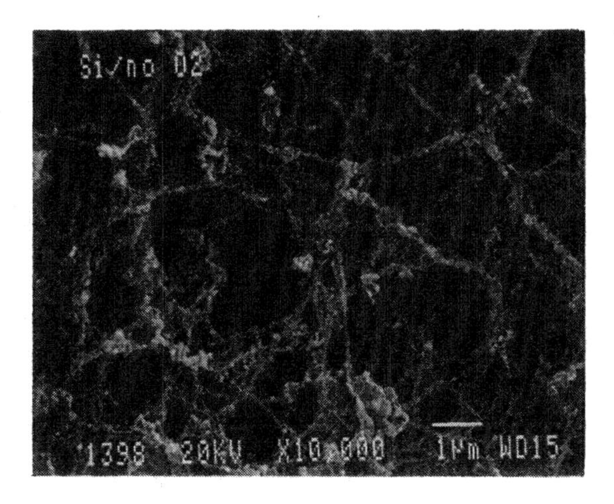

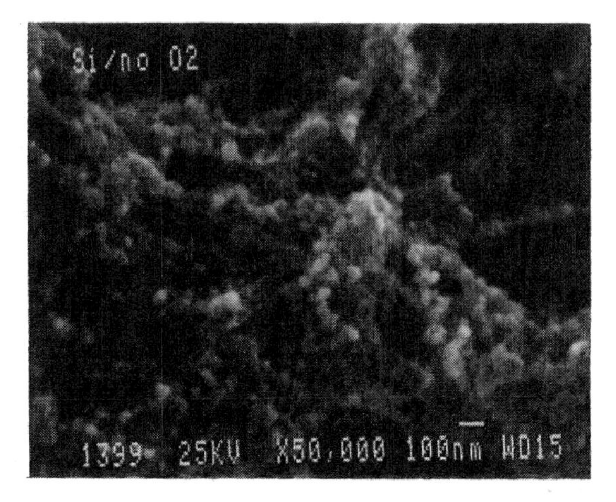

Figure 8. SEM micrographs of the web-like agglomeration of the silicon nanocrystals.

individual particles are spherical and uniform in size, about 10 nm in diameter and are connected in a web-like structure.[17]

The Raman spectrum of the Si nanoparticles prepared in the absence of oxygen, displayed in Figure 9a, shows a sharp peak at about 510 cm^{-1}. This is close to the Raman allowed optical phonon characteristic of microcrystalline silicon at 520 cm^{-1}.[83] This peak shifts to higher frequency by increasing the amount of O_2 added during the particles' synthesis as can be seen from the Raman spectrum displayed in Figure 9b. The downshift of this band compared to the bulk sample is attributed

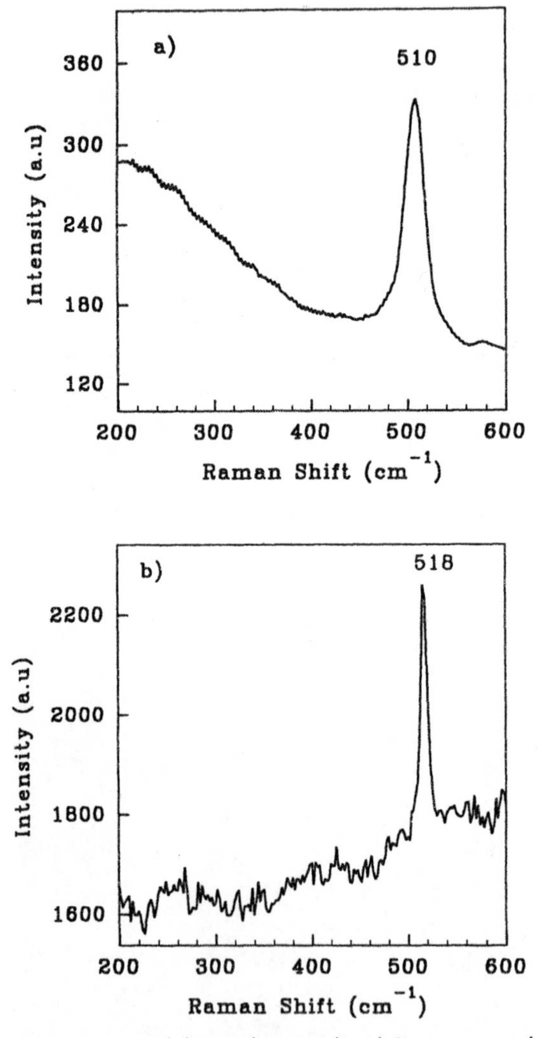

Figure 9. Raman spectrum of the surface-oxidized Si nanocrystals: (**a**) no oxygen added, and (**b**) with 100 torr O_2 in the preparation.

to size and strain effects.[63,84,85] We note that porous Si samples prepared under various etching conditions show Raman shifts which have been correlated to the particle size. For example, Raman shifts of 517 cm^{-1}, 515 cm^{-1}, 507 cm^{-1}, and 501 cm^{-1} were correlated to particle sizes of 7 nm, 4.8 nm, 2.5 nm, and ~ 1.6 nm, respectively.[86] Based on the Raman shift of our sample, the average particle size can be estimated as ~ 4 nm. This is considerably smaller than the particle size shown by the SEM micrographs and this suggests that particles smaller than the SEM is able to detect may be present in the sample.

The crystallinity of the particles is verified by the X-ray diffraction spectrum shown in Figure 10. The spectrum conclusively shows crystalline Si lines 111, 220, and 311 at scattering angles of 28°, 47°, and 56°, respectively. Such crystalline lines are not present in the X-ray diffraction pattern of silica nanoparticles prepared in the presence of O_2. The silica particles exhibit a completely amorphous structure similar to that of fumed silica as shown in Figure 5.

The surface oxidation of the Si nanoparticles is studied by FTIR. Figure 11 shows the dependence of the IR absorption spectrum on the annealing temperature of the nanoparticles. The spectra of the as-deposited particles (Figure 11a) are measured 20 min after removing the particles from the preparation chamber. Spectra b and c are measured after annealing the as-deposited sample in air for 60 min at 288 °C and 60 min at 593 °C, respectively. The FTIR of the as-deposited sample (Figure 11a) shows a broad absorption band at about 1080 cm^{-1} accompanied by weak

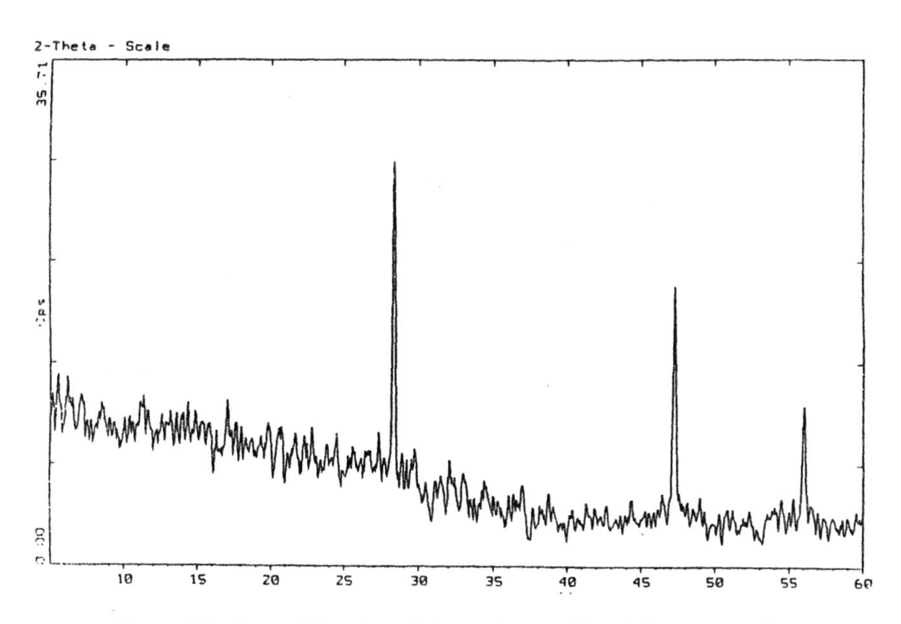

Figure 10. X-ray diffraction of the surface-oxidized Si nanocrystals.

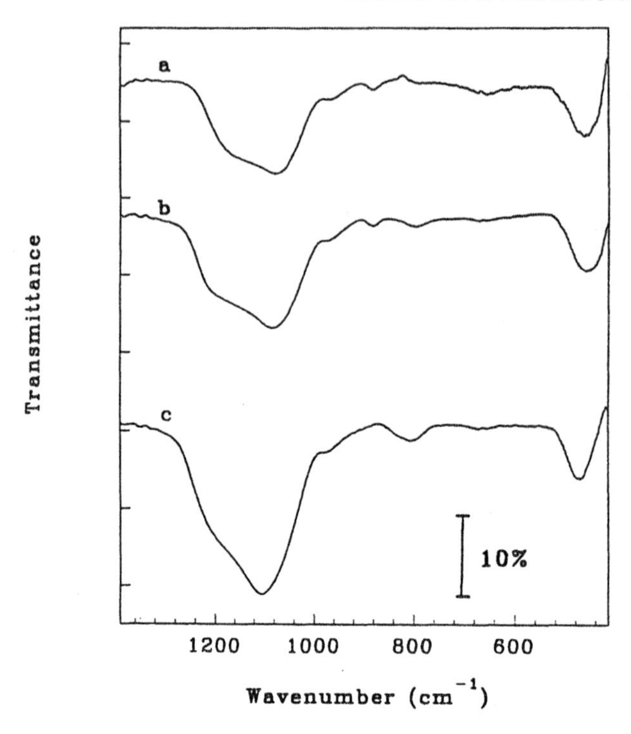

Figure 11. FTIR of Si nanocrystals: (**a**) as deposited particles. Spectra **b** and **c** are measured after annealing the as deposited sample in air for 60 min at 288 °C and 60 min at 593 °C, respectively.

bands at 958 cm^{-1}, 880 cm^{-1}, and 454 cm^{-1}. The main band at 1080 cm^{-1} is due to Si–O–Si stretching and is usually observed in thin films of amorphous silicon oxide (SiO_x with x varying from 0–2).[87,88] Other peaks observed in the IR of amorphous SiO_x films are a weak band at 800 cm^{-1} and a moderate peak at 460 cm^{-1} which are assigned to bending and rocking vibrations, respectively, of the Si–O–Si bonds.[88] In addition, a shoulder at 956 cm^{-1} is also observed and attributed to the stretching of Si–O–Si bridge of SiO_x.[89,90] Another peak at 880 cm^{-1} has also been observed in the IR of amorphous silicon films[87] and is also observed in our sample. The origin of this peak has not been clearly identified. Previous studies have attributed this feature to various sources: (1) to a Si_2O_3 phase within the amorphous Si film;[91] (2) to the combination of Si–(O_2Si_2), Si–(O_3Si), and Si–(O_4) structures;[87] (3) to the Si–N bond;[92] and to the motion of the nonbridging Si–O bonds or change in the Si–O–Si angle.[93] In addition, Knolle and Maxwell, from both experimental and theoretical analysis, assign the 880 cm^{-1} peak to the interaction between Si and nonbridging O atoms in SiO_2.[94] This last explanation is best supported by our results, as explained below.

The observed IR peaks of the as-deposited Si sample are assigned to a surface oxide layer of SiO_x that apparently forms after the particles are removed from the reaction chamber. Annealing the particles for 60 min at 288 °C (Figure 11b) results in the appearance of a weak band at 800 cm^{-1} and an increase in both the broad band at 1080 cm^{-1} and the 460 cm^{-1} band. These changes are attributed to developing the oxide layer (SiO_x) by both increasing its oxygen content (i.e. x increases toward 2) and by increasing its thickness through the slow oxidation of the Si core. It should be noted that no significant shift of the broad band at 1080 cm^{-1} is observed even by when the annealing time is increased to 90 min at 288 °C. However, after annealing the particles for 60 min at 427 °C, a noticeable decrease in the 880 cm^{-1} band and a high-frequency shift of the main broad band to 1086 cm^{-1} occur. These changes can clearly be produced by annealing the particles for shorter times but at higher temperatures. For example, by continuously annealing the particles for 60 min at 593 °C, (spectrum 11c), further shift of the broad band to 1106 cm^{-1} is observed. It should be noted that spectrum c is very similar to the IR spectrum of silica nanoparticles prepared by using a O_2/Ar gas mixture.[14] In particular, the band at 800 cm^{-1} is very pronounced in the silica nanoparticles as is the broad band at 1100 cm^{-1} which has nearly the same frequency as in the spectrum of Figure 11c. This is taken as evidence for complete oxidation of the Si nanoparticles (including the core particles) achieved by annealing the particles at quite high temperatures such as 593 °C.

To further support the conclusion that the changes in the IR spectra are directly related to the oxidation of the Si nanoparticles, we prepared several Si samples by using variable amounts of O_2 in the reaction chamber.[17] As the partial pressure of O_2 increases in the reaction chamber, the following features can be observed in the IR spectra: (1) the broad band at ~1080 cm^{-1} and the 460 cm^{-1} band shift to higher frequency, (2) the shoulder peak at 956 cm^{-1} becomes stronger, (3) the peak at 881 cm^{-1} disappears, and (4) the weak peaks at 800 cm^{-1} and 460 cm^{-1} become stronger. Figure 12 illustrates the magnitude of the shift in the broad band at 1080 cm^{-1} and in the 460 cm^{-1} peak as a function of O_2 partial pressure during the particles' formation. These peak shifts are strongly related to both the oxygen content and the thickness of the SiO_x oxide layer. It has been shown both experimentally and by theoretical analysis that the thicker the SiO_2 oxidation layer or the higher the oxygen content of the SiO_x layer, the higher the frequency of the Si–O stretching and rocking vibrations.[87,88,95,96] The disappearance of the 880 cm^{-1} peak at higher O_2 partial pressure (more than 20 torr in our experiments) is consistent with the assignment of Knolle and Maxwell, who attributed this peak to the interaction between core Si and nonbridging O atoms in the SiO_2 layers.[94] Oxidation eliminates the nonbonding Si atoms and thus weakens the interaction between Si and non-bridging O atoms. Thus the 880 cm^{-1} peak gradually decreases and eventually disappears at higher O_2 partial pressure or higher annealing temperatures.

Another result relating to the oxidation of the Si nanoparticles is that complete oxidation can be achieved by focusing the 532 nm second harmonic of the Nd:YAG

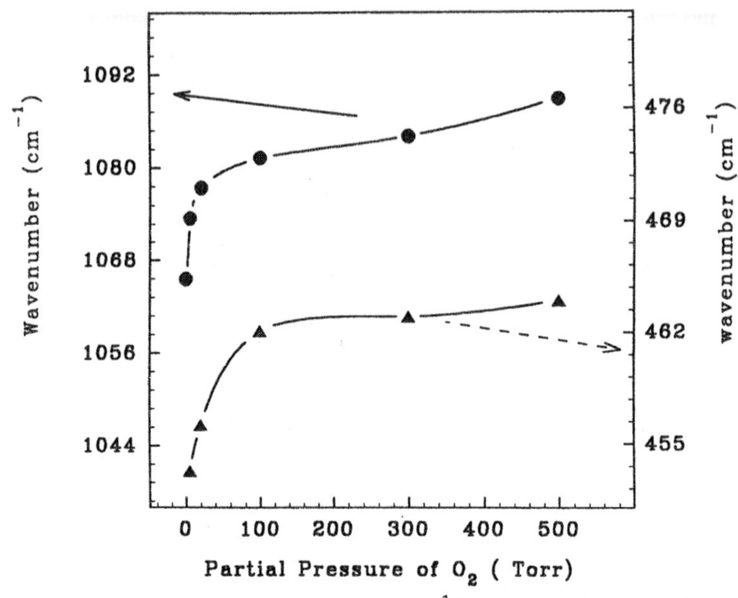

Figure 12. FTIR shifts of the 1080 and 460 cm^{-1} bands as a function of O_2 partial pressure during the particles' formation.

laser (30 mJ/pulse) on the Si particles in air.[17] Bright red orange luminescence is observed upon irradiation of the particles with a single laser pulse (10 ns). This is accompanied by a mild sound, and the yellow Si particles turn white.[97] The observed emission can be reproduced by irradiation of fresh Si sample, but after the particles turn white, further irradiation does not produce any red emission. We measured the IR of the Si particles before and after the irradiation with the YAG laser and the results are shown in Figure 13. The IR spectrum measured after irradiation shows clearly all the features associated with complete oxidation of the Si nanoparticles. These features include the disappearance of the 880 cm^{-1} peak, the enhancement of the 470 cm^{-1} and the 800 cm^{-1} peaks, and the high-frequency shift and enhancement of the broad band at 1080 cm^{-1}. These results indicate that the incident laser pulse efficiently breaks the thin SiO_x layer around the Si core and therefore, induces immediate oxidation of the bare particles in air.

From the FTIR study, it can be concluded that the as-deposited Si nanoparticles are coated with a surface layer of SiO_x when exposed to air. The oxygen content can be increased to approach $x = 2$ by annealing the particles at lower temperatures (300 °C) in air. We estimate the thickness of the SiO_x layer (in the as-deposited particles) as 1–2 nm. The thickness of this layer can be increased by further oxidation of the core Si particles, which takes place efficiently only at relatively high temperatures (e.g. 10 min at 593 °C), or by rapid breaking of the oxide surface

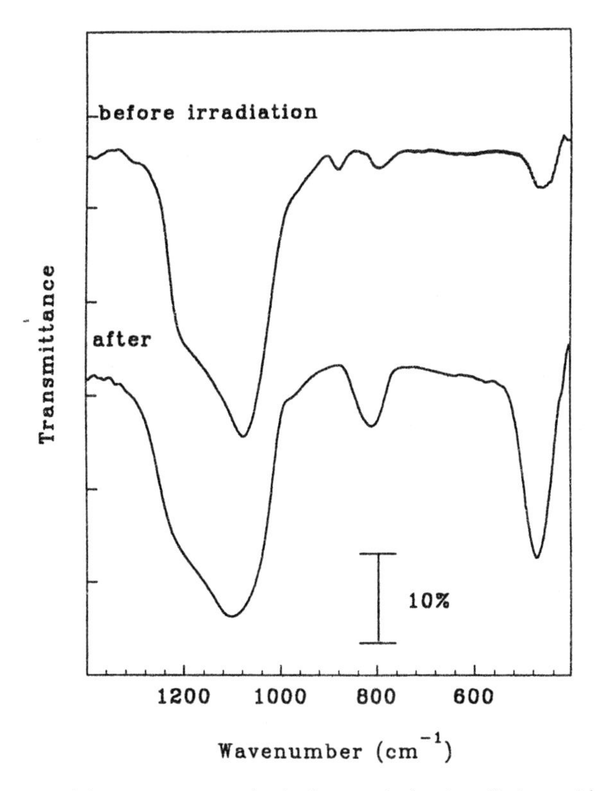

Figure 13. FTIR of the Si nanoparticles before and after irradiation with the 532 nm second harmonic of the Nd:YAG laser.

layer by, for example, irradiation with intense laser pulses. Under ambient temperature, further oxidation of the Si core appears to be very slow and inefficient. These conclusions are consistent with previous results which showed that the oxide growth rate of small Si particles is smaller than the rate of oxidation of Si wafers.[98] One explanation of this phenomenon is that stress in the oxide layer may suppress the oxidation of the small particles. Our results are also in full agreement with the conclusions of Hayashi et al. from their study of the surface oxidation of thermally evaporated Si nanoparticles (10 nm).[88]

Figure 14a shows the UV-vis absorbance of the Si nanoparticles suspended in methanol. The spectrum shows the absorption features associated with indirect band gap transitions, particularly the yellow absorption tail which extends from 400 nm across the visible and the stronger absorption from 370 to 240 nm. The indirect band gap of semiconductors can be obtained from the following relation,[99]

$$(\alpha\, h\nu)^{1/2} = B\,(h\nu - E_g) \tag{13}$$

where α is the reciprocal absorption length, $h\nu$ is the photon energy, B is a constant and E_g is the band gap. Using the yellow absorption tail in the UV-vis spectrum, we plot $(\alpha\, h\nu)^{1/2}$ vs. $h\nu$ as shown in Figure 14b. The plot shows an approximate straight line which extrapolates to give 1.78 eV as the calculated band gap of the Si nanoparticles. This corresponds to 700 nm emission. The blue shift from the bulk Si indirect band gap (1.1 eV) is a direct manifestation of the quantum confinement effect.[68,70,75]

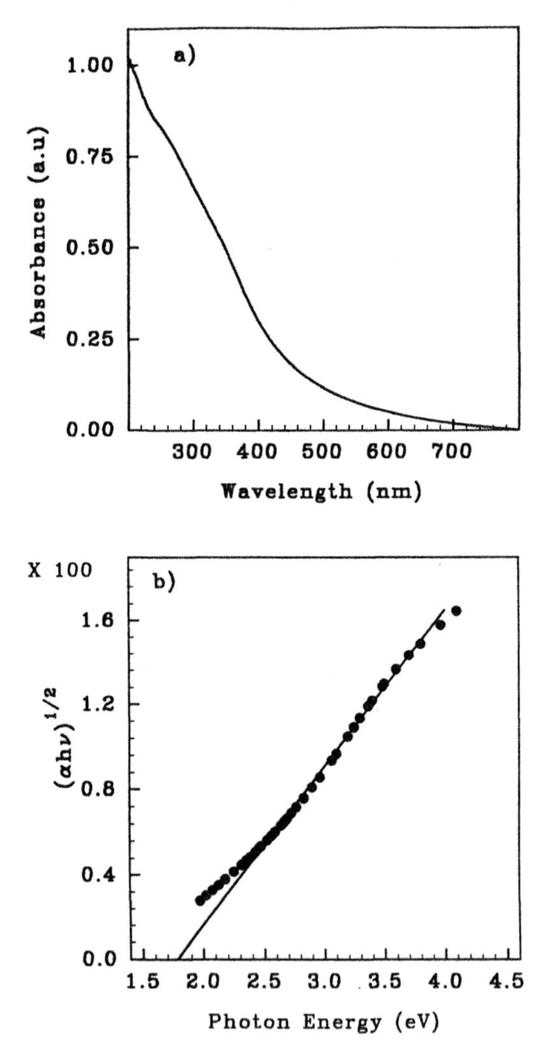

Figure 14. (**a**) UV-vis absorbance of Si nanocrystals suspended in methanol. (**b**) Plot of $(\alpha\ h\nu)^{1/2}$ vs. $h\nu$ for the absorption tail of the Si nanocrystals.

The dispersed luminescence spectra of the Si nanoparticles excited by the 363.8 and 514.5 nm Ar ion cw laser lines are shown in Figure 15. In both spectra a broad emission is seen in the red region. This emission can be seen with the naked eye in the presence of normal room light. The luminescence quantum yield, estimated by using Rhodamine B as a relative standard, is about 1.3%. The red emission curve can be fit to a Gaussian shape. For the 363.8 nm excitation the width (FWHM) is 184 nm and the maximum is at 740 nm; for the 514.5 nm excitation the width is 190 nm and the maximum is at 760 nm. The 20 nm blue shift in the emission maximum of the 363.8 nm excitation may arise as a result of excitation of higher energy states which are not accessible with longer wavelength excitations. With the quantum confinement interpretation,[68,70,75] this suggests that different sizes of nanocrystals are simultaneously present in the sample. Accordingly, the longer wavelength excitation selects larger particles which in turn emit further to the red than the smaller particles selected by the shorter wavelength excitation.

The insert in Figure 15 shows a weak blue emission feature that is observed when 363.8 nm excitation is used. This feature peaks at 450 nm and appears similar to the blue emission observed when SiO_2 nanoparticles are excited. This blue emission is more pronounced when pulsed excitation is used and will be discussed further below.

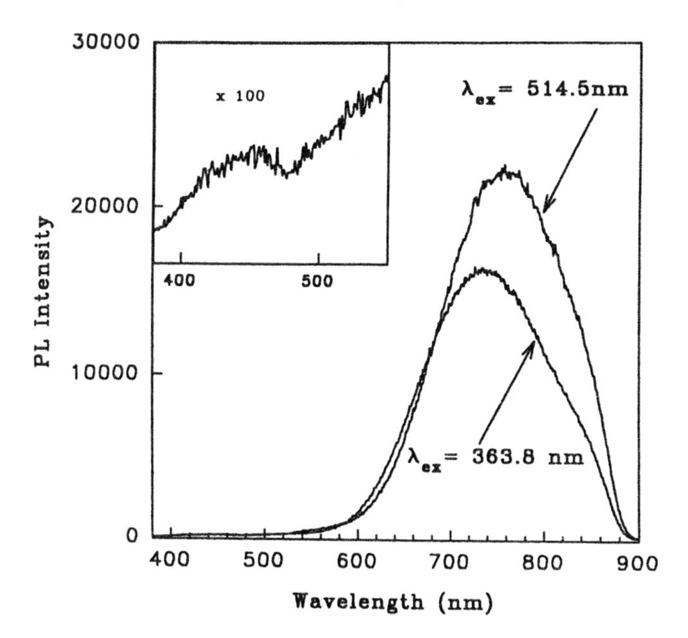

Figure 15. Dispersed emission of the Si nanocrystals excited by the 363.8 and 514.5 nm Ar ion laser.

Dispersed luminescence spectra obtained with 340 nm pulsed laser excitations are shown in Figure 16. The spectra differ by the position of the boxcar gate, which ranges from 0 delay with respect to the laser excitation pulse to 5 μs delay. The 0 delay spectrum enhances the blue emission component, since the lifetime associated with it is short (less than 20 ns) compared to the lifetimes of the red emission. The shape of the blue emission appears similar to that from SiO_2 nanoparticles and it is probably due to the oxidized surface layer of the Si nanoparticles.[14] The cw excitation experiments, similar to time-integrated pulsed ones, weigh the long-lived red emission more heavily compared to the short-lived blue emission than do gated experiments with short delays. This makes the blue band hard to detect. By using the 266 nm excitation, the red emission shifts to shorter wavelength as compared to the emission from the 340 nm excitation.[17] This again, could be attributed to preferential excitation of the smaller Si particles by the 266 nm pulse.

As the boxcar delay increases, the blue emission decreases in intensity due to its relatively short lifetime and the red to blue intensity ratio increases. Also, the maximum of the red emission band shifts to longer wavelengths as the delay increases. This is due to the lifetime lengthening that occurs as emission wavelength increases. This lifetime lengthening can also be observed in the time and frequency resolved emission spectra. These decays are multiexponential, but clearly have both long and short components.[17] They can be fit to the biexponential form,

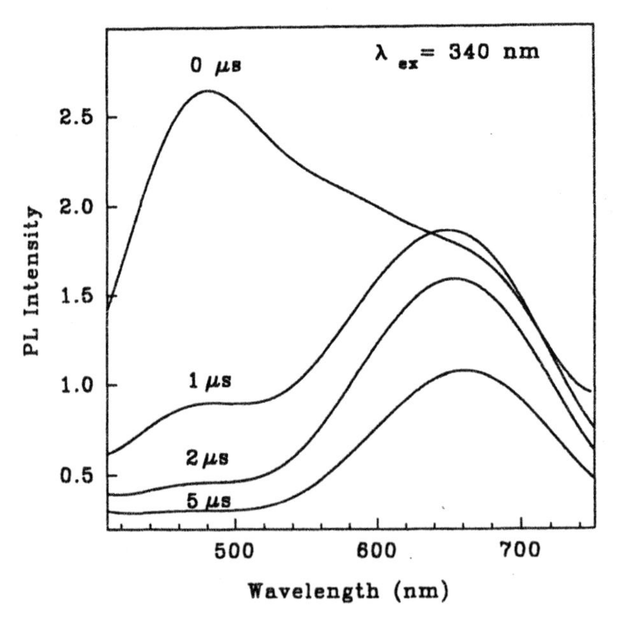

Figure 16. Dispersed emission of the Si nanocrystals obtained with 340 nm pulsed laser excitation at different boxcar gate delays from the excitation pulse.

$$I(t) = I_s^\circ \exp(-t/\tau_s) + I_l^\circ \exp(-t/\tau_l) \tag{14}$$

where the subscripts 's' and 'l' refer to the short and long-lived components, respectively. A decay at 640 nm is shown in Figure 17 together with the biexponential fit. The short lifetime, about 12 μs, does not depend on emission wavelength; the long lifetime ranges from 90 to over 130 μs, increasing with emission wavelength. The dependence of τ_l on emission energy is shown in Figure 18a, and the energy dependence of the intensity ratio, I_l°/I_s°, is shown in Figure 18b. The long lifetime and the intensity ratio increases with emission wavelength both contribute to the red shift with boxcar delay seen in the dispersed spectra (Figure 16).

The red PL from the Si nanocrystals can be quenched by electron acceptor molecules such as 1,4-dinitrobenzene and 4-nitrophthalonitrile. Interestingly, no detectable PL quenching has been observed with water. Similar results have been reported for porous Si.[100] The PL spectra taken form colloid solutions of the Si nanocrystals in ethanol after successive additions of 1,4-dinitrobenzene and 4-nitrophthalonitrile quenchers and excitations with 355 nm are shown in Figures 19a and 19b, respectively. The systematic loss of the PL intensity with increasing the aromatic quencher concentration is evident. The quenching mechanism can be

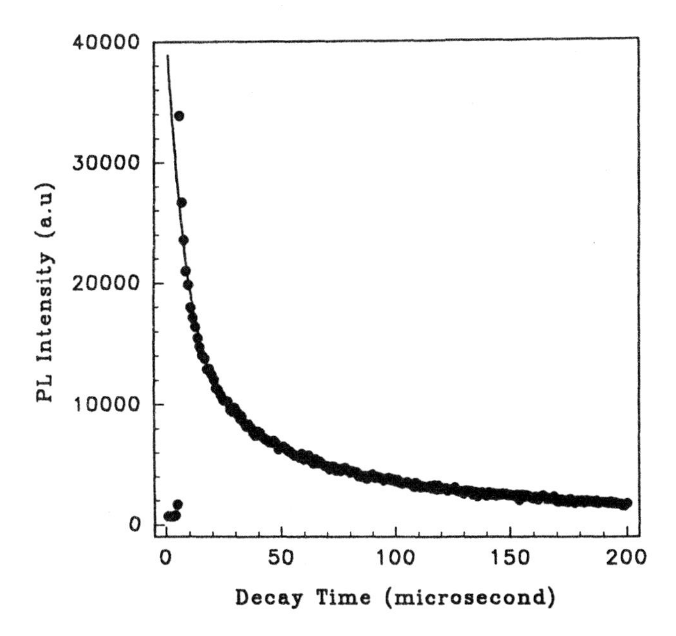

Figure 17. Decay of the PL from the Si nanocrystals at 640 nm (solid line represents the biexponential fit).

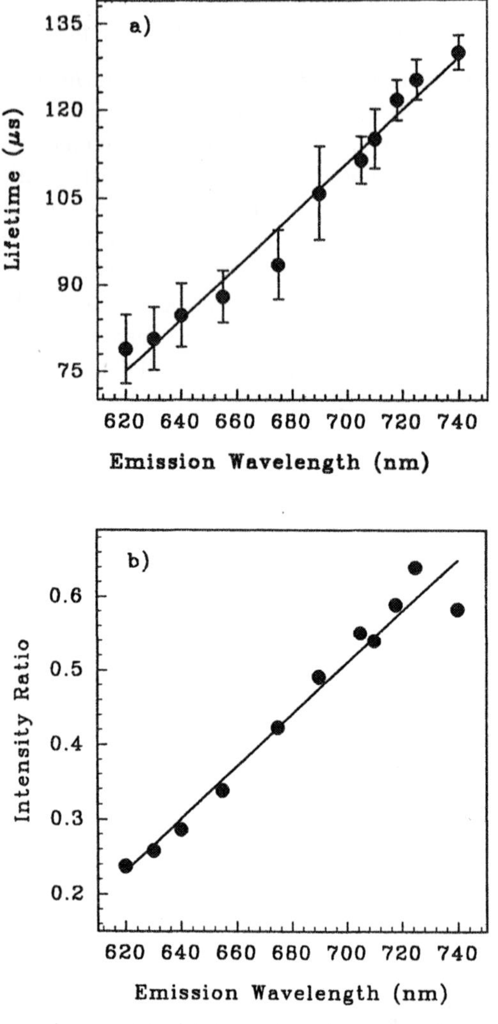

Figure 18. (**a**) Dependence of the lifetime of the longer-lived component of the red emission on the emission wavelength. (**b**) Dependence of the intensity ratio $\mathring{I}_l/\mathring{I}_s$ on the emission wavelength.

interpreted as energy or electron transfer. Recent studies by Sailor and coworkers identified an energy transfer pathway for quenching of porous Si by aromatic molecules in which low-lying triplet states are involved.[100,101] However, the results of Fauchet and coworkers provide evidence for quenching of porous Si via an electron or hole transfer mechanism.[102] Work is in progress in our laboratory to

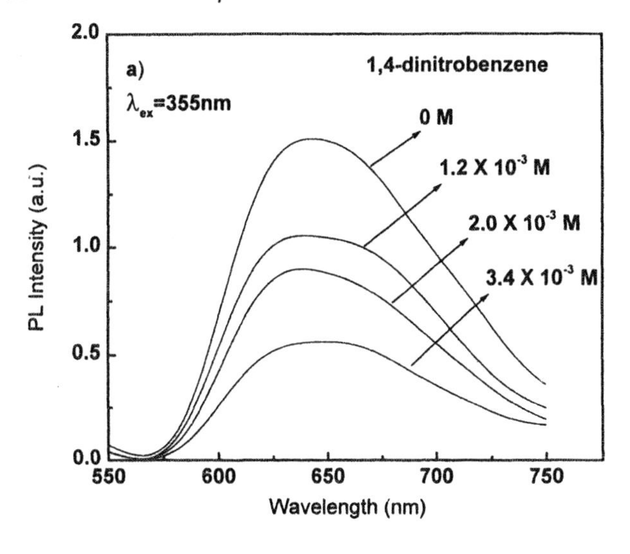

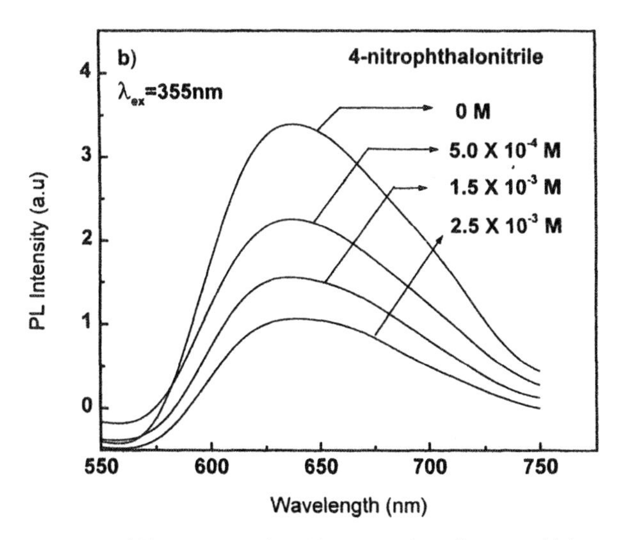

Figure 19. PL spectra of Si nanocrystals with successive aliquots of (**a**) 1,4-dinitrobenzene and (**b**) 4-nitrophthalonitrile in ethanol.

measure the PL quenching rate constant by a wide variety of aromatic molecules in order to examine the correlation between the quenching rate and both the ionization potential and the triplet energy of the quencher. This work will determine the contributions of the electron transfer and energy transfer pathways to the PL quenching of Si nanocrystals.

On the Mechanism of Photoluminescence of the Si Nanocrystals

The web-like aggregates of coalesced Si nanocrystals, passivated by a SiO_x barrier layer, exhibit intense red PL upon excitation with visible or UV light. The freshly prepared Si nanoparticles do not show the emission unless a passivated layer stabilizes the nanocrystal surface by saturating all the dangling bonds. FTIR of the freshly prepared sample shows weak peaks due to the stretching, bending, and rocking vibrations of the Si–O–Si bonds, indicating the presence of the surface oxidized layer SiO_x, but no red or blue emission is observed until more complete development of the oxide layer has been achieved. This appears to be essential for efficient passivation of the nanocrystal surface. Part of this development appears to be the increase in the oxygen content of the oxide layer to approach the SiO_2 composition; another part may be an increase in its thickness to form an efficient barrier layer. This is apparently a slow process under normal ambient conditions since the emission can be observed only 1–2 days after removing the nanoparticles from the reaction chamber to the ambient air. It is also important to note that fully oxidized Si nanoparticles with SiO_2 composition and completely amorphous phase do not show the red PL upon UV or visible excitations.

The red PL from the Si particles shifts to longer wavelength with decreasing the excitation energy. This emission is characterized by a multiexponential decay having a long component (80–130 μs at 300 K) that increases in lifetime and intensity with emission wavelength (630–740 nm). These features of the red emission are similar to those observed for porous Si[59–66] and from Si nanocrystals.[67–75] Because of such strong similarity between the PL properties of porous Si and Si nanocrystals, Brus et al. concluded that the luminescent chromophore is the same in the two systems.[68,70,75] They assigned the red PL to single-crystalline quantum confinement in which both electron and hole are in totally symmetric volume-confined states with no significant trapping within the optical absorption band gap.[70,75] The barriers to carrier mobility between crystallites make the radiationless processes much less important in nanocrystals. This results in weak PL efficiency in Si nanocrystals and porous Si.[75]

In addition to the red PL, a weak blue PL is observed with pulsed laser excitation (266 and 340 nm). Its wavelength does not shift with excitation wavelength and it can also be observed under high-intensity cw 363 nm laser excitation. This blue emission has a fast decay (less than 20 ns) and it appears similar to the blue PL observed from amophorous silica (SiO_2) nanoparticles.[14] The features observed for the blue PL are similar to those reported by Kanemitsu using surface-oxidized Si crystallites prepared by laser breakdown of silane.[71,72,74] Kanemitsu also observed the slow decay red PL with very similar features to our present results and to those previously reported by Brus. However, in contrast to Brus conclusions, Kanemitsu proposed a model in which the red PL originates from surface localized states while the blue PL originates from the nanocrystal core.[74] According to this model, the shift of the red PL to longer wavelength with time delay is due to the relaxation of the carriers or excitons to lower energy states.

In comparing the PL results from different laboratories, it is important to point out the differences between the particles investigated in each study. For example, the particles used by Brus and Kanemitsu were prepared by high-temperature pyrolysis and laser breakdown of disilane and silane, respectively. In both studies, the surface-oxidized Si crystallites were around 3 nm in diameter. In Brus' study, larger crystallites with Si cores of > 3 nm and an oxidized layer of 0.5–1.0 nm emit in the 800–1000 nm range and smaller crystallites (with Si core diameters of 1–2 nm) emit in the 600–750 nm range. Our particles were prepared by laser vaporization of Si and the method does not involve hydrogen. As indicated by the FTIR, Raman, XPS and X-ray diffraction data, our particles appear to be only a crystalline Si core coated with an oxide SiO_x layer.[17] The SEM data indicates the presence of large particles (~10 nm) aggregated in a web-like morphology, while the Raman shift suggests an average particle size of ~4 nm. If the estimation of the particle size from the Raman shift is reliable, a significant number of smaller particles (~4 nm) may exist in our sample to produce the observed red PL. The red PL from these aggregated solid particles at 300 K excited by 363.8 nm light shows a Gaussian-shaped band centered at ~740 nm. The PL reported by Brus et al.[75] from 3 nm particles suspended in ethylene glycol glass at 15 K and excited by 350 nm shows a band centered at ~700 nm as compared to 740 nm observed in our study. This shift in the PL may reflect larger size particles used in our experiments.

Simple three-dimensional quantum confinement calculations suggest the particle size must be less than 5 nm for visible emission. To be consistent with this, our sample must contain particles significantly smaller than the 10 nm ones indicated by the SEM data. The resolution of the SEM micrographs may be inadequate for the smaller particles or it may be that higher level calculations that include electron–hole interactions are needed to give a better correlation between band gap and Si particle size.

In view of the many different mechanisms proposed to explain the visible PL from porous Si and Si nanocrystals,[59–75] we note that many of the observed features associated with the red PL from our particles can be explained by the quantum confinement mechanism. For example, the failure to observe the red PL from the freshly prepared particles and the long annealing time in ambient air required before emission can be observed, can be explained by the need to provide an efficient surface passivation which permits radiative recombination in the crystallites by removing competitive nonradiative relaxation at the surface states. Also, the shift of the PL towards shorter wavelength by increasing excitation energy is explained by the selective excitation of the smaller particles which emit more to the blue. Finally, the increase in the lifetime of the longer lived component of the red emission with emission wavelength is consistent with larger particles having longer lifetimes. In addition, we note that the features associated with the fast-decay blue emission can't be explained with the confinement mechanism. For example, this emission does not shift with the excitation energy and its short lifetime does not depend on the emission wavelength. Moreover, this blue emission is similar to that observed from silica particles and several models related to surface defects and surface adsorbed OH groups have been proposed to explain it.[14,103,104]

The surprisingly long annealing time required in ambient air before efficient PL can be observed from the Si nanoparticles deserves further comment. This long passivation time may suggest that slow continuous oxidation of the Si core takes place in ambient air and this process results in a gradual decrease of the core size. Accordingly, the red PL is observed only when the size of the Si core becomes comparable to the quantum confinement regime. This suggestion appears to be consistent with the absence of red PL from the completely oxidized Si nanoparticles. To test the validity of this proposal, one must make systematic measurements of PL, Raman shift, and particle size over long particle oxidation times. If extended oxidation reduces the core size, one would expect to see a blue shift in the red PL along with a significant peak narrowing as the Si core size continuously decreases. Such measurements are currently in progress in our lab.

In summary, the Si nanocrystals prepared by the LVCC method aggregate into a novel web-like porous microstructure. From the Raman shift of the Si nanoparticles, the average particle size is estimated as ~ 4 nm. FTIR of the freshly prepared particles shows weak peaks due to the stretching, bending, and rocking vibrations of the Si–O–Si bonds, indicating the presence of a surface oxidized layer SiO_x ($x < 2$). Under ambient temperature, further oxidation of the Si core appears to be very slow and inefficient. Annealing at higher temperatures facilitates the oxidation.

The Si nanocrystals show luminescence properties that are similar to those of porous Si and Si nanoparticles produced by other techniques. The nanoparticles do not luminesce unless, by exposure to air, they acquire the SiO_x passivated coating. They show a short-lived blue emission characteristic of the SiO_2 coating and a biexponential longer lived red emission. The short lifetime component of the red emission, about 12 μs, does not depend on emission wavelength. The longer lived component has a lifetime that ranges from 80 to over 130 μs (at 300 K), increasing with emission wavelength. The results are consistent with the quantum confinement mechanism as the source of the red photoluminescence.

C. Zinc Oxide Nanoparticles Prepared by Physical and Chemical Methods

Zinc oxide, a semiconductor, has been prepared as nanoparticles in transparent colloidal solutions by Bahnemann et al.,[105] Henglein and coworkers,[106,107] Spanhel and Anderson,[108] and others. These particles exhibit quantum size effects:[109,110] their bandgap absorption and emission are blue-shifted with respect to bulk ZnO and their visible emission, in the 500 nm region, shows wavelength and lifetime dependence on size. Surface alterations cause noticeable photophysical changes.

In this section, we compare the properties of ZnO nanoparticles prepared by the LVCC technique and by sol–gel and reversed micelle methods. In the sol–gel Method, we used the method of Henglein and coworkers.[106,107,111] Using a reversed micelle technique, ZnO nanoparticles, coated with stearate, were prepared in a water-in-oil microemulsion.[112]

SEM and TEM analyses of the ZnO particles synthesized using 20% O_2 in He at a total pressure of 800 torr and a top plate temperature of −100 °C have indicated that the particle diameters are typically between 10 and 20 nm. Examples of the SEM micrographs are shown in Figure 20.

(a)

(b)

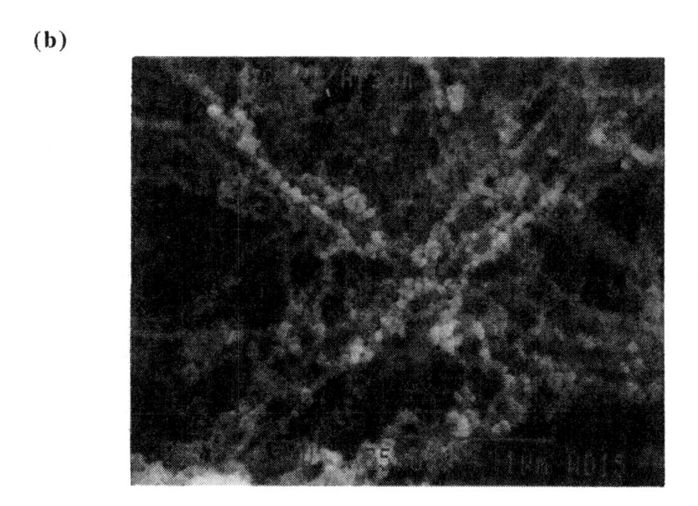

Figure 20. SEM micrographs of the ZnO nanocrystals synthesized in (**a**) He and (**b**) Ar.

The X-ray diffraction patterns of ZnO powder samples obtained by the three different synthetic techniques are similar in peak positions and relative intensities as shown in Figure 21. They are also the same as the pattern for bulk, crystalline ZnO. Thus the nanoparticles have crystalline structures identical to the bulk structure, which is of the hexagonal, wurzite type. The peak intensities, which correspond to the degree of crystallinity, are strongest for the particles made in methanol and weakest for those coated with stearate. Particle diameters, estimated from peak widths, are about 5 nm for those made from a one-day-old methanol solution and about 6.5 and 6.8 nm for those made by the laser and reversed micelle techniques, respectively. These sizes are consistent with those obtained from the absorption spectra.

FTIR results indicate that the phonon spectra of the ZnO nanoparticles are the same as bulk ZnO.[13] As shown in Figure 22a, the FTIR spectrum of the LVCC sample shows a broad band at 443 cm^{-1} with a shoulder at 550 cm^{-1}. The ZnO surface is known to be reactive towards CO_2 which results in the formation of carboxylate species due to the reaction with atmospheric CO_2 as indicated by the small bands at 1363 and 1579 cm^{-1}.[113] The Raman spectrum, shown in Figure 22b, shows only one unambiguous band at 438 cm^{-1}.

Bandgap absorption occurs when semiconductor electrons are excited from the valence band to the conduction band. The onset of this absorption occurs at the bandgap energy, E_g, which for bulk ZnO is 3.3 eV or 380 nm. For ZnO nanoparticles

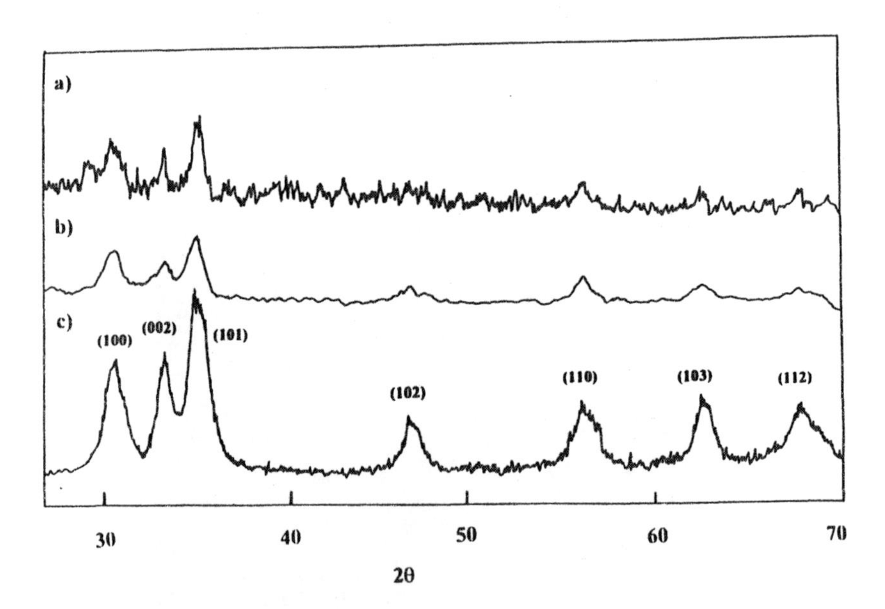

Figure 21. X-ray diffraction patterns of ZnO nanoparticles prepared by (**a**) reversed micelle, (**b**) LVCC and (**c**) sol–gel methods.

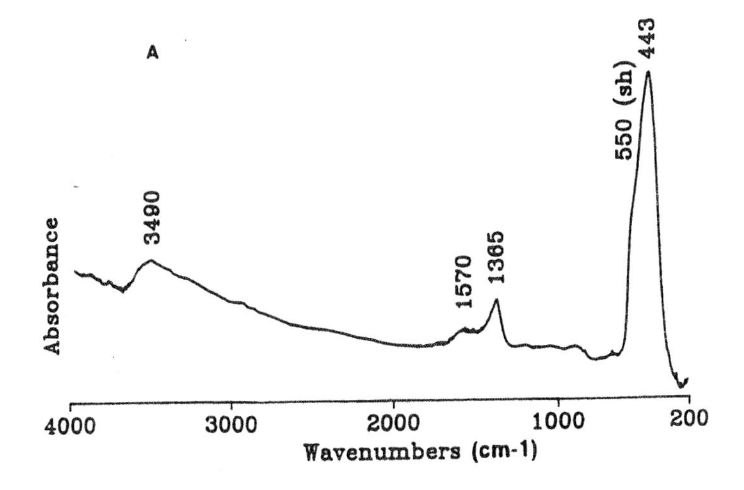

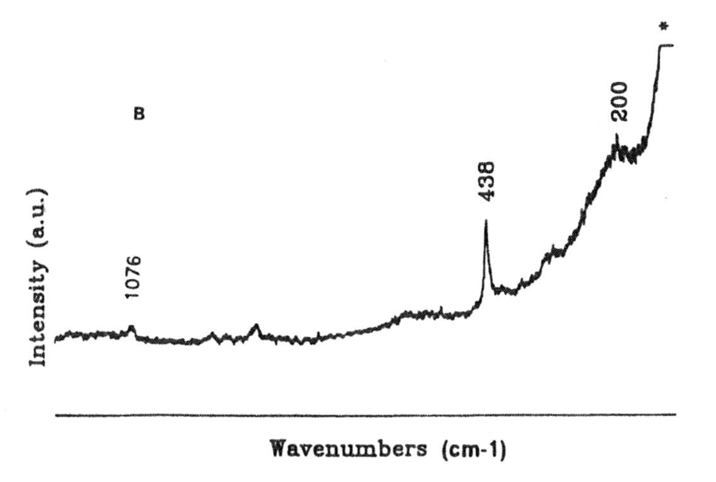

Figure 22. (**A**) FTIR and Raman (**B**) spectra of ZnO nanoparticles prepared by the LVCC method.

the electron levels become discrete and E_g increases: this increase correlates with decreasing particle size. The bandgap energy is calculated from the absorption coefficients in the onset region[99,114] and then the shift in bandgap from bulk ZnO is used to estimate particle size.[115]

The bandgap and size of ZnO particles produced by the LVCC technique depend on the O_2 partial pressure in the chamber. The bandgap wavelength changes from 370 to 380 nm as the O_2 partial pressure increases from 20 to 300 torr. From 300 to 800 torr O_2 partial pressure, the bandgap wavelength does not change. Figure 23a shows absorption spectra (in methanol) for nanoparticles prepared at two different O_2 pressures. These results indicate that the smallest ZnO particles produced by the LVCC method are made at low O_2 pressures.

One minute after zinc acetate is mixed with NaOH in the sol–gel preparation method, a bandgap absorption is seen with an onset at 328 nm as shown in Figure 23b. This shift of 52 nm from the bulk onset corresponds to particles of about 4 nm diameter. With time the particles grow until at 15 min after mixing, the growth is nearly complete. The bandgap wavelength shifts to 350 nm and the corresponding particle diameter is about 5.5 nm.

In the reversed micelle method, particle formation occurs in the nanoscale environment generated by surfactant zinc stearate molecules, water, and toluene. The particle size is dependent on the $[OH^-]/[Zn^{2+}]$ ratio. The bandgap wavelength shifts from 373 to 365 nm when the ratio increases from 10 to 40; the particle size decreases with increasing OH^- concentration. Figure 23c shows absorption spectra at three different OH^- concentrations. By comparing bandgap shifts, it appears that the stearate coated ZnO particles produced by this method are comparable in size to the ZnO particles made by the LVCC method, but are larger than those made in methanol by the sol–gel technique.

When the ZnO nanoparticles, produced by the sol–gel technique, are irradiated for 2 h with 266 nm laser radiation, a blue shift in the bandgap absorption occurs. This blue shift persists up to 1 day after illumination. It has been previously reported for ZnO and other colloids[106,107] and may be due to laser-induced dissolution of ZnO particles which results in smaller particles that absorb further to the blue.

Bulk ZnO shows two broad luminescence peaks: one, attributed to bandgap emission, is at 380 nm and the other is at 500 nm. The nanoparticle samples all show these two emissions when excited with 266 nm light, but the absolute and relative intensities and the positions of these two bands vary with the method of particle preparation. For the laser-prepared particles, the blue maximum at 384 nm is less intense than the green one at 515 nm. For the stearate-coated particles, the blue peak is again at 384 nm but is now relatively stronger than the green one which is red-shifted to 540 nm. The absolute intensity of the emission from stearate coated particles is about 10 times greater than that from the laser prepared ones. The freshly prepared colloidal ZnO sol is turbid and shows only green emission; the blue emission apparently is reabsorbed by the sample. After the sol ages for 4 h, weak blue emission appears; this emission then grows more intense and shifts to the blue. After aging for 24 h, the blue peak, now at 335 nm, is more intense than the green one.

Different explanations have been proposed for the green emission. Henglein and coworkers[106,107] have attributed it to surface anion vacancies, while Bahnemann and coworkers[105] have explained it by the tunneling of surface-bound electrons to preexisting, trapped holes. The trapped state explanation better fits our alkaline colloids and stearate-coated particles which are unlikely to have surface anion

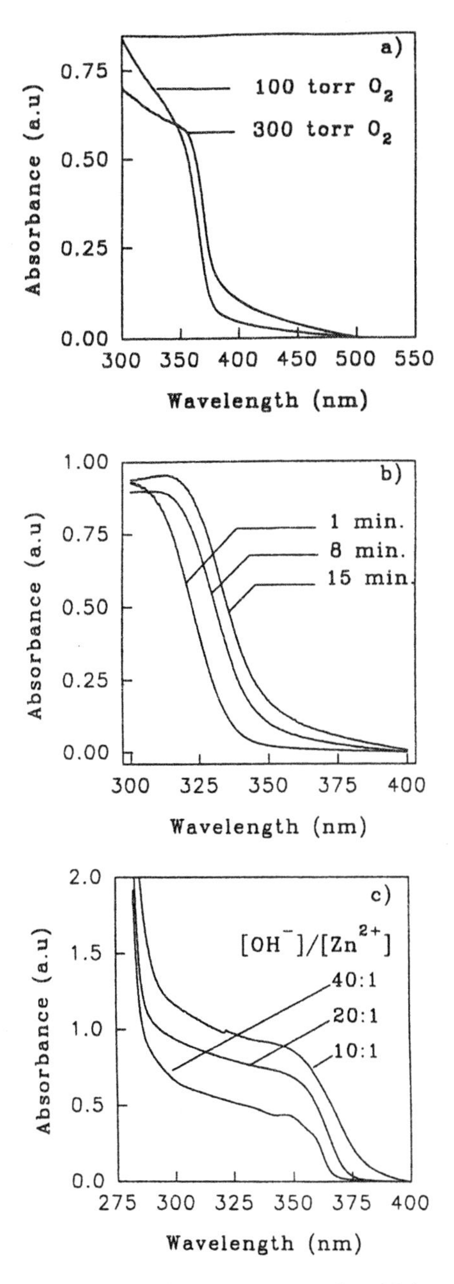

Figure 23. Absorption spectra of ZnO nanoparticles: (**a**) (in methanol) prepared by the LVCC method at two different O_2 pressures, (**b**) (in methanol) prepared by the sol–gel method and (**c**) (in toluene) prepared by the reversed micelle method.

vacancies. The intense green emission from the stearate-coated particles is similar to the strong red emission that arises when Si nanoparticles are passivated.[75] In both cases nonradiative surface relaxation may be reduced by surface modification.

The blue bandgap emission has a picosecond lifetime[105,106] which cannot be measured by our nanosecond lasers. The green emission has been shown to be multiexponential,[116] with at least two components shorter than our 10 ns laser pulses. Our wavelength selected time decays can be fit to a biexponential form where the fast component includes all components with lifetimes less than 10 ns and the slow component lifetime depends on emission wavelength. For our particles, the fast component amplitude is much greater (~103 times) than the slow component amplitude; the integrated intensities for the two components, however, are more comparable.

The slow component lifetime at an emission wavelength of 520 nm is 0.85 μs for the laser prepared ZnO particles and 1.42 μs for the stearate-coated particles. The longer lifetime for the coated particles may reflect the ability of the coating to reduce nonradiative relaxation. The slow component lifetime increases with emission wavelength. For the coated particles, this lifetime increases from 1.35 μs at 480 nm to 1.50 μs at 600 nm. Figure 24a shows a series of emission spectra from laser prepared particles; the spectra have different boxcar delay times. A similar series is shown in Figure 24b for stearate-coated particles. The increase in slow component lifetime with emission wavelength is shown by the red shift in the green maximum with increasing boxcar delay. This red shift, similar to that seen for CdS nanoparticles,[117] can be explained by the trapped electron-trapped hole model. In this model, larger particles with longer electron-hole distances, emit at longer wavelengths with slower rates. This also explains the 10 nm red shift that is observed when the excitation wavelength for laser-prepared particles is changed to 340 nm from 266 nm. The longer wavelength excitation preferentially excites larger particles which then preferentially emit to the red.

In summary, the ZnO nanoparticles produced by LVCC technique form a web-like agglomeration and, like the solution-produced particles, show quantum size effects and interesting photoluminescence properties. The particles produced by the sol–gel method are smallest, but all show bandgap shifts and size-dependent dispersed and time-resolved emission spectra. Of particular interest is the intense luminescence from the stearate-coated particles produced by the reversed micelle method; the surface coating may inhibit nonradiative relaxation processes.

D. Gallium Oxide Nanoparticles

The X-ray diffraction pattern of the gallium oxide nanoparticles prepared by the LVCC method shows several weak crystalline lines similar to the β-Ga_2O_3 pattern. The crystallinity of the as-deposited nanoparticles is not very high and probably some amorphous phase exists. The FTIR shows a strong absorption between 800 and 500 cm^{-1} with double peaks centered at 700 and 500 cm^{-1}. The IR of the

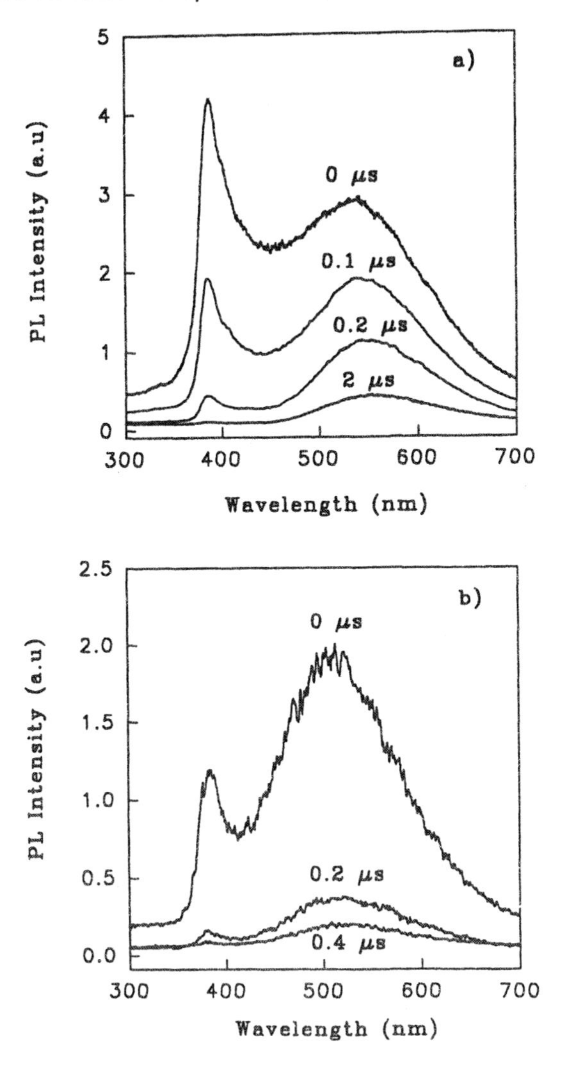

Figure 24. Dispersed emission of the ZnO nanoparticles excited by 266 nm for the nanoparticles prepared by (**a**) laser vaporization and (**b**) reversed micelle.

amorphous gallium oxide reported in the literature consists of a strong broad band centered at 550 cm^{-1} and a medium broad band at 305 cm^{-1}, while the polycrystalline β-Ga_2O_3 is characterized by two strong, broad bands at 430 and 680 cm^{-1}.[118] The FTIR of the nanoparticles, shown in Figure 25, is very close to that of the polycrystalline β-Ga_2O_3, thus confirming the existence of this phase in the nanoparticles.

The metallic gallium nanoparticles prepared in the absence of O_2 do not show IR absorption in the range of 800 to 400 cm^{-1}. However, after exposing the particles

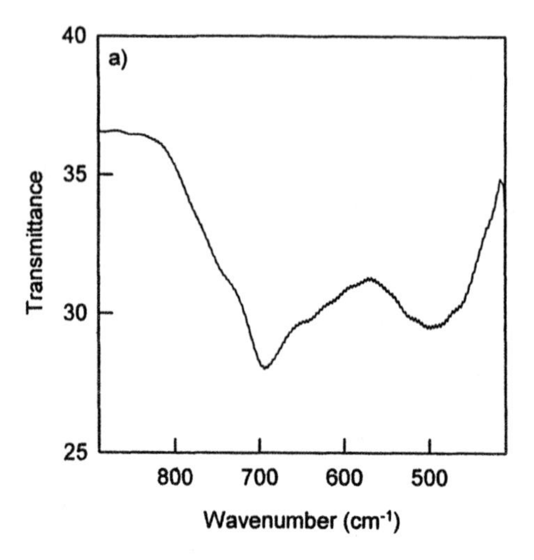

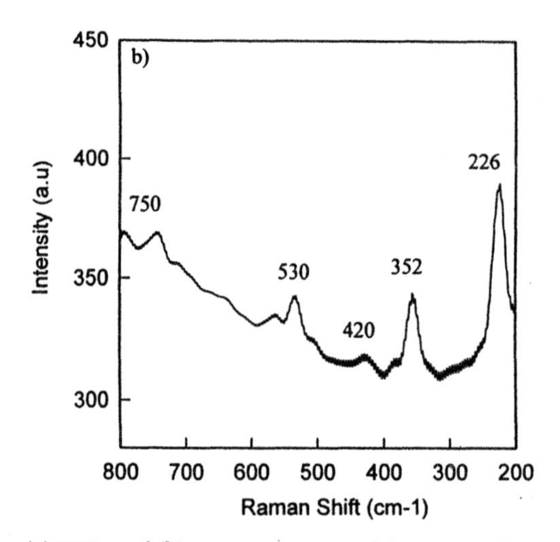

Figure 25. (**a**) FTIR and (**b**) Raman spectra of the Ga oxide nanoparticles.

to air for several days, the absorption peaks between 800 and 400 cm^{-1} appear. This is apparently due to the formation of an oxide layer on the surface of Ga nanoparticles. The oxide formation can be enhanced by annealing the particles at higher temperatures in air. For example, by annealing the Ga nanoparticles at 700 °C for 60 min, a complete conversion to the β-Ga_2O_3 phase can be achieved.

The UV-vis absorbance of the gallium oxide nanoparticles suspended in methanol shows a long, weak absorption tail in the visible region which increases sharply in the UV region. The absorption spectrum is similar to that of β-Ga_2O_3. Extrapolation of the absorption band gives a optical bandgap of 4.4 eV which is very close to that of the bulk β-Ga_2O_3.[119] This result indicates that the Ga oxide nanoparticles are probably too large for quantum confinement and therefore the bandgap energy is similar to the bulk value.

The dispersed PL spectra of the Ga oxide nanoparticles excited by the 266 nm pulsed laser lines are shown in Figure 26. The spectra differ by the position of the boxcar gate, which ranges from 0 delay with respect to the laser excitation pulse to 4 μs delay. As the boxcar delay increases, the PL intensity decreases and the emission maximum shifts to longer wavelengths. This is due to the lifetime lengthening that occurs as the emission wavelength increases. The time-resolved PL decay can be fit to a biexponential function with short- and long-lived components. The short and long lifetimes range from 0.2 to 1.2 μs and 4.9 to 10 μs, respectively, depending on the emission wavelength. Both the short and long lifetimes increase with emission wavelength.

A striking feature of Ga oxide nanoparticles is that they exhibit very efficient PL. The quantum yield measured, using a standard fluorescein dye as a reference, is about 0.24 at room temperature. Because of such high PL efficiency, it should be possible to use Ga oxide nanoparticles in light-emitting diodes (LEDs) and possibly in laser diodes.

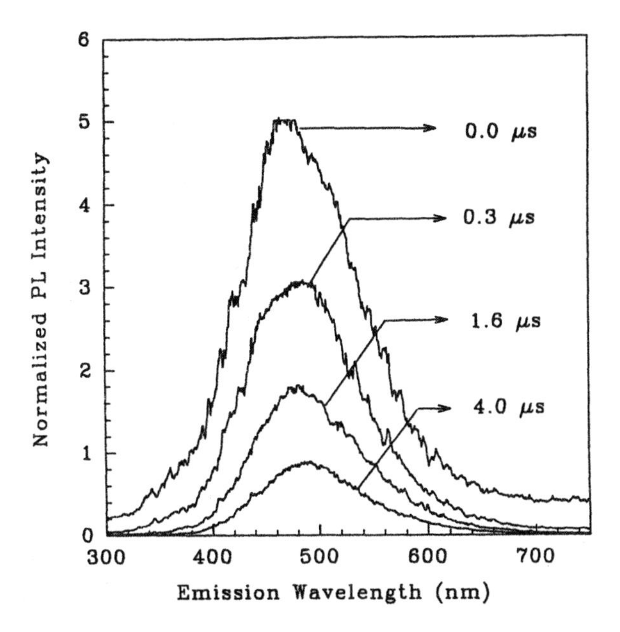

Figure 26. Dispersed emission of the Ga oxide nanoparticles obtained with 266 nm excitation at different boxcar gate delays from the excitation pulse.

With respect to the mechanism of PL from the Ga oxide nanoparticles, it appears the bright blue emission is related to the presence of the β-Ga_2O_3 phase in the nanoparticles. This is consistent with the reported blue PL from high-purity β-Ga_2O_3 upon excitation with 253.7 nm.[120] Also, it has been reported that this blue PL can be obtained from Ga_2O_3 after heating above 1100 °C in air where it is believed that β-Ga_2O_3 is formed at such a high temperature.[121] It is also known that the PL from β-Ga_2O_3 is strongly temperature-dependent.[122,123] At lower temperature, the emission is close to the bandgap of β-Ga_2O_3. The lifetime associated with this emission is very short (10 ns),[123] as compared to the μs lifetimes measured for our nanoparticles. The large energy migration in the PL spectra indicates that the electron or the hole relaxes to the self-trapped states at higher temperatures.

E. Magnetic Nanoparticles

Nanoparticles of iron and iron oxides have been prepared by the laser vaporization of iron in a He atmosphere containing variable concentrations of O_2. The color of the samples goes from very dark brown, almost black (the color of FeO and Fe_3O_4) to reddish brown (the color of γ-Fe_2O_3) with increasing oxygen pressure. SEM/TEM micrographs were used to study the particle morphology. TEM bright field images showed the samples to consist of small particles with a mean diameter of about 60 °A. The diffuse TEM electron diffraction patterns (shown in Figure 27)

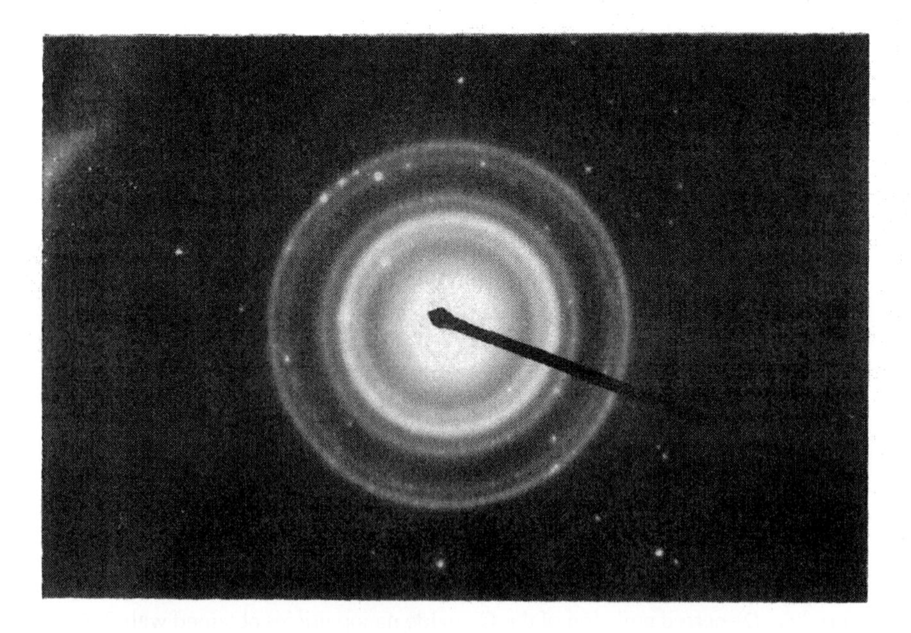

Figure 27. TEM electron diffraction from iron oxide nanoparticles.

indicate the presence of FeO, Fe_3O_4, and γ-Fe_2O_3 phases in all the samples. The diffraction patterns from different samples were much the same. In particular, no lines indicating α-Fe were observed in the Fe sample. This could arise from the Fe core being very small or having an amorphous structure and hence showing very broad lines.

The magnetic behavior of these particles was studied using DC- and AC-susceptibility measurements.[124] The results from samples prepared using different oxygen concentrations are shown in Figure 28. All samples exhibited superparamagnetism with blocking temperatures ranging from 50 K to above room temperature. Magnetic anisotropy constants were found to be 1 order of magnitude higher than the

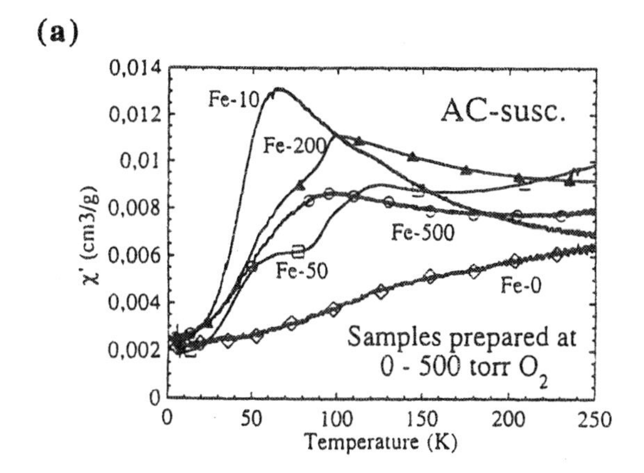

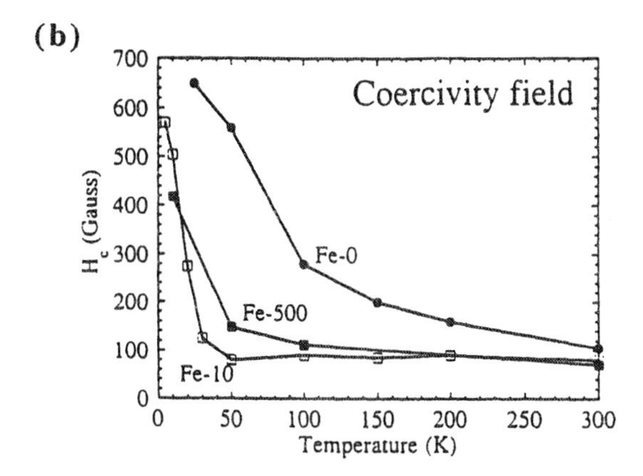

Figure 28. (**a**) In-phase AC susceptibility and (**b**) coercivity field measured for the iron oxide nanoparticles as functions of temperature.

known bulk values. The mean particle size estimated from the magnetic data was found to be in perfect agreement with the TEM data which showed a mean particle diameter of about 60 °A.

F. Nanoparticles of Other Metal Oxides

Several other metal oxide nanoparticles have been synthesized using the LVCC technique. This includes Bi_2O_3, PdO, NiO, AgO, MgO, SnO, In_2O_3, CuO, Al_2O_3, ZrO_2, MoO_3, V_2O_5, TeO_2, Sb_2O_3, SnO_2, and Cr_2O_3. It is important to note that by controlling the experimental parameters such as the partial pressure of oxygen, the partial pressure of the metal (by varying the vaporization laser power), and total pressure of the carrier gas (He or Ar) it is possible to control the stoichiometry of the oxides formed.[12,16] Several suboxides such as MoO_x ($x < 3$), V_2O_x ($x < 5$), and PdO_x ($x < 1$) were prepared and characterized using IR, Raman, and XPS measurements. Some of the IR and Raman characteristics of the oxide nanoparticles are presented below.

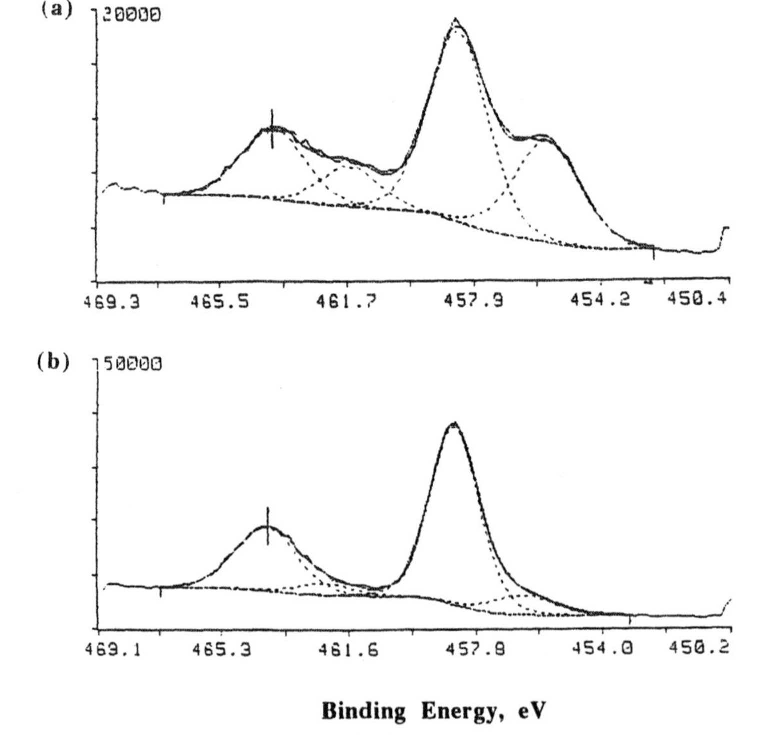

Figure 29. XPS data of titanium oxide nanoparticles synthesized in **(a)** Ar and **(b)** He. **(a)** Peaks appear at 455.7, 458.4, 461.6, and 463.9 eV, respectively. **(b)** Peaks appear at 456.4, 458.4, 462.3, and 464.0 eV, respectively.

Titanium Oxide

Figure 29 shows the XPS results for TiO_2 particles synthesized using O_2/Ar and O_2/He mixtures containing the same concentration of O_2 (20%). The XPS data show two small peaks at 456 and 462 eV which are due to the $2p_{3/2}$ and $2p_{1/2}$ core level of Ti metal.[12] As expected for oxide formation, these peaks are shifted to 458 and 464 eV, respectively in the TiO_2 particles. The TiO_2/Ti ratios in the samples synthesized in Ar and He are estimated as 65 and 94%, respectively. Therefore, it appears that more efficient oxidation takes place in He than in Ar and this may be attributed to inefficient cooling of the hot metal atoms in He which could lead to higher reaction probability with O_2. Furthermore, the diffusion coefficient of O_2 is greater in He than in Ar and this is related to the efficiency of the oxidation process.

Figure 30 displays representative Raman spectra of TiO_2 nanoparticles prepared using Ar or He as a carrier gas. As shown in Figure 30a, the particles synthesized in Ar exhibit the characteristic Raman peaks of the nanoscale anatase phase of TiO_2 (the low-temperature phase).[125] The Raman line at 144 cm^{-1} is the most intense line

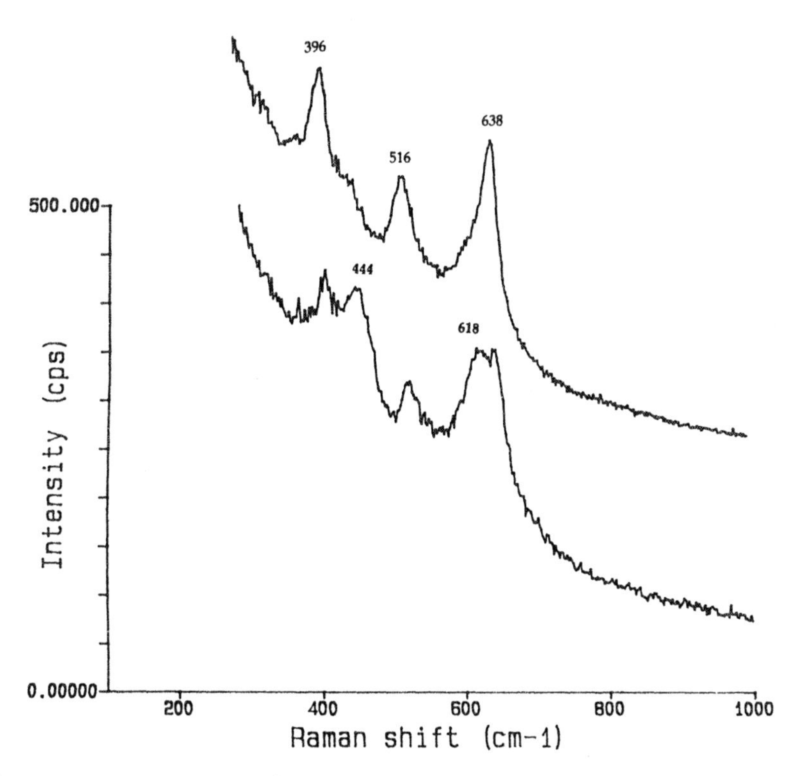

Figure 30. Raman spectra of TiO2 nanoparticles synthesized in O2/Ar (upper spectrum) and in O2/He (lower spectrum).

in the Raman spectrum of the anatase natural crystal[125] and this line is observed at 142 cm^{-1} in the spectrum of the nanoparticles shown in Figure 30a. Other Raman lines of the anatase nanoparticles are observed near 396, 516, and 638 cm^{-1} in good agreement with the corresponding lines reported for the bulk materials. For example, the major Raman lines observed for natural polycrystalline, for a single crystal and for synthetic powder of anatase TiO_2, are reported at 392, 510, and 633 cm^{-1}, 400, 515, and 640 cm^{-1}, and 393, 512, and 635 cm^{-1}, respectively.[125] Figure 30b compares the Raman spectra of the TiO_2 nanoparticles synthesized in He and in Ar and shows that the former particles exhibit Raman features characteristic of both

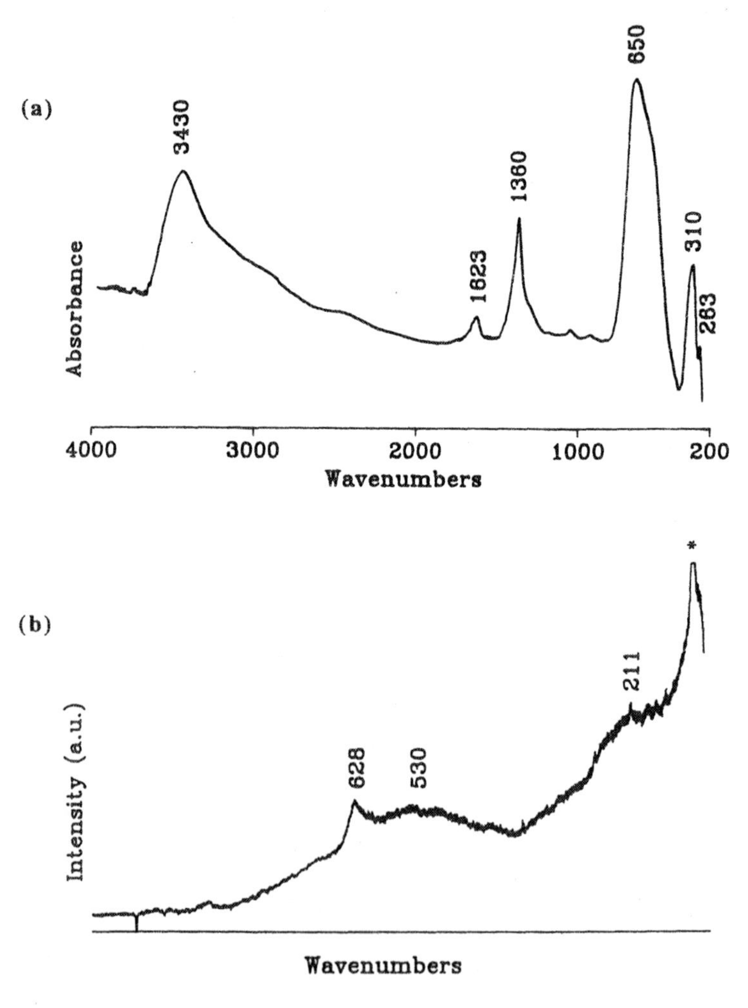

Figure 31. (**a**) FTIR and (**b**) Raman spectra of the tin oxide nanoparticles.

the anatase and the rutile phases, thus indicating a mixture of the two forms. Previous Raman measurements on nanoscale TiO_2 particles showed that both the rutile and anatase phases were present in the sample with the rutile phase being the most dominant.[126] It is well known that the rutile phase can be obtained by high-temperature annealing of the anatase particles.[125] This is consistent with the inefficient collisional cooling of He which may result in significant formation of the high-temperature rutile phase. However, it is remarkable that by using Ar as a carrier gas it is possible to exclusively synthesize the anatase form of TiO_2.

Tin Oxide

Compared with literature data,[127,128] the FTIR spectrum (Figure 31a) reveals an SnO_2 rutile structure rather than SnO. The band at 1360 cm^{-1} is assigned to the ν (C–O) stretching vibration originating from chemisorbed atmospheric CO_2 on the surface.[113] The Raman spectrum (Figure 31b) shows only one band at 628 cm^{-1} close to the value of the A1g vibrational mode of the single crystal at 632 cm^{-1}.[129] The other bands which should appear at 472 and 773 cm^{-1} are not visible. Broad features at ~ 530 and 221 cm^{-1} could indicate an intermediate oxidation state SnO_x ($x < 2$).[130] Another explanation can be found in surface effects associated with crystallites size or particles shape effects.[128] The XPS, displayed in Figure 32, shows two peaks corresponding to Sn (11%) and SnO_2 (89%) for the $3d_{5/2}$ photoelectrons which appear at 483.0 and 484.4 eV, respectively.

Vanadium Oxide

The most intense band in the FTIR spectrum (Figure 33a) is located at 992 cm^{-1}. In pure crystalline V_2O_5, this band appears at 1020 cm^{-1} due to the (V=O) stretching mode,[131] while the characteristic vibrations of the bridging oxygen in V–O–V are at 830, 450, and 200 cm^{-1}. The V–O–V asymmetric stretching

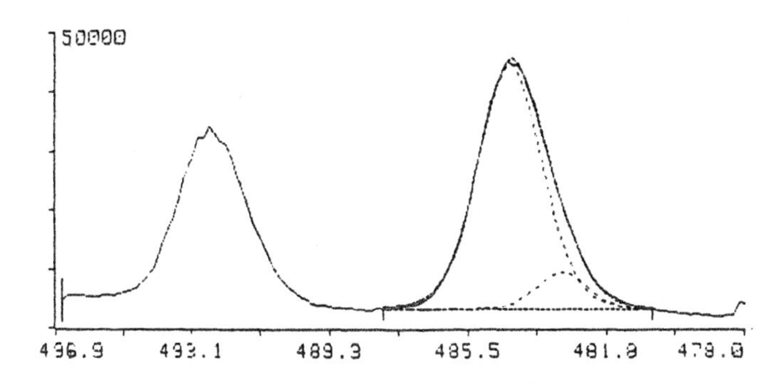

Figure 32. XPS spectrum of tin oxide nanoparticles.

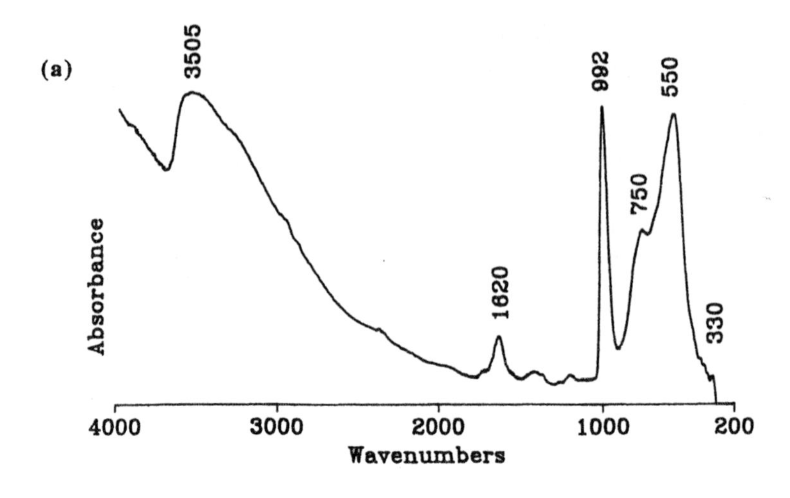

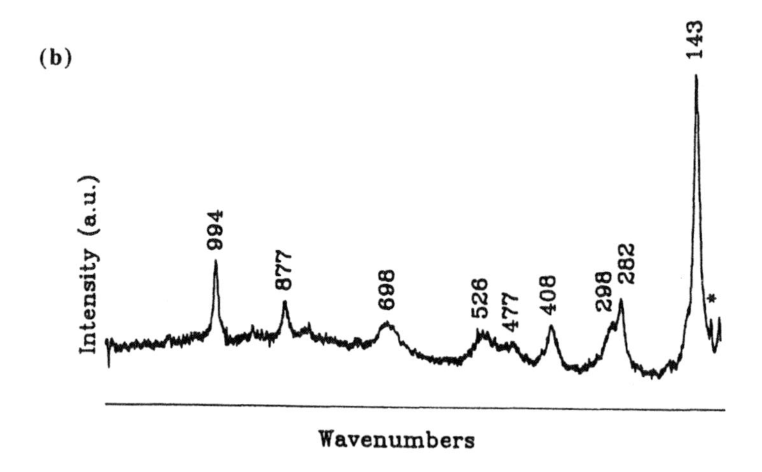

Figure 33. (a) FTIR and (b) Raman spectra of the vanadium oxide nanoparticles.

vibration absorbs at ~ 550 cm^{-1}, while the mode at ~ 350 cm^{-1} is the corresponding bending vibration.[131]

The spectrum of the vanadium oxide nanoparticles resembles the one of the amorphous compound because of the broad bands.[131] Moreover, the shift of the intense band from 1020 to 992 cm^{-1} may mean that either VO_4 vanadate species would be present in the oxide[131–133] or there would exist a V=O···V interlayer interaction.[134] However, the absorption maximum of this band depends also on the crystalline state of the oxide.[135]

Compared to the Raman spectrum of a V_2O_5 single crystal,[131] the spectrum of the nanoparticles is fairly good (Figure 33b). Obviously, the bands are broader, indicating that the order of the crystalline state is not perfect. The rather intense band at 877 cm^{-1} can be explained either by the presence of vanadate species or by another vanadium oxide such as V_2O_4.[136] However, the XPS, shown in Figure 34, exhibit two peaks for the $2p_{3/2}$ and $2p_{1/2}$ photoelectrons of the V_2O_5 which appear at 517.4 and 524.7 eV, respectively.

Bismuth Oxide

The FTIR spectrum (Figure 35a) shows three bands (554, 351, and 281 cm^{-1}) which could be assigned to the Bi–O vibrations.[137] But the strong and sharp bands at 1468, 1387, and 844 cm^{-1} are due to the carbonate ion.[113,127,138] The reactivity of the bismuth oxide surface toward atmospheric CO_2 leads to the formation of carbonate groups whose numbers increase with increasing the specific surface area of the oxide nanoparticles.

According to the Raman spectra of different bismuth oxides and the calculated frequencies for ideal bismuth oxide structures,[139] the nanoparticles may be compared to an α-Bi_2O_3 structure (Figure 35b). However, the band at ~450 cm^{-1}, which should exist in both a and b forms, does not appear in the spectrum. The XPS data, shown in Figure 36, show peaks corresponding to Bi (18%) and Bi_2O_3 (82%) for the $4f_{7/2}$ and $4f_{5/2}$ photoelectrons which appear at 162 eV (Bi-$4f_{5/2}$), 163.7 eV (Bi_2O_3-$4f_{5/2}$), 156.8 eV (Bi-$4f_{7/2}$), and 158.5 eV (Bi_2O_3-$4f_{7/2}$), respectively.

Tellurium Oxide

The FTIR spectrum (Figure 37a) of the nanoparticles is comprised of two main bands at 632 cm^{-1}, with a shoulder at 717 cm^{-1}, and at 370 cm^{-1} with a shoulder

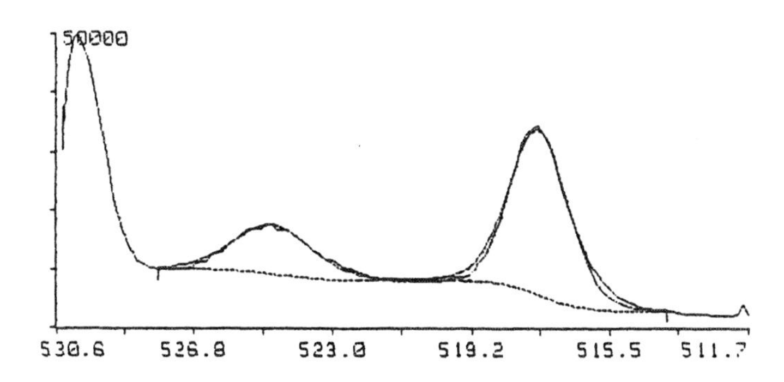

Figure 34. XPS spectrum of vanadium oxide nanoparticles.

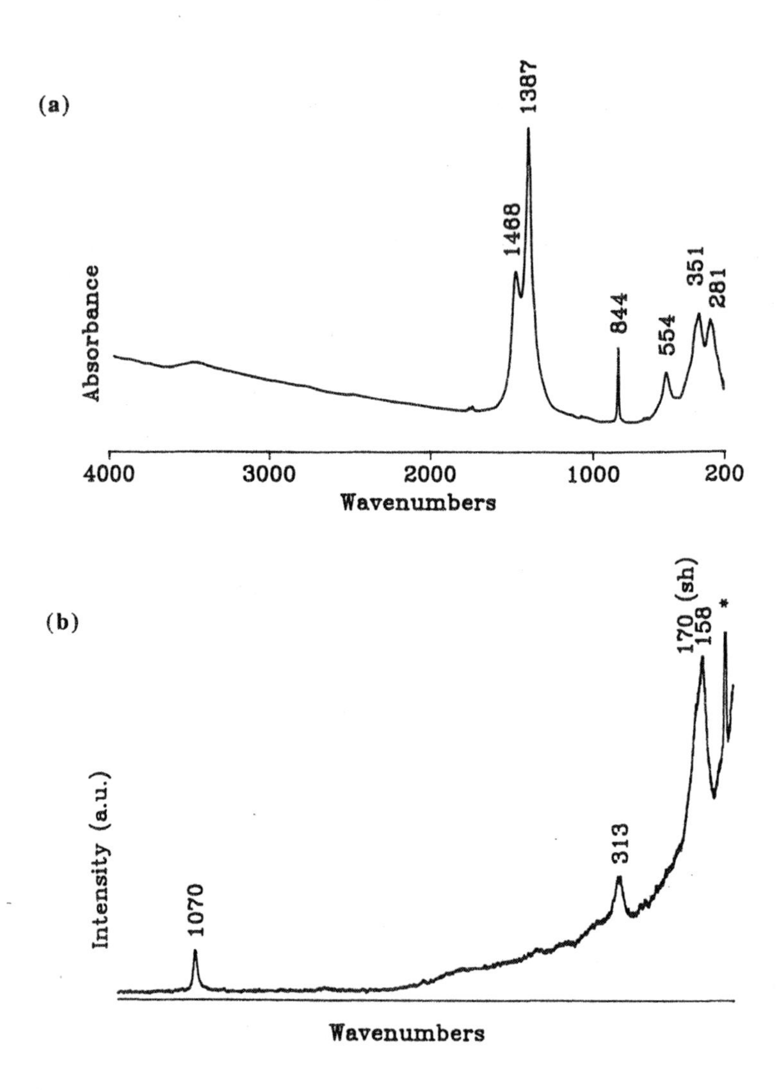

Figure 35. (**a**) FTIR and (**b**) Raman spectra of the bismuth oxide nanoparticles.

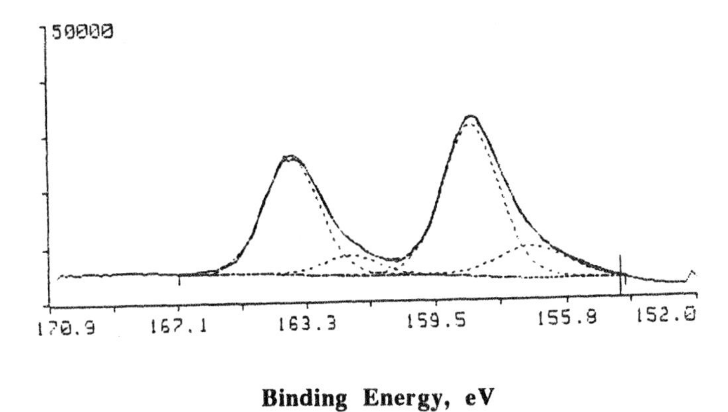

Figure 36. XPS spectrum of bismuth oxide nanoparticles.

at 492 cm^{-1}. It is quite comparable to the spectrum of the bulk TeO_2 reported in the literature.[127,137] Two strong and sharp Raman bands appear at low frequency, 120 and 139 cm^{-1} as shown in Figure 37b, which indicate a relatively well organized structure (Figure 37b). The XPS, shown in Figure 38, shows peaks corresponding to Te (14%) and TeO_2 (86%) for the $3d_{5/2}$ photoelectrons which appear at 572.5 and 574.8 eV, respectively.

G. Nanoparticles of Oxide Mixtures and Mixed Oxides

One of the significant advantages of the laser vaporization method is the ability to synthesize mixtures of nanoparticles of controlled composition or particles of mixed metals or metal oxides. This can be achieved by sequential or simultaneous vaporization of different metal targets. Using sequential vaporization we prepared several homogeneous mixtures of oxides such as CuO/ZnO, TiO_2/Al_2O_3, and TiO_2/Al_2O_3/ZnO. In these experiments, laser vaporization is alternated on the different metal targets of interest and by adjusting the number of laser shots on each target we were able to control the composition of the oxides mixture.[12,16]

The FTIR of the mixture of Cu/Zn oxide nanoparticles is shown in Figure 39. As mentioned in Section IV.C, the broad absorption band of zinc oxide is centered at 450 cm^{-1}. As for copper oxides, a broad band located at ~ 500 cm^{-1} for CuO and a narrower one at 620 cm^{-1} for Cu_2O have been reported in the literature.[127,137] In the mixed sample, it is hard to tell the difference between the pure ZnO spectrum (Figure 22a) and the Cu/Zn oxide one (Figure 39a). The bands at 1596 and 1360 cm^{-1} along with the group of weak bands around 1100 and 1030 cm^{-1} resemble the ones observed in malachite[127] and could be assigned to carbonate ions bonded to the basic surface sites Zn^{2+} and Cu^{2+} as well as to carbonates.[127]

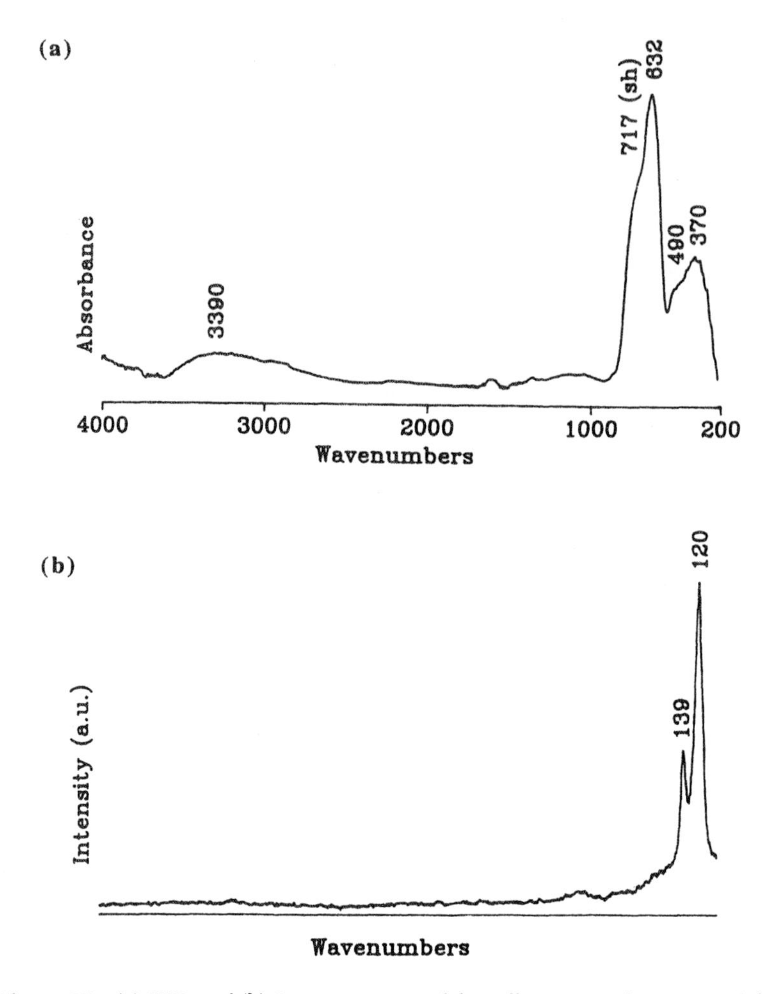

Figure 37. (**a**) FTIR and (**b**) Raman spectra of the tellurium oxide nanoparticles.

The broad band at 438 cm^{-1} which clearly appears in Zn oxide Raman spectrum becomes broader and shifts toward lower frequency (Figure 39b). Moreover, other bands at 580, 317, and 273 cm^{-1} indicate that this sample is no longer a pure zinc oxide.

The XPS data, shown in Figure 40, exhibits one peak in the Cu $2p_{3/2}$ region corresponding to CuO, and two peaks in the Zn $2p_{3/2}$ region corresponding to Zn (18%) and ZnO (82%). The separation of the latter two peaks is about 2 eV, while the separation between Zn and ZnO peaks reported in the literature is about 0.2 eV.

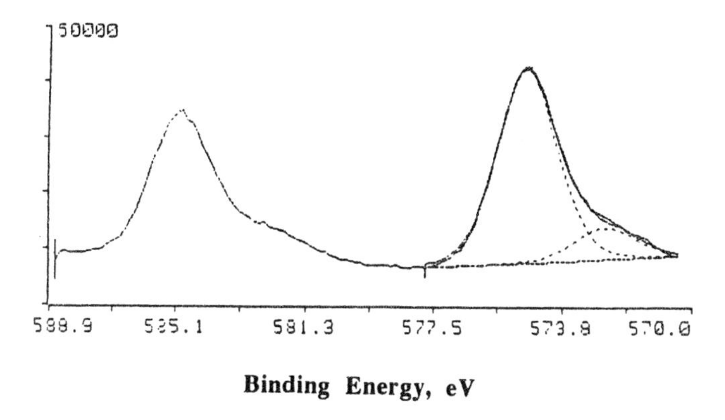

Figure 38. XPS data of tellurium oxide nanoparticles.

This difference may be due to the formation of the carbonate ions on the surface of the ZnO nanoparticles.

Mixed composite particles have been prepared by simultaneous vaporization of two or more metals. In this case beam splitters are used to focus the laser on the different targets. Figure 41a displays an SEM micrograph obtained for the mixed Si/Al oxide. The particles appear to have a uniform size and shape. Energy dispersed analysis (shown in Figure 41b) indicates that single particles contain both Si and Al with a composition that is identical throughout the sample. Other examples of composite particles prepared by this method include Si/B oxide, Mg/Fe oxide, and Cd/Te oxide.

H. Metal Carbides and Other Carbonaceous Particles

The synthesis of metal carbide nanoparticles has been demonstrated by carrying out laser vaporization in the presence of a convenient source of carbon such as methane, ethylene, or isobutene. Thus, silicon carbide nanoparticles were prepared by laser vaporization of Si in a He/isobutene mixture.[13] Surface analysis using FT-IR has indicated that the nanoparticles consist of SiC cores covered with pure silica layers.[13]

Figure 42 displays the Raman spectrum obtained for the nanoparticles synthesized by laser vaporization of Ti in He carrier gas containing 20% ethylene. The three Raman peaks shown in Figure 42 appear in close proximity to the Raman lines characteristic of TiO_2 (Figure 30). Since the force constants of the Ti–O and Ti–C bonds are close, one expects small Raman shifts between Ti–O and Ti–C bonds. Laser desorption mass spectrometry of the same sample shows peaks corresponding to $(TiC)_2^+$, $(TiC_2)_2^+$, and $(TiC_2)_4^+$. Complete characterizations of this and other nanoparticles are currently under active investigation in our laboratory.

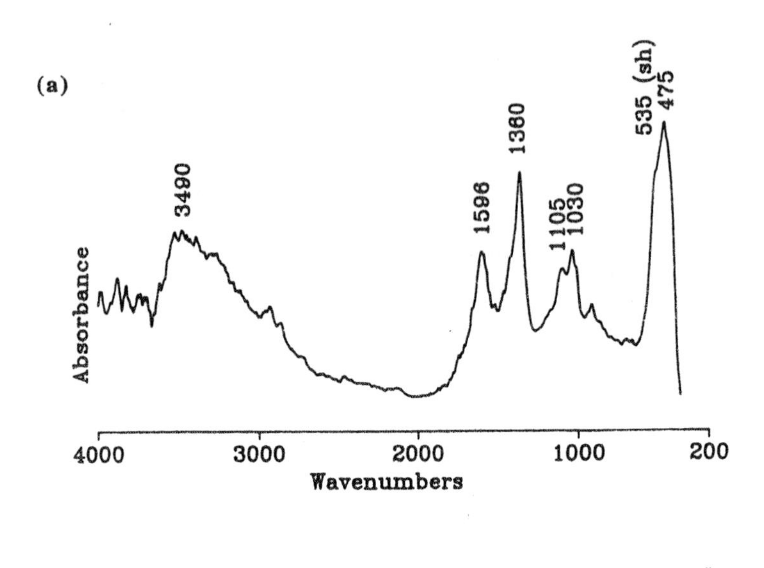

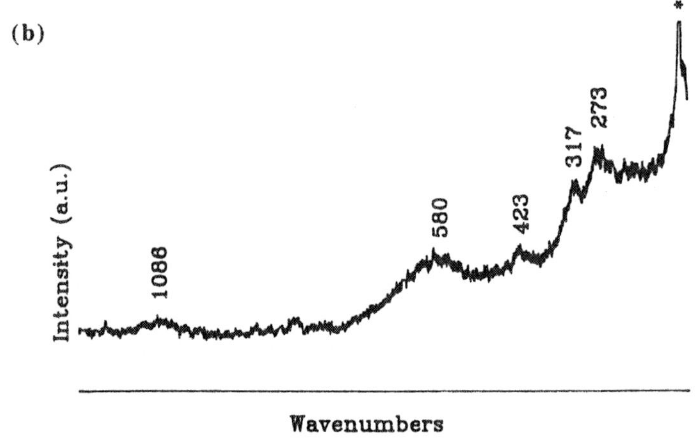

Figure 39. (**a**) FTIR and (**b**) Raman spectra of the Cu/Zn oxide nanoparticles.

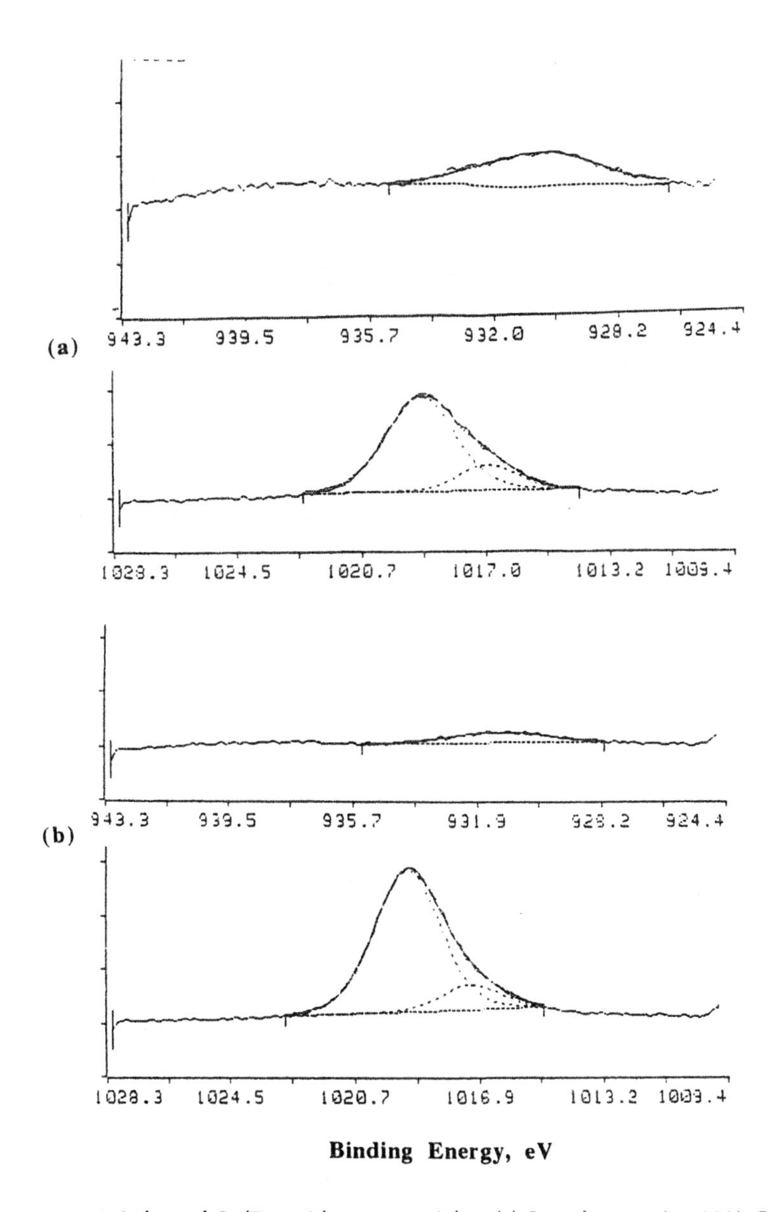

Figure 40. XPS data of Cu/Zn oxide nanoparticles. (**a**) Sample contains 23% CuO, peaks appear at 930.6 eV (CuO-$2p_{3/2}$), 1016.9 eV (Zn-$2p_{3/2}$), and 1019.0 eV. (**b**) Sample contains 16% CuO, peaks appear at 931.3 eV (CuO-$2p_{3/2}$), 1017.2 eV (Zn-$2p_{3/2}$), and 1019.2 eV.

(a)

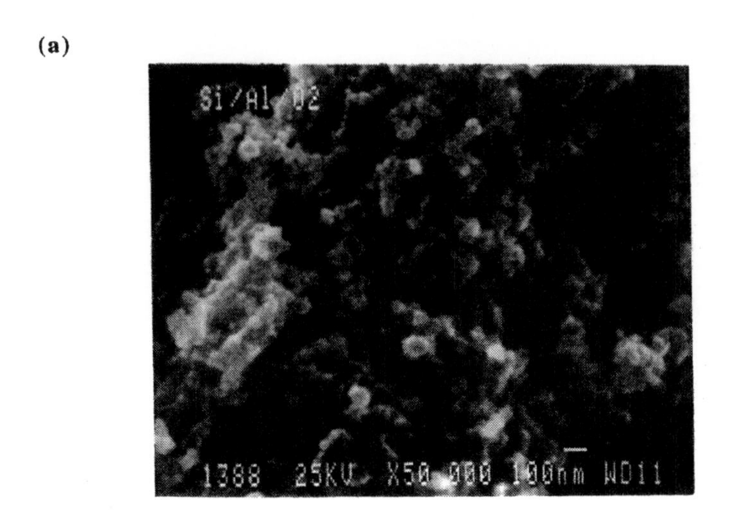

(b)

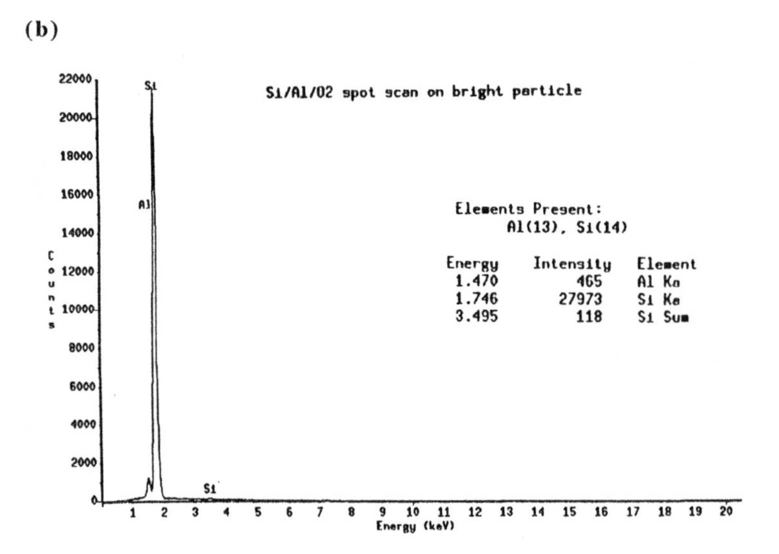

Figure 41. (**a**) SEM of Al/Si oxide and (**b**) DES of Al/Si oxide.

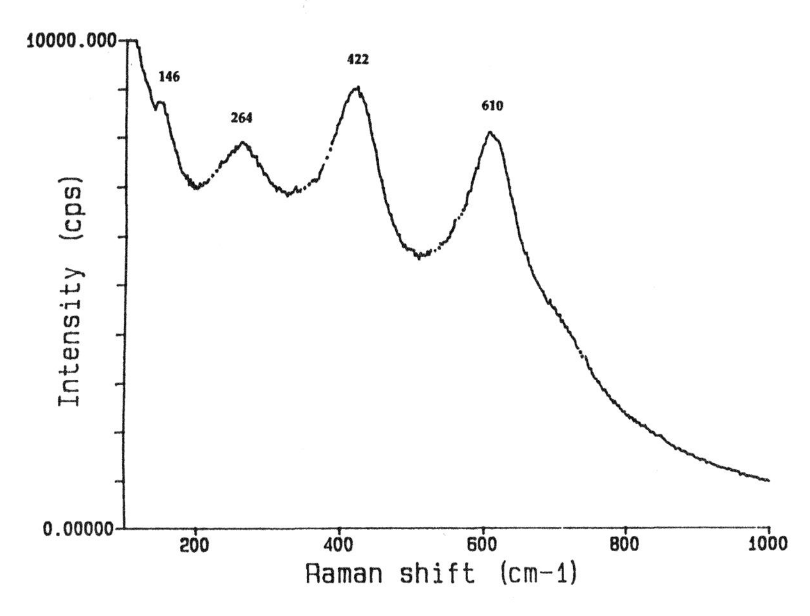

Figure 42. Raman spectrum of the nanoparticles synthesized by laser vaporization of Ti in 20% ethylene/He mixture.

When Mo was vaporized in the presence of a He/isobutene mixture, the mass spectra of the nanoparticles showed peaks corresponding to molybdenum carbide clusters of the stoichiometry $(MoC_4)_n$ with $n = 1$–4 as shown in Figure 43. These clusters have also been synthesized by laser irradiation of $Mo(CO)_6$ as recently reported by Jin et al.[140] Interestingly, in our laser vaporization experiment, the mass spectra of the particles also revealed abundant ions corresponding to higher fullerenes (C_{80}–C_{150}) with small peaks corresponding to C_{60} and C_{70}.[141] This suggests some catalytic effect for the Mo nanoparticles in the formation of higher fullerenes.

I. Heat and Solvent Effects on the Morphology of the Nanoparticles

It is important to study the effects of composition (chemical composition and oxide or carbide/carbon particle ratios), and conditions (pressure, carrier gas flow rate, temperature) on the nanoparticles' size, on the geometry of the porous aggregates, and the degree of porosity, i.e. particle/pore volume ratios.

The thermal stability of the porous structures can be investigated by heating the porous aggregates deposited on inert supports in vacuum between 400 and 1000 K and examining by SEM the possible thermal coagulation of the porous structures, as well as thermal composition changes in mixed particles. We have performed some preliminary experiments on FeO nanoparticles prepared by the LVCC method and the results from TEM and SEM are shown in Figure 44. The primary particle size of the as-deposited sample was 10–20 nm as shown in Figure 44a. After heating the sample to 600 °C in vacuum for 40 min, coagulation to produce larger particles was clearly evident as shown in Figure 44b. Continuous heating at 600 °C for 2 h produced even larger elongated particles as shown in Figure 44c. Measuring the rate at which such changes occur over a range of temperatures will yield the activation energies for the coagulation process. This information is important in relation to the stability of porous aggregates, for example, in a range of interstellar environments.

The stability of porous structures can also be affected by the condensation and subsequent evaporation of organic solvents and water onto the aggregates. We performed some preliminary experiments in this direction. We observed that even prolonged exposure to atmospheric humidity did not affect the porous structures.

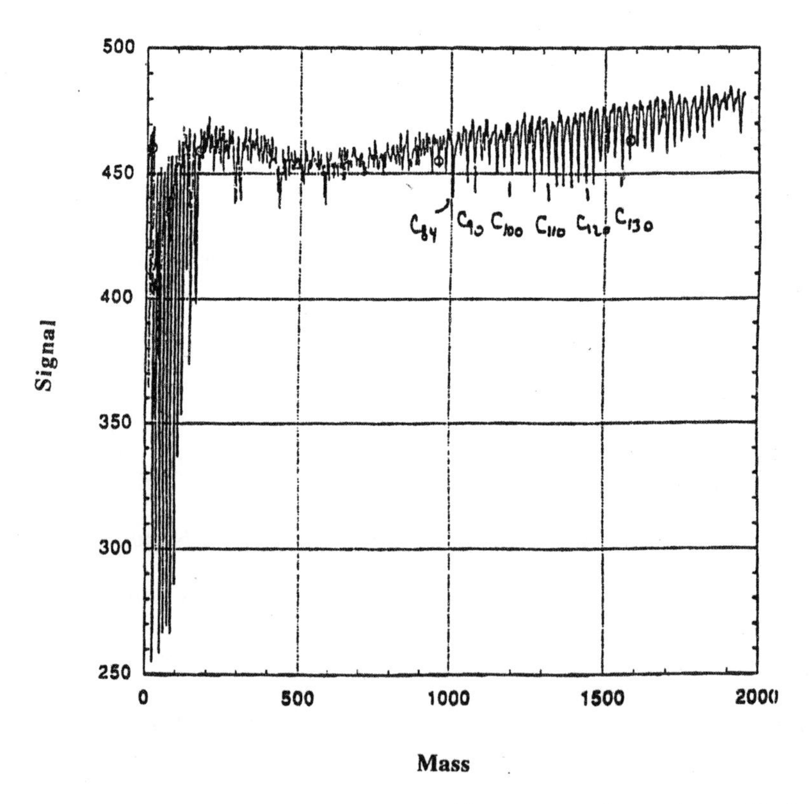

Figure 43. Laser desorption mass spectrum of the nanoparticles synthesized by laser vaporization of Mo in isobutene/Ar mixture. The mass spectrum shows the formation of Mo_nC_{4n} (n = 1–4) and C_n (n = 84–160).

However, exposure to liquid water caused the aggregates to coalesce, whereas an organic solvent such as hexane left the particles in a more dense fibrous form as compared to the original morphology. Nonaqueous polar solvents such as methanol and acetonitrile caused the particles to agglomerate into smaller masses which still retain some of the open structure of the as-deposited morphology. Examples of these effects are shown in Figure 45.

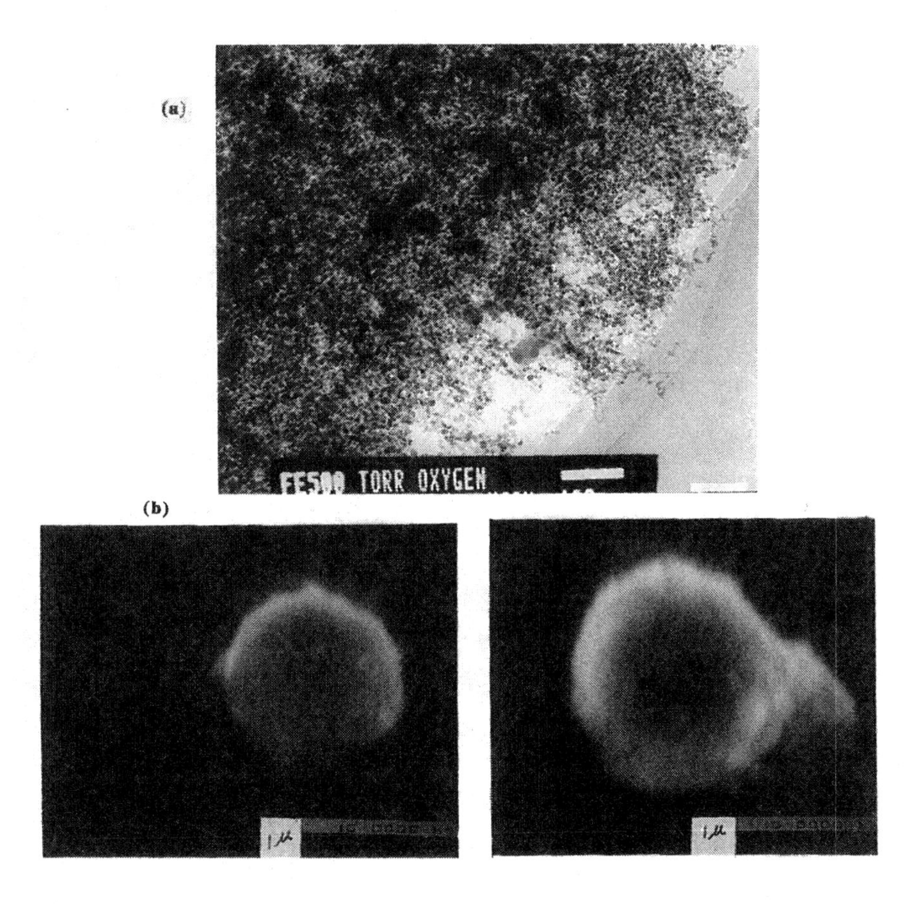

Figure 44. Temperature effect on the morphology of Fe oxide nanoparticles, (**a**) no heating, (**b**) heating at 600 °C for 40 min, and (**c**) heating at 600 °C for 120 min.

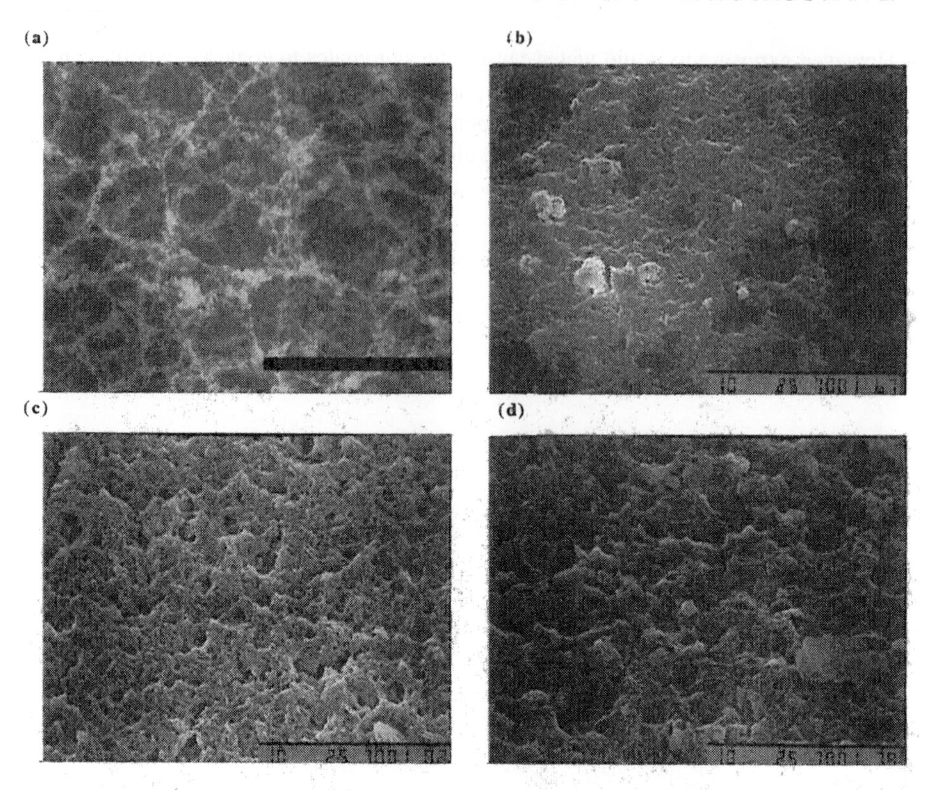

Figure 45. SEM micrographs showing the morphology of silica nanoparticles (**a**) as-deposited, and after exposure to (**b**) water, (**c**) hexane, and (**d**) methanol.

V. CONCLUSIONS

It has been shown that the LVCC method can be used to produce a wide variety of nanoparticles and porous aggregates where the growth of the particles, state of aggregation, and composition of the composite can be controlled. The method can also be used to produce simulated interstellar grains for astrophysical studies.

ACKNOWLEDGMENTS

The authors gratefully acknowledge financial support from the NASA Microgravity Materials Science Program (Grant NAG8-1276). We thank Dr. M .I. Baraton (LMCTS - CNRS, Limoges, France) for the FTIR and Raman measurements and analyses and Dr. Ismat Shah (DuPont, Wilmington, Delaware) for the XPS measurements.

REFERENCES

1. *Nanomaterials: Synthesis, Properties and Applications*; Edelstein, A. S.; Cammarata, R. C., Eds.; Institute of Physics Publishing: Bristol and Philadelphia, 1996.
2. *Nanotechnology: Molecularly Designed Materials*; Chow, G. M.; Gonsalves, K. E., Eds.; ACS Symposium Series 622: Washington DC, 1996.
3. *Nanophase Materials: Synthesis, Properties, Applications*; Hadijipanyis, G. C.; Siegel, R. W., Eds.; Kluwer Academic Publications: London, 1994.
4. Siegel, R. W. *Nanostructured Materials* **1994**, *4*, 121; Siegel, R. W. *Annu. Rev. Mater. Sci.* **1991**, *21*, 559.
5. Steigerwald, M. L.; Brus, L. E. *Acc. Chem. Res.* **1990**, *23*, 183.
6. Ozin, G. A. *Adv. Materials* **1992**, *4*, 612.
7. Kamat, P. V. *Chem. Rev.* **1993**, *93*, 267.
8. Henglein, A. *J. Phys. Chem.* **1993**, *97*, 5457; Henglein, A. *Chem. Rev.* **1989**, *89*, 1861.
9. Gleiter, H. *Adv. Materials* **1992**, *4*, 474.
10. Hoffmann, M. R.; Martin, S. T.; Choi, W.; Bahnemann, D. W. *Chem. Rev.* **1995**, *95*, 69.
11. Alivisatos, A. P. *J. Phys. Chem.* **1996**, *100*, 13226.
12. El-Shall, M. S.; Slack, W.; Vann, W.; Kane, D.; Hanley, D. *J. Phys. Chem.* **1994**, *98*, 3067; El-Shall, M. S.; Slack, W.; Hanley, D.; Kane, D. In *Molecularly Designed Ultrafine/Nanostructured Materials*; Gonsalves, K. E.; Chow, G.; Xiao, T. D.; Cammarata, R. C., Eds.; *Mater. Res. Soc. Proc.* **1994**, *351*, 369 (Pittsburgh, PA).
13. El-Shall, M. S.; Graiver, D.; Pernisz, U.; Baraton, M. I. *Nanostructured Materials* **1995**, *6*, 297; Baraton, M. I.; El-Shall, M. S. *Nanostructured Materials* **1995**, *6*, 301.
14. El-Shall, M. S.; Li, S.; Turkki, T.; Graiver, D.; Pernisz, U. C.; Baraton, M. E. *J. Phys. Chem.* **1995**, *99*, 17805.
15. El-Shall, M. S.; Graiver, D.; Pernisz, U. C. "Silica Nanoparticles" U.S. patent # 5580655. Issued on 12/03/1996.
16. El-Shall, M. S.; Li, S.; Graiver, D.; Pernisz, U. C. In *Nanotechnology: Molecularly Designed Materials*; Chow, G. M.; Gonsalves, K. E., Eds.; ACS Symposium Series 622: Washington DC, 1996, Chap. 5, pp. 79–99.
17. Li, S.; Silvers, S. J.; El-Shall, M. S. *J. Phys. Chem.* **1997**, *101*, 1794.
18. Abraham, F. F. *Homogeneous Nucleation Theory*; Academic Press: New York, London, 1974.
19. *Nucleation*; Zettlemayer, A. C., Ed.; Marcel Dekker: New York, 1969.
20. Frenkel, J. *Kinetic Theory of Liquids*; Dover Publications: New York, 1955, Chap. 7.
21. *Nucleation and Atmospheric Aerosols*; Fukuta, N.; Wagner, P. E., Eds.; A. DEEPAK Publishing: 1992.
22. Reiss, H. In *Advances in Chemical Reaction Dynamics*; Rentzepis, P. M.; Capellos, C., Eds.; NATO ASI series, 1985, Vol. 184.
23. *Atmospheric Aerosols and Nucleation*; Wagner, P. E.; Vali, G., Eds.; Lecture Notes in Physics; Springer: Berlin, 1988, Vol. 309.
24. Weast, R. C. *Handbook of Chemistry and Physics*; Chemical Rubber Co.: Boca Raton, FL, 1973.
25. El-Shall, M. S. *J. Chem. Phys.* **1989**, *90*, 6533; Wright, D.; Caldwell, R.; Moxley, C.; El-Shall, M. S. *J. Chem. Phys.* **1993**, *98*, 3356.
26. Reiss, H.; Marvin, D. C.; Heist, R. H. *J. Colloid Interface Sci.* **1977**, *58*, 125; McGraw, R.; Reiss, H. *J. Colloid Interface Sci.* **1979**, *72*, 172; Reiss, H. *Science* **1987**, *238*, 1368.
27. Kane, D.; Daly, G. M.; El-Shall, M. S. *J. Phys. Chem.* **1995**, *99*, 7867; Kane, D.; El-Shall, M. S. *Chem. Phys. Letters* **1996**, *259*, 482.
28. Reiss, H. *J. Chem. Phys.* **1950**, *18*, 840; Wilemski, G. *J. Chem. Phys.* **1975**, *62*, 3763; Mirabel, P.; Clavelin, J. L. *J. Chem. Phys.* **1978**, *68*, 5020; Jaecker-Voirol, A.; Mirabel, P.; Reiss, H. *J. Chem. Phys.* **1987**, *87*, 4849.
29. Ruth, V.; Hirth, J. P.; Pound, G. M. *J. Chem. Phys.* **1988**, *88*, 7079.

30. Pound, G. M. *J. Phys. Chem. Ref. Data* **1972,** *1*, 119.
31. Langer, J. S.; Turski, L. A. *Phys. Rev.* **1973,** *A8*, 3230.
32. Lothe, J.; Pound, G. M. *J. Chem. Phys.* **1968,** *48*, 1849.
33. Reiss, H.; Katz, J. L.; Cohen, E. R. *J. Chem. Phys.* **1968,** *48*, 5553.
34. Anderson, J. B. In *Molecular Beams and Low Density Gas Dynamics*; Wagner, P. P., Ed.; Dekker: New York, 1974, pp. 1–91; Hagena, O. *ibid*, p. 93.
35. Miller, D. R. In *Atomic and Molecular Beam Methods*; Scoles, G., Ed.; Oxford University Press: Oxford, 1988, pp. 14–53; Gentry, W. R. *ibid*, pp. 55–82.
36. Mandich, M. L.; Reents, W. D., Jr. In *Atomic and Molecular Clusters*; Bernstein, E. R., Ed.; Elsevier: Amsterdam, 1990, pp. 69–357.
37. Uyeda, R. *J. Cryst. Growth* **1974,** *24*, 69.
38. Hahn, H.; Averback, R. S. *J. Appl. Phys.* **1990,** *67*, 1113.
39. Mandich, M. L.; Bondybey, V. E.; Reents, W. D. *J. Chem. Phys.* **1987,** *86*, 4245.
40. Haggerty, J. S. In *Laser-Induced Chemical Processes*; Steinfeld, J. I., Ed.; Plenum: New York, 1990; Fantoni, R.; Borsella, E.; Piccirillo, S.; Enzo, S. *Proc. SPIE* **1990,** *1279*, 77; Curcio, F.; Ghiglione, G.; Musci, M.; Nannetti, C. *Appl. Surf. Sci.* **1989,** *36*, 52; Rice, G. W.; Woodin, R. L. *J. Am. Ceram. Soc.* **1988,** *71*, C181.
41. Tam, A.; Moe, G. G.; Happer, W. *Phys. Rev. Lett.* **1975,** *35*, 1630; Kizaki, Y.; Kandori, T.; Fujiyama, Y. *J. Appl. Phys.* **1985,** *24*, 800; Yabuzaki, T.; Sato, T.; Ogawa, T. *J. Chem. Phys.* **1980,** *73*, 2780; Rice, G. W. *J. Am. Ceram. Soc.* **1986,** *69*, C-183.
42. Duncan, M. A.; Dietz, T. G.; Smalley, R. E. *J. Am . Chem. Soc.* **1981,** *103*, 5245; Beck, S. M. *J. Chem. Phys.* **1989,** *90*, 6306; Harano, A.; Kinoshita, J.; Koda, S. *Chem. Phys. Lett.* **1990,** *172*, 219.
43. Chaiken, J.; Casey, M. J.; Villarica, M. *J. Phys. Chem.* **1991,** *96*, 3183; Zafiropulos, V.; Kollia, Z.; Fotakis, C.; Stockade, J. A. D. *J. Chem. Phys.* **1993,** *98*, 3067; Matsuzaki, A.; Horita, H.; Hamada, Y. *Chem. Phys. Lett.* **1992,** *190*, 331.
44. El-Shall, M. S.; Edelstein, A. S. In *Nanomaterials: Synthesis, Properties and Applications*; Edelstein, A. S.; Cammarata, R. C., Eds.; Institute of Physics Publishing: Bristol, Philadelphia, 1996, Chap. 2.
45. Dietz, T.; Duncan, M.; Liverman, M.; Smalley, R. *J. Chem. Phys.* **1980,** *73*, 4816; Dietz, T.; Duncan, M.; Powers, D.; Smalley, R. *J. Chem. Phys.* **1981,** *74*, 6511.
46. Smalley, R. E. *Laser Chem.* **1983,** *2*, 167; Powers, D. E.; Hansen, S. G.; Geusic, M. E.; Puiu, A. C.; Hopkins, J. B.; Dietz, T. G.; Duncan, M. A.; Langridge-Smith, P. R. R.; Smalley, R. E. *J. Phys. Chem.* **1982,** *86*, 2556.
47. Powers, D. E.; Hansen, S. G.; Geusic, M. E.; Michalopolous, D. L.; Smalley, R. E. *J. Chem. Phys.* **1983,** *78*, 2866.
48. Heath, J. R.; Liu, Y.; O'Brien, S. C.; Zhang, Q-L.; Curl, R. F.; Tittel, F. K.; Smalley, R. E. *J. Chem. Phys.* **1985,** *83*, 5520.
49. Granqvist, C. G.; Buhrman, R. A. *J. Appl. Phys.* **1976,** *47*, 2200; Ichinose, N.; Ozaki, Y.; Kashu, S. *Superfine Particle Technology* (Engl. Translation); Springer: London, 1992.
50. Gleiter, H. *Prog. Mater. Sci.* **1989,** *33*, 223.
51. Birringer, R.; Herr, U.; Gleiter, H. *Trans. Japan Inst. Met. Suppl.* **1986,** *27*, 43.
52. Kane, D.; El-Shall, M. S. *J. Chem. Phys.* **1996,** *105*, 7617; El-Shall, M. S.; Rabeony, H. M.; Reiss, H. *J. Chem. Phys.* **1989,** *91*, 7925.
53. Katz, J. L. *J. Chem. Phys.* **1970,** *52*, 4733; Katz, J. L.; Mirabel, P.; Scoppa, C. J.; Virkler, T. L. *J. Chem. Phys.* **1976,** *65*, 392; Fisk, J. A.; Chakarov, V. M.; Katz, J. L. *J. Chem. Phys.* **1996,** *104*, 8657.
54. Iler, R. K. *The Chemistry of Silica, Solubility, Polymerization, Colloid and Surface Properties, and Biochemistry*; Wiley-Interscience: New York, 1979, Chapter I.
55. Griscom, D. L. *J. Non-Cryst. Solids* **1985,** *73*, 51.
56. Stathis, J. H.; Kastner, M. A. *Phys. Rev. B.* **1987,** *35*, 29727.
57. Skuja, L. N.; Streletsky, A. N.; Pakovich, A. B. *Solid State Commun.* **1984,** *50*, 1069.

58. Sato, S.; Ono, H.; Nozaki, S.; Morisaki, H. *Nanostructured Materials* **1995**, *5*, 589; Nozaki, S.; Sato, S.; Ono, H.; Morisaki, H. *Mat. Res. Soc. Symp. Proc.* **1997**, *452*, 159.
59. Canham, L. T. *Appl. Phys. Lett.* **1990**, *57*, 1046.
60. Cullis, A. G.; Canham, L. T. *Nature* **1991**, *353*, 335.
61. Light Emission from Silicon; Iyer, S. S.; Collins, R. T.; Canham, L. T., Eds.; *Mater. Res. Soc. Proc.* **1992**, *256* (Pittsburgh, PA).
62. Microcrystalline Semiconductors: Materials Science and Devices; Fauchet, P. M.; Tsai, C. C.; Canham, L. T.; Shimizu, I.; Aoyahi, Y., Eds.; *Mater. Res. Soc. Proc.* **1992**, *283* (Pittsburgh, PA).
63. Sui, S.; Leong, P. P.; Herman, I. P.; Higashi, G. S.; Temkin, H. *Appl. Phys. Lett.* **1992**, *60*, 2086.
64. Cullis, A. G.; Canham, L. T.; Williams, G. M.; Smith, P. W.; Dosser, O. D. *J. Appl. Phys.* **1994**, *75*, 493.
65. Canham, L. T.; Cullis, A. G.; Pickering, C.; Dosser, O. D.; Cox, T. I.; Lynch, T. P. *Nature* **1994**, *368*, 133.
66. Chin, R. P.; Shen, Y. R.; Petrova-Koch, V. *Science* **1995**, *270*, 776.
67. Brus, L. E. *Nature* **1991**, *353*, 301.
68. Littau, K. A.; Szajowski, P. J.; Muller, A. J.; Kortan, A. R.; Brus, L. E. *J. Phys. Chem.* **1993**, *97*, 1224.
69. Takagi, H.; Ogawa, H.; Yamazaki, Y.; Ishizaki, A.; Nakagiri, T. *Appl. Phys. Lett.* **1992**, *256*, 117.
70. Wilson, W. L.; Szajowski, P. F.; Brus, L. E. *Science* **1993**, *262*, 1242.
71. Kanemitsu, Y.; Ogawa, T.; Shiraishi, K.; Takeda, K. *Phys. Rev. B* **1993**, *48*, 4883.
72. Kanemitsu, Y. *Phys. Rev. B* **1993**, *48*, 12357.
73. Brus, L. *J. Phys. Chem.* **1994**, *98*, 3575.
74. Kanemitsu, Y. *Phys. Rev. B* **1994**, *49*, 16845.
75. Brus, L. J.; Szajowski, P. F.; Wilson, W. L.; Harris, T. D.; Schuppler, S.; Citrin, P. H. *J. Am. Chem. Soc.* **1995**, *117*, 2915.
76. Fojtik, A.; Weller, H.; Fiechter, S.; Henglein, A. *Chem. Phys. Lett.* **1987**, *134*, 477.
77. Heath, J. R. *Science* **1992**, *258*, 1131.
78. Heinrich, J.; Curtis, C.; Credo, G.; Kavanagh, K.; Sailor, M. *Science* **1992**, *255*, 66.
79. Cannon, W. R.; Danforth, S. C.; Flint, J. H.; Haggerty, J. S.; Marra, R. A. *J. Am. Ceram. Soc.* **1982**, *65*, 324.
80. Matsunawa, A.; Katayma, S.; Arata, Y. *J. High Temp. Soc.* **1987**, *13*, 69.
81. Takagi, H.; Yamazaki, Y.; Ishizaki, A.; Nakagiri, T. *Appl. Phys. Lett.* **1990**, *56*, 2379.
82. Furukawa, S.; Miyasato, T. *Jpn. J. Appl. Phys.* **1988**, *27*, 2207.
83. Okada, T.; Iwaki, T.; Yamamoto, K.; Kasahara, H.; Abe, K. *Solid State Comm.* **1984**, *49(8)*, 809.
84. Tsu, R.; Shen, H.; Dutta, M. *Appl. Phys. Lett.* **1992**, *60*, 1112.
85. Kaneko, K.; Onisawa, M.; Kita, Y.; Minemura, T. *Jpn. J. Appl. Phys.* **1993**, *32*, 4907.
86. Prokes, S. M. In *Nanomaterials: Synthesis, Properties and Applications*; Edelstein, A. S.; Cammarata, R. C., Eds.; Institute of Physics Publishing: Bristol, Philadelphia, 1996, pp. 349–457; Prokes, S. M. *Bull. Am. Phys. Soc.* **1993**, *38*, 157.
87. Nakamura, M.; Mochizuki; Usami, K. *Solid State Comm.* **1984**, *50(12)*, 1079.
88. Hayashi, S.; Tanimoto, S.; Yamanoto, K. *J. Appl. Phys.* **1990**, *68*, 5300.
89. Koropecki, R. R.; Arce, R. *J. Appl. Phys.* **1986**, *60*, 1802.
90. Lucovsky, G.; Yang, J.; Chao, S. S.; Tyler, J. E.; Czubatyi, W. *Phys. Rev. B* **1983**, *28*, 3225.
91. Ritter, E. *Opt. Acta* **1962**, *19*, 197.
92. Anderson, D. A.; Moddel, G.; Paesler, M. A.; Paul, W. *J. Vac. Sci. Technol.* **1979**, *16*, 906.
93. Shabalov, A. L.; Feldman, M. S. *Thin Solid Films* **1983**, *110*, 215.
94. Knolle, W. R.; Maxwell, H. R., Jr.; Beneson, R. E. *J. Appl. Phys.* **1980**, *51*, 4385.
95. Boyd, I. W.; Wilson, J. I. B. *J. Appl. Phys.* **1987**, *62*, 3195.
96. Boyd, I. W. *Appl. Phys. Lett.* **1987**, *51*, 418.
97. El-Shall, M. S.; Graiver, D.; Pernisz, U. C., "Silicon Nanoparticles", U.S. patent application # DC 4304, Filed 11/22/1995.

98. Okada, R.; Iijima, S. *Appl. Phys. Lett.* **1991**, *58*, 1662.
99. Hagfeldt, A.; Gratzel, M. *Chem. Rev.* **1995**, *95*, 49.
100. Lauerhaas, J. M.; Credo, G. M.; Heinrich, J. L.; Sailor, M. J. *J. Am. Chem. Soc.* **1992**, *114*, 1911.
101. Fisher, D. L.; Harper, J.; Sailor, M. J. *J. Am. Chem. Soc.* **1995**, *117*, 7846.
102. Rehm, J. M.; McLendon, G. L.; Fauchet, P. M. *J. Am. Chem. Soc.* **1996**, *118*, 4490.
103. Tamura, H.; Ruckschloss, M.; Wirschem, T.; Veprek, S. *Appl. Phys. Lett.* **1994**, *65*, 1537.
104. Morisaki, H.; Hashimoto, H.; Ping, F. W.; Nozawa, H.; Ono, H. *J. Appl. Phys.* **1993**, *74*, 2977.
105. Bahnemann, D. W.; Kormann, C.; Hoffmann, M. R. *J. Phys. Chem.* **1987**, *91*, 3789.
106. Koch, U.; Fojtik, A.; Weller, H.; Henglein, A. *Chem. Phys. Lett.* **1985**, *122*, 507.
107. Haase, M.; Weller, H.; Henglein, A. J. *Phys. Chem.* **1988**, *92*, 482.
108. Spanel, L.; Anderson, M. J. *J. Am. Chem. Soc.* **1991**, *113*, 2826.
109. Brus, L. E. *J. Chem. Phys.* **1984**, *80*, 4403.
110. Brus, L. *J. Phys. Chem.* **1986**, *90*, 2555.
111. Li, S.; Silvers, S. J.; El-Shall, M. S. In *Advances in Microcrystalline and Nanocrystalline Semiconductors - 1996*; Fauchet, P. M.; Collins, R. W.; Alivisatos, P. A.; Shimizu, I.; Shimada, T.; Vial, J. C., Eds.; Materials Research Society Symposium Proceedings Series, 1997, 452. In press.
112. Joselevich, E.; Willner, I. *J. Phys. Chem.* **1994**, *98*, 7628.
113. Busca, G.; Lorenzelli, V. *Materials Chem.* **1982**, *7*, 89; Gervasini, A.; Auroux, A. *J. Thermal Analysis* **1991**, *37*, 1737; Lavalley, J. C.; Saussey, J.; Bovet, C. *J. Mol. Struct.* **1982**, *80*, 191.
114. Hoffman, A. J.; Mills, G.; Yee, H.; Hoffmann, M. R. *J. Phys. Chem.* **1992**, *96*, 5546.
115. Wang, Y.; Herron, N. *Phys. Rev. B* **1990**, *42*, 7253.
116. Kamat, P. V.; Patrick, B. *J. Phys. Chem.* **1992**, *96*, 6329.
117. Chestnoy, N.; Harris, T. D.; Hull, R.; Brus, L. E. *J. Phys. Chem.* **1986**, *90*, 3393.
118. Palik, E. D.; Ginsburg, N.; Holm, R. T.; Gibson, J. W. *J. Vac. Sci. Technol.* **1978**, *15*, 1488.
119. Passlack, M.; Schubert, E. F.; Habson, W. S.; Hong, M.; Moriya, N.; Chu, S. N. G.; Konstadinidis, K.; Mannaerts, J. P.; Zydzik, G. J. *J. Appl. Phys.* **1995**, *77*, 686.
120. Herber, W. C.; Minnier, H. B.; Brown, J. J. *J. Electrochem Soc.* **1969**, *116*, 1019.
121. Wanmaker, W. L.; Vrugt, J. W. *J. Electrochem Soc.* **1969**, *116*, 871.
122. Blasse, G.; Bril, A. *J. Phys. Chem. Solids* **1970**, *31*, 707.
123. Harwig, T.; Kellendonk, F.; Slappendel, S. *J. Phys. Chem. Solids* **1978**, *39*, 675.
124. Jonsson, B. J.; Turkki, T.; Strom, V.; El-Shall, M. S.; Rao, K. V. *J. Appl. Phys.* **1996**, *79*, 5063.
125. Parker, J. C.; Siegel, R. W. *J. Mater. Res.* **1990**, *5*, 1246; Capwell, R. J.; Spagnolo, F.; DeSesa, M. A. *Appl. Spectrosc.* **1972**, *26*, 536.
126. Melendres, C. A.; Narayanasamy, A.; Maroni, V. A.; Sigel, R. W. *J. Mater. Res.* **1989**, *4*, 1246.
127. Nyquist, R. A.; Kagel, R. O. *Infrared Spectra of Organic Compounds*, 2nd edn.; Academic Press: New York, 1973.
128. Ocana, M.; Serna, C. J.; Matijevic, E. *Materials Lett.* **1991**, *12*, 32; Ocana, M.; Serna, C. J. *Spectrochim. Acta* **1991**, *47A*, 765.
129. Scott, J. F. *J. Chem. Phys.* **1970**, *53*, 852.
130. Geurts, J.; Rau, S.; Richter, W.; Schmitte, F. J. *Thin Solid Films* **1984**, *121*, 217.
131. Abello, L.; Husson, E.; Repelin, Y.; Lucazeau, G. *Spectrochim. Acta* **1983**, *39A*, 641.
132. Iordanova, R.; Dimitriev, Y.; Dimitriov, V.; Klissurski, D. *J. Non-Crystalline Solids* **1994**, *167*, 74.
133. Carrazan, S. R. G.; Rives, V. *Materials Chem. Phys.* **1991**, *28*, 227.
134. Ferrer, E. G.; Baran, E. *J. Spectrochim. Acta* **1994**, *50A(2)*, 375.
135. Hardcastle, F. D.; Wachs, I. E. *J. Phys. Chem.* **1991**, *95*, 5031.
136. Wachs, I. E.; Chan, S. S. *Appl. Surf. Sci.* **1984**, *20*, 181.
137. Wada, K. *The Cumulative Index to IR Spectra of Inorganic Compounds*; Reports of the Central Customs Laboratory: 1978, Vol. 18, pp. 1–4.
138. Zecchina, A.; Coluccia, S.; Spoto, G.; Scarano, D.; Marchese, L. *J. Chem. Soc., Faraday Trans.* **1990**, *86*, 703.
139. Hardcastle, F. D.; Wachs, I. E. *J. Solid State Chem.* **1992**, *97*, 319.

140. Jin, C.; Haufler, R. E.; Hettich, R. L.; Barshick, C. M.; Compton, R. N.; Puretzky, A. A.; Dem'yanenko, A. V.; Tuinman, A. A. *Science* **1994**, *263*, 68.
141. El-Shall, M. S. In *Science and Technology of Atomically Engineered Materials*; Jena, P.; Rao, B. K.; Khanna, S., Eds.; World Scientific: 1996, pp. 67–75.

ENDOHEDRAL METALLOFULLERENES

Amer Lahamer, Z. C. Ying, R. E. Haufler,

R. L. Hettich, and R. N. Compton

I. GENERAL REVIEW AND INTRODUCTION

The story of the all-carbon fullerene molecule began with the initial calculations by Ozawa[1] and by Bochvar and Galpern[2] who considered the possible structures of caged carbon molecules such as C_{60}. Their calculations did indeed suggest that C_{60} might be stable but could only dream of its existence and eventual isolation.

Advances in Metal and Semiconductor Clusters
Volume 4, pages 179–203

ISBN: 0-7623-0058-2

Unfortunately these papers were published in non-English journals and went unnoticed by the western scientific community until the discovery and recognition of C_{60}, the mother of all fullerenes, some 20 years later. The story continues with the observation by the Exxon group of Cox, Rolfing, and Kaldor[3] of a slightly enhanced peak in the carbon cluster mass spectrum at a mass corresponding to 60 carbon atoms. The mass spectrum also exhibited only even numbered carbon cluster ions, a signature that we now know to be characteristic of fullerenes. The ions were produced by laser ablation of graphite into a nozzle jet which constitutes the now famous "Smalley source" provided to them by Rick Smalley. The fullerene era in chemistry essentially began in 1985 following a series of Nobel Prize winning experiments (1985) by Smalley, Kroto, Curl and students at Rice University.[4] The ensuing recognition and attention directed to fullerene research has been unequaled in science. The ability of the fullerenes to undergo exohedral chemical reactions as well as to provide a central "cage" into which other atoms and ions can reside has captured the imagination of chemists and physicists alike. Very soon after their initial discovery of C_{60}, the Rice group suggested that an atom or small molecule could be trapped inside the empty fullerene cage.[5,6] In fact, the studies of this group on laser "desorption" of successive C_2 dimers from endohedral metal fullerenes to the point that the metal atom is "squeezed" out of the cage, provided initial compelling evidence that C_{60} was indeed a "caged" soccer ball shaped molecule. These studies are often referred to as the "shrink wrapping" experiments.

Research on the physics and chemistry of endohedral fullerenes is expanding at a rapid pace. This interest is due to both the promise of useful new materials and the new science to be discovered in the study of caged atoms. In this chapter, we will briefly review endohedral fullerene research over the past several years before discussing our studies of endohedral metallofullerenes. This is not intended to be a comprehensive review of the field but rather an introduction to our contribution.

The simplest endohedral fullerene consists of a single electron or positron "inside" a fullerene cage designated by $e@C_{60}$ or $e^+@C_{60}$. The endohedral electron is believed to exist in a positive energy state (i.e. unbound) for small fullerenes (e.g. C_{60}) as a consequence of the so-called "endohedral effect".[7] An electron is therefore unstable inside C_{60}. On the other hand, a positron could reside on the inside of the fullerene cage prior to electron–positron annihilation. Positron annihilation in C_{60} has been studied by a number of authors.[8] Studies of the pressure dependence indicate that the positron is annihilating with an electron near the interstitial octahedral sites of the fcc lattice.[9] The attachment of electrons to the "outside" (i.e. occupying the t_{1u} orbital) of the C_{60} molecule has been reported by a number of authors.[10,11] The possibilities of excited state endohedral binding in which the electron density would be located to a large extent on the inside of the cage has not yet been explored. However, it is possible that resonance features observed in the electron attachment studies[10,11] could be due to states in which excess electron density is localized inside the cage.

Studies of muons implanted in fullerenes provide evidence that the muonioum atom can also exist inside the fullerene cage; i.e. Mu@C_{60}. Muonium is a light exotic isotope of hydrogen, i.e. Mu = μ^+e^-. This experimental evidence that the Mu is located at the center of the cage is supported by theoretical calculations.[12] Observations of the temperature independence of the depolarization rate of the fullerene (6, 5 isomer of $C_{61}H_2$) Mu@$C_{61}H_2$ provide evidence that external functionalization does not significantly alter the π-electron density inside the cage.[13]

A limited number of studies have considered entrapping hydrogen atoms in fullerenes. The insertion of a deuteron and deuterium molecules into C_{60} (i.e. D@C_{60} and D_2@C_{60}) was reported following experiments involving collisions between C_{60}^+ and D_2.[16]

A. Noble Gas Endohedral Fullerenes

The first experiments that suggested the presence of endohedral fullerenes containing noble gas inclusion were conducted at Technische Universität in Berlin in 1991.[14] In these experiments, C_{60} and C_{70} cations with kinetic energies of several thousand electron volts collided with helium atoms. Mass analysis of the product ions indicated the presence of He@C_{n-2x} cations (n = 60 and 70, x = 1–6). The parent He@C_n (n = 60 and 70) cations were observed in subsequent experiments in this and other laboratories.[15,16,17] In one experiment, the He@C_{60} cations were neutralized by reaction with $(CH_3)_3N$.[16]

Other rare gas atoms were found to be inserted into the fullerene cage during collisions. In addition to He, noble gas atoms of Ne and Ar as well as D and D_2 atoms were intercolated into the C_{60} cage.[16] Insertion of two helium atoms[16] or one helium and one krypton atom[18] into a single C_{60} cage was also achieved by collisions methods.

Sauders and Cross at Yale University discovered in 1992 that He@C_{60} actually existed in the soot produced by the well-known arc vaporization method.[19] Using a highly sensitive mass spectrometer, they found helium desorbed thermally from the C_{60} material at temperatures around 800 °C. The desorption activation energy was estimated to be 80 kcal/mol (3.5 eV). This energy is certainly too high for helium atoms attached to the outside of the fullerene cage, where the interaction is expected to be a weak van der Waals force. A logical explanation is that an endohedral species is formed during arc vaporization of graphite in a helium atmosphere. The helium atoms escaped the fullerene cages and were detected by the mass spectrometer during heating.

The insertion of a small atom into the fullerene cage needs to overcome an energy barrier. Estimation, using the benzene ring as a model for the fullerene's hexagonal faces, put the value around 10 eV.[20] This number is much higher than the value obtained from the thermal-desorption experiments. Sauders and Cross therefore proposed a "window" mechanism, where temporary bond breaking allowed trapping of the helium atoms with the observed low activation energy.[19] To test this idea, they heated C_{60} to 600 °C under three atmospheres of ^{3}He and Ne for 1 h and found that a small amount (~1 ppm) of ^{3}He and Ne was indeed inserted into the fullerene cage. The low activation energy obtained from the thermal desorption experiments[19] is consistent with the threshold energy for endohedral He@C_{60} formation by collision (3 eV)[21] and an early theoretical study.[22] A recent calculation has suggested alternative mechanisms.[23]

The concentration of the endohedral material was subsequently improved in experiments using a vessel with pressures up to 2700 atm.[24] After the high-pressure, high-temperature treatment, the concentration of G@C_{60} (G = He, Ne, Ar, and Kr) in the product reached 0.1–0.3%, a 3-order-of-magnitude improvement over the previous experiments.[19] As expected, xenon is more difficult to be inserted into the fullerene cage. The concentration of Xe@C_{60} in the product was only 0.008%. Noble gas atoms were also inserted into C_{70}.

The successful synthesis of noble gas endohedral fullerene molecules and the observation of a lower concentration of Xe@C_{60} has prompted calculations of the binding energies of these endohedral complexes. Calculations using a semiempirical Lennard–Jones potential have shown that endohedral complexes are more stable than the corresponding exohedral species for He, Ne, Ar, and Kr inside C_{60}.[25,26] Among these four species, Ar@C_{60} is the most stable one with a binding energy of 0.34 eV. On the other hand, Xe atoms appear to be too large to form stable endohedral species inside C_{60} cages. Using the same method, Dicamillo et al. have calculated the binding energies for noble gas atoms trapped inside C_{84}.[27] In this case the larger cage calculations show that Xe@C_{84} has the largest binding energy of all the rare gases, about 0.74 eV (see Figure 1).

Further enhancement of the endohedral molecule's concentration was achieved using high-performance liquid chromatography (HPLC). Successful separation of Ar@C_{60} from other fullerene species has been reported independently by groups at Yale[28] and Oak Ridge National Laboratory.[27]

Other methods, such as hot-atom chemistry[29] and neutron irradiation,[30,31] have been used to produce endohedral fullerenes doped with noble gas atoms, particularly radioactive ones. These radioactive endohedral fullerenes are useful in studies involving nuclear spin[32,33] and have potential applications for high-speed biological nuclear magnetic resonance imaging.

While experimentalists are still working to obtain pure noble gas endohedral fullerenes of sufficient quantities for electronic and ionic property measurements, theorists have performed a number of calculations of their properties. Rotation,[34] vibrational,[34,35] magnetic,[36] plasmon,[37] and other properties[38] have been calculated.[38]

B. Rare Earth Metal Endohedral Fullerenes

The first experiments of endohedral metallofullerenes were conducted at Rice University within a week after the discovery of C_{60} in 1985.[5] A piece of graphite

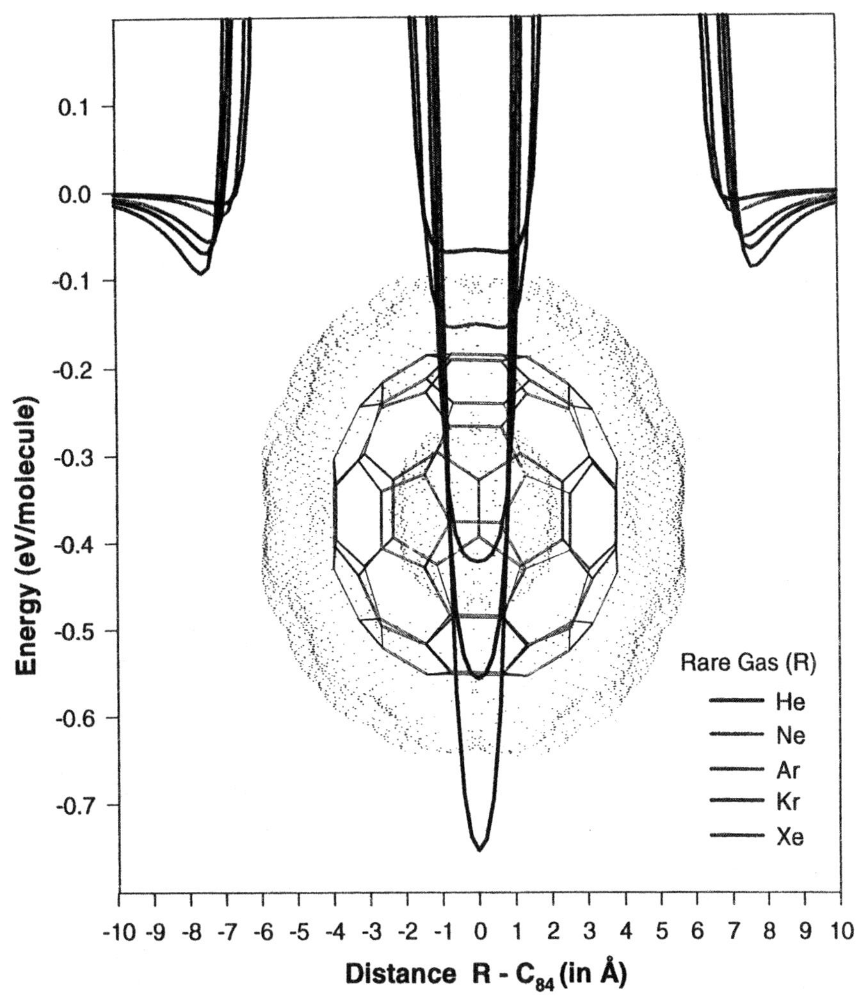

Figure 1. Endohedral and exohedral van der Waals interaction calculated for rare gas–C_{84} molecules (see ref. 27).

that had been boiled in a saturated water solution of $LaCl_3$ was used as the laser ablation target in a "Smalley source". The positive ions were detected by a time-of-flight mass spectrometer, which shows the presence of $La@C_n^+$ ($n = 44–76$), with $La@C_{60}^+$ being an intense peak. Interestingly, C_{60}^+ and C_{70}^+ were the only bare carbon clusters that were observed. The first measurement of electronic properties for these lanthanum-doped fullerene molecules was conducted at Exxon. The ionization potential (IP) of the lanthanum-doped fullerene molecules was estimated to be less than 6.42 eV, substantially lower than those of the undoped fullerene molecules.[39] The Exxon group has also shown that two or more metal atoms could be attached to the fullerene structure.[40]

Several years later, macroscopic quantities of metallofullerenes were produced at Rice University.[6] Laser ablation of a solid rod of La_2O_3/graphite mixture was carried out at 1200 °C in an argon atmosphere, resulting in a black-brown deposit on a quartz tube. Vacuum sublimation of the deposit produced a black film on a Cu disk. Its content was analyzed using Fourier transform mass spectrometer (FTMS), which displayed La@C_n (n = 60, 70, 74, and 82) and other doped and empty fullerenes. The metallofullerenes exhibited no reactivity towards O_2 and NH_3, and photodecomposed under high-power laser irradiation only by successive C_2 loss. The authors therefore suggested that the La atoms were trapped inside the fullerenes cage. They called the new species a "superatom" and suggested the notation for endohedral species of M@C_n. In a parallel experiment, hot toluene extract of the laser-ablation deposit was analyzed using FTMS. This time, La@C_{82} was the only metallofullerene which survived.

A number of other endohedral metallofullerene molecules were made in the early 1990s using either the laser ablation or arc discharge method at various laboratories. Rare earth metals have been found to be easily trapped inside the fullerene cage. The list includes molecules with a single atom inside, RE@C_n (RE = Y, La, Ce, Pr, Nd, Sm, Eu, Gd, Tb, Dy, Ho, Er, Yb, and Lu; $n \geq 60$),[41–44] with two atoms inside, RE_2@C_n (RE = Y, La, Ce, Tb, Ho, and Lu; $n \geq 80$),[41,43–48] and with three atoms inside, RE_3@C_n (RE = Sc and La; $n \geq 82$).[48–50] The endohedral fullerenes with atoms other than rare earth elements include U@C_n (n = 28–70),[45] Ca@C_{60},[46] Sr@C_{60},[44,47] and Ba@C_{60}.[48] Whether a metal atom can be trapped inside a fullerene cage by the arc-discharge method appears to depend on its electronegativity.[49]

With macroscopic quantities of (impure) metallofullerene materials available, experiments have been conducted to measure their molecular properties.[50] The most important questions are the location and charge state of the metal atom.

An early evidence for the endohedral geometry came from the fact that La@C_n and Y@C_n do not react with water or oxygen, suggesting that the fullerene cage acts to protect the metal atom inside.[6,41] An absence of fragmentation following collisions of 200-eV, La_2@C_{80}^+ ions on a Si(100) surface provides another piece of evidence that the metal atoms are located inside the fullerene cages.[51] Measurements using the extended X-ray absorption fine structure (EXAFS) technique provides strong evidence that Y@C_{82} and Y_2@C_{82} form two 6-carbon-atom shells at 2.6 and 2.9 Å, consistent with endohedral geometry with the metal atoms located off the cage center.[52]

Three groups of scientists at IBM and in Japan used electron paramagnetic resonance (EPR) and electron spin resonance (ESR) to measure the charge state of the metal atoms for several rare earth endohedral fullerenes. They found that the lanthanum atom in La@C_{82} is in the +3 state,[53] as well as yttrium in Y@C_{82}.[54] Another interesting observation was that RE@C_{82} (RE = Sc, Y, and La) in toluene solution exists as multiple species.[55,56] In studies of $Sc_x@C_{82}$ (x = 1–3), it was shown that the Sc atom in $Sc@C_{82}$ is in the +3 state and that the three Sc atoms in $Sc_3@C_{82}$ form an equilateral triangle.[45] An absence of EPR signal for $Sc_2@C_{82}$ suggested pairing of the Sc valence electrons. A later study of the linewidths of ESRs hyperfine components suggested that the Sc atom in $Sc@C_{82}$ is in the +2 state.[57] This new result is consistent with calculations using the molecular orbital and spin-unrestricted Hartree–Fock method.[58,59] The theory also predicts that Ce in the $Ce@C_{82}$ is also in the +2 state,[60] instead of +3 as for La in the $La@C_{82}$.[59]

Beginning in 1993, pure samples of certain rare earth metallofullerenes became available. The production of the metallofullerenes was produced by the arc discharge of metal/graphite mixture. The separation of the rare earth metallofullerene molecules from the undoped fullerene molecules and of rare earth metallofullerene molecules of different numbers of metal and/or carbon atoms was accomplished by using HPLC. The first successful report was on three di-scandium species: $Sc_2@C_n$ (n = 74, 82, and 84).[61] Then $La@C_{82}$, a single-metal-atom endohedral species, was isolated.[62] Pure samples of these and other rare earth metallofullerenes, such as $Y@C_{82}$,[63] $Ce@C_{82}$,[64] $Nd@C_{82}$,[65] $Gd@C_{82}$,[66] and $Sc_3@C_{82}$,[67,68] can now be produced in several laboratories in the world.

One of the first experiments using pure rare earth endohedral fullerene samples was on valence level of $La@C_{82}$, as measured by ultraviolet photoemission spectroscopy.[69] The data showed peaks at 0.9 and 1.6 eV below the Fermi level for $La@C_{82}$, in comparison with that of C_{82}. This observation was attributed to charge transfer of three electrons from the metal atom to the fullerene cage. A later photoemission study showed that the bonding of $La@C_{82}$ in the solid state is significantly stronger than that for C_{82} and the solid material is nonmetallic.[70]

Another interesting experiment has been reported for X-ray diffraction of solid $Y@C_{82}$.[71] It confirmed through analysis of the diffraction data that the metal atom is indeed located inside the fullerene cage, but at a location off the center, which was predicted in 1992 for $La@C_{82}$.[72] The X-ray diffraction study of $Sc_2@C_{84}$ has pointed out that the two metal atoms are in the endohedral geometry with the Sc–Sc and Sc–C distances to be 3.9 and 2.4 Å, respectively.[73] The theory has predicted that the two Sc atoms inside $Sc_2@C_{84}$ oscillate with considerable amplitude, while the two La atoms inside $La_2@C_{80}$ can circulate inside the fullerene cage with a small barrier.[74] The availability of the pure rare earth metallofullerene materials also allowed refinement of the EPS and ESR measurements that were made earlier using mixed samples. Spin relaxation,[75] magnetic properties,[66] and adsorption geometry on surfaces[65] are further examples of measurements for pure rare earth metallofullerene materials.

It should be noted that, while C_{60} has received the greatest attention of the fullerenes, the most studied rare earth endohedral fullerene species are not $M@C_{60}$ but $M@C_{82}$ (M = Y and La). In experimental studies of rare earth endohedral fullerenes, $M@C_{60}$ was observed together with $M@C_{74}$ and $M@C_{82}$ with approximately the same abundance in macroscopic samples produced using either the laser-ablation or arc-discharge technique, as long as the samples had not been exposed to the commonly used solvents (toluene and pyridine).[6,41,76] In the solvent-extracted samples, however, $M@C_{60}$ disappeared and the most abundant endohedral fullerene became $M@C_{82}$. It appears that either $M@C_{60}$ is insoluble in toluene and pyridine or it is unstable in these solvents. A recent study has shown that many metallofullerenes $M@C_{60}$ could be extracted using aniline under an air atmosphere.[77] The air sensitivity of $M@C_{60}$ is an unsettled issue. It has been reported that $M@C_{60}$ does not react with oxygen, water, or nitrous oxide.[41,78] On the other hand, $La@C_{76}$, extracted using toluene under anaerobic conditions, is found to be air-sensitive; its ESR intensity decreases to less than 5% of the original value after air exposure after 5 days.[79] From our own experience, the sample films prepared by vacuum sublimation of the laser-ablation soot, which were never exposed to any solvent, did not exhibit observable changes after exposed to air for a few days. The $La@C_{60}$ signal as seen from FTMS, however, did decrease substantially after 2-month exposure to air, while the $La@C_{82}$ signal showed much smaller degradation.

II. "SUPER ATOMS"

The Rice University group[6] christened the first endohedral metallofullerenes as "super atoms." The analogy to an atom was posed since a metal atom chemically bound in a carbon cage is expected to give up one or more electrons to the empty orbitals of the fullerene molecule forming a monopole charge-transfer salt. In the rare cases where electrons from the cage are partially located on the endohedral metal atom, the structure could be called a "super anti-atom." Roger and Wästberg[80] have taken this analogy a step further to make a connection with the electronic structure of the super atom to that of the quasiatomic semiconductor hetrostructure in solid-state physics.[81]

The semiconductor hetrostructure super atom consists of a spherical core, doped with a controlled number of donors and surrounding impurity-free matrix with a higher electron affinity. The electronic structure for the "super atom" is very different from that of an ordinary atom as a result of the absence of the $1/r$ singularity in the centrosymmetric Coulomb potential. For example, self-consistent local-density-functional calculations for the $Al_{0.35}Ga_{0.65}As$-GaAs super atom gave an energy ordering 1s, 2p, 3d, 2s as opposed to the conventional atomic 1s, 2s, 2p, 3s ordering. Calculations[80] suggest that $La@C_{60}$ could be considered as a super atom candidate. Further tests of this super atom concept will occur when electronic spectroscopy is carried out on these new molecules.

As emphasized in the preceding brief review, the separation of metallofullerenes of C_{60} and C_{70} from arc-discharge soot has proven to be very difficult owing to their apparent high chemical sensitivity. Although $M@C_{60}$ and $M@C_{70}$ are the most abundant endohedral fullerenes in the soot, they are apparently absent or scarce following most common separation techniques (HPLC). The first experiment of the Rice group showed that $M@C_{60}$ and $M@C_{70}$ were the most abundant ionic species generated in direct laser ablation of the metal doped graphite. Our interests are in the characterization of $M@C_{60}$. This species represents the simplest molecular prototype of a caged super atom. Experimental studies of electronic and ionic properties of $M@C_{60}$ are more amenable to comparison with theoretical calculations.

III. PRODUCTION OF $M@C_n$ (n = 60, 70, ... ; M = Li, K, Ca, Sr, Ba, Sc, La, and Eu)

In this section we discuss the laser-ablation method for the production of endohedral metallofullerenes. Laser ablation of graphite in a heated rare gas was first employed by Smalley for the production of fullerenes.[6] A schematic of our experimental setup is shown in Figure 2. Four major components were employed in these laser-ablation experiments: a pulsed Nd:YAG laser, a target graphite rod containing the desired metal atom, an oven, and an inert carrier gas in a quartz tubing of about 1 in. in diameter and about 36 in. in length. The laser source used for our experiments was the second harmonic (532 nm) of a Nd:YAG (Quantel International 571-C) operated in the nano-second mode with a repetition rate of 10 Hz. The incident beam (about 0.15 J per pulse) was focused to a spot size of a few millimeters in diameter on one end of the target rod. The target rod was situated inside the quartz tube surrounded by an oven with an inert carrier gas. The rod was connected to a motor which

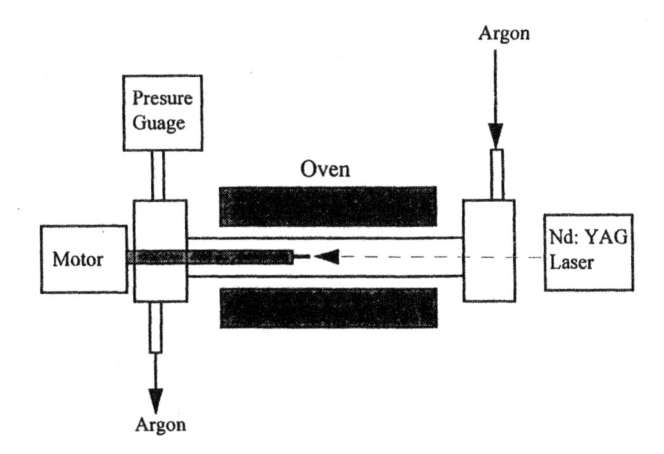

Figure 2. Laser ablation apparatus. The collected material was analyzed by a linear TOF mass spectrometer (Comstock Inc.) and by a FTMS (Finnigan FTMS-2000) mass spectrometer. Both spectrometers used the laser desorption technique for generating the positive and negative ions. The FTMS used the third harmonic of a Nd-YAG laser (Spectra Physics DCR-11) with a wavelength of 355 nm, while the TOF used a nitrogen (Laser Photonic Megaplus) laser with 337 nm wavelength. The resolution, m/Δm, was up to 2000 for the TOFMS and about 6000 for the FTMS.

provided a rotational motion of about 10 rpm. Rotation of the target was necessary in order to expose different areas of the target for the laser beam producing a uniform consumption of the target material. Before these rods were ready for ablation, they were first prepared by mixing graphite (Dylon Industries), graphite cement (Dylon Industries) with either an oxide or a carbonate form of the metal desired to be entrapped with elemental proportion (metal/carbon) between 0.1 and 1%. The mixture was cured in a stainless steel mold for about 4 h at about 120 °C after which they were baked at 1100 °C in a low flow rate of argon gas at about 100 mtorr for about 4 h. Once this process was completed they were machined to fit on a thin connecting carbon rod to the motor. A typical procedure for an ablation run was to evacuate the quartz tube to about 10 mtorr and purge it several times with the carrier gas which usually was argon or helium. The oven was set to 1100 °C while adjusting the argon to a slow flow rate at a pressure of about 138 torr. The ablation process usually took between 10 and 20 min. As the material was being ablated, the carrier gas transports the fullerenes to the cooler side of the quartz tube where the material was deposited. After the oven cools off, the material usually was sublimed onto a tip to be used in the time-of-flight mass spectrometer (TOFMS) or a sample disc for use in the Fourier transform ion cyclotron resonance mass spectrometer (FTMS). The tip or the disk was situated on an aluminum rod cooled with circulating chilled water. This sublimation was usually carried out at 600 °C and at 138 torr for about 30 min.

The calcium and lanthanum targets were also ablated with the free electron laser (FEL) of Vanderbilt in Nashville, Tennessee with the same experimental arrangement as that used for the Nd: Yag laser. The FEL laser wavelength was 7 μm with an energy of 13 mJ per macropulse. The 2.5 μs macropulse consists of 1000 micropulses each of which has a duration of about 1 ps. The collected material was analyzed by the TOF mass spectrometer, the calcium result of which is shown later in Figure 10.

A. TOFM Spectra

We describe below a series of experiments which were successful in producing a variety of endohedral metallofullerenes. The high quality of the laser desorption time-of-flight (LDTOF) mass spectra is a result of employing the new technique of pulsed field extraction. An ion draw-out pulse is applied to the TOF backing plate which corrects for the TOF of ions initially created on the surface with differing kinetic energies. Ions of the same mass with higher initial kinetic energies will be extracted with lower kinetic energy (i.e. they will "fall" through a lower voltage) allowing the ions initially produced with lower energies to "catch-up".

Li@C_n

To our knowledge the entrapment of Li inside C_n has not been reported using the laser ablation technique. There have been reports of Li entrapment using ion implantation technique.[95] In Figure 3, a peak at a mass of 727 amu corresponding to $Li@C_{60}$ and a peak at mass of 847 amu corresponding to $Li@C_{60}$ were clearly present. One possible explanation of the low abundance of these species is that during the laser ablation which takes place under high temperatures, the small lithium atoms might have escaped during laser desorption.

K@Cn

Endohedral potassium fullerenes were observed by accident. The intention of the original experiment was to use different compounds that contain iron for the purpose of entrapment of iron. All of our efforts to trap the iron were not successful; however when using potassium ferrocinide as a source of iron, a mass of 759 amu corresponding to the $K@C_{60}$ molecules was present in the TOF mass spectra (Figure 4).

Ca@Cn

Figure 5 shows the TOF mass spectrum with a clear signal of $Ca@C_{60}$. A peak of mass 968 amu which corresponds to a calcium dimer trapped inside C_{74} is also present which might suggest a minimum size for entrapment of two calcium atoms. There have been many reports of calcium entrapment and isolation of calcium metallofullerenes by several groups.[46]

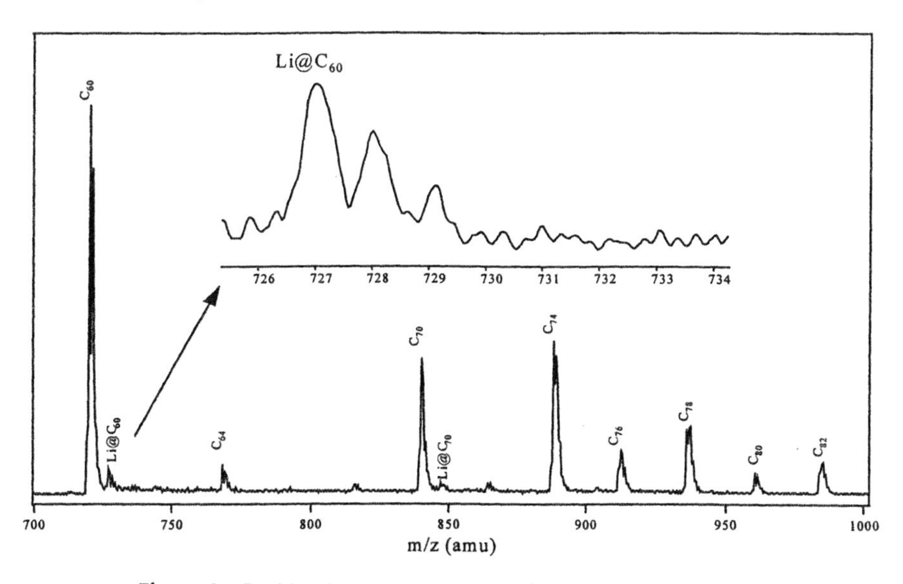

Figure 3. Positive ion TOF spectrum of lithium/graphite target.

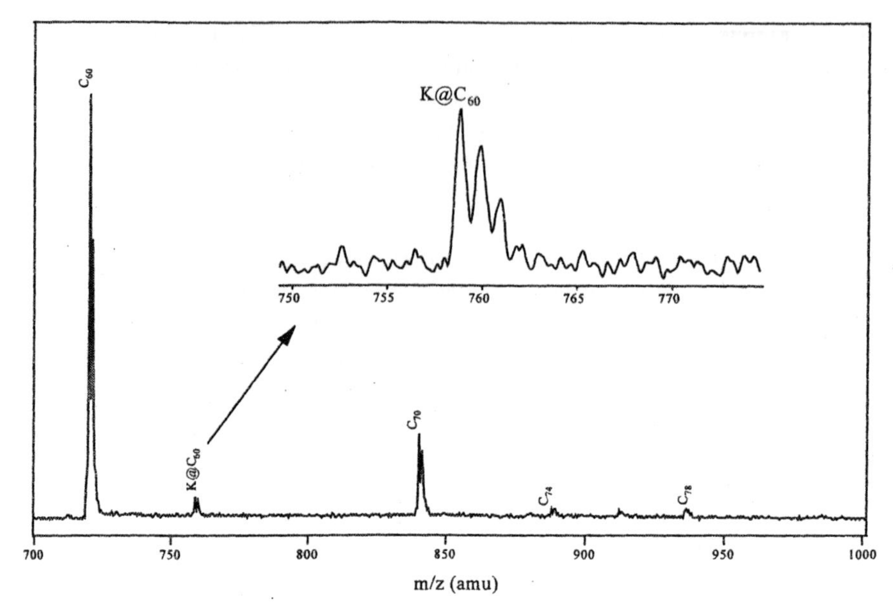

Figure 4. Positive ion TOF spectrum of potassium ferrocinide/graphite target.

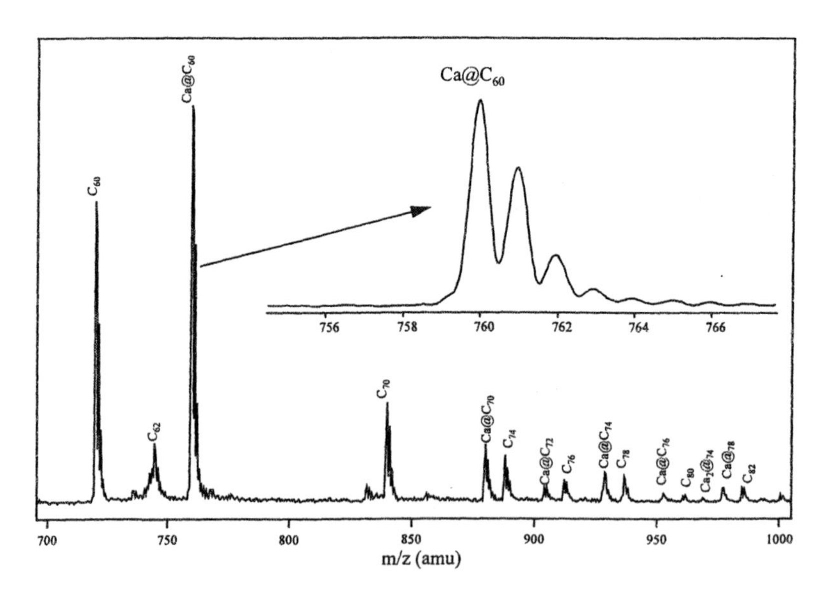

Figure 5. Positive ion TOF spectrum of calcium/graphite target.

Sr@Cn

Figure 6 is a typical TOF mass spectra where Sr@C_{60} and Sr@C_{70} molecules are present. Other reports of successful entrapment and isolation of this metallofullerene have also been reported.[44,47]

Ba@Cn

Various barium to carbon composition ratios as well as different parent compounds of barium (oxides and carbonates) have been attempted with somewhat similar results. One of the successful results were obtained when 0.1% ratio of barium (barium oxide) and carbon (graphite) was used. Figure 7 shows the presence of Ba@C_{60} and higher metallofullerenes (Ba@C_n, $n > 60$) in the LDTOF mass spectra.

La@Cn

Several laser ablations of the lanthanum/graphite target have been performed with interesting results. For example, in addition to La@C_n, a peak of a mass of 1158 amu was observed in the negative ion mode spectra (Figure 8b) which might correspond to LaC@C_{84} molecule. Figure 8a shows TOF mass spectra in the positive mode.

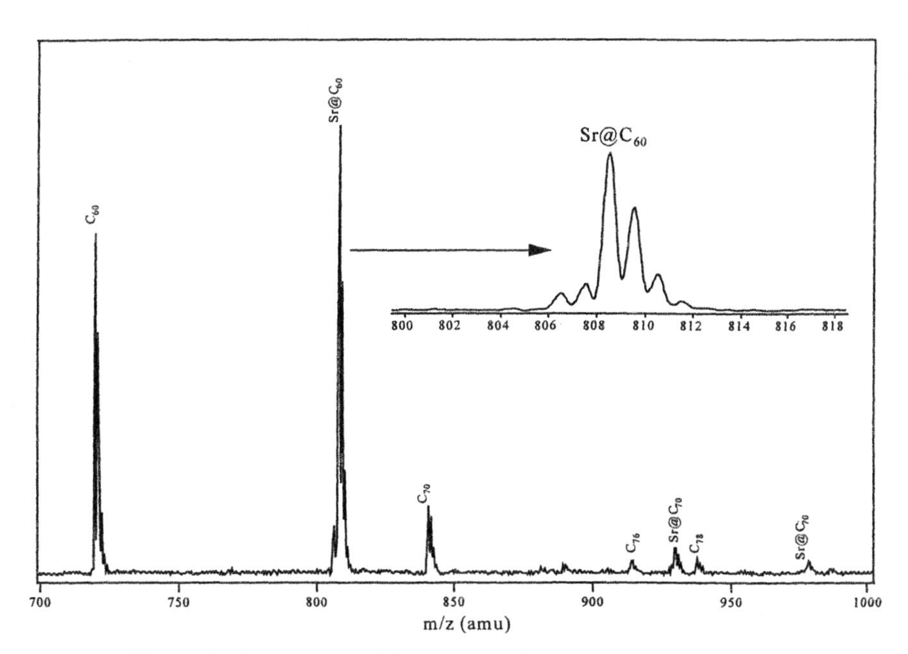

Figure 6. Positive ion TOF spectrum of strontium/graphite target.

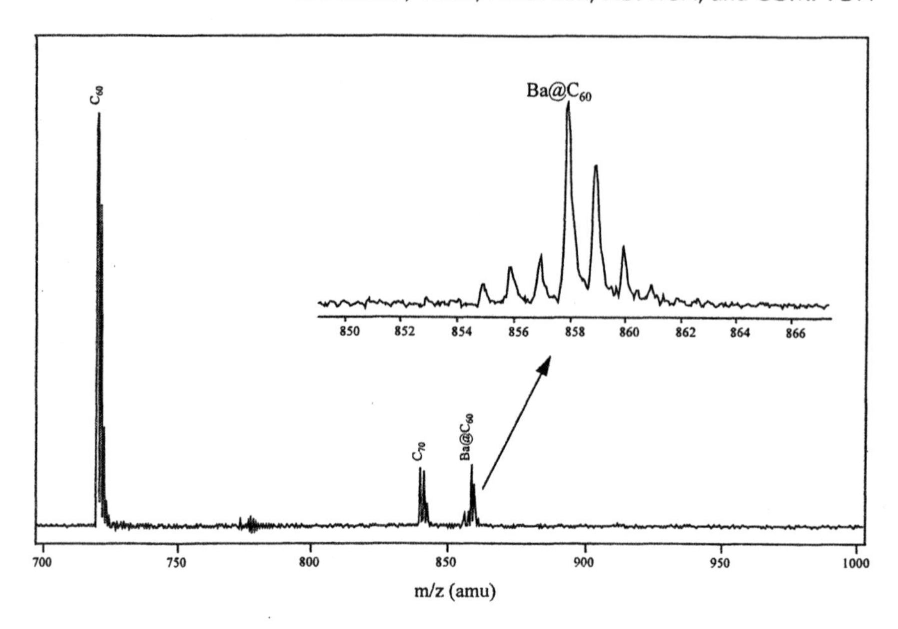

Figure 7. Positive ion TOF spectrum of barium/graphite target.

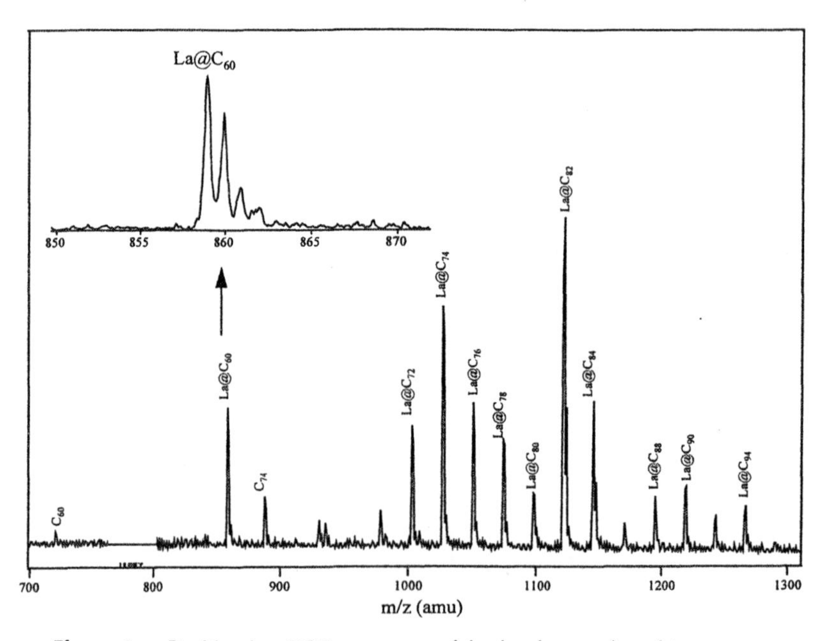

Figure 8a. Positive ion TOF spectrum of the lanthanum/graphite target.

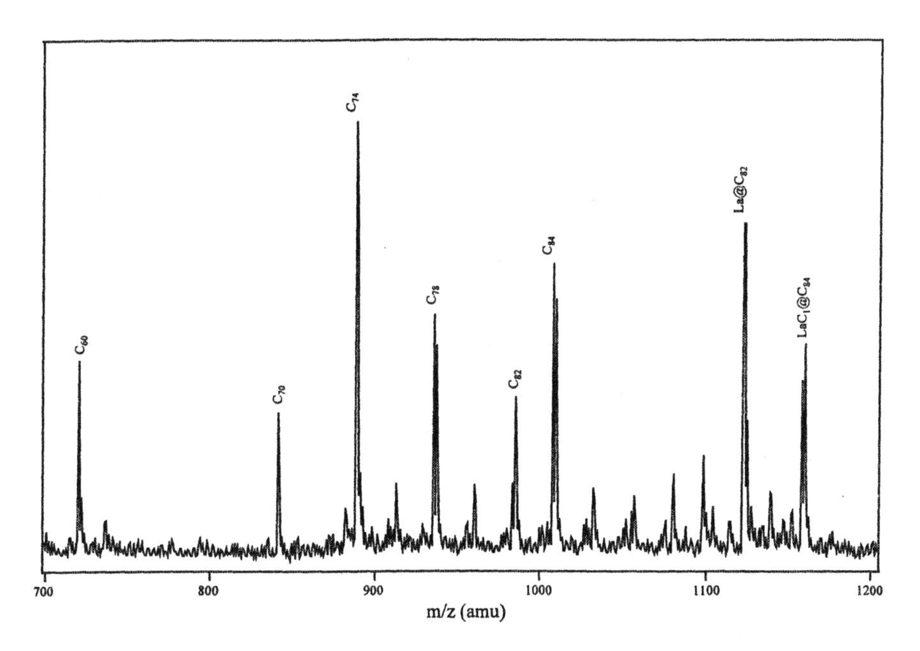

Figure 8b. Negative ion TOF spectrum of the lanthanum/graphite target.

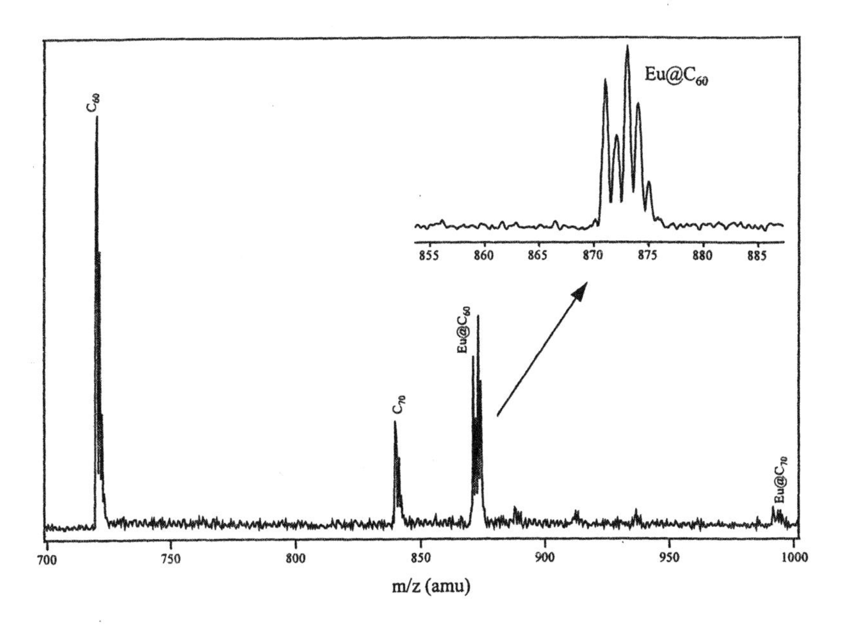

Figure 9. Positive ion TOF spectrum of europium/graphite target.

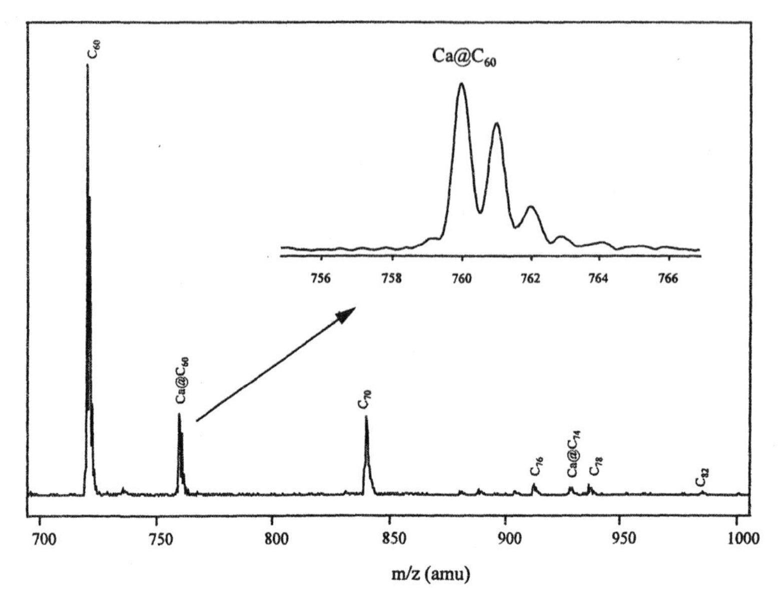

Figure 10. Positive ion TOF spectrum of calcium/graphite target ablated by the Vanderbilt University FEL.

Eu@C$_n$

Figure 9 shows the TOF mass spectra of europium/graphite target with Eu@C_{60} having the expected mass distribution.

Ca@C$_n$

Figure 10 shows the TOF mass spectra of calcium metallofullerenes produced by FEL laser ablation of the calcium/graphite target. The Nd:YAG laser preparation of calcium metallofullerenes showed higher yield of Ca@C_{60}.

B. FTM Spectra

The FTMS experiments were performed with a Finnigan FTMS-2000 Fourier transform ion cyclotron resonance mass spectrometer, equipped with a 3 Tesla superconducting magnet and a Spectra Physics DCR-11 pulsed Nd:YAG laser. Samples were placed on a stainless steel probe disk, which was inserted into the FTMS vacuum chamber and used for the laser desorption experiments. The third harmonic of the Nd:YAG laser (355 nm) was focused to approximately 10^7 W/cm^2 onto the surface of the sample. A single laser shot was sufficient to desorb and simultaneously ionize part of the sample, generating positive or negative ions which can be detected with the FTMS instrument. For most spectra, multiple laser shots were signal-averaged prior to Fourier transform analysis to improve the quality of the mass spectra. More details of the principles of FTMS operation in general and laser desorption FTMS in particular have been outlined previously.[82,83]

The laser-desorbed ions can be trapped in the FTMS cell for times ranging from a few milliseconds to several minutes prior to ion detection. By trapping the ions in the FTMS cell in the presence of a reagent gas, it is possible to study ion–molecule reactions, such as charge transfer. Ion fragmentation can be studied by using collision-induced dissociation experiments, which are conducted by isolating an ion of interest in the FTMS cell (i.e. all other ions are ejected) and then accelerating that ion into a collision gas such as argon (maintained in the cell at a static pressure of about 10^{-6} torr).

Characterization of La@C$_n$

LD-FTMS of the lanthanum metallofullerene samples generated by the experimental technique outlined above revealed the presence of some C_{60}^+ and C_{70}^+, along with abundant peaks corresponding to La@C_n^+ for n = 60, 74, 82. Weaker peaks corresponding to La@C_n^+ for n = 66–100 were also observed. As we will discuss below, the lower ionization potentials for the endohedral La@C_n species relative to C_n overrepresent their actual abundance in the mass spectrum. Three FTMS experiments were conducted to examine the La@C_n^+ species: (1) ion–molecule reactions with oxygen-containing species, (2) collision-induced dissociation reactions, and (3) charge exchange reactions to bracket the ionization potentials and electron affinities. The first two types of reactions were used to interrogate the structure of the La@C_n^+ species to verify that the metal atom was inside the cage (endohedral) and not attached to the outside of the cage (exohedral). The third type of reaction (charge exchange) was used to bracket the ionization potentials of these endohedral species to ascertain the effect of the "trapped" heteroatom on the fullerene ionic structure.

Ion–molecule reactions of La@C_n^+ with N_2O were examined by trapping the La@C_n^+ species in a static pressure of N_2O (1×10^{-6} torr) for times ranging up to 5 s. Experiments were conducted by isolating individual endohedral species, such as La@C_{60}^+, and examining its reactivity with N_2O, as well as simply letting the entire La@C_n^+ packet react with N_2O. No reactions were observed in any case. Because atomic lanthanum ions are known to be quite reactive with oxygenated species, these results indicate that the lanthanum atom in the La@C_n^+ is inaccessible for reaction with N_2O, and support the assumption that these species are endohedral fullerenes. For comparison, metal exohedral complexes MC_{60}^+ for M = Fe, Co, Ni, Cu, Rh, La, Y, and VO have been generated and studied.[84,85,86] These complexes are quite reactive. The exohedral complexes of the early transition metals (such as La, Y) were found to react readily with oxygen-containing compounds to generate the MO^+ species.

The collisional dissociation of La@C_n^+ for n = 60, 74, 82 revealed primarily sequential C_2 loss, as shown in Figure 11 for La@C_{82}^+ and summarized in Table 1. These low energy (E_{com} < 50 eV) collisional dissociation experiments in FTMS

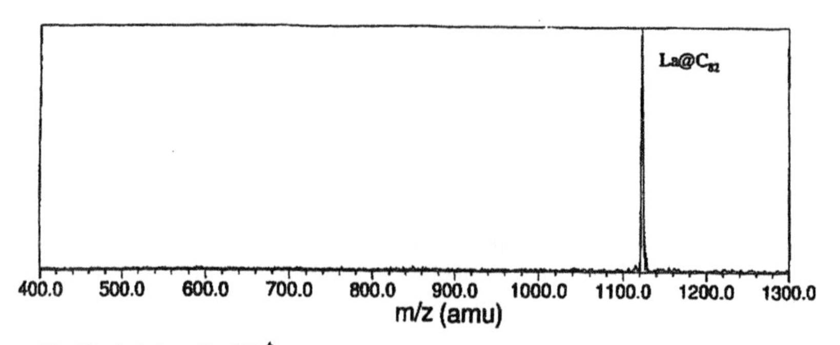

Fig. 12a. Isolation of La@C82+

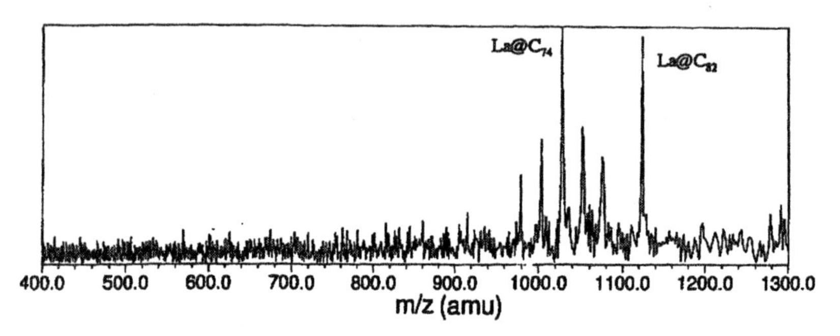

Figure 11. (**a**) Isolation of La@C_{82}^{+}. (**b**) Collision-induced dissociation of La@C_{82}^{+}.

involve multiple collisions prior to ion fragmentation. Note that the empty fullerenes are not observed as fragment ions at these energies (i.e. C_{60}^{+} is not observed as a fragment ion of La@C_{60}^{+}). These results further support the conclusions from the ion–molecule reactions outlined above which imply that the lanthanum atom is inside the cage (endohedral) and not attached to the outside of the fullerene (exohedral). This collisional dissociation fragmentation is similar to that observed for the photodissociation of La@C_{60}^{+}.[87] In contrast, the exohedral MC_{60}^{+} species fragment to give M^{+} and C_{60}^{+}.[85] For the early transition metals such as La and Y, the

Table 1. Collision-Induced Dissociation of La@C_{n}^{+}

Parent Ion	*Fragment Ions*
La@C_{60}^{+}	La@C_{58}^{+}, La@C_{58}^{+}, (No C_{60}^{+})
La@C_{74}^{+}	La@C_{70}^{+}, (No C_{70}^{+})
La@C_{82}^{+}	La@C_{78}^{+}, La@C_{76}^{+}, La@C_{74}^{+}
	La@C_{72}^{+}, La@C_{70}^{+}

M^+ fragment ion is usually the most abundant ion, due to the lower IP of the metal relative to that of C_{60}.

Ionization potentials for La@C_n species can be bracketed by examining the following charge-exchange reaction:

$$\mathrm{La@C}_n^+ + A \longrightarrow \mathrm{La@C}_n + A^+$$

If the IP (La@C_n) > IP (A), then the reaction shown above will be observed; however, if IP (La@C_n) < IP (A), then no reaction will occur. Thus, the ionization potentials for La@C_n can be determined by using a series of different reagent species, A, whose ionization potentials are well known. La@C_n^+ was observed to charge exchange with decamethyl-ferrocene (I.P. = 5.8 ± 0.1 eV[88]) and *N,N,N,N*-tetramethyl-phenylenediamine (I.P. = 6.2 ± 0.05 eV[88]), shown in Figure 12, suggesting that I.P. (La@C_n) > 6.2 eV. Experimentally, charge exchange is observed as a decrease in the La@C_n^+ signal and an increase in the ionized reagent with increasing reaction time. La@C_n^+ is unreactive with ferrocene (I.P. = 6.75 ± 0.01 eV[88]), implying that I.P. (La@C_n) < 6.75 eV. Photoionization results of La@C_n indicate that these species can be ionized with a ArF excimer laser, implying that

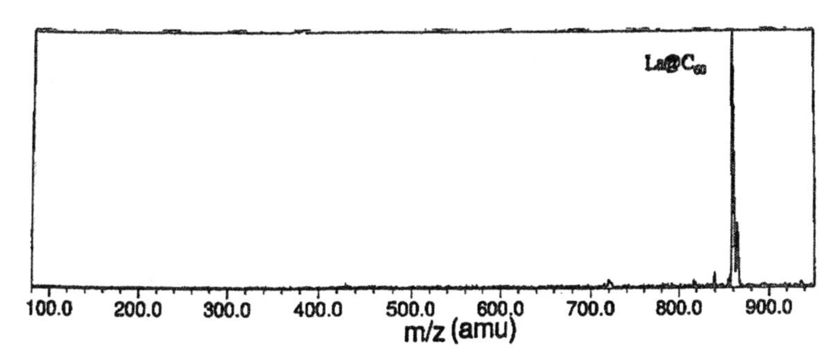

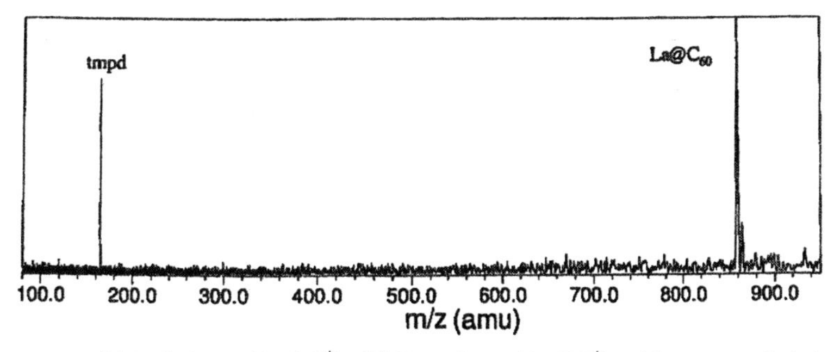

Figure 12. **(a)** Isolation of La@C_{60}^+. **(b)** Reaction of La@C_{60}^+ with tetramethylphenylenediamine.

I.P. (La@C_n) < 6.42 eV.[39] These experimental results imply that the I.P. of La@C_n is between 6.20 and 6.4 eV.

Control Experiments

Control experiments were conducted to verify that the charge transfer reactions could be used to determine ionization potentials. C_{60}^+ (generated by the same laser desorption conditions as the La@C_n^+ ions) was observed to charge transfer with both *N,N,N,N*-tetramethylphenylenediamine (I.P. = 6.2 ± 0.05 eV) and ferrocene (I.P. = 6.75 ± 0.009 eV), which is expected since the I.P. of C_{60} is 7.6 eV.[89]

Note that the I.P. of La@C_n is between that of C_{60} (7.6 eV) and La (5.577 eV).[88] Similarly, the electron affinities of La@C_n species were bracketed by reacting La@C_n^- with a variety of reagents. La@C_n^- was observed to charge exchange with fluroanil (*EA* = 2.8 eV), but not TCNQ (*EA* = 3.3 eV).[90] This places *EA*(La@C_n) between 2.8 and 3.3 eV. For comparison, the electron affinity of C_{60} is 2.666 eV.[91] Theoretical calculations predict an electron affinity for La@C_{60} of 3.8 eV and an IP of 6.9 eV.[92] Recent calculations yield somewhat lower values for La@C_{82} (*EA* = 3.22 eV and IP = 6.19 eV).[60]

These ionization potential and electron affinity bracketing measurements support the overall picture that the endohedral La atom transfers its three valence electrons to the lowest unoccupied molecular orbital (LUMO) of C_{60}. As a result, La@C_n consists of a positively charged La ion inside a negatively charged C_{60} shell. The molecular orbitals of C_{60} remain roughly the same, but are shifted downward in energy as a result of the Coulombic interactions between La^{3+} and C_{60}^{3-}.

IV. DISCUSSION AND CONCLUSIONS

The laser ablation of graphite containing certain metal atoms in a furnace with rare gases is demonstrated in this study to be an efficient technique for the production of fullerenes and metallofullerenes of the type MC_n where $n = 60, 70, 74, 80, 82, 84$ and other higher fullerenes. For some species, two and three atoms are found to be associated with higher ($n \geq 74$) fullerenes. In addition, LaC was found attached to or inside C_{84}. As observed by all other researchers, the metallofullerenes consisting of C_{60} and C_{70} are very fragile, i.e. air- and solvent-sensitive. Our studies of the reactions with toluene suggest that La is released from La@$C_{60,70}$. Based upon the laser desorption time of flight mass spectrometer results, $MC_{60,70}$ are the most abundant metallofullerenes present in the laser furnace soot. This is also observed in the arc discharge studies when the soot is extracted under anaerobic conditions. Although there are reports of separation of M@C_{60} and M@C_{70} from their siblings in certain solvents (e.g. aniline)[77] it is generally observed that these metallofullerenes do not survive solvation. From careful studies of extraction of these species in solvents containing various levels of impurities (i.e. water and oxygen) we find that the lifetime of M@$C_{60,70}$ in these solvents increases with the purity of the solvent. However, we could never reach a level of purity in which HPLC could be used to separate these specific metallofullerenes.

Freiser et al.[85] have found that co-depositing C_{60} and certain metal atoms (M) in a vacuum produces exohedral bonded MC_{60} molecules. These species are very reactive and easy to dissociate. It is also well known that superconducting fulleride salts are produced from 3:1 mixtures of alkali atoms with C_{60} fullerenes.[93] These superconductors are likewise very air-sensitive. Given this information, it is reasonable to suspect that the $C_{60,70}$ metallofullerenes produced in the laser furnace might be exohedrally bonded. Our studies support endohedral bonding of $M@C_{60,70}$. We list below some of the arguments that support endohedral bonding of $M@C_{60,70}$:

1. Only certain atoms are observed to form $M@C_{60}$. If co-deposition were occurring, in principle any complex could form. In particular, $Mg@C_{60}$ was not observed whereas all other group II metals were effective at producing these metallofullerenes species.
2. The $M@C_{60}^{+}$ ions could be photodissociated in both the LDMS and FTMS losing C_2 and not the metal atom, M. This is similar to the "shrink wrapping" observations demonstrating the endohedral nature of $La@C_{60}$.[87]
3. Studies of the reactivity of $M@C_{60}^{+/-}$ with other species shows electron exchange occurs before the loss of M.[85]
4. The measured IP and EA for $La@C_{60}$ are consistent with calculations for that of endohedral metallofullerenes.
5. If exohedral MC_{60} is produced, it would be reasonable to expect that some level of M_2C_{60} would occur, which it does not.

Based upon these arguments, we conclude that endohedral metallofullerenes are primarily produced in the laser ablation furnace reactor.

The mechanism or mechanisms responsible for the insertion of metal atoms into carbon cages are not known. One scenario would be that the carbon cage grows around the metal atom by the addition of carbon in the form of atoms, dimers, small fragments, and larger carbon clusters. The type of the metal atom that successfully forms the endohedral metallofullerene would be atoms that form strong noncovalent bonds with the growing fullerene precursors. Those metal atoms that form strong covalent metal–carbide bonds would not successfully form endohedral metal fullerenes since further carbon–carbon bonding would not occur and fullerene growth is thwarted. Thus, metal atoms that successfully form the endohedral metallofullerenes are those metal atoms that do not form strong metal–carbide bonds at the temperature where the fullerenes are growing (1000 to 5000 °C, depending on the technique—an arc discharge or laser ablation).

The metal atom present in the carbon vapor undergoes collisions with the polyatomic carbon fullerene precursors. As the fullerene precursors increase in size, there are more internal degrees for dissipation of the energy of collision with other carbon atoms, polyatomic carbon clusters, and metal atoms. According to the pentagon road mechanism for fullerene growth, the growing fullerene precursors have curvature and the "inside" and "outside" of the fullerenes begins to evolve. As the growing fullerene precursors begin to approach the size of a complete fullerene, the surface area available for sticking of the metal atoms on the inside is very large. That is, the average metal carbon atom distance is smaller for endohedral than exohedral species. Thus the sticking coefficient for the metal atom with the "inside" part of the growing fullerene precursor is higher than the "outside" part. Those atoms that do stick to the outside of the fullerene would also be fragile and likely to be fragmented during sublimation and harvesting.

A second scenario is one in which a negative ion and a positive ion recombine by inserting the positive ion into the negatively charged fullerene. Fullerenes possess high electron affinities and are present in the laser ablation or arc discharge. Similarly, the metal atoms which are presently known to insert into fullerenes all have low ionization potentials and they would likewise exist as singly and multiply charged cations in the laser or arc discharge soot. The presence of the metal atoms would also provide more electrons to the discharge thereby enhancing the negative ion yield of fullerenes. The long range coulomb attraction of the C_{60}^- anion and M^+ cation could lead to insertion through one of the six-member rings of the cage. The smaller size of the cation and the high degree of vibrational excitation of the anion would combine to provide a facile insertion into the cage.[11] For singly charged ions, the relative kinetic energy of the two singly charged ions at the point of touching would be about 5 eV which is necessary to overcome the barrier to insertion. The excess charges would also combine to lower the barrier. The cation would be trapped inside the cage by vibrational excitation and possible further charge transfer to the cage. This model was suggested by the recent observations of Ohtsuki et al.[94] These authors detected $^7Be@C_{60}$ following a recoil process of a nuclear reaction. Computer simulations of Be^{2+} colliding with C_{60}^{2-} at a collision energy of 5 eV demonstrated the insertion through a six-member ring within 60 fs. Experiments in our laboratory are being designed to investigate the reactivity of singly charged fullerene anions with doubly charged metal cation. Rate constants for the formation of $M@C_{60}^+$ could then be determined. This model is consistent with the recent observations of ion insertion of metal ions in fullerenes on surfaces.[95,96] Nitrogen ion implantation in solid C_{60} has produced a nitrogenated fullerene in which a single N atom is believed to be encapsulated.[97] EPR measurements show a isotropic hyperfine interaction suggestive of $N@C_{60}$.

ACKNOWLEDGMENTS

The first author wishes to acknowledge the financial support from the U.S. Department of Energy (DOE) Faculty Research Participation Program at the Oak Ridge National Laboratory, Oak Ridge, Tennessee during his 1996–1997 sabbatical leave from Berea College, Berea, Kentucky. This research was also supported in part by the Molecular Design Institute funded by the office for Naval Research through the Georgia Institute of Technology and the Vanderbilt University Free Electron Laser Center. We acknowledge the assistance of Dr. Richard Haglund and Dr. Oguz Yavas during the FEL studies. We thank Dr. Barbara DiCamillo for the use of Figure 1.

REFERENCES

1. Osawa, E. *Kagaku* **1970**, *25*, 854–863 (in Japanese). See also Yoshida, Z.; Osawa, E. *Aromaticity* (in Japanese); Kyoto: Kadjakudojin, 1971, pp. 174–178.
2. Bochvar, D. A.; Galpern, E. G. *Rep. Soviet Acad. Sci.* **1973**, *209*, 610 (in Russian).
3. Cox, D. M.; Rolfing, Kalder, A. *J. Chem. Phys.* **1984**, *81*, 3322.
4. Kroto, H. W.; Heath, J. R.; O'Brien, S. C.; Curl, R. F.; Smalley, R. E. *Nature* **1985**, *318*, 165.
5. Heath, J. R.; O'Brien, S. C.; Zhang, Q.; Liu, Y.; Curl, R. F.; Kroto, H. W.; Tittel, F. K.; Smalley, R. E. *J. Am. Chem. Soc.* **1985**, *107*, 7779.
6. Chai, Y.; Guo, T.; Jin, C.; Haufler, R. E.; Chibante, L. P. F.; Fure, J.; Wang, L.; Alford, J. M.; Smalley, R. E. *J. Phys. Chem.* **1991**, *95*, 7564.
7. Cioslowski, J.; Nanayakkara, A. *J. Chem. Phys.* **1992**, *96*, 8354; Cioslowski, J.; Ragharachari, K. *J. Chem. Phys.* **1993**, *98*, 8734.
8. Puska, M. J.; Nieminen, R. M. *Rev. Modern Phys.* **1994**, *66*, 3841.
9. Jean, Y. C.; Lu, X.; Lou, Y.; Bhavathi, A.; Sundar, C. S.; Lyy, Y.; Hor, T. H.; Chu, C. W. *Phys. Rev. B* **1992**, *45*, 12126.
10. Jaffke, T.; Illenberger, E.; Lezius, M.; Matejcik, S.; Smith, D.; Mark, T. D. *Chem. Phys. Lett.* **1994**, *226*, 213.
11. Huang, J.; Carman, H. S., Jr.; Compton, R. N. *J. Phys. Chem.* **1995**, *99*, 1719.
12. Percival, P. W.; Wlodck, S. *Chem. Phys. Lett.* **1992**, *196*, 317.
13. Cristofolini, Ricco, M.; De Renzi, R.; Dalcanale, E.; Marilla, L. *Chem. Phys. Lett.* **1995**, *234*, 260.
14. Weiske, T.; Böhme, D. K.; Hrušák, J.; Krätschmer, W.; Schwarz, H. *Angew. Chem. Int. Ed. Engl.* **1991**, *30*, 884.
15. Ross, M. M.; Callahan, J. H. *J. Phys. Chem.* **1991**, *95*, 5720.
16. Caldwell, K. A.; Giblin, D. E.; Hsu, C. S.; Cox, D.; Gross, M. L. *J. Am. Chem. Soc.* **1991**, *113*, 8519.
17. Weiske, T.; Wong, T.; Krätschmer, W.; Terlouw, J. K.; Schwarz, H. *Angew. Chem. Int. Ed. Engl.* **1992**, *31*, 183.
18. Weiske, T.; Schwarz, H.; Giblin, D. E.; Gross, M. L. *Chem. Phys. Lett.* **1994**, *227*, 87.
19. Saunders, M.; Jiménes-Vázquez, H. A.; Cross, R. J.; Poreda, R. J. *Science* **1993**, *259*, 1428.
20. Hrušák, J.; Böhme, D. K.; Weiske, T.; Schwarz, H. *Chem. Phys. Lett.* **1992**, *193*, 97.
21. Sprang, H.; Halikow, A.; Campbell, E. E. B. *Chem. Phys. Lett.* **1994**, *227*, 91.
22. Murry, R. L.; Scuseria, G. E. *Science* **1994**, *263*, 791.
23. Ratchkovskii, S.; Thiel, W. *J. Am. Chem. Soc.* **1996**, *118*, 7164.
24. Saunders, M.; Jiménez-Vázquez, H. A.; Cross, R. J.; Mroczkowski, S.; Gross, M. L.; Giblin, D. E.; Poreda, R. J. *J. Am. Chem. Soc.* **1994**, *116*, 2193.
25. Pang, L.; Brisse, F. *J. Phys. Chem.* **1993**, *97*, 8562.
26. Son, M.-S.; Sung, Y. K. *Chem. Phys. Lett.* **1995**, *245*, 113.
27. DiCamillo, B. A.; Hettich, R. L.; Guichon, G.; Compton, R. N.; Saunders, M.; Jiménez-Vázquez, H. A.; Khong, A.; Cross, R. J. *J. Phys. Chem.* **1996**, *100*, 9197.
28. Saunders, M.; Khong, A.; Shimshi, R.; Jiménez-Vázquez, H.; Cross, R. J. *Chem. Phys. Lett.* **1996**, *248*, 127.
29. Jiménes-Vázquez, H. A.; Cross, R. J.; Saunders, M.; Poreda, R. J. *Chem. Phys. Lett.* **1994**, *229*, 111.
30. Braun, T.; Rausch, H. *Chem. Phys. Lett.* **1995**, *237*, 443.

31. Gadd, G. E. et al. *Chem. Phys. Lett.* **1997**, *270*, 108.
32. Saunders, M.; Jiménes-Vázquez, H. A.; Cross, R. J.; Mroczkowski, S.; Freedberg, D. I.; Anet, F. A. L. *Nature* **1994**, *367*, 256.
33. Driehuys, B.; Gates, G. D.; Happer, W. *Phys. Rev. Lett.* **1995**, *74*, 4943.
34. Joslin, C. G.; Gray, C. G.; Goddard, J. D.; Goldman, S.; Yang, J.; Poll, J. D. *Chem. Phys. Lett.* **1993**, *213*, 377.
35. Bug, A. L. R.; Wilson, A.; Voth, G. A. *J. Phys. Chem.* **1992**, *96*, 7864.
36. Cioslowski, J. *Chem. Phys. Lett.* **1994**, *227*, 361.
37. Puska, M. J.; Nieminen, P. M. *Phys. Rev.* **1993**, *A 47*, 1181.
38. Saunders, M.; Cross, R. J.; Jiménez-Vázquez, H.; Shimshi, R.; Khong, A. *Science* **1996**, *271*, 1693.
39. Cox, D. M.; Trevor, D. J.; Reichmann, K. C.; Kaldor, A. *J. Am. Chem. Soc.* **1986**, *108*, 2457.
40. Cox, D. M.; Riechmann, K. C.; Kaldor, A. *J. Chem. Phys.* **1988**, *88*, 1588.
41. Weaver, J. H.; Chai, Y.; Kroll, G. H.; Jin, C.; Ohno, T. R.; Haufler, R. E.; Guo, T.; Alford, J. M.; Conceicao, J.; Chibante, L. P. F.; Jain, A.; Palmer, G.; Smalley, R. E. *Chem. Phys. Lett.* **1992**, *190*, 460.
42. McElvany, S. W. *J. Phys. Chem.* **1992**, *96*, 4935.
43. Gillan, E. G.; Yeretzian, C.; Min, K. S.; Alverez, M. M.; Whetten, R. L.; Kaner, R. B. *J. Phys. Chem.* **1992**, *96*, 6869.
44. Moro, L.; Ruoff, R. S.; Becker, C. H.; Lorents, D. C.; Malhotra, R. *J. Phys. Chem.* **1993**, *97*, 6801.
45. Guo, T.; Diener, M. D.; Chai, Y.; Alford, M. J.; Haufler, R. E.; McClure, S. M.; Ohno, T.; Weaver, J. H.; Scuseria, G. E.; Smalley, R. E. *Science* **1992**, *257*, 1661.
46. Wang, L. S.; Alford, J. M.; Chai, Y.; Diener, M.; Zhang, J.; McClure, S. M.; Guo, T.; Scuseria, G. E.; Smalley, R. E. *Chem. Phys. Lett.* **1993**, *207*, 354.
47. Kubozono, Y.; Noto, T.; Ohta, T.; Maeda, H.; Kashino, S.; Emura, S.; Ukita, S.; Sogabe, T. *Chem. Lett.* **1996**, 453.
48. Kubozono, Y.; Maeda, H.; Takabashi, Y.; Hiraoda, K.; Nakai, T.; Kashino, S.; Emura, S.; Ukita, S.; Sogabe, T. *J. Am. Chem. Soc.* **1996**, *118*, 6998.
49. Guo, T.; Smalley, R. E.; Scuseria, G. E. *J. Chem. Phys.* **1993**, *99*, 352.
50. Bethune, D. S.; Johnson, R. D.; Salem, J. R.; de Vries, M. S.; Yannoni, C. S. *Nature* **1993**, *366*, 123.
51. Yeretzian, C.; Hansen, K.; Alvarez, M. M.; Min, K. S.; Gillan, E. G.; Holczer, K.; Kaner, R. B.; Whetten, R. L. *Chem. Phys. Lett.* **1992**, *196*, 337.
52. Park, C.-H.; Wells, B. O.; DiCarlo, J.; Shen, Z.-X.; Salem, J. R.; Bethune, D. S.; Yannoni, C. S.; Johnson, R. D.; de Vries, M. S.; Booth, C.; Bridges, F.; Pianetta, P. *Chem. Phys. Lett.* **1993**, *213*, 196.
53. Johnson, R. D.; de Vries, M. S.; Salem, J.; Bethune, D. S.; Yannoni, C. S. *Nature* **1992**, *355*, 239.
54. Shinohara, H.; Sato, H.; Saito, Y.; Ohkohchi, M.; Ando, Y. *J. Phys. Chem.* **1992**, *96*, 3571.
55. Suzuki, S.; Kawata, S.; Shiramaru, H.; Yamauchi, K.; Kikuchi, K.; Kato, T.; Achiba, Y. *J. Phys. Chem.* **1992**, *96*, 7159.
56. Hoinkis, M.; Yannoni, C. S.; Bethune, D. S.; Salem, J. R.; Johnson, R. D.; Crowder, M. S.; de Vries, M. S. *Chem. Phys. Lett.* **1992**, *198*, 461.
57. Kato, T.; Suzuki, S.; Kikuchi, K.; Achiba, Y. *J. Phys. Chem.* **1993**, *97*, 13425.
58. Nagase, S.; Kobayashi, K.; Kato, T.; Achiba, Y. *Chem. Phys. Lett.* **1993**, *201*, 475.
59. Nagase, S.; Kobayashi, K. *Chem. Phys. Lett.* **1993**, *214*, 57.
60. Nagase, S.; Kobayashi, K. *Chem. Phys. Lett.* **1994**, *228*, 106.
61. Shinohara, H.; Yamaguchi, H.; Hayashi, N.; Sato, H.; Ohkohchi, M.; Ando, Y.; Saito, Y. *J. Phys. Chem.* **1993**, *97*, 4259.
62. Kikuchi, K.; Suzuki, S.; Nakao, Y.; Nakahara, Y.; Wakabayashi, T.; Shiromaru, T.; Saito, K.; Ikemoto, I.; Achiba, Y. *Chem. Phys. Lett.* **1993**, *216*, 67.
63. Kikuchi, K.; Nakao, Y.; Suzuki, S.; Achiba, Y.; Suzuki, T.; Maruyama, Y. *J. Am. Chem. Soc.* **1994**, *116*, 9367.

64. Ding, J.; Weng, L.-T.; Yang, S. *J. Phys. Chem.* **1996**, *100*, 11120.
65. Lin, N.; Ding, J. Q.; Yang, S. H.; Cue, N. *Phys. Lett.* **1996**, *A* 222, 190.
66. Funasaka, H.; Sajurai, K.; Oda, Y.; Yamamoto, K.; Takahashi, T. *Chem. Phys. Lett.* **1995**, *232*, 273.
67. Shinohara, H.; Inakuma, M.; Hayashi, N.; Sato, H.; Saito, Y.; Kato, T.; Bandow, S. *J. Phys. Chem.* **1994**, *98*, 8597.
68. van Loosdrecht, P. H. M.; Johnson, R. D.; de Vries, M. S.; Kiang, C.-H.; Bethune, D. S.; Dorn, H. C.; Burbank, P.; Stevenson, S. *Phys. Rev. Lett.* **1994**, *73*, 3415.
69. Hino, S.; Matsumoto, K.; Hasegawa, S.; Iwasaki, K.; Yakushi, K.; Morikawa, T.; Takahashi, T.; Seki, K.; Kikuchi, K.; Suzuki, S.; Ikemoto, I.; Achiba, Y. *Phys. Rev. B* **1993**, *48*, 8418.
70. Poirier, D. M.; Knupfer, M.; Weaver, J. H.; Andreoni, W.; Lassonen, K.; Parridello, M.; Bethune, D. S.; Kikuchi, K.; Achiba, Y. *Phys. Rev.* **1994**, *B 49*, 17403.
71. Takata, M.; Umeda, B.; Nishibori, E.; Sakata, M.; Salto, Y.; Ohno, M.; Shinohara, H. *Nature* **1995**, *377*, 46.
72. Laasonen, K.; Andreoni, W.; Parrinello, M. *Science* **1992**, *258*, 1916.
73. Takata, M.; Nishibori, E.; Umeda, B.; Sakata, M.; Yamamota, E.; Shinohara, H. *Phys. Rev. Lett.* **1997**, *78*, 3330.
74. Kobayashi, K.; Nagase, S.; Akasake, T. *Chem. Phys. Lett.* **1996**, *261*, 502.
75. Okade, N.; Ohba, Y.; Suzuki, S.; Kawata, S.; Kikuchi, K.; Achiba, Y.; Iwaizumi, M. *Chem. Phys. Lett.* **1995**, *235*, 564.
76. Ross, M. R.; Nelson, H. H.; Callahan, J. H.; McElvany, S. W. *J. Phys. Chem.* **1992**, *96*, 5231.
77. Kubozono, Y.; Maeda, H.; Takayashi, Y.; Hiraoka, H.; Nakai, T.; Kashino, S.; Emura, S.; Ukita, S.; Sogabe, T. *J. Am. Chem. Soc.* **1996**, *118*, 6998.
78. Huang, R.; Lu, W.; Yang, S. *J. Chem. Phys.* **1995**, *102*, 189.
79. Bandow, S.; Kitagawa, H.; Mitani, T.; Inokuchi, H.; Saito, Y.; Yamaguchi, H.; Hayashi, N.; Sato, H.; Shinohara, H. *J. Phys. Chem.* **1992**, *96*, 9609.
80. Rosen, A.; Wästberg, B. *Z. Phys.* **1989**, *D. 12*, 387–390.
81. Inoshita, T.; Ohnishi, S.; Oshiyama, A. *Phys. Rev. Lett.* **1989**, *57*, 2560–2563.
82. Marshall, A. G.; Grosshans, P. B. *Anal. Chem.* **1991**, *63*, 215A.
83. Hettich, R. L.; Jin, C. In *Laser Ablation*; Miller, J. C., Ed.; Springer Series in Materials Science; Springer-Verlag: Berlin, 1994, Vol. 28, p. 135.
84. Roth, L. M.; Huang, Y.; Schwedler, J. T.; Cassady, C. J.; Ben-Amotz, D.; Kahr, B.; Freiser, B. S. *J. Amer. Chem. Soc.* **1991**, *113*, 6298.
85. Huang, Y.; Freiser, B. S. *J. Amer. Chem. Soc.* **1991**, *113*, 9418.
86. McElvany, S. W. *J. Phys. Chem.* **1992**, *96*, 4935.
87. Weiss, E. D.; Elkind, J. L.; O'Brien, S. C.; Curl, R. F.; Smalley, R. E. *J. Amer. Chem. Soc.* **1988**, *110*, 4464.
88. Lias, S. G.; Bartmess, J. E.; Liebman, J. F.; Holmes, J. L.; Levin, R. D.; Mallard, W. G. *J. Phys. Chem. Ref. Data, Suppl. 1* **1988**, *17*.
89. Zimmerman, J. A.; Eyler, J. R.; Bach, S. B. H.; McElvany, S. W. *J. Chem. Phys.* **1991**, *94*, 3556.
90. Hettich, R.; Jin, C.; Compton, R. *Int. J. Mass Spec. Ion. Proc.* **1994**, *138*, 263.
91. Brink, C.; Andersen, L. H.; Hvelplund, P.; Mathur, D.; Voldstad, J. D. *Chem. Phys. Lett.* **1995**, *233*, 52.
92. Rosen, A.; Wastberg, B. *J. Amer. Chem. Soc.* **1988**, *110*, 8701.
93. Haddon, R. C. et al. *Nature (London)* **1991**, *350*, 320.
94. Ohtsuki, T.; Masumoto, K.; Ohno, K.; Maruyma, Y.; Kawazoe, Y.; Sueki, K.; Kikuchi, K. *Phys. Rev. Lett.* **1996**, *77*, 3522.
95. Tellgmann, R.; Krawez, N.; Lin, S.-H.; Hertel, I. V.; Campbell, E. E. *Nature* **1996**, *382*, 382.
96. Shimshi, R.; Cross, R. J.; Saunders, M. *J. Am. Chem. Soc.* **1997**, *119*, 1163.
97. Almeida Murphy, T.; Pawlik, Th.; Weidinger, A.; Hohne, M.; Alcala, R.; Spaeh, J. M. *Phys. Rev. Lett.* **1997**, *77*, 1075.

ENDOHEDRAL METALLOFULLERENES: STRUCTURES AND ELECTRONIC PROPERTIES

Hisanori Shinohara

Advances in Metal and Semiconductor Clusters
Volume 4, pages 205–226

ISBN: 0-7623-0058-2

I. INTRODUCTION

Endohedral metallofullerenes (endo-fullerenes, for short) are novel forms of fullerene-related materials which can encage various kinds of metal atom(s) in a variety of fullerene cages.[1,2] In this chapter, recent advances on the production, separation (isolation), and various spectroscopic characterization of endo-fullerenes are presented in order to clarify their structural and electronic properties. Endohedral metallofullerenes are normally produced by the DC arc-discharge method with metal/graphite composite rods as positive electrodes. The endo-fullerenes so far produced are centered on rare earth metallofullerenes such as Sc, Y, La, Ce, Pr, and Gd which are based mainly on the C_{82} fullerene.

The purification of endo-fullerenes had been difficult, mainly because that the content of the metallofullerenes in soot extracts was very limited and, furthermore, the solubility in normal HPLC solvents was even lower than that of various higher hollow fullerenes. It took almost 2 years for metallofullerenes to be completely isolated by high-performance liquid chromatography (HPLC)[3,4] since the first extraction of La@C_{82} by the Rice group.[5] The success of the purification was a big breakthrough for a further progress on metallofullerene studies. By using the two (or multi)-stage HPLC (high performance liquid chromatography) technique, these metallo-fullerenes have been completely purified and isolated, such as C_{60} and C_{70}.

It has been revealed by electron spin resonance (ESR) and theoretical calculations that substantial electron transfers are taking place from the encaged metal atom to the carbon cage: intra-fullerene electron transfers. Recent synchrotron X-ray diffraction studies[6,7] show that a metal atom is not in the center of the fullerene but very close to the carbon cage, indicating the presence of a strong metal–cage interaction. Up to the present, various spectroscopic and structural analyses including UPS, TEM, EXAFS, STM, ESR, ^{13}C NMR, and X-ray diffraction have been performed to elucidate their novel molecular and solid-state properties.

In the present review, we first present the production and sample purification procedures of the metallofullerene and then discuss the various structural and spectroscopic characterization of the metallofullerenes based on scanning tunneling microscopy (STM), ^{13}C NMR, and synchrotron X-ray diffraction.

II. PRODUCTION, EXTRACTION, AND PURIFICATION OF ENDOHEDRAL METALLOFULLERENES

A. Production and Extraction of Metallofullerenes

We will discuss the production and extraction of scandium fullerenes in particular, because scandium fullerenes exist as mono-, di-, and even tri-scandium fullerenes, which is quite unique as compared with other rare earth metallofullerenes such as lanthanum and yttrium endo-fullerenes. The production of the mono-, di-, and tri-scandium fullerenes was found to be sensitive to the mixing ratio of scandium molecules show no characteristic bright (or dark) spots (which may correspond to the position of scandium atoms) on and around the carbon cage, and all images are essentially the same as those of hollow C_{84} molecules.[16] The two scandium atoms are trapped inside the C_{84} cage. An intriguing and important observation in Figure 4 is that the nearest neighbor distance of the $Sc_2@C_{84}$ molecules (11.7 Å) is slightly smaller than that of the hollow C_{84} molecules (12.0 Å) on the surface. An intermolecular interaction among the $Sc_2@C_{84}$ molecules must play a significant role in the shrinkage of the close-packing geometry on the Si surface. The hyperfine structure analysis of ESR studies on the scandium fullerenes[8,9] indicated that the scandium fullerene has a relatively large permanent dipole moment due to the electron transfer from the encaged metal atoms to the carbon cage. An ab initio calculation[20] indicates that each encaged scandium atom is in the +2 oxidation state in $Sc_2@C_{84}$, leading to the electronic structure of $(Sc^{2+})^2@C_{84}^{4-}$.

Figure 5 shows the height distribution of the $Sc_2@C_{84}$ molecules adsorbed on the Si(100)2 × 1 surface, in reference to that of C_{60}. The distribution has the

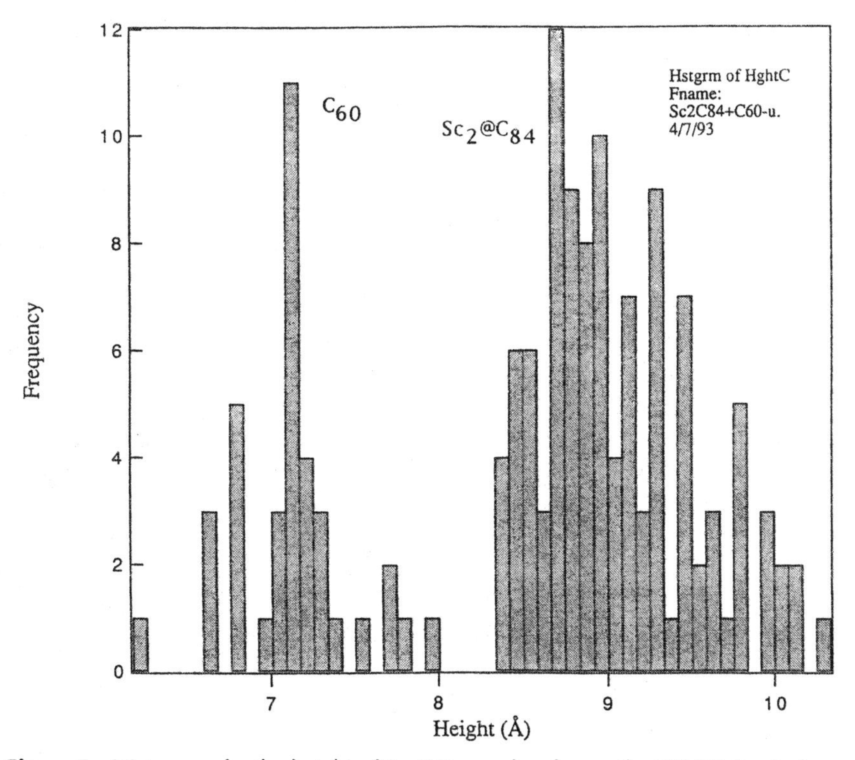

Figure 5. Histogram for the height of $Sc_2@C_{84}$ molecules on the Si(100) 2 × 1 clean surface as determined by the present STM observation. The height of $Sc_2@C_{84}$ molecules is scaled with respect to that of C_{60} (7.1 Å).

maximum at the height of 8.5–8.7 Å which corresponds to the long axes of the D_{2d}(No.23) and D_2(No.22) isomers of C_{84}.[21] It has been known that there are 24 structural isomers of empty C_{84} which satisfy the so-called "isolated pentagon rule" (IPR). Although, we are not able to observe the details of the $Sc_2@C_{84}$ atomic configuration at present, we can exclude the other possible C_{84} isomers such as C_{84} (D_{6h}), C_{84} (T_d), and C_{84} based on the present STM analysis.

B. $Y@C_{82}$ on Cu(111)1 × 1 Surface

A typical large scale (400 Å × 400 Å) STM image of the very small amount (23 molecules/(1000 Å)^2) of $Y@C_{82}$ on Cu(111)1 × 1 surface at room temperature shows a preferential adsorption to the terrace edges (Figure 6). The $Y@C_{82}$ molecules are sublimated from the tantalum boat onto the copper surface and impinge on the terrace of the surface with a kinetic energy corresponding to ca. 400°C. The $Y@C_{82}$ molecules are mobile on the surface and segregate to the terrace edges. The impinging $Y@C_{82}$ molecules migrate to the edges following adsorption since the bonding to the substrate surface is relatively weak. The C_{60} adsorption on the Cu(111)1 × 1 surface showed a similar mobile tendency.[22] This is in sharp contrast to the adsorption of fullerenes on Si(100) and Si(111) surfaces in which the fullerenes such as C_{60},[15] C_{70},[23] and C_{84}[19] do not freely migrate.

One of the most important findings of the present study is that the $Y@C_{82}$ fullerenes predominantly form clusters, $(Y@C_{82})_n$ (n = 2–6), and in particular dimers, $(Y@C_{82})_2$, on the Cu(111) surface even at the very initial stage of adsorption.[24] A typical STM observation (Figure 6) images three dimers and one trimer of $Y@C_{82}$ at a terrace edge. To further explore the dimer and cluster formation of $Y@C_{82}$, a statistical analysis of the cluster size distribution of $Y@C_{82}$ was performed, as compared with that of C_{60}, on the copper surface.

Figure 7 shows a histogram for the cluster size distribution of $Y@C_{82}$ and C_{60} on the same copper surface. The total number of $Y@C_{82}$ and C_{60} molecules that were actually counted in the present analysis are 1108 and 1716, respectively, which is sufficient to perform a statistical analysis. Moreover, the average numbers of adsorbed $Y@C_{82}$ and C_{60} fullerenes at the step edge were 3.4 molecules/1000 Å and 4.1 molecules/1000 Å, respectively, which means that the adsorption conditions were very similar in both cases. This allows us to make a fair statistical comparison of these experimental results. The adsorption rate was kept very low, typically 3000–5000 molecules per minute for an area of 1000 Å × 1000 Å, in order not to induce unnecessary multiple fullerene–fullerene collisions on the surface.

The distribution has the maximum at dimer and shows that more than 60% of the $Y@C_{82}$ molecules on the Cu(111) surface exist as dimers or larger clusters. However, a similar histogram for the cluster size distribution of hollow C_{60} molecules on a similar surface shows a totally different trend: more than 70% of the C_{60} molecules exist as monomers on the surface, and the formation of the C_{60} dimers and the clusters is a minor process under the similar experimental conditions results show that the two Sc atoms are indeed trapped within the C_{84} cage. A similar equivalence of trapped atoms has been reported on an endohedral tri-scandium fullerene, $Sc_3@C_{82}$, based on an analysis of the hyperfine structure of the corresponding ESR measurements.[8,32,33]

An endohedral structural model for $Sc_2@C_{84}$ (III), which is consistent with the present ^{13}C NMR result, is presented in Figure 11; the two Sc atoms are encaged along the D_{2d} axis of the C_{84}-D_{2d} (No.23) fullerene in a symmetric position with respect to the center of the C_{84} cage. A recent ab initio calculation[20] has predicted that such an endohedral structure is the most stable structure. It also predicts that each Sc atom donates two valence electrons to the C_{84} cage, leading to a formal electronic structure $(Sc^{2+})^2@C_{84}^{4-}$. However, at least at room temperature, some dynamical averaging of the Sc ions might be taking place around the optimum scandium position such as in Figure 11. This is mainly due to Coulomb repulsion between the two Sc^{2+} cations within the cage. This dynamical picture for the endohedral structure of $Sc_2@C_{84}$ (III) is also consistent with the present NMR results. In fact, our preliminary synchrotron X-ray powder diffraction study on $Sc_2@C_{84}$ (III)[7] shows strong evidence that this is the better structural description than the static picture.

The fact that in $Sc_2@C_{84}$ the two Sc atoms are preferentially trapped within one of the two major isomers of the corresponding hollow C_{84} fullerene, C_{84}-D_{2d} (No.23), will be an important clue for elucidating the growth mechanism of endohedral metallofullerenes.

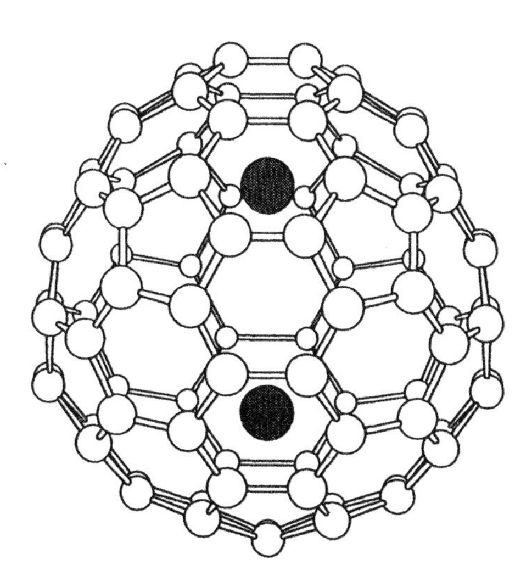

Figure 11. A molecular structure for $Sc_2@C_{84}$ (III) consistent with the ^{13}C NMR results (see text).

V. SYNCHROTRON X-RAY DIFFRACTION STUDY OF METALLOFULLERENES

Synchrotron X-ray diffraction has been a powerful technique for obtaining structural information on various types of fullerenes. The synchrotron powder X-ray diffraction method has great advantages over the conventional laboratory X-ray measurement: intense X-ray and high spectral resolution. We have applied this useful technique to the structural analysis of endohedral metallofullerenes.

A. $Y@C_{82}$

In the preceding sections, we have discussed the structural and electronic aspects of several endo-fullerenes in terms of STM and ^{13}C NMR, and presented evidence that these metallofullerenes are "endohedral" metallofullerenes. Previous experimental evidence including extended X-ray absorption fine structure (EXAFS)[35,36] and transmission electron microscopy (TEM)[37] indicate that the metal atoms are inside the fullerenes. Theoretical calculations also suggest that such metallofullerenes are endohedral. However, the first conclusive evidence of the endohedral nature of a metallofullerene was obtained by a synchrotron X-ray diffraction study: the yttrium atom is encapsulated within the C_{82} fullerene for $Y@C_{82}$ and is strongly bound to the carbon cage.[6]

To collect an X-ray powder pattern with good counting statistics, the synchrotron radiation X-ray powder experiment with an imaging plate detector was carried out at Photon Factory BL-6A2 (Institute for High Energy Physics, Tsukuba). The X-ray powder data of the hollow C_{82} fullerene is also measured as a reference under a similar experimental condition. The powder pattern of $Y@C_{82}$ is also similar to that of the hollow C_{82}. The space group is assigned to P_21, which is monoclinic for both $Y@C_{82}$ and C_{82}. The experimental data are analyzed in an iterative way of combination of Rietveld analysis[38] and the maximum entropy method (MEM).[39,40]

First, the structure factors of the individual reflections are derived with a help of the Rietveld method. The central position of the C_{82} sphere, the radius of the sphere, and the position of yttrium atom are refined in the Rietveld analysis. The reliability factors (R-factors) based on the Bragg intensities, R_I, are 14.4 and 12.7% for $Y@C_{82}$ and C_{82}, respectively. Second, the MEM analysis[41] is carried out with a computer program, MEED.[42] The MEM can produce an election density distribution map from a set of X-ray structure factors without using any structural model. It is well known that the MEM map is consistent with the observed structure factors and least biased with respect to unobserved structure factors. By the MEM analysis, the R_I become as low as 1.5 and 1.4% for $Y@C_{82}$ and C_{82}, respectively.

A series of the iterative steps involving the Rietveld analysis, using the revised structural model based on the previous MEM map, and the MEM analysis are carried out until no significant improvement is obtained. Eventually, the R_I factor improved from 14.4 to 5.9% (RWP = 3.0%). In Figure 12, the best fit of the Rietveld

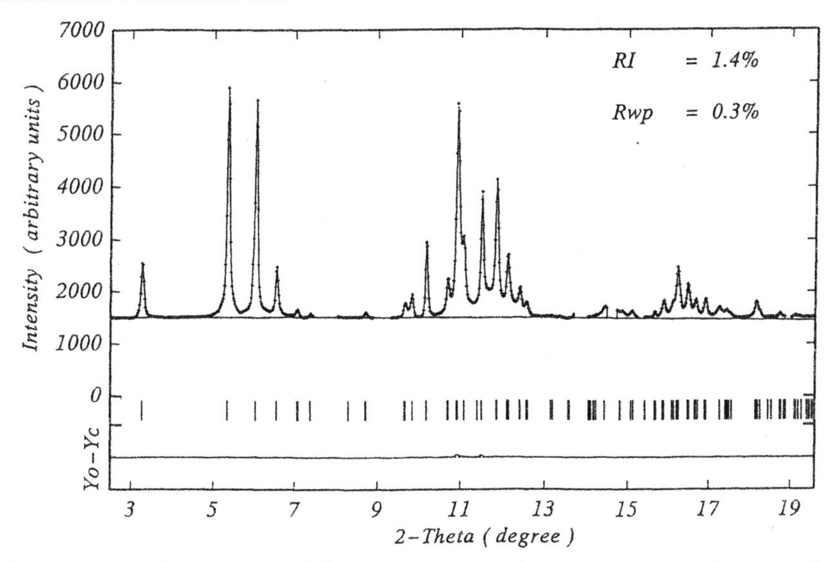

Figure 12. Synchrotron X-ray diffraction patterns and the corresponding fitting results of Y@C_{82} based on the calculated intensities from the MEM electron density.

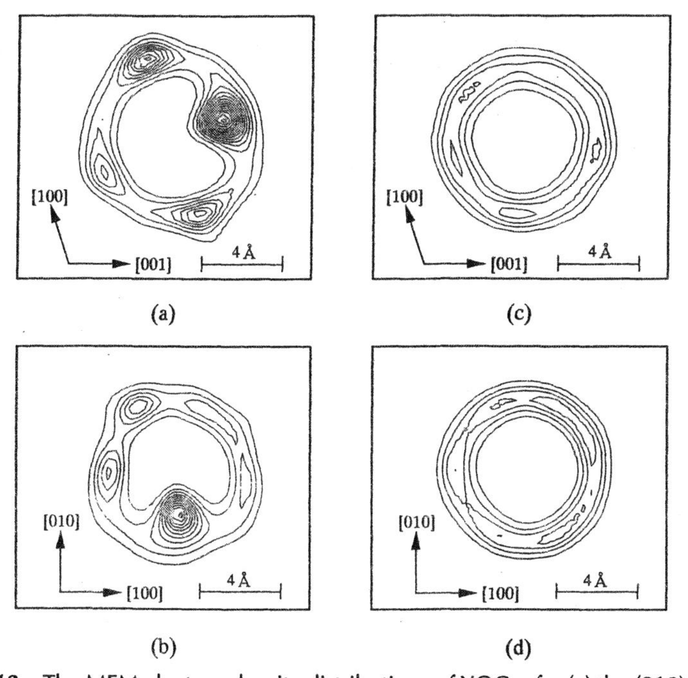

Figure 13. The MEM electron density distributions of Y@C_{82} for (**a**) the (010) and (**b**) (001) sections and those of C_{82} for (**c**) the (010) and (**d**) (001) sections. The counter lines are drawn from 0.0 [e/A^3] with a 0.5[e/A^3] step width. The endohedral nature of Y@C_{82} is clearly shown in (**a**) and (**b**).

analysis of Y@C_{82} is shown. The positional coordinates determined from the final MEM map coincide with that of Rietveld analysis. To visualize the endohedral nature of the Y@C_{82}, the MEM electron density distributions of Y@C_{82} and C_{82} are shown in Figure 13. These figures correspond to a section of Y@C_{82} and C_{82}. There exist remarkably high densities just inside the C_{82} cage (Figure 13a,b). Evidently, the density maxima at the interior of the C_{82} cage is the yttrium atom.

The cage structure of Y@C_{82} differs from that of the hollow C_{82} fullerene. There are many local maxima along the cage in Y@C_{82}, whereas electron densities of the C_{82} cage is relatively uniform (Figure 13c,d). This suggests that in Y@C_{82} the rotation of the C_{82} cage is very limited around a certain axis even at room temperature, while that in C_{82} is almost free. The MEM election density map further reveals that the yttrium atom does not reside in the center of the C_{82} cage but is very close to the carbon cage as suggested theoretically. The ESR and theoretical studies suggest the presence of a strong charge transfer interaction between the Y^{3+} ion and the C_{82}^{3-} cage, which may cause the aspherical electron density distribution of atoms. The Y–C distance calculated from the MEM map is 2.9(3) Å which is slightly longer than a theoretical prediction of 2.55–2.65 Å.[26,30] The present study also reveals that the Y@C_{82} molecules are aligned along the [001] direction in a "head-to-tail" (....Y@C_{82}....Y@C_{82}....Y@C_{82}....) order in the crystal, indicating the presence of a strong dipole–dipole and charge transfer interactions among the Y@C_{82} fullerenes. The ability to achieve a molecular alignment of the endohedral metallofullerene in a certain direction in the crystal or clean surfaces may in future lead to materials with novel solid-state properties.

B. Sc_2@C_{84}

In addition to mono-metallofullerenes, various metallofullerenes supposed to encapsulate two or three metal atoms within fullerene cages, such as La_2@C_{80}, Y_2@C_{82}, Sc_2@C_{74}, Sc_2@C_{82}, Sc_2@C_{84}, and Sc_3@C_{82}, have been successfully synthesized and purified. Among them the di-metallofullerenes are especially interesting, because they might exhibit novel solid-state properties due to their very small band gaps.[1]

In the series of the procedure as in Y@C_{82}, the C_2 axis was finally refined to the [5 3 7] direction and an anisotropic thermal vibrational model was introduced for the Sc atoms to express the elongated feature parallel to the C_2 axis in the MEM map. Eventually, the positional parameters of all carbon atoms and Sc atoms were included as refined parameters in the Rietveld analysis maintaining the D_{2d} symmetry as determined by the ^{13}C NMR study.[11] Under this D_{2d} symmetry constraint of the molecule, the carbon atomic sites are reduced from 84 to 11 for one 4b site and ten 8d sites. In the final stage, the number of structural parameters refined are 54. The R_I factor finally became to 7.9% (R_{wp} = 5.3%). In Figure 14a, the best fit of the Rietveld analysis of Sc_2@C_{84} is shown. The final MEM map was obtained by using the observed structure factors evaluated from the Rietveld analysis with R_I = 7.9%.

clearly exhibits six- and five-membered rings. This configuration of the rings is exactly the same as that of the $C_{84}(D_{2d})$ cage. The two density maxima can be seen through the carbon cage. The number of electrons around each maximum inside the cage is 18.8 which is very close to that of a divalent scandium ion Sc^{2+} (19.0). Evidently, each of the two density maxima at the interior of the C_{84} cage corresponds to a scandium atom, indicating that $Sc_2@C_{84}$ is endohedral. A theoretical study has predicted that the formal electronic structure of $Sc_2@C_{84}$ is well represented by $(Sc^{2+})^2@C_{84}^{4-}$, where two 4s electrons of each Sc atom are transferring to the C_{84} cage.[20] The positive charge of the Sc atom from the MEM charge density is +2.2 which is in good accordance with the theoretical value.[20]

The charge density of the present $Sc_2@C_{84}$ molecule shows a distinct feature which is in very sharp contrast to that of $Y@C_{82}$.[6] The 5/6 ring structure of C_{84} can be clearly seen though the present experiment was performed at room temperature, whereas only averaged charge density of C_{82} was obtained for $Y@C_{82}$. This suggests that rotations of the C_{84} cage are virtually quenched even at room temperature in a $Sc_2@C_{84}$ solid, while the $Y@C_{82}$ molecule is still in a hindered rotation.

In Figure 16a, the section of the MEM charge density including the encaged Sc atoms is shown for (010) plane. Two density maxima can be clearly seen inside the section of the carbon cage and are located in a symmetric position with respect to the center of the cage. The Sc–Sc distance in C_{84} derived from the MEM charge density is 3.9(1) Å which is a bit smaller than that of a theoretical value of 4.029 Å.[20] The nearest Sc–C distance is 2.4(2) Å (2.358 Å theoretical value[20]) (Figure 16b). One of the most intriguing observations in Figure 16a is that the charge density of the Sc atoms shows a salient tear-drop feature as if the two Sc atoms (ions) are in a stretching vibration within the C_{84} cage. The C–C distance of the double bond adjacent to the Sc atom is 1.9(3) Å, which is considerably longer than a theoretical value of 1.434(8) Å, indicating some significant distortion of the polar regions of the C_{84} cage. Such anomalous large C–C distance (1.90(15) Å) has been also reported for polymeric fullerene, RbC_{60}.[43] Although a further study will be required for such an unusual distance with the higher resolution data, the present result suggests a strong indication of the elongation of C–C distance. Such an indication is also recognized even in the Riedvelt analysis. This might be closely related to the thermal motion of the Sc atoms in the C_{84} cage, and to the existence of some localized interaction caused by the charge transfer between the encapsulated Sc atoms and the C_{84} cage.

A rotational vibration of the Sc atoms along the C–C double bonds, which is predicted by a recent ab initio calculation,[44] is not obvious in the present MEM charge density. The theoretical calculation was done at 0 K. Therefore, low-temperature experiments should be required to confirm the existence of the rotational vibration mode of the Sc atoms. The variety of thermal behavior of metal atoms inside the carbon cage may be lead to the discovery of novel physical property of the material.

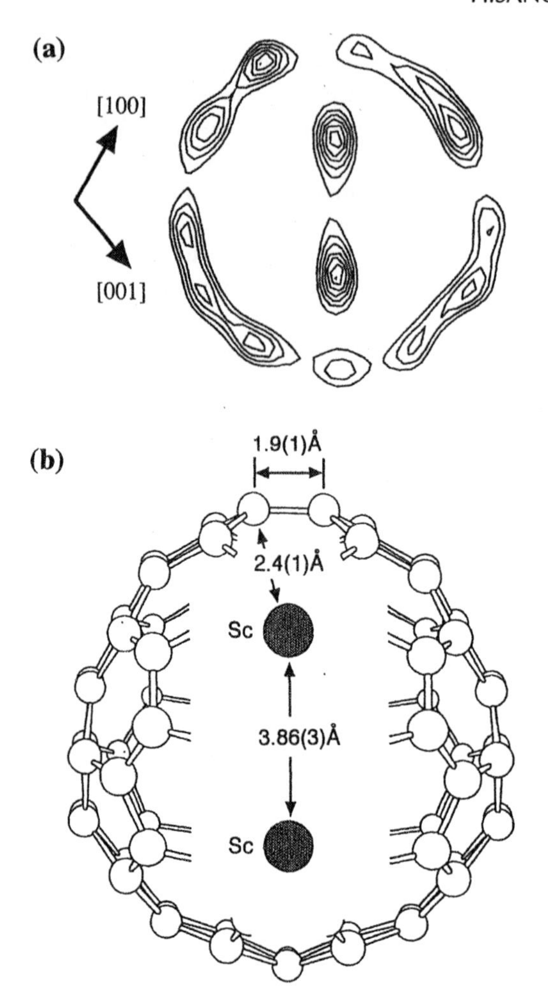

Figure 16. (**a**) The MEM electron density distribution of $Sc_2@C_{84}$ for (010) section. The contour lines are drawn with 0.3 eÅ^{-3} intervals from 1.0 eÅ^{-3}. The endohedral nature of $Sc_2@C_{84}$ is clearly shown in these figures. (**b**) The schematic representation of the Sc–Sc, Sc–C, and double-bonded C–C distances obtained from the MEM charge density.

VI. FUTURE PROSPECTIVE IN ENDOHEDRAL METALLOFULLERENES

Endohedral metallofullerenes are novel forms of fullerene-related materials which we were able to obtain, in their purified form, only a few years ago. Since then, as we have seen in this chapter, numerous investigations in various aspects of the metallofullerenes have been intensively done by many research groups. These studies have revealed unique structural and novel electronic properties of the metallofullerenes. Even so, there still remain many potentially important and intriguing aspects of them to be clarified. For instance, the solid-state properties, such as electric conductivity and magnetic behavior, are not well known. Physiological and medical applications of the endohedral metallofullerenes will be important in relation to tracer chemistry in biological systems, and definitely await future studies.[45] In any case, endohedral metallofullerenes will continue to provide us many stimuli in basic science and technology.

ACKNOWLEDGMENTS

I express my thanks to Dr. T. Sugai, M. Ohno, T. Kimura, T. Nakane and M. Inakuma, E. Yamamoto (Nagoya University) for experimental help. The author is indebted to the STM collaboration with Prof. T. Sakurai (IMR, Tohoku University) and Dr. T. Hashizume (Hitachi Ltd.). The synchrotron X-ray collaboration with Dr. M. Takata and Prof. M. Sakata (Nagoya University) is greatly acknowledged. I also thank the Japanese Ministry of Education, Science, Sports and Culture for Grants-in-Aid for Scientific Research on Priority Areas (No.05233108) and Scientific Research (A) (2) (No.08554020).

REFERENCES

1. Shinohara, H.; Takata, M.; Sakata, M.; Hashizume, T.; Sakurai, T.; *Materials Sci. Forum* **1996**, *232*, 207.
2. Bethune, D. S.; Johnson, R. D.; Salem, J. R.; de Vries, M. S.; Yannoni, C. S. *Nature* **1993**, *366*, 123.
3. Shinohara, H.; Yamaguchi, H.; Hayashi, N.; Sato, H.; Ohkohchi, M.; Ando, Y.; Saito, Y. *J. Phys. Chem.* **1993**, *97*, 4259.
4. Kikuchi, K.; Suzuki, S.; Nakao, Y.; Nakahara, N.; Wakabayashi, T.; Shiromaru, H.; Saito, I.; Ikemoto, I.; Achiba, Y. *Chem. Phys. Lett.* **1993**, *216*, 67.
5. Chai, Y.; Guo, T.; Jin, C.; Haufler, R. E.; Chibante, L. P. F.; Fure, J.; Wang, L.; Alford, J. M.; Smalley, R. E. *J. Phys. Chem.* **1991**, *95*, 7564.
6. Takada, M.; Umeda, B.; Nishibori, E.; Sakata, M.; Saito, Y.; Ohno, M.; Shinohara, H. *Nature* **1995**, *377*, 46.
7. Takada, M.; Nishibori, E.; Umeda, B.; Sakata, E. M.; Yamamoto, E.; Shinohara, H. *Phys. Rev. Lett.* **1997**, *78*, 3330.
8. Shinohara, H.; Sato, H.; Ohchochi, M.; Ando, Y.; Kodama, T.; Shida, T.; Kato, T.; Saito, Y. *Nature* **1992**, *357*, 52.
9. Inakuma, M.; Ohno, M.; Shinohara, H. In *Fullerenes: Recent Advances in the Chemistry and Physics of Fullerenes and Related Materials*; Kadish, K.; Ruoff, R., Eds.; Electrochemical Society, Reno, NV, 1995, Vol. 2, p. 330.
10. Shinohara, H.; Inakuma, M.; Hayashi, N.; Sato, H.; Saito, Y.; Kato, T.; Bandow, S. *J. Phys. Chem.* **1994**, *98*, 8597.
11. Yamamoto, E.; Tansho, M.; Tomiyama, T.; Shinohara, H.; Kawahara, H.; Kobayashi, Y. *J. Am. Chem. Soc.* **1996**, *118*, 2293.
12. Wilson, R. J.; Meijer, G.; Bethune, D. S.; Johnson, R. D.; Chambliss, D. D.; de Vries, M. S.; Hunziker, H. E.; Wendt, H. R. *Nature* **1990**, *348*, 621.

13. Wragg, J. L.; Chamberlain, J. E.; White, H. W.; Kraetschmer, W.; Huffman, D. R. *Nature* **1990**, *348*, 623.
14. Li, Y. Z.; Patrin, J. C.; Chander, M.; Weaver, J. H.; Chibante, L. P. F.; Smalley, R. E. *Science* **1991**, *252*, 547.
15. Hashizume, T.; Wang, X.D.; Nishina, Y.; Shinohara, H.; Saito, Y.; Kuk, Y.; Sakurai, T. *Jpn. J. Appl. Phys.* **1992**, *31*, L880.
16. Wang, X. D.; Hashizume, T.; Shinohara, H.; Saito, Y.; Nishina, Y.; Sakurai, T. *Jpn. J. Appl. Phys.* **1992**, *31*, L983.
17. Sakurai, T.; Wang, X. D.; Xue, Q. K.; Hasegawa, Y.; Hashizume, T.; Shinohara, H. *Prog. Surf. Sci.* **1996**, *51*, 263.
18. Shinohara, H.; Hayashi, N.; Sato, H.; Saito, Y.; Wang, X. D.; Hashizume, T.; Sakurai, T. *J. Phys. Chem.* **1993**, *97*, 13438.
19. Wang, X. D.; Hashizume, T.; Shinohara, H.; Saito, Y.; Nishina, Y.; Sakurai, T. *Phys. Rev. B* **1993**, *47*, 15923.
20. Nagase, S.; Kobayashi, K. *Chem. Phys. Lett.* **1994**, *231*, 319.
21. Manolopoulos, D. E.; Fowler, P. W. *J. Chem. Phys.* **1992**, *96*, 7603.
22. Hashizume, T.; Motai, K.; Wang, X. D.; Shinohara, H.; Saito, Y.; Maruyama, Y.; Ohno, K.; Kawazoe, Y.; Nishina, Y.; Pickering, H. W.; Kuk, Y.; Sakurai, T. *Phys. Rev. Lett.* **1993**, *71*, 2959.
23. Wang, X. D.; Xue, Q.; Hashizume, T.; Shinohara, H.; Nishina, Y.; Sakurai, T. *Phys. Rev. B* **1994**, *49*, 7754.
24. Shinohara, H.; Inakuma, M.; Kishida, M.; Yamazaki, S.; Hashizume, T.; Sakurai, T. *J. Phys. Chem.* **1995**, *99*, 13769.
25. Shinohara, H.; Sato, H.; Saito, Y.; Ohkohchi, M.; Ando, Y. *J. Phys. Chem.* **1992**, *96*, 3571.
26. Nagase, S.; Kobayashi, K. *Chem. Phys. Lett.* **1993**, *214*, 57.
27. Watanabe, H.; Inoshita, T. *Optoelectron. Dev. Technol.* **1986**, *1*, 33.
28. Rosen, A.; Waestberg, B. *Z. Phys. D* **1989**, *12*, 387.
29. Saito, S. In *Clusters and Cluster-Assembled Materials*; Averback, R. S.; Bernholc, J.; Nelson, D. L., Eds.; MRS: Pittsburgh, 1990, p. 115.
30. Nagase, S.; Kobayashi, K. *Chem. Phys. Lett.* **1994**, *228*, 106.
31. Kikuchi, K.; Nakahara, N.; Wakabayashi, T.; Suzuki, S.; Shiromaru, H.; Miyake, Y.; Saito, K.; Ikemoto, I.; Kainosho, M.; Achiba, Y. *Nature* **1992**, *357*, 142.
32. Kato, T.; Bandow, S.; Inakuma, M.; Shinohara, H. *J. Phys. Chem.* **1995**, *99*, 856.
33. Yannoni, C. S.; Hoinkis, M.; de Vries, M. S.; Bethune, D. S.; Salem, J. R.; Crowder, M. S.; Johnson, R. D. *Science* **1992**, *256*, 1191.
34. Fowler, P. W.; Manolopoulos, D. E. *An Atlas of Fullerenes*; Clarendon: Oxford, 1995, pp. 258–259.
35. Park, C. H.; Wells, B. O.; Dicarlo, J.; Shen, Z. X.; Salem, J. R.; Bethune, D. S.; Yannoni, C. S.; Johnson, R. D.; de Vries, M. S.; Booth, C.; Bridges, F.; Pianetta, P. *Chem. Phys. Lett.* **1993**, *213*, 196.
36. Kikuchi, K.; Nakao, Y.; Achiba, Y.; Nomura, M. In *Fullerenes: Recent Advances in the Chemistry and Physics of Fullerenes and Related Materials*; Kadish, K.; Ruoff, R., Eds.; Electrochemical Society: Pennington, 1994, Vol. 1, pp. 1300–1308.
37. Beyers, R.; Kiang, C-H.; Johnson, R. D.; Salem, J. R.; de Vries, M. S.; Yannoni, C. S.; Bethune, D. S.; Dorn, H. C.; Burbank, P.; Harich, K.; Stevenson, S. *Nature* **1994**, *370*, 196.
38. Rietveld, H. M. *J. Appl. Cryst.* **1968**, *2*, 65.
39. Collins, D. M. *Nature* **1982**, *298*, 49.
40. Bricogne, G. *Acta Cryst.* **1988**, *A44*, 517.
41. Sakata, M.; Sato, M. *Acta Cryst.* **1990**, *A46*, 263.
42. Kumazawa, S.; Kubota, Y.; Tanaka, M.; Sakata, M.; Ishibashi, Y. *J. Appl. Cryst.* **1993**, *26*, 453.
43. Stephens, P. W. et al. *Nature* **1994**, *370*, 636.
44. Kobayashi, K.; Nagase, S. Private communication.
45. Kobayashi, K.; Kuwano, M.; Sueki, K.; Kikuchi, K.; Achiba, Y.; Nakahara, H.; Kananishi, N.; Watanabe, M.; Tomura, K. *J. Radio Anal. Nucl. Chem.* **1995**, *192*, 81.

SYNTHESIS AND PROPERTIES OF ENCAPSULATED AND INTERCALATED INORGANIC FULLERENE-LIKE STRUCTURES

R. Tenne, M. Homyonfer, and Y. Feldman

Advances in Metal and Semiconductor Clusters
Volume 4, pages 227–252

ISBN: 0-7623-0058-2

I. INTRODUCTION

Nanoclusters of various inorganic layered compounds, like metal dichalcogenides MX_2 (M = Mo, W; X = S, Se), are known to be unstable in the planar form and form a hollow cage—inorganic fullerene-like (IF-MX_2) structures like nested fullerenes and nanotubes.[1–3] Not surprisingly, nanoparticles of hexagonal boronitrides with graphite-like structure were shown to behave similarly.[4,5] As would be anticipated from the W-S phase diagram, spontaneous crystallization of hollow cage WS_2 polyhedra from amorphous WS_3 soot at room temperature was also observed within a few years time.[6,7] Irradiation of a given precursor with intense electron[6,8,9] and ion beams[10] accelerates the crystallization of fullerene and fullerene-like nanoparticles to a time scale measured in minutes. However, this is a rather aggressive process which must be carried out in a controlled atmosphere. Alternatively, IF-MoS_2 could be synthesized by application of an electric pulse from the tip of a scanning tunneling microscope (STM).[11]

Intercalation of carbon nanotubes with alkali metal atoms from the vapor phase was recently described.[12] The intercalated films were found to arrange in stage–1 (n = 1) superlattice, i.e. alkali metal layers were stacked between each two carbon layers. The composite nanostructures were found to disintegrate when exposed to air and complete shattering of the nanotubes (exfoliation) was obtained upon immersion in water. The intercalation of 2H-MoS_2, 2H-WS_2, and other metal dichalcogenide compounds was discussed in detail,[13–15] but staging was not observed in either of the former compounds, i.e. the alkali atoms were found to distribute randomly in the lattice. Here too, deintercalation occurs upon exposure to air and exfoliation upon immersion in water.

A new method for the synthesis of large quantities of IF structures was recently described.[16] This method affords also intercalation of the IF structures with various metal atoms, which has a remarkable influence on the solubility of these structures in aprotic solvents. The formation of stable suspensions from the intercalated IF materials permits deposition of thin films with a range of potential applications. The use of such films as the photosensitive element in solar cells and for the fabrication of inert STM tips is described in the present work.

II. SYNTHESIS OF DOPED OXIDE NANOPARTICLES AND INTERCALATION OF IF

A. General Considerations

Scheme 1 illustrates a flow chart of the processes used for synthesizing alkali-metal intercalated IF structures and for fabricating the films, from e.g. Na intercalated IF-WS_2. The first step in this hierarchy consists of synthesizing oxide powder in a modified vacuum deposition apparatus (see Figure 1a). For that purpose, a W (or other metal) wire was heated and evaporated in the presence of water vapor.

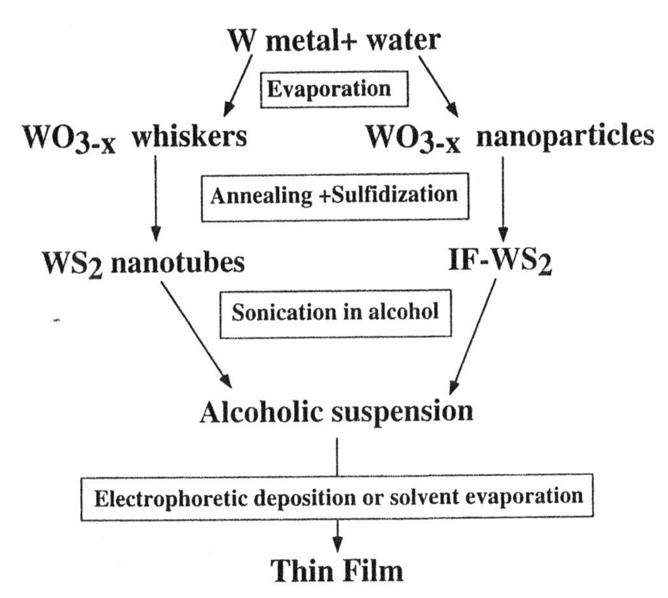

Scheme 1. Flow chart illustrating the various steps in the fabrication of the photosensitive IF-WS_2 (MoS_2) films. To obtain the suspensions, a 50 mg portion of IF particles (nanotubes) was mixed with 50 mL ethanol, and the mixture was sonicated for 5 min.

Metal or metal oxides could be evaporated from tungsten boats, but this would usually lead to mixed tungsten–metal oxide (and sulfide after sulfidization) nanoparticles. To intercalate the IF particles with metal atoms, metal-doped oxide nanoparticles were first prepared. This was done by dissolving 10^{-3}–10^{-2} M of alkali metal salts, like NaCl, etc., in water and co-evaporating it together with the heated wire to form nanoparticles of alkali-doped metal oxides, which accrued on the walls of the bell jar. Further experimental details of the apparatus and the work protocol are described in the legend to Figure 1. In the second step of Scheme 1, the oxide powder was converted into a metal sulfide with nested fullerene or nanotube structures (generically known as inorganic fullerene-like material—IF) by annealing the oxide in a reducing atmosphere with H_2S gas.[3]

Stable alcoholic suspensions of the IF particles were obtained by mixing a few mg of the intercalated IF powder in 3 mL of ethyl alcohol and sonication of the mixture (third step in Scheme 1). The solubility of the IF particles was found to be proportional to the amount of intercalant, whereas IF particles containing > 8% of alkali metal atoms were totally stable in suspension; nonintercalated IF particles did not form stable suspensions at all.

Next, deposition of intercalated IF films from the suspensions was accomplished (fourth step in Scheme 1), using two alternative routes: solvent evaporation or electrophoretic deposition onto gold substrates. In general, electrophoretic deposition resulted in more adhesive films and was therefore preferred in the present study.

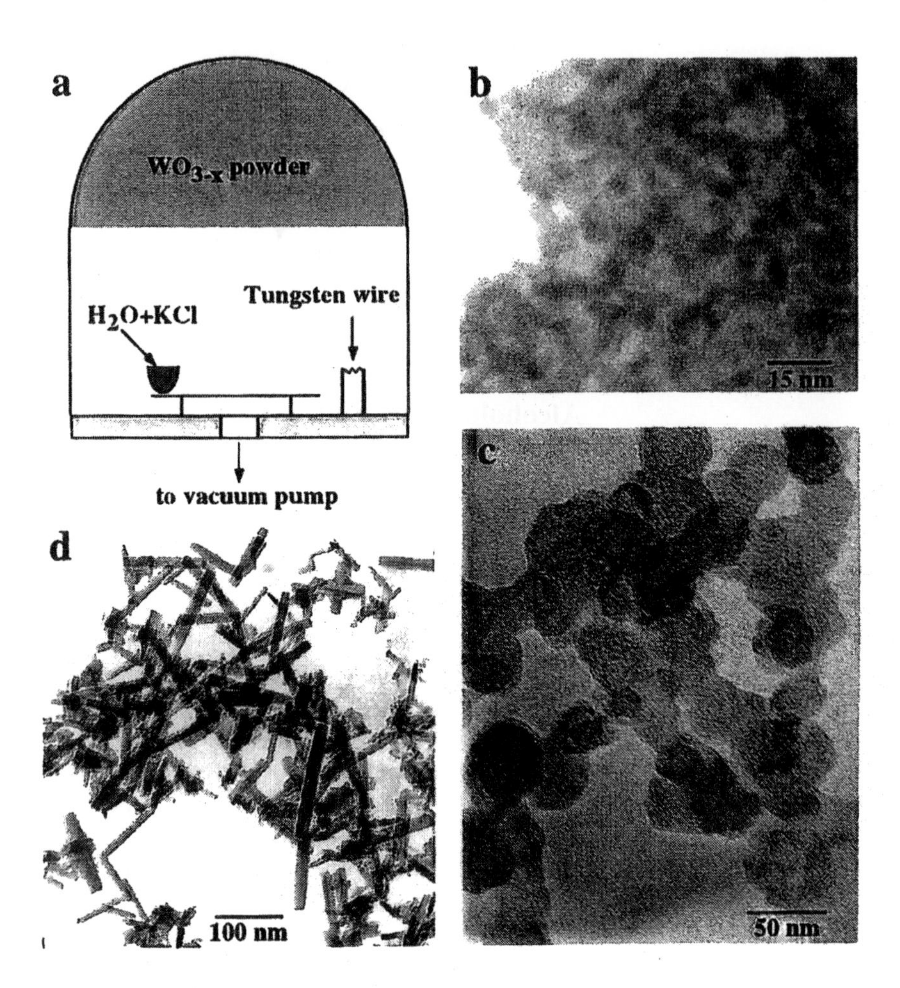

Figure 1. (**a**) Schematic representation of the evaporation apparatus used for the synthesis of the oxide nanoparticles and whiskers. In the first stage, the bell jar was evacuated to ca. 10^{-5} torr. The tungsten (molybdenum) wire was cleaned by heating it close to its melting temperature for a few minutes, after which the system was allowed to cool to ambient temperatures. The bell jar was opened and a beaker with about 3 mL of water was inserted. Upon evacuation to about 10^{-3} torr, the water in the beaker froze. At this point, the gate valve was closed and the W wire heated again. After a few minutes, the wall of the bell jar was covered with a blue oxide powder which was collected for characterization and further processing. The size of the oxide particles and their shape could be varied according to the process parameters, such as pumping speed, vacuum, etc. TEM images of the oxide precursors: (**b**) small (< 10 nm); and (**c**) large (ca. 50 nm) spherical particles; (**d**) whiskers.

B. Synthesis of Oxide Nanoparticles with Alkali Metal Dopants

Using the evaporation method described above, the particle sizes (7–50 nm) and shapes could be varied by varying the experimental conditions. Figures 1b and c show TEM images of the oxide nanoparticles. Their composition was found from X-ray photoelectron spectroscopy (XPS) to vary between $WO_{2.9}$ and $WO_{2.5}$. Tungsten oxide is known to have stable nonstoichiometric phases with O/W ratio varying in this range[17] (vide infra). The alkali metal content varied between 4 and 11 atomic %. An X-ray diffraction (XRD) study indicated that the oxide precursor was mainly amorphous. Using a lower vacuum (10^{-3} torr) and higher pumping speed, which produced more water vapor, a blue powder consisting mainly of oxide nanowhiskers with an average length of 300 nm was deposited on the bell jar (Figure 1d). Careful control of the evaporation conditions was imperative for maintaining a high yield of the oxide nanoparticles and nanowhiskers. In particular, the metal cleaning process prior to the evaporation was found to be very important for the success of the process.

C. Synthesis of Alkali Metal-Intercalated IF

Growth Mechanism

The sulfidization process of oxide nanoparticles and the production of IF materials of various kinds has been described,[2,3] and is schematically presented in Figure 2. The sulfidization starts at the outermost surface of the oxide and the oxide core is progressively consumed and converted into sulfide. Gas-phase reaction between

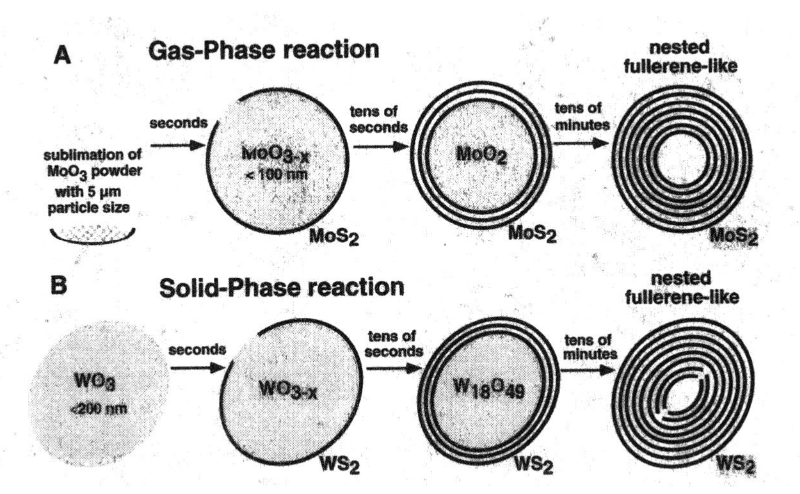

Figure 2. Schematic representation of the growth model of the inorganic fullerene-like MoS_2 (**A**) WS_2 (**B**) nested polyhedra from oxide nanoparticles.

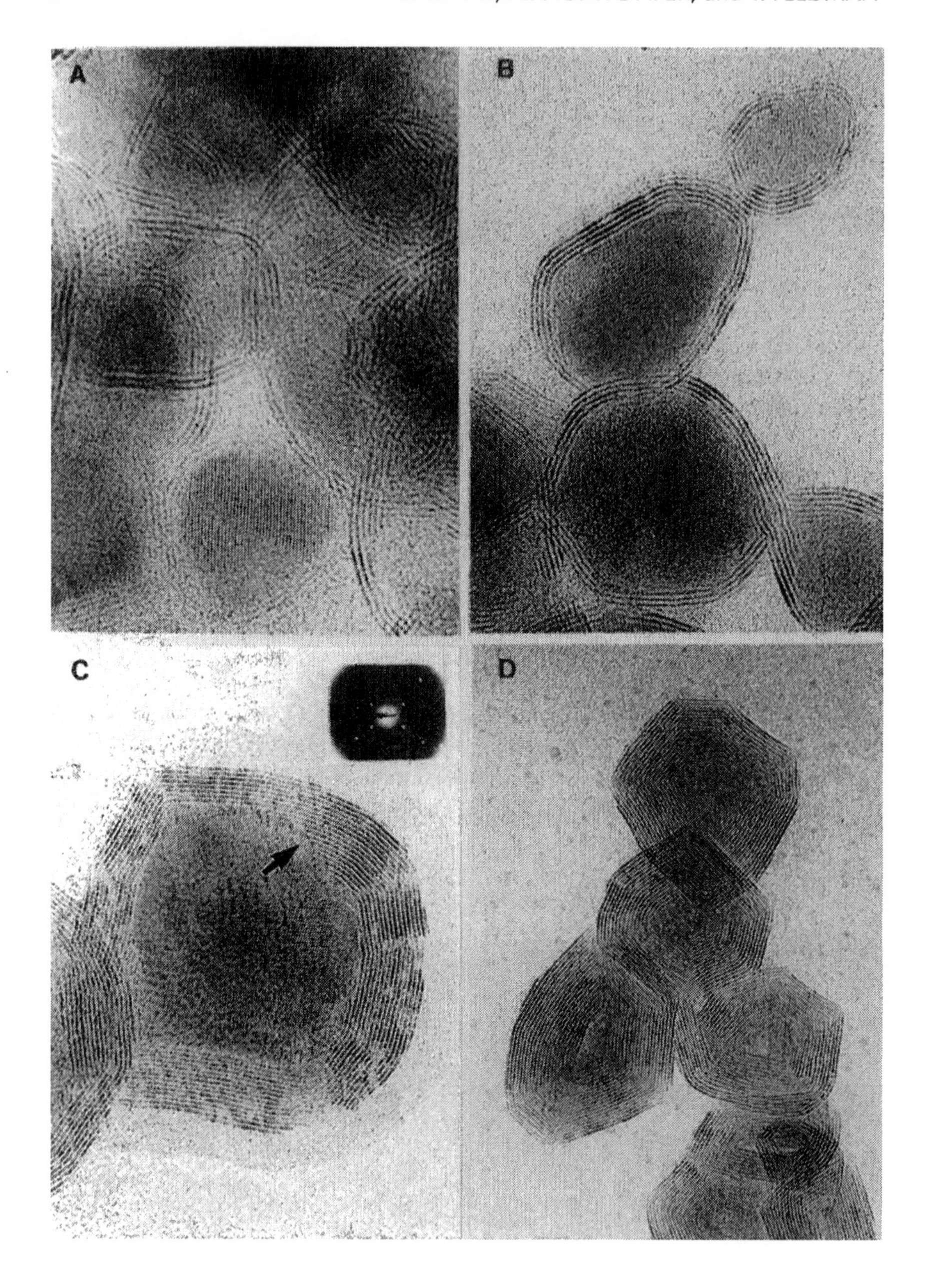

Figure 3. TEM micrographs showing the gradual transformation of molybdenum oxide nanoparticles into IF-MoS_2 (**A–D**) nested polyhedra. The electron diffraction pattern inset of Figure 2C is consistent with (111) of MoO_2.

MoO_3 and H_2S was undertaken for the synthesis of IF-MoS_2 (Figure 3)[2], while for IF-WS_2 material, a solid–gas reaction between WO_3 and H_2S has been adopted (Figure 4). Interestingly, the alkali metal-doped molybdenum oxide powder obtained in the present synthetic apparatus is much less volatile than the nonintercalated nanoparticles studied before.[2] Therefore, the simpler solid–gas reaction between MoO_3 powder and H_2S gas was adopted for the synthesis of intercalated IF-MoS_2. It was found that during the early stages of the sulfidization process, reduction and crystallization of the oxide core takes place (Figure 2). In particular, using XRD, $W_{18}O_{49}$ was observed as a stable intermediate phase in the sulfidization of (nonintercalated) tungsten oxide nanoparticles (Figure 5).[3] In the intermediate stages of the oxide to sulfide conversion, IF particles encapsulating oxide nanoparticles are obtained. Such nanoparticles may find various applications in the future. In the case of molybdenum oxide, the reduced core of the nanoparticle was identified to be MoO_2. It is likely that this or a similar suboxide phase is also formed during the present sulfidization process.

One can discriminate between two time domains in the reaction. In the first instant of the reaction, which takes less than 1 min, the oxide core is already fully reduced, while only 1 or 2 layers of sulfide have been formed, which nonetheless wrap the entire nanoparticle and maintain its integrity (Figures 2–4). Furthermore, only one growth front of the sulfide shell is observed in each nanoparticle, and the process is being completed within an hour or two. This suggests that facile diffusion of hydrogen, oxygen, and possibly water occur through the advancing sulfide membrane and the oxide core, while intercalation of sulfur in the van der Waals planes and its diffusion along these planes is the rate-limiting process for the conversion of the reduced oxide into (IF) sulfide.

Alkali Metal Intercalated IF

The sulfidization of oxide nanoparticles led to the production of copious amounts of IF particles of various sizes and topologies. When the precursor was alkali metal-doped oxide nanoparticles, intercalated IF particles were obtained after sulfidization. IF structures with a nearly spherical shape or polyhedral topology were obtained from the quasi-spherical oxide particles (Figure 6a), while nanotubes, which were closed at both ends, were obtained from the oxide whiskers (Figure 6b). The size of the IF particles retained the size of the oxide precursor particles. An abundance of T-bars, closed from three ends (Figure 6c) and all showing negative curvature, [18–21] were obtained by a similar procedure. The number of closed sulfide layers in the IF structures could be controlled through the firing time. Because the reaction can be interrupted at any time, macroscopic amounts of fullerene-like structures and nanotubes with different numbers of sulfide layers and an oxide core including single, double, or multiple layers (with oxide core) or fully converted (hollow) IF could be synthesized. The concentration and structure of the intercalated atoms in the host lattice were studied in detail (vide infra).

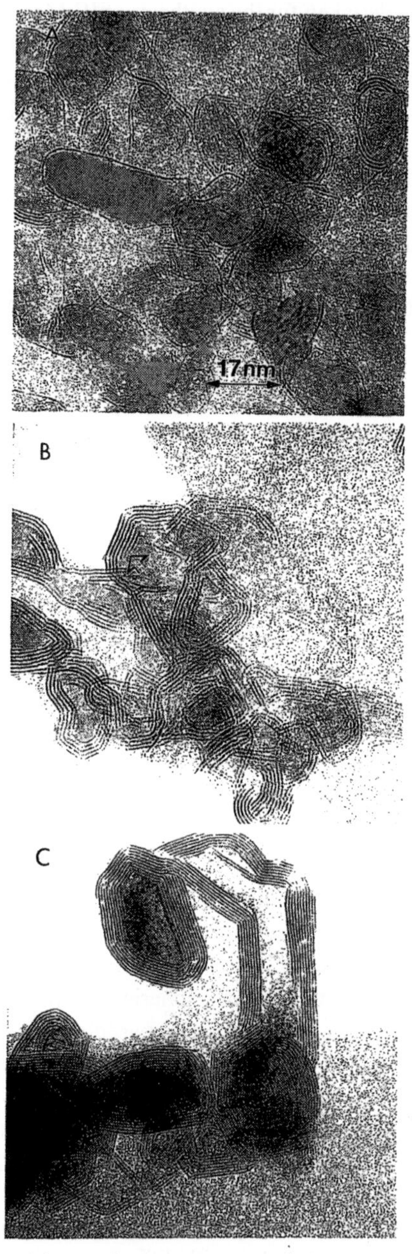

Figure 4. TEM images of the gradual conversion of tungsten oxide into IF-WS_2. The interlayer spacing of 0.62 nm is clearly visible in band **C**.

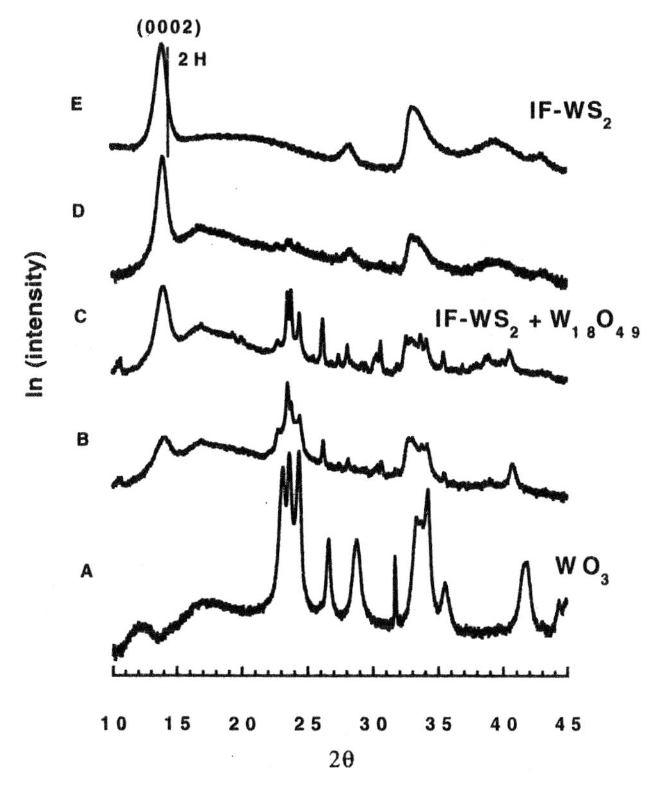

Figure 5. The transformation of WO_3 nanoparticles into IF-WS_2 followed by X-ray powder diffraction: (**A**) WO_3 precursor; (**B**) the same powder after 2 min annealing; (**C**) after 8 min annealing; (**D**) 15 min annealing. Annealing conditions are 850 °C. Rates of gas flow are: forming gas—130 cc/min; and H_2S—2 cc/min.

In general, the intercalation process is not limited to the metallic atoms and intercalation of solvent molecules from solutions cannot be avoided.[22] However, in the present process, water molecules, if at all incorporated into the oxide particles, would be outgassed during the sulfide synthesis. Indeed, no evidence for the intercalation of solvent molecules into the host lattice was found. Furthermore, since no prismatic edge planes (1120) exist in these closed structures, intercalation of solvent molecules, which makes the 2H platelets susceptible to solvent intercalation and exfoliation,[22,23] was avoided.

The above method is rather versatile and is not limited to IF of Mo or W chalcogenides. Figure 7 presents the lattice image of VS_2 fullerene-like particles which were obtained with the same apparatus. IF structures of γ-In_2S_3, PtS_2, etc. were also obtained but are not shown. The starting material for IF-VS_2 was V_2O_5 which was heated in a molybdenum boat in the presence of water vapor. This

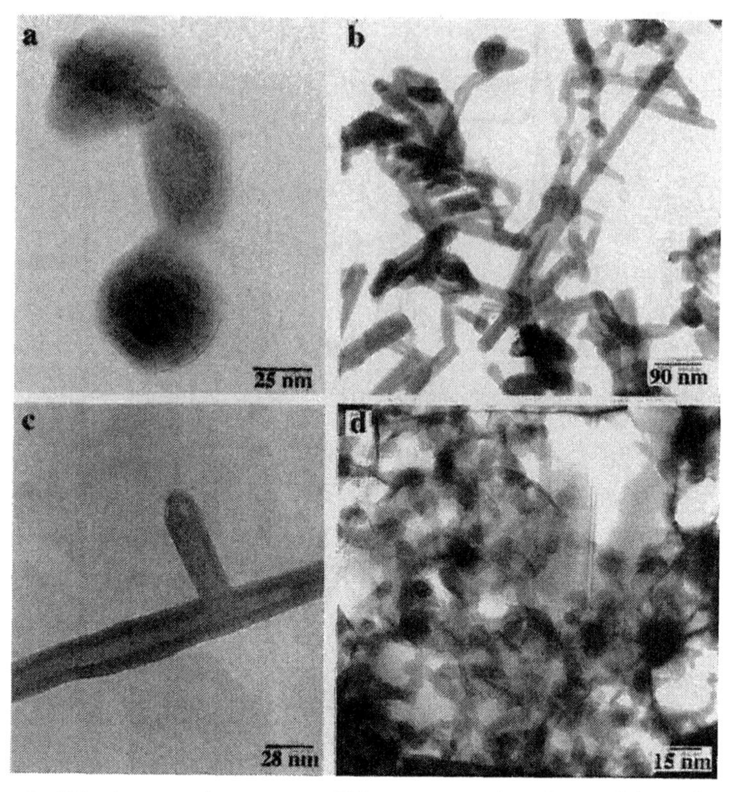

Figure 6. TEM images of tungsten–sulfide nanoparticles obtained from the oxide precursors shown in Figure 1: (**a**) oblate and quasi-spherical IF particles; (**b**) nanotubes; (**c**) T-bar exhibiting negative curvature (Schwartzite). (**d**) shows a full-fledged IF-WS_2 film on a Au substrate. The IF film was deposited by first evaporating a gold film (35 nm) on mica and annealing it to 250 °C for 12 h, which resulted in (111) textured Au crystallites with a typical size of 1 μm. The IF films were obtained by applying a bias of 20 V between the gold cathode and a Pt foil which were immersed in the (intercalated) IF suspension. IF films up to 500 nm in thickness were obtained by this procedure.

explains the observation that some of the nanoparticles were mixed IF-$V_xMo_{1-x}S_2$. It is likely that Na intercalation facilitates the synthesis of IF-VS_2, since it has been established previously that Na intercalation promotes 2H-VS_2 formation.[11] The precursor for In_2O_3 was In shot-heated in a Mo boat. This demonstrates the versatility of the present procedure, which in fact can be used for the synthesis of IF structures from virtually any metal chalcogenide having a layered-type structure.

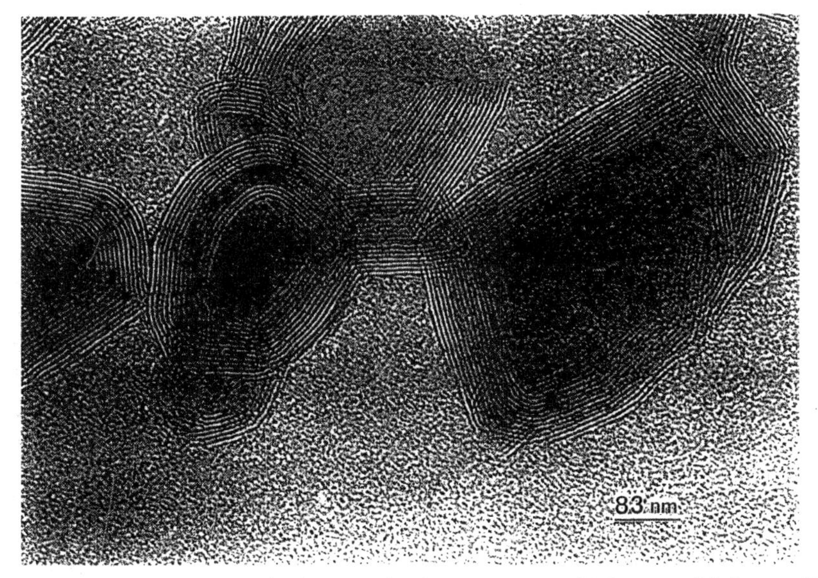

Figure 7. TEM micrograph showing the lattice image of VS_2 nested fullerene-like particles obtained from the respective oxide nanoparticles which were produced by the apparatus of Figure 1a. Lattice spacing (c) is 0.62 nm.

D. IF Suspensions and Thin Films

The IF powder synthesized in the second step formed very stable suspensions in various alcohols upon sonication (third step of Scheme 1). IF-MS_2 powders, which were synthesized according to previously reported procedures,[2, 3] did not form stable suspensions even after prolonged sonication. These results indicate that the intercalation of alkali metal atoms in the van der Waals gap of the IF particles led to a partial charge transfer from the alkali metal atom to the host lattice which increased the polarizability of the nanostructures, enabling them to disperse in polar solvents. The transparency of the suspensions and their stability increased with the amount of alkali metal intercalated into the IF structures. Suspensions prepared from IF powder (both fullerene-like particles and nanotubes) which contained large amount of intercalant (> 5%) were found to be virtually indefinitely stable. The optical absorption of the IF suspensions was very similar to that of thin films of the same nanoparticles. Details of the optical properties of IF films are deferred to a forthcoming publication.[24]

In the next step, IF films were deposited on a gold substrate by electrophoretic deposition. Given the chemical affinity of sulfur to gold, it is not surprising that electrophoretic deposition led to relatively well-adhering IF films. Furthermore, some selectivity with respect to the IF sizes and number of MS_2 layers in the films was achieved by varying the potential of the electrode. The thickness of the film was controlled by varying the electrophoresis time. Figure 6d presents a TEM image of a very thin film (tens of nm) of IF particles together with the gold substrate. Since the nonintercalated IF particles do not form stable suspensions, films of such material were obtained by electrophoresis from vigorously sonicated dispersions of the IF powder. Alternatively, IF films were prepared by evaporating the solvent from a dip-coated metal substrate; this method, however, resulted in poorly adhering films.

Using XPS and inductive coupled plasma (ICP) analysis, the alkali content of the IF particles was found to vary from 4–10 atom %, depending on the parameters of preparation of the oxide precursor and the subsequent sulfidization process. X-ray diffraction patterns of IF-WS_2 and 2H-WS_2 powders were measured using CuK_α radiation (Figure 8). The most pronounced evidence for the formation of fullerene-like particles is the shift of the (0002) Bragg diffraction peak towards lower angles and the simultaneous broadening of this peak. The shift reflects the strain-relief mechanism associated with folding of S–W–S layers.[2, 3] The peak broadening is due to the reduced domain size for coherent X-ray scattering in the direction perpendicular to the basal (0001) plane. The inherent average microstrain within IF was estimated using the Williamson–Hall analysis of the (0002l) peaks (l = 1–3) to be $1.5*10^{-2}$. The expansion along the c-axis of the IF particles (ca. 2–4%) may also reflect the random distribution of the intercalated atoms, at least in a subgroup of the IF particles. However, the variation between the lattice spacing (d_{0002}) of the IF particles, which are doped with different alkali atoms, is small (6.35 Å for Na compared to 6.45 Å for K). The appearance of superstructure Bragg diffraction peaks at small angles (Figure 8) strongly indicates the presence of an ordered IF phase with staging.[9,25] The three (000l) Bragg reflections allude to a sixth stage (n

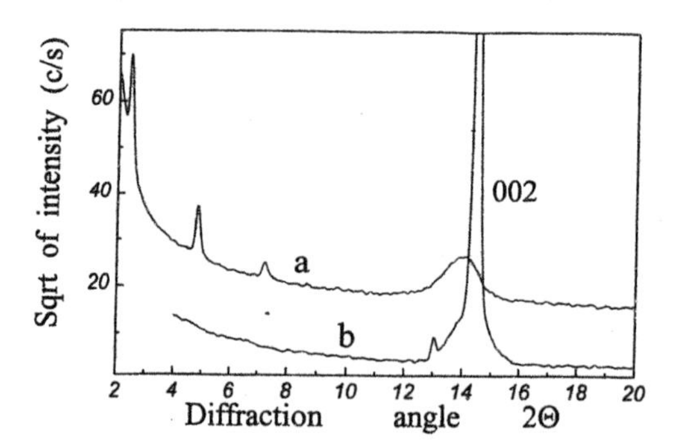

Figure 8. (**a**) Powder diffraction of IF-WS_2 with sodium as intercalant; (**b**) WS_2 platelets (the peak at 13° is assigned to the CuK_β). To preclude parasitic scattering, the specimens were deposited on background-free single-crystal quartz plates. Step size was 0.005° and exposure time was 6 s/point. The square root intensity presentation was used to increase the dynamic range of the figure.

= 6) in IF-WS_2. The repeat distance between the adjacent intercalated layers varies from 36.7 Å for Na to 38.9 Å with K. These results suggest that two populations of fullerene-like particles exist: one having ordered dopant atoms (staging) with n = 6, the other with randomly distributed alkali atoms. Since the number of MS_2 layers in the IF particles varies from about 4 to 10, those particles with fewer than 6 layers cannot exhibit staging, and consequently a random distribution of the alkali metal atoms is thermodynamically favored in this case.

It was reported[12] that intercalation of alkali metal atoms into the lattice of 2H-MoS_2 did not lead to staging, and furthermore, the intercalated material was found to be extremely sensitive to air and moisture. Finally, two additional weak diffraction peaks with lattice spacing of 2.81 and 3.03 Å were observed. These peaks were assigned to a microphase of $NaNO_3$, which might be formed through the reaction of forming gas (N_2-95%/H_2-5%) with oxygen during sulfidization of the oxides.

III. PROPERTIES AND APPLICATIONS OF INTERCALATED IF-WS2 FILMS

A. Photovoltaic Properties of the Intercalated IF

The optical absorption spectra of intercalated 2H-MoS_2 did not show appreciable changes for alkali metal concentrations less than 30%, where a transition into a metallic phase at room temperature and a further transition into a superconductor at about 3–7 K was reported.[10,22] Since the concentration of the intercalating metal atoms did not exceed 10% here, no changes in the optical transmission spectra were anticipated nor were found to occur. The intercalation of alkali atoms in the IF particles also induces n-type conductivity of the host.

The prevalence of dangling bonds on the prismatic faces of 2H-WS_2 crystallites leads to rapid recombination of photoexcited carriers. Consequently, the performance of thin-film photovoltaic devices of layered compounds has been disappointing. The absence of dangling bonds in IF material suggested to us that this problem could be alleviated here. Therefore the photocurrent response of IF-WS_2 films in selenosulfate solutions was examined and compared to that of 2H-WS_2 films. The response was found to be very sensitive to the density of dislocations in the film. Curve **a** of Figure 9 shows the quantum efficiency (number of collected charges/number of incident photons) of a typical IF-WS_2 film with a low density of dislocations as a function of the excitation wavelength. On the other hand, films having nested fullerene-like particles with substantial amounts of dislocations exhibited a poor photoresponse and substantial losses at short wavelengths (curve **b**), which indicates that the dislocations impair the lifetime of excited carriers in the film.[26] Films of nanotubes also showed substantial photoresponse (curve **c**). Finally, films made of 2H-WS_2 platelets (each about 1 μm in size), which are known to have many recombination centers on the prismatic edges ($11\bar{2}0$), did not exhibit

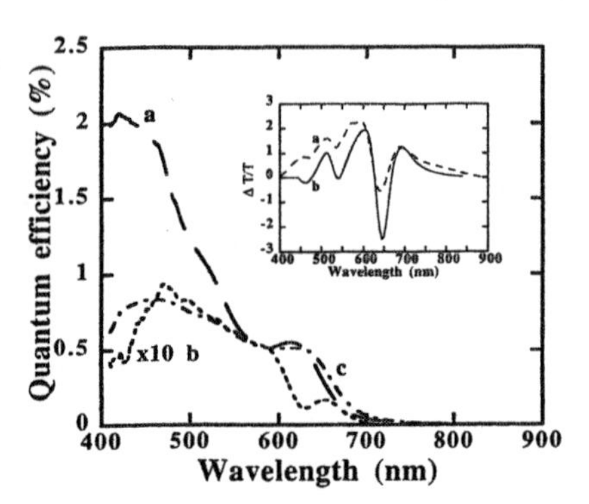

Figure 9. Photoresponse spectra of thin films of IF-WS_2: (**a**) IF film with low density of dislocations; (**b**) film consisting of IF particles with a high density of dislocations; (**c**) nanotubes. The inset presents EET spectra of: (**a**) dislocation-free IF film and (**b**) nanotubes. The EET spectra were obtained by measuring the transmission spectrum while superimposing an ac modulation of 0.45 V on the photoelectrode.

any measurable photoresponse under comparable conditions. The photocurrent decreased with negative bias, reaching zero at −1.0 V vs. the Pt foil counterelectrode, thus affirming the *n*-type conductivity of the alkali metal-intercalated IF particles. The photoresponse of the IF films did not show any degradation after 48 h of continuous illumination. Electrolyte electrotransmission (EET) spectra of the IF films were also recorded.[27] The inset of Figure 9 shows such a spectrum, which clearly reveals the direct excitonic transitions of the film at 2.02 (A exciton) and 2.4 eV (B exciton), respectively.[24]

B. Scanning Probe Microscopy of IF

STM was used to probe the photoresponse of individual IF particles electrodeposited on a gold film. Initially, STM measurements were made in the dark. Both I/V spectroscopy, and topographic images were made. The I/V spectroscopy yielded either ohmic behavior (not shown) corresponding to exposed gold regions on the surface, or a currentless region centered around 0/V bias corresponding to the "bandgap" of the individual IF particle (curve **a** of Figure 10). These data must be interpreted in the framework of a metal–semiconductor–metal structure. The log of the current increases linearly with voltage, behavior which is associated with Poole conduction.[28] The slight current enhancement in the positive sample bias region is expected for the *n*-type IF under study. The I-V curve could be used to determine tunneling conditions which do not sweep the IF particles away by the tip: the bias was set to 1.5 V with current setpoint < 1 nA. Photoresponse was then measured by illuminating the junction with modulated 650 nm light. The laser was applied after observing no photo effect when using 0.5 watt, mildly focused white light. We do not expect thermal effects such as tip expansion to be present under

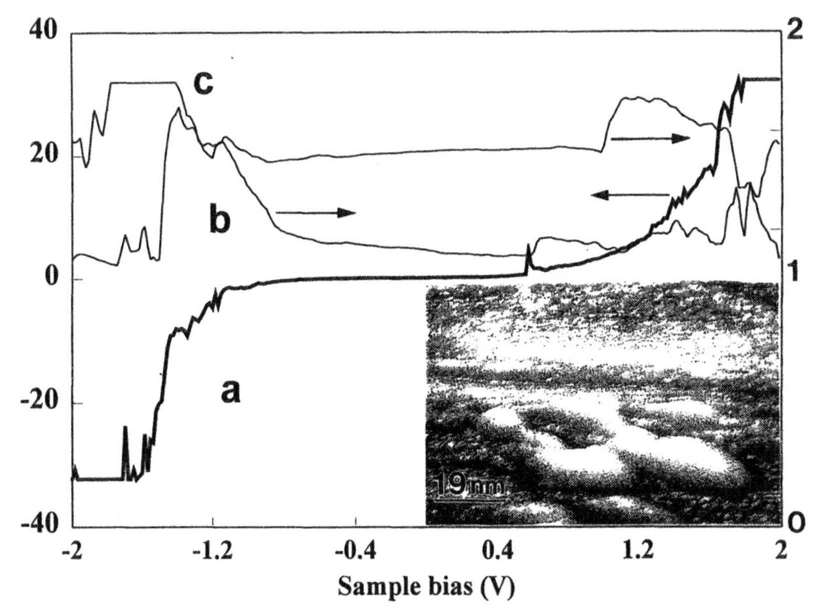

Figure 10. Scanning tunneling current spectroscopy of fullerene-like WS_2 films deposited on gold substrate, using a Pt-Ir tip. Curve **a** is an I-V plot in the dark showing a bandgap of about 1.6 eV; curve **b** shows background ac signal without illumination; and curve **c** shows an I-V plot under modulated 650 nm illumination, attenuated by a factor of 4. The modulation signal represents the lock-in output in arbitrary units, whereas the I-V signal is in nA. See text for further description. The peaks at negative sample bias in curves **b** and **c** are an artifact due to capacitive pickup. Prior to the measurements, the sample was wetted with selenosulfate solution, then dried for about 30 min. Inset: topography taken with a sample bias of 50 mV; a current setpoint of 0.7 nA, under cw laser illumination, showing a group of fullerene-like structures on a gold terrace. This illumination induces strong tunneling current over the IF, allowing it to be seen as protrusions on the surface (contrast to curve **a**).

our working conditions of relatively low-power laser light.[29] Using the lock-in output as a measure of the photoresponse, two effects can be observed by comparing curves **b** and **c** of Figure 10. First, the signal level is raised over the entire voltage scan. Second, significant enhancement of the signal was observed at positive sample bias. Note that curve **c** in Figure 10 was attenuated by a factor of 4 to keep it on scale. The peak at negative bias seen in both curves is an artifact due to capacitive pickup.

The current enhancement under illumination was exploited to improve the STM imaging of the individual fullerene-like particles. Using a bias voltage of 50 mV, conditions which would not allow observation of the IF in the dark, the sample was imaged under CW illumination. The resulting image is shown in the inset of Figure

10. The contrast between the IF particles and the gold substrate stems from the photosensitivity of the former, which produced large current contrast even under small bias. This effect also indicates that the response is due to photocurrent, not photovoltage which would have a much smaller current dependence on distance. Choice of small bias sets the tip-surface distance and hence impedance to a relatively low value, which increases sensitivity to the photocurrent. The photo-assisted STM imaging was obtained only after the film was wetted by a drop of a selenosulfate solution. Recently, a few authors discussed the influence of electrolyte solutions in mediating the charge transfer in films consisting of semiconducting nanoparticles (e.g. refs. 30, 31).

We must also consider the effect of the tip proximity in the STM experiments: tip-induced band-bending due to a surface space-charge region can lead to a reduced current for the lower bias values. Although the Na concentration is high enough to lead to a very small space-charge region if there was full charge transfer from the alkali metal atoms to the host, the actual charge transfer is only partial. This may explain the difference between measured bandgap (1.6 eV) and published values (1.3 eV)[32]. Coulomb charging cannot be a significant factor in determining the bandgap as this effect would only increase the measured bandgap by a few meV for our 30-nm-radius IF particles. The asymmetry observed in curve **c** of Figure 10, is due to the *n*-type nature which led to the asymmetry in curve **a** of Figure 10 as discussed above. Here, the illumination aids to flatten the bands and enhance the effect.

The photocurrent observed in the IF film can be explained by separation of photogenerated charges by a space-charge layer in the semiconductor, which however is not likely to be very large in these nanostructures. A second possible explanation may be that the photocurrent flow is determined by preferential trapping/transfer of either photogenerated electrons or holes. In this scenario, the free charge defines the position of the Fermi level upon illumination.[31] There are a number of possible explanations for the modest quantum efficiencies exhibited by the illuminated films: potential barrier between the nanoparticles or between the nanoparticles and the back contact impede the flow of the carriers from the front surface into the back contact; insufficient doping of the IF film leads to internal resistance within the nanoparticles; surface trapping of carriers by imperfections and dislocations; etc. Photosensitive films of this kind can be used for various purposes, including the photoremediation of water, photochemical storage of solar energy, etc. The d-d nature of the photoexcitation process prohibits degradation of the film photoresponse over extended periods of time.[33]

We found that the relatively low affinity of the IF to any surface made their imaging with scanning force microscopy (SFM) problematic. Using contact-mode SFM with a Si or Si_3N_4 cantilever/tip, the fullerene-like particles were brushed aside and the images were blurry and irreproducible. This was true both at ambient and low-humidity conditions, and using cantilevers with lateral (torsional) spring constants of 3 to 250 N/m, with the lowest workable loading force, typically 5–20 nN. Using the noncontact mode, the IF were clearly seen, which proves that this imaging problem is related to the affinity between the tip and surface. To reduce the interaction between the tip and the IF, a thin layer of IF-WS_2 particles was deposited from the alcoholic suspension using the Si cantilever/tip as a cathode. Formation of geometrically and chemically well-defined tips is of great interest to the SFM community. Prior to this work, Ducker et al.[34] successfully attached 1 μm colloidal spheres to the cantilever of an SFM tip and measured force laws in electrolyte solution. A geometrically well-defined tip with radius 2 orders of magnitude smaller allow us to investigate a completely different range of forces. Dai et al. have recently reported attachment of carbon nanotubules to the apex of an SFM tip.[35] With the new tip produced by attaching IF to the cantilever, clear images of the IF film surface were obtained in contact mode (Figure 11). This "IF-tip" was used to image several surfaces without degradation, before eventually falling off. We have yet to determine the reason for the stability of this composite Si/IF tip compared to evaporated or electrodeposited IF film. Presumably, the

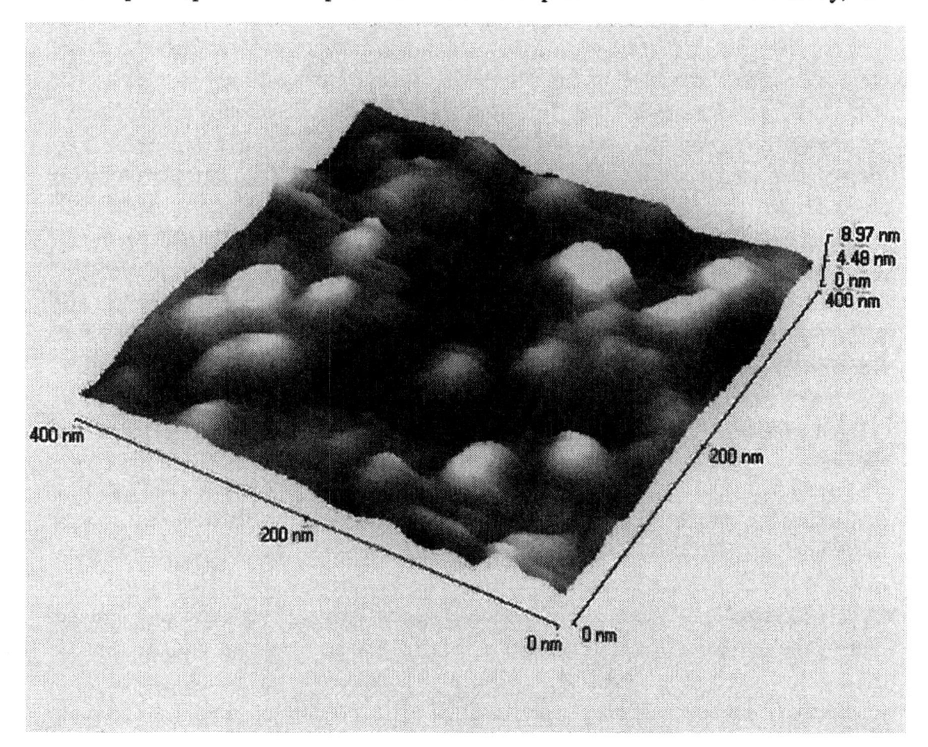

Figure 11. Contact-mode SFM image of the IF-WS_2 film used in the experiments of Figure 4, obtained with the Si/IF composite tip, which was prepared by electrophoretic deposition of IF film on the Si tip. To prepare the IF coated tip, the Si cantilever was inserted into the IF suspension with average particle size of 30 nm. A Pt foil served as anode and a bias of 20 V was applied between the cathode and anode for ca. 1 min.

stability of this composite tip is related to the presence of a high electric field at the sharp tip, which makes a more stable bond between the IF particle and the Si tip. By imaging a Nb thin-film surface with sharp features (General Microdevices), we estimate this tip radius to be 20 nm, which coincides with a typically sized IF particle.

IV. SYNTHESIS OF IF-MoS_2 USING ELECTRICAL PULSES FROM STM

A. Synthetic Procedure

For sample preparation, a 35-nm-thick polycrystalline gold film was evaporated onto a quartz substrate at room temperature and then annealed for 12 h at 250 °C. The textured film consisted of (111) oriented gold grains, ca. 0.25 μm in size. Subsequently, a discontinuous film consisting of a-MoS_3 nanoparticles was deposited on the gold substrate using a modification of a previous procedure. The film was not uniform: in certain areas a-MoS_3 nanoparticles mostly having oval shape were obtained, while in other areas a continuous film, consisting of Mo and S but with varying stoichiometry was deposited. The film was prepared by electrochemical deposition from a bath of 0.1 M ammonium thiomolybdate and 0.05 M Na_2SO_4 dissolved in an aqueous solution at pH 5 and a bias of 0.23 V vs. saturated calomel electrode (SCE). The size of the nanocrystallites depended strongly on the bath temperature, with the typical size varying between 130, 40, and 5 nm at 10, 0, and −10 °C, respectively. The films which were used in the present work were prepared at 0 °C. Ambient STM provided with a Pt/Ir tip was used in the experiments. The surface of the sample was carefully scratched at certain points which served as markers for the identification of the STM-treated zones during the subsequent TEM analysis.

STM imaging of the a-MoS_3/Au surface was done with a tunneling current of about 0.5 nA and a positive tip bias of 1.5–2.7 V, compared with 0.2 V used for a pure gold film. Before applying the electric pulses, an area of ca. 50×50 nm^2 was defined on the sample. Pulses ranging between 4 and 9 V (tip positive) were applied to induce crystallization of the amorphous nanoparticles, while the tip was positioned a few angstroms above the sample. The duration of the pulses varied from 10 to 1000 ms. The system automatically stopped the scanning prior to the pulse. During the pulse, the feedback loop was disengaged and a sample-hold loop fixed the tip in its position. After the pulse was applied, the feedback system was reactivated and the scanning was resumed. The predefined area was scanned repetitively to verify that apart from the oblate objects which appeared as a result of the pulse, no other modifications on the sample surface had occurred. Usually a few series of scan/pulse/scan routines were applied for the same predefined area. Subsequently, another 50×50 nm^2 area was defined on the same sample and the same procedure was repeated, and so forth. In several cases, well-defined nanoscopic features appeared following the pulses, while in other cases no change in the STM image was noticed after a series of pulses. This irregularity is not surprising in view of the heterogeneity of the sample surface as pointed out above. Quite often, a pulse caused an instability in the successive scans. This instability was attributed to a cluster of sulfur atoms, which detached from the a-MoS_3 particle (film) during crystallization, and sticked to the tip (vide infra).

Since a-MoS_3 is insulating, STM imaging of the nanoparticles or the continuous film was not possible. For this reason, TEM and later on HRTEM were the most useful means for the direct imaging of the nanoparticles and the film. Following the electrical pulse, typically 20 nm round objects were identified by the STM (Figure 12). The contour of the STM tip across such nanoparticle was quite spherical indeed. Atomically resolved images of these objects could not be obtained with the STM, due likely to the curvature of the objects, which implies that tunneling from more than one atom of the STM tip to the sample (and vise versa) takes place. The STM images resemble the ones reported before,[6] but it was not possible to decide whether the synthesis yielded nanoparticles with fullerene-like structure at this stage. However, the appearance of such objects after the electrical pulses served as a criterion for the success of such experiments, and only such samples were further analyzed by TEM etc. To get a better idea of the structure of the nanoparticles, the treated sample was carefully transferred onto a polymer coated Cu grid and a TEM analysis was carried out. Figure 13a shows a TEM image of a group of amorphous oval a-MoS_3 nanoparticles deposited on the textured gold film. The average size of the nanoparticles is approximately 30 nm. Smaller a-MoS_3 nanoparticles are often observed (see inset of Figure 13a). A TEM image of a typical group of IF-MoS_2 particles, which were obtained after a series of electrical pulses were applied to the original film, is shown in Figure 13b. Note that in most cases, the core of the nanoparticles remained amorphous. Thus encapsulation of IF-MoS_2 with a-MoS_3 core were obtained in this synthetic procedure. In the inset of Figure 13b a 15 nm nanoparticle is shown, which has been fully converted from a-MoS_3 into IF-MoS_2. Growth fronts of the crystalline phase inside the amorphous core can be discerned (marked by arrows). The growth fronts are very similar to the ones observed during the conversion of oxide nanoparticles into IF (see Section II.c). Pulse experiments on top of the continuous nonstoichiometric Mo–S film or with pulse voltage smaller than 4 V did not lead to crystallization of fullerene-like structures (see also refs. 36,37). Contrarily, pulses of 5 V and above produced fullerene-like objects, reproducibly. Complete destruction of the film surface and usually loss of an image was obtained with a pulse voltage larger than 8 V. The topology of the nested fullerene-like objects resembled structures reported earlier.[1–6]

Contrary to this behavior, a completely different mode of crystallization was observed for areas which were covered by a continuous thin film of a-MoS_3. Small 2H-MoS_2 platelets, which were embedded in the amorphous matrix, were formed as a result of the electrical pulse.[11] The size of the crystalline domain is rather small (2–3 nm thick) in this case. The dissipation of thermal energy, deposited by the

Figure 12. STM images of assortments of IF-MoS_2 nanoparticles produced by electrical pulse from the tip of the STM. Pulse voltage was 5 V in these experiments. The nature of each of the IF particles was verified through the semiconductor behavior of the I-V curve.

electrical pulse in the continuous Mo–S film, was probably very fast and hence the energy density was below the required threshold for sulfur abstraction and crystallization of the material. Attempts to use the same method to obtain IF structures from large area MoS_2 crystals were not successful. This result clearly indicates that IF-MoS_2 could occur only in discontinuous nanoscopic a-MoS_3 domains of a diameter smaller than ca. 50 nm.

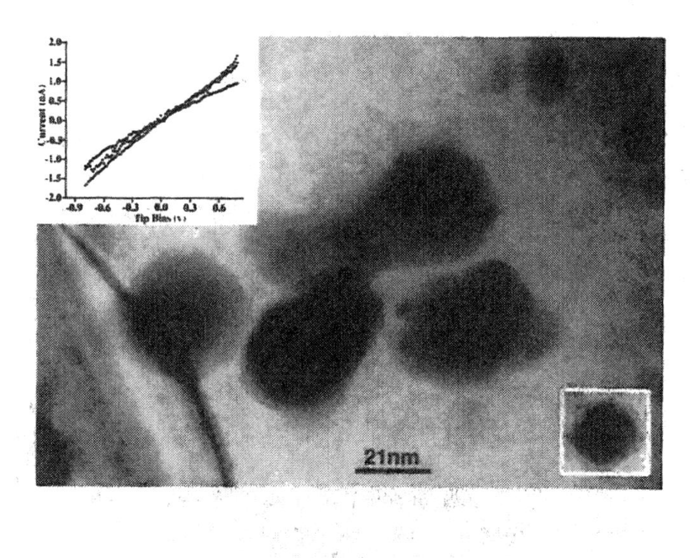

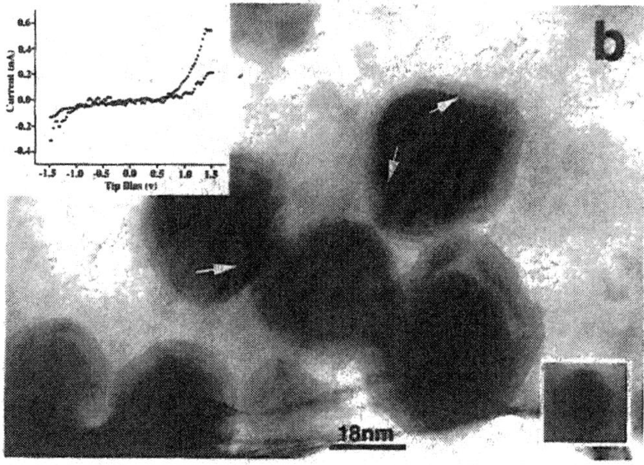

Figure 13. (**a**) TEM image of a group of a-MoS_3 nanoparticles. In the insets are shown an amorphous particle of ca. 15 nm and a typical I-V curve obtained with this film. (**b**) Crystalline IF-MoS_2 nanoparticles with an a-MoS_3 core. The insets show a fullerene-like nanoparticle of ca. 15 nm and a typical I-V curve obtained while the tip was fixed a few angstroms above one of the nanoparticles.

B. Characterization of the Encapsulated IF Nanoparticles

Prior to the application of the electrical pulses, an Ohmic behavior was observed for the film (inset of Figure 13a). The I-V curve which was measured while the tip was fixed above a fullerene-like nanoparticles (inset of Figure 13b) exhibited a typical semiconductor-like behavior with a statistically averaged bandgap of −1.1 eV (± 0.05 eV).[6] The value of the bandgap is ca. 0.1 eV smaller than the bandgap of the bulk (2H) material and was confirmed by direct optical measurements.[24] The shape of the I-V curve, which changed from a metallic-like behavior to a semiconductor-like behavior was another criteria used to judge the success of an experiment. The van der Waals surface of MoS_2 is known to be very inert against oxidation, which is probably the reason for the fact that atomic resolution for the 2H polytype and meaningful values for the bandgap can be reproducibly obtained with ambient STM.[6]

Unfortunately, it is not possible to determine the accurate tip position with respect to the nanoparticle, while the electrical pulse was applied on the sample. Furthermore, the size of the IF particles did not seem to depend on the pulse potential (between 5 and 8 V), which indicates that their size is determined by the size of the precursor (a-MoS_3) nanoparticles. This observation is not unique to the present process since the size of the IF particles, which were produced from an oxide precursor, was also found to depend solely on the size of the precursor nanoparticles.[6] This analogy is not surprising in view of the fact that in both cases the crystallization goes from outside-inwards.

In order to get more information about the process, EDS probe mounted on HRTEM (1–5 nm probe) was used to analyze the samples prior to and after the electrical pulses. The high resolution of the microscope, coupled with the small thickness of the film and substrate, permitted chemical analysis of the nanoparticles one by one. Apart from the contributions of the gold substrate, the Cu grid, and the constituting elements of the nanoparticles, no other foreign element could be detected. Table 1 shows the calculated atomic ratios of the various elements of the two typical nanoparticles. While the ratio of S/Mo was found to be 3.0 for the amorphous nanoparticles, the ratio varied between 1.8 and 2.2 for the shell of the nanoparticles with IF structure, in accordance with the hypothesized mechanism.

Table 1. Compositional EDS Analysis of a Single MoS_x Nanoparticle Before and After the Application of an Electrical Pulse from the STM Tip

Element	*X-ray Line*	*Mole % (before the pulse)*	*Mole % (after the pulse)*
S	K	25.8	15.9
Cu	K	49.7	58.5
Mo	L	8.6	7.3
Au	M	15.9	18.3

Note that the contribution of the Cu grid stems from the large cross section of this metal with respect to X-ray emission.

C. Growth Mechanism

Modification of materials by pulse from the STM tip have been published.[38–41] The main mechanisms proposed for the process are field emission of electrons and atoms, deposition of thermal energy,[39] and vibrational excitation of surface atoms.[41] Alternatively, the reaction can be initiated by pulse-induced cleavage of oxygen and water molecules and formation of O· and OH· radicals. In the absence of data for the heat conductivity and heat capacity of MoS_3, the temperature rise cannot be calculated. Rough estimates for the temperature rise in amorphous metals under a similar pulse were as high as 1000 °C.[39]

In order to crystallize, each MoS_3 molecule in the nanoparticle must first lose one sulfur atom: $MoS_3 \rightarrow MoS_2 + S$ ($\Delta G^0 = -6.2$ kcal/mol[42]). This reaction is well-documented in the literature.[43] Since the reaction is carried out in ambient conditions, an alternative pathway involving sulfur abstraction by oxygen is possible: $MoS_3 + O_2 \rightarrow MoS_2 + SO_2\uparrow$ ($\Delta G^0 = -77.6$ kcal/mol[42]) and $MoS_3 + 3/2O_2 \rightarrow MoS_2 + SO_3\uparrow$ ($\Delta G^0 = -94$ kcal/mol[42]). Although all processes are exothermic, the oxygen induced reactions are appreciably more favorable, and the evolution of gaseous SO_2 (SO_3) drives the reaction to completion. However, the latter reactions might be kinetically hindered by shortage of oxygen, particularly for the crystallization of the inner MoS_2 layers.

In Section IIC, a slow conversion of tungsten (molybdenum) oxide into fullerene-like sulfide was discussed. The rate-determining step in this process was attributed to the intercalation of sulfur in the topmost sulfide layers and its slow diffusion towards the growth front inside the nanoparticle. The fast process observed in the present experiment suggests a different growth mechanism for the fullerene-like particles.

The outermost layers of the fullerene-like nanoparticles are found to be always complete. Also, the size and morphology of the fullerene-like nanoparticles do not seem to be influenced by the duration of the STM pulse. These facts indicate that the reaction starts at some point on the outer surface of the amorphous MoS_3 nanoparticle and is only terminated when the MoS_2 layers enveloping the surface are completed. These observations could possibly suggest a self-sustained, self-extinguishing mechanism which is reminiscent of the metathesis reactions between alkali sulfide and molybdenum (tungsten) halides.[44] Although the exothermicity of the metathesis reactions is appreciably larger, the present reaction produces heat in constricted domains of 15–40 nm. The heat cannot be effectively dissipated to the gold substrate due to the small contact area between the spherical nanoparticle and the flat gold substrate film. Radiative emission to the ambient atmosphere is also rather slow. Therefore, the energy density which is released during the electrical pulse is likely to be even higher than the metathesis reaction and the reaction can therefore self-sustain itself. The local heating of the nanoparticle provides also the necessary energy to form the point defects, which are responsible for the bending of the otherwise flat surface of the hexagonal network.[1,45] Once the fullerene-like shell is completed, the reaction is switched off automatically. The threshold bias of 4 V is required for the initial abstraction of sulfur atoms through one of the above mechanisms. These atoms may remain on the STM tip surface and induce current instability, which is often observed after the pulse.

On the other hand the crystallization of the continuous film is interrupted very shortly after the pulse, and the domain size of the 2H-polytype nanocrystallites is very small (ca. 2–3 nm) in this case, which suggests that the local heating of this sample due to the electrical pulse is confined to a very small domain and excludes a self-sustained reaction here. Most of the heat is shunted through the gold film which serves as a substrate for the Mo–S film. The large contact area between these two films facilitates the heat transfer across the interface. The size of the 2H platelets can also be influenced by the irregular composition (S/Mo ratio) in the film. There is no reason to believe that the reaction mechanism for the sulfur abstraction in the continuous film is very different from the mechanism in the discontinuous nanocrystallites, and hence the same mechanism for crystallization should pertain for both cases. This argument indicates that the slow dissipation of electrical (thermal) energy from the amorphous nanoparticle leads to a very strong local heating of the nanoparticle which is responsible for the thermally activated process: $MoS_3 \rightarrow$ IF-MoS_2 + S.[43]

V. SUMMARY

Very short electrical pulses from an STM tip were applied to amorphous MoS_3 nanoparticles and induced fast crystallization of an amorphous precursor into fullerene-like MoS_2 particles in ambient conditions. A mechanism for this unique process has been proposed. A Schottky-like junction between the fullerene-like MoS_2 nanoparticle and the supporting gold film has been established and is currently being investigated.

ACKNOWLEDGMENTS

We are indebted to Dr. Yu. Rosenberg of Tel-Aviv University Materials Science Center; Dr. S.R. Cohen of the Surface Analysis Laboratory of the Weizmann Institute; B. Alperson and Dr. G. Hodes of the Department of Materials and Interfaces of the Weizmann Institute; M. Hershfinkel and Prof. V. Volttera of the Physics Department and Prof. M. Talianker of the Materials Engineering Department, all from Ben-Gurion University in Beer Sheva; Drs. J. Sloan and J.L. Hutchison of the Department of Materials in the University of Oxford. This work was supported by grants from the Minerva Foundation, Munich (Germany); NEDO (Japan); Israeli Ministry of Science; the ACS-PRF Foundation (USA).

REFERENCES

1. Tenne, R.; Margulis, L.; Genut, M.; Hodes, G. *Nature* **1992**, *360*, 444.
2. (a) Feldman, Y.; Wasserman, E.; Srolovitz, D.J.; Tenne, R. *Science* **1995**, *267*, 222; (b) Srolovitz, D.J.; Safran, S.A.; Homyonfer, M; Tenne, R. *Phys. Rev. Lett.* **1995**, *74*, 1799.
3. Feldman, Y.; Frey, G.L.; Homyonfer, M.; Lyakhovitskaya, V.; Margulis, L.; Cohen, H.; Hodes, G.; Hutchison J.L.; Tenne, R. *J. Am. Chem. Soc.* **1996**, *118*, 5362.
4. Stephan, O.; Ajayan, P.M.; Colliex, C.; Redlich, Ph.; Lambert, J.M.; Bernier P.; Lefin, P. *Science* **1994**, *266*, 1683.
5. Chopra, N.G.; Luyken, R.J.; Cherrey, K.; Crespi, V.H.; Cohen, M.L.; Louie, S.G.; Zettl, A. *Science* **1995**, *269*, 966.
6. Hershfinkel, M.; Gheber, L.A.; Volterra, V.; Hutchison, J.L.; Margulis, L.; Tenne, R., *J. Am. Chem. Soc.* **1994**, *116*, 1914.
7. Margulis, L.; Tenne, R.; Iijima, S. *Micro., Microanal. Microstruc.* **1996**, *7*, 87.
8. Ugarte, D. *Nature* **1992**, *359*, 707.
9. José-Yacamán, M.; Lopez, H.; Santiago, P.; Galván, D.H.; Garzón I.L.; Reyes, A. *Appl. Phys. Lett.* **1996**, *69*, 1065.
10. Chadderton L.T.; Fink D.; Gamaly Y.; Moeckel H.; Wang L.; Omichi H.; Hosoi F. *Nucl. Instr. Methods Phys. Res.* **1994**, *B 91*, 71.
11. Homyonfer, M.; Mastai, Y.; Hershfinkel, M.; Volterra, V.; Hutchison, J.L.; and Tenne, R. *J. Am. Chem. Soc.* **1996**, *118*, 7804.
12. Zhou, O.; Fleming, R. M.; Murphy, D.W.; Chen, C.H.; Haddon, R.C.; Ramirez A.P.; Glarum, S.H. *Science* **1994**, *263*, 1744.
13. Brec, R.; Rouxel, J. *Intercalation in Layered Materials*; Dresselhaus, M.S., Ed.; *NATO ASI Series B: Physics*; Plenum Press: New York, 1986, Vol. 148, pp. 75–91.
14. Friend, R.H.; Yoffe, A.D. *Adv. Phys.* **1987**, *36*, 1.
15. Somoano, R.B.; Hadek, V.; Rembaum, A. *J. Chem. Phys.* **1973**, *58*, 697.
16. Homyonfer, M.; Alperson, B.; Rosenberg, Yu.; Sapir, L.; Cohen, S.R.; Hodes, G.; Tenne, R. *J. Am. Chem. Soc.* In press.
17. Wells, A.F. *Structural Inorganic Chemistry*, 3rd edn.; Oxford University Press: Oxford, 1962, p. 468.
18. Iijima, S.; Ichihashi, T.; Ando,Y. *Nature* **1992**, *356*, 776.
19. Terrones, H.; Terrones, M.; Hsu, W.K. *Chem. Soc. Rev.* **1995**, 341.
20. Dunlap, B. *Phys. Rev. B.* **1992**, *46*, 1933.
21. Homyonfer, M.; Feldman, Y.; Margulis, L.; Tenne, R. Submitted.
22. Somoano R.B.; Woollam, J.A. *Intercalated Layered Materials*; Lévy, F., Ed.; D. Reidel Publishing: Dordrecht, 1979, pp. 307–319.
23. Jonson, P.; Frindt, R.F.; Morrison, S.R. *Mater. Res. Bull.* **1986**, *21*, 457.
24. Frey, G.L., et al. To be published.
25. Dresselhaus M.S.; Dresselhaus, G. *Adv. Phys.* **1981**, *30*, 139.
26. Lewerenz, H.J.; Heller, A.; DiSalvo, F. *J. Am. Chem. Soc.* **1980**, *102*, 1877.
27. Bordas, J. In *Physics and Chemistry of Materials with Layered Structures*; Lee, P.A.; Reidel, D., Eds.; Publishing: Dordrecht, 1976, pp. 145–230.
28. Geddes, N.J.; Sambles, J.R.; Parker, W.G.; Couch, N.R.; Jarvis, D.J. *J. Phys. D. Appl. Phys.* **1990**, *23*, 95.
29. Hamers, R.J.; Merkert, K. *Phys. Rev. Lett.*. **1990**; *64*, 1051.
30. Gerischer, H.; Heller, A. *J. Phys. Chem.* **1991**, *95*, 5261.
31. Hodes, G.; Howell, I.D.; Peter, L.M. *J. Electrochem. Soc.* **1992**, *139*, 3136.
32. Kam, K.K.; Parkinson, B.A. *J. Phys. Chem.* **1982**, *86*, 433.
33. Tributsch, H. *Structure and Bonding* **1982**, *49*, 127.
34. Ducker, S.A.; Senden, T.J.; Pashley, R.M. *Nature* **1992**, *353*, 239.

35. Dai, H.; Hafner, J.H.; Rinzler, A.G.; Colbert, D.T.; Smalley, R.E. *Nature* **1996,** *384*, 147.
36. Schimmel, Th.; Fuchs, H.; Lux-Steiner, M. *Phys. Stat Sol. (a)* **1992**, *131*, 47.
37. Akari S.; Möller S.; Dransfeld K. *Appl. Phys. Lett.* **1991**, *59*, 243.
38. Staufer, U.; Wiesendanger, R.; Eng, L.; Rosenthaler, L.; Hidber, H.R.; Güntherodt, H.-J.; Garcia, N. *Appl. Phys. Lett.* **1987**, *51*, 244.
39. Staufer, U.; Wiesendanger, R.; Eng, L.; Rosenthaler, L.; Hidber, H.R.; Güntherodt, H.-J.; Garcia, N. *J. Vac. Sci. Technol. A.* **1988**, *6*, 537.
40. Parkinson, B.A. *J. Am. Chem. Soc.* **1990**, *112*, 7498.
41. Shen, T.C.; Wang, C.; Abeln, G.C.; Tucker, J.R.; Lyding, J.W.; Avouris, P.; Walkup, R.E. *Science* **1995**, *268*, 1590.
42. *Standard Potentials in Aqeuous Solution*; Bard, A.J.; Parsons, R.; Jordan, J., Eds; Marcel Dekker: New York, 1985.
43. *Constitution of Binary Alloys, First Supplement*; Elliot R.P., Ed.; McGraw-Hill: New York, 1965, p. 632.
44. Bonneau P.R.; Jarvis Jr., R.F.; Kaner, R.B. *Nature* **1991**, *349*, 510.
45. Zhang Q.L. et al. *J. Phys. Chem.* **1986**, *90*, 525; Kroto H.W. *Nature* **1987**, *329*, 529.

AB INITIO CALCULATIONS ON MET-CARS:

A COMPARISON OF DIFFERENT LEVELS OF THEORY ON MODEL COMPOUNDS

Robert G. A. R. Maclagan and Gustavo E. Scuseria

I. INTRODUCTION

In 1992, Guo, Kerns, and Castleman[1] reported the discovery of a new class of molecular clusters containing metal (Ti) and carbon atoms. The first "met-car" discovered was $Ti_8C_{12}^+$. They proposed a pentagonal dodecahedron structure with

Advances in Metal and Semiconductor Clusters
Volume 4, pages 253–261

ISBN: 0-7623-0058-2

T_h symmetry. In one view the structure consists of eight Ti atoms in a cubic arrangement with six C_2 units capping each square face. When charge transfer is taken into account, which is likely since C_2 has a large electron affinity, the dominant structure would be $Ti_8^{6+}(C_2)_6^{6-}$ or $(Ti_8)^{12+}(C_2)_6^{12-}$. It was also suggested that Ti_8C_{12} could be electronically a closed shell state. Other M_8C_{12} cluster compounds including those of V, Cr, Fe, Nb, Zr, Mo, and Hf have been reported.[2,3] Binary metal met-car clusters have also been reported.[4] Other met-car compounds different from the M_8C_{12} species have been observed.[5]

In response to Castleman's discovery, quantum chemical calculations were submitted for publication in 1992 including SCF calculations by Rohmer et al.,[6] Hay,[7] and Lin and Hall,[8] and density functional calculations by Dance (LDA),[9] Reddy et al. (LCGTO and DVM),[10] Methfessel et al. (Muffin tin),[11] and Grimes and Gale (LDA).[12,13] In addition qualitative discussions appeared from Pauling,[14] Ceulemans and Fowler,[15] and Rantala et al.[16]

It was soon realized that an undistorted dodecahedral structure is unlikely. Ceulemans and Fowler[15] argued that a D_{3d}, rather than a T_h structure would have a closed shell structure. Chen et al.[17] found a D_{2d} structure to be lower in energy than the T_h structure. Dance[9] reported a T_d structure to be lower with C_2 units parallel to the longest diagonal of the underlying face, rather than parallel to the Ti–Ti bonds as in the T_h structure. In this structure there are two distinct sets of Ti atoms: the outer, large tetrahedron, designated *THN* by Rohmer et al.,[18] and an inner, small tetrahedron, designated *thn*. The *THN* Ti atoms are σ-coordinated to the C_2 species, while the *thn* Ti atoms are side-on coordinated.

In addition, the closed-shell ground-state structure was questioned. Rohmer et al.[6] showed that two ^3T_g states obtained by the promotion of pairs of electrons from the $4a_g$ and both the $4a_g$ and $1a_u$ orbitals to the $4t_g$ orbital were lower in energy than the ^1A_g state. Hay's calculations[7] found a ^9A_g structure to be even lower in energy. In a later report Rohmer et al.[19] found a $^5A_2$ state with a T_d structure was the ground state, with a $^1A_1$ D_{2d} state 1.54 eV higher in SCF calculations, but at the C.I. level of theory, the $^1A_1$ state of the structure with T_d symmetry is lower in energy than the $^5A_2$ state. Electron correlation effects are obviously important in determining the electronic structure of these clusters. The calculations of Rohmer et al.[38] indicate that the ground state of Ti_8C_{12} involves a "localized" state with four ferromagnetically coupled *d* electrons accommodated in the small (*thn*) tetrahedron.

The conclusions drawn from quantum chemical calculations as to geometry and electronic structure (including spin state) on species including metal atoms, like the met-car species, can vary significantly according to the quantum chemical level of theory employed. However the highest levels of theory require more computing resources than are usually available or practical, particularly for a series of structures for several species or electronic states. We have undertaken a series of studies to compare various levels of theory to determine if lower levels of theory can give results in reasonable agreement with higher levels of theory, so that comparative studies on met-car structures could be performed which would be both practical and reliable.

II. CHOICE OF LEVEL OF THEORY

In studying the effect of the level of theory employed the metal atom basis set was a (14s,11p,6d)/[10s,8p,3d] formed from Wachter's (14s,9p,5d) basis set[20] with the addition of Hay's diffuse 3d function[21] and Hood et al. 4p functions.[22] For carbon the standard DZP (9s,5p,1d)/[4s,2p,1d] basis set[23,24] was used. The calculations were performed using the following programs: MOLPRO 94[25] (for the HF, CCSD(T) and MRCI calculations), and both GAUSSIAN 94[26] and a development version of Gaussian[27] (for the DFT calculations). All DFT calculations were carried out using unrestricted Kohn–Sham theory as implemented in Gaussian. The MRCI or second-order CI calculations included all single and double excitations out of any configuration with a CSF whose coefficient was larger than 0.05. The density functionals used were local spin density approximation (LSDA), and Becke's three-parameter hybrid method[29] which uses the Lee, Yang, and Parr correlation functional (B3LYP).[30] In addition, calculations were also performed using Becke's 1988 exchange functional[31] combined with Perdew and Wang's 1991 gradient corrected correlation functional (BPW91)[32] which are not reported in Tables 1 and 2 (see next section).

The highest level of theory we have used is the multireference configuration interaction (MRCI) method. In comparing the various methods, we will use it as an approximation to the exact result. The Hartree–Fock based methods we have examined are the restricted open-shell Hartree–Fock (ROHF) method, MRCI, and the coupled cluster approximation CCSD(T)[33,34] where triple excitations were included. The three density functional approximations were the LSDA, BPW91, and B3LYP methods.

III. RESULTS AND DISCUSSION

In Table 1 are reported the calculated equilibrium bond lengths r_e for various states of the diatomic species TiC, VC, CrC, NiC, and Ti_2. In Table 2 are listed the dissociation energies D_e for the same states. The calculations on TiC,[35] VC,[36] and CrC[37] have been described in more detail elsewhere.

Calculations of Benard et al.[38] in 1995 gave Ti–Ti distances of 2.932 and 3.052 Å and Ti–C distances of 1.964 and 2.258 Å for Ti_8C_{12}, and V–V distances of 2.893 and 3.184 Å and V–C distances of 1.962 and 2.263 Å for V_8C_{12}. These were optimized at the RHF level of theory using a (13s,8p3d)/[5s,3p,3d] basis set for the metal atoms and a (9s5p)/[3s,2p] basis set for the carbon atoms. The M–C distances calculated for Ti_8C_{12} and V_8C_{12} are much longer than in any of the states of the diatomics studied. The same is true of Ti–Ti distances. They are similar to those we have obtained in preliminary calculations on TiC_2. For all the MC states the HF

Table 1. Calculated Bond Lengths r_e (Å) of Various M_nC_m Species

		Method				
Molecule	*State*	*HF*	*LSDA*	*B3LYP*	*CCSD(T)*	*MRCI*
TiC	$^3\Sigma^+$	1.617	1.657	1.679	1.703	1.733
	$^1\Sigma^+$	1.539	1.602	1.641	—	1.790
VC	$^2\Delta$	1.509	1.572	1.592	1.651	1.645
	$^2\Sigma^+$	1.483	1.565	1.564	1.614	1.646
CrC	$^3\Sigma^-$	1.488	1.557	1.642	1.879	1.676
	$^5\Sigma^-$	1.587	1.623	1.716	1.700	1.756
	$^7\Sigma^-$	—	1.949	1.948	—	1.959
	$^9\Sigma^-$	2.113	2.073	2.132	2.112	2.165
NiC	$^1\Sigma^+$	1.649	1.602	1.622	—	1.670
Ti_2	$^1\Sigma_g^+$	2.384	1.858	1.938	—	2.065
	$^3\Sigma_u^+$	1.955	—	2.008	—	—
	$^3\Delta_g$	2.031	1.861	1.910	—	2.420
	$^7\Sigma_u^+$	2.363	2.343	2.297	—	2.510

bond lengths are shorter by 0.1–0.2 Å than the MRCI values. The LSDA distances tend to be shorter than the B3LYP values which in turn are shorter than the MRCI values. The LSDA method r_e values are 0.02–0.08 Å less than the B3LYP values. Of the three density functional methods, the B3LYP method gave the best agreement with the MRCI values, with the r_e values 0.03–0.06 Å less than the MRCI values for the ground states of the MC species. The bond distances calculated with the BPW91 method are either poorer than the LSDA method or in between the values from the LSDA and B3LYP methods. For Ti_2 the density functional bond lengths are significantly less than the MRCI values. For the $^3\Delta_g$ state the B3LYP bond length is 0.5 Å less than the MRCI result.

The shorter values for r_e calculated by the density functional methods are reflected in large values calculated for the harmonic vibrational frequency. For the TiC $^3\Sigma^+$ state the B3LYP frequency is 988 cm^{-1} compared with the MRCI value of 704 cm^{-1}. The LSDA method gives the poorest estimate of the harmonic vibrational frequencies. With large structures like the met-cars, differences in zero point vibrational energies could be significant in determining the minimum energy structure or electronic state. The energy differences quoted by Rohmer et al.[18] suggest that errors in vibrational frequencies are unlikely to affect the minimum energy structure of Ti_8C_{12} having T_d symmetry, but could affect which electronic state is lowest.

Except for the $^7\Sigma^-$ state of CrC, all MC states are predicted to be unbound at the HF level of theory. The density functional methods overestimate the dissociation energy with the LSDA method giving more binding than the B3LYP method which tends to give more than the MRCI method. The LSDA method predicts the incorrect

Table 2. Dissociation Energies D_e (eV) of Various M_nC_m Species

		Method				
Molecule	*State*	*HF*	*LSDA*	*B3LYP*	*CCSD(T)*	*MRCI*
TiC	$^3\Sigma^+$	−1.24	6.06	3.60	3.05	2.82
	$^1\Sigma^+$	−1.83	6.30	3.53	—	2.66
VC	$^2\Delta$	−1.17	6.52	3.87	2.92	2.92
	$^2\Sigma^+$	−1.84	5.96	3.17	2.58	2.64
CrC	$^3\Sigma^{-}$ [a]	−4.67	4.96	2.77	3.19	3.00
	$^5\Sigma^-$	−4.04	4.25	2.46	1.89	2.00
	$^7\Sigma^-$	−4.25	1.45	1.57	—	1.54
	$^9\Sigma^-$	1.08	2.46	1.69	1.70	1.51
NiC	$^1\Sigma^+$	−3.52	5.82	3.06	—	2.31
Ti_2	$^1\Sigma_g^+$	−13.69	1.20	0.73	—	−0.38
	$^3\Sigma_u^+$	−6.70	—	0.27	—	—
	$^3\Delta_g$	−9.58	4.80	1.13	—	−0.88
	$^7\Sigma_u^+$	−5.92	3.69	0.33	—	−2.53

Note: [a]The $^3\Sigma^-$ state dissociates to the Cr $4s3d^5$ ^5S state, not the ground $4s3d^5$ ^7S state.

ground state, putting the triplet above the singlet state. This is mainly due to the LSDA method also giving the wrong ground state of Ti: sd^3 ^5F, instead of s^2d^2 ^3F. The singlet is calculated to be 0.225 eV below the triplet, instead of 0.81 eV observed experimentally. The same incorrect assignment is obtained with the BPW91 method. For VC, all methods give the correct order of the $^2\Delta$ and $^2\Sigma$ states. The ROHF method gives the ground state of CrC to be the $^9\Sigma^-$ state, whereas the other methods give the ground state to be the $^3\Sigma^-$ state. The CCSD method also gives the $^9\Sigma^-$ to be the ground state, but the CCSD(T) method gives what appears to be the correct order.

The calculations on Ti_2 display the same uncertainties as to the ground state as did the calculations of Bauschlicher et al.[41] They concluded that the ground state is the $^3\Delta_g$ state, although at the MRCI and ACPF levels of theory the $^7\Sigma_u^+$ was lower. Only with the MCPF modified coupled pair functional (MCPF) method was the $^3\Delta_g$ state lower. For Zr_2 they concluded that the $^1\Sigma_g^+$ state was the ground state. They also concluded that inner shell correlation effects are important. In our calculations the density functional calculations favor the $^3\Delta_g$ assignment, but the MRCI calculation (which does not include 3s, 3p correlation) favors the $^1\Sigma_g^+$ state. The HF calculation favors the $^7\Sigma_u^+$ state. The nondensity functional methods do not give bonding, but there is bonding predicted with the density functional methods. In the Ti–C met-cars, the longer Ti–Ti bond length may make some conclusions from these calculations on Ti_2 less applicable.

Many of the excited states of the diatomic species are not well described by a single reference state. For example the coefficients of the $7\sigma^2 3\pi^4 8\sigma^2$ and $7\sigma^2 3\pi^4 9\sigma^2$

configurations are 0.77 and 0.44 for the TiC $^1\Sigma^+$ state. In the CrC $^7\Sigma^-$ state, the dominant configuration has a coefficient of 0.740. This is also reflected in very large values for the τ_1 diagnostic in CCSD calculations. For the $^2\Delta$ state of VC it was 0.250. For the $^7\Sigma^-$ state of CrC it was 0.55.

How well the various methods predict the dipole moment or the charge on the metal atom varies from species to species. For the ground state of TiC, the value of the dipole moment calculated from the B3LYP calculations of 3.16 D is greater than the value of 2.73 D calculated by the MRCI method. By contrast for the ground state of VC, the value of dipole moment calculated by the B3LYP method of 5.86 D is less than the value of 7.36 D calculated by the MRCI method.

IV. ECP BASIS SETS

The basis sets used in the above study are reasonably large with polarization functions included. However to treat clusters the size of the basis set needs to be as small as possible. One widely used alternative are effective core potentials (ECP) which can also take into account relativistic effects. In Table 3 we report the optimized bond lengths calculated at the B3LYP/LanL2DZ level of theory for some states of the diatomic species discussed earlier, together with the B3LYP and MRCI values calculated with the larger basis set. The LanL2DZ basis set uses the D95 basis set[39] for C and the Los Alamos ECP plus DZ[40] on the metal atoms. For the MC species, with the exception of the $^3\Sigma^-$ state of CrC, the bond length calculations using effective core potentials give closer agreement to the MRCI results. The dissociation energies D_e for the same states are given in Table 4. For TiC and CrC the B3LYP calculations with effective core potentials give slightly poorer agreement with the MRCI values for D_e, but for VC and NiC they are better.

Table 3. Calculated Bond Lengths r_e (Å) of Various M_nC_m Species—Use of ECP Basis

		Method		
Molecule	*State*	*B3LYP*	*MRCI*	*B3LYP(ECP)*
TiC	$^3\Sigma^+$	1.668	1.733	1.685
	$^1\Sigma^+$	1.604	1.790	1.621
VC	$^2\Delta$	1.592	1.645	1.628
	$^2\Sigma^+$	1.564	1.646	1.584
CrC	$^3\Sigma^-$	1.642	1.676	1.717
	$^5\Sigma^-$	1.716	1.756	1.775
	$^9\Sigma^-$	2.132	2.165	2.173
NiC	$^1\Sigma^+$	1.622	1.670	1.659
Ti_2	$^1\Sigma_g^+$	1.938	2.065	1.826
	$^3\Delta_g$	1.910	2.420	1.833

Table 4. Dissociation Energies D_e (eV) of Various M_nC_m Species—Use of ECP Basis

		Method		
Molecule	*State*	*B3LYP*	*MRCI*	*B3LYP(ECP)*
TiC	$^3\Sigma^+$	3.60	2.82	3.71
	$^1\Sigma^+$	3.53	2.66	3.65
VC	$^2\Delta$	3.87	2.92	3.63
	$^2\Sigma^+$	3.17	2.64	2.66
CrC	$^3\Sigma^-$	2.77	3.00	2.50
	$^5\Sigma^-$	2.46	2.00	2.09
	$^9\Sigma^-$	1.69	1.51	1.59
NiC	$^1\Sigma^+$	3.06	2.31	2.61
Ti_2	$^1\Sigma_g^+$	0.73	–0.38	0.81
	$^3\Delta_g$	1.13	–0.88	0.61

V. CONCLUSIONS

The calculations on the diatomic species give us the basis for judging the reliability of calculations on met-cars. Our calculations using the HF (SCF) method often did not predict bonding and gave bond lengths for the MC species which were too small. MRCI and CASSCF calculations showed that a single reference configuration was inadequate. For this reason, the HF method results are suspect. Correlation is important in determining the ground-state structure of met-cars. The LSDA method gave better results than the HF method. The B3LYP density functional method gives significantly better results. Since for any comparative study a MRCI calculation is unlikely to be practicable, our results suggest that the B3LYP method is the method of choice. Our calculations also suggest that the use of effective core potentials such as the LanL2DZ basis give results in good agreement to those with a larger, all electron basis set with the added advantage that some relativistic effects are taken into account.

ACKNOWLEDGMENT

This work was supported by the Air Force Office of Scientific Research (Grant No. F49620-95-1-0203).

REFERENCES AND NOTES

1. Guo, B. C.; Kerns, K. P.; Castleman, A. W. *Science* **1992**, *255*, 1411.
2. Guo, B. C.; Wei, S.; Purnell, J.; Buzza, S.; Castleman, A. W. *Science* **1992**, *256*, 515.
3. Pilgrim, J. S.; Duncan, M. A. *J. Am. Chem. Soc.* **1993**, *115*, 6958.
4. Cartier, S. F.; May, B. D.; Castleman, A. W. *J. Am. Chem. Soc.* **1994**, *116*, 5295.
5. Pilgrim, J. S.; Duncan, M. A. *J. Am. Chem. Soc.* **1994**, *115*, 9724.

6. Rohmer, M. M.; de Vaal, P.; Benard, M. *J. Am. Chem. Soc.* **1992**, *114*, 9696.
7. Hay, P. J. *J. Phys. Chem.* **1993**, *97*, 3081.
8. Lin, Z.; Hall, M. B. *J. Am. Chem. Soc.* **1992**, *114*, 10054.
9. Dance, I. *J. Chem. Soc., Chem. Commun.* **1992**, 1779.
10. Reddy, B. V.; Khanna, S. N.; Jena, P. *Science* **1992**, *258*, 1640.
11. Methfessel, M.; van Schilfgaarde, M.; Scheffler, M. *Phys. Rev. Lett.* **1992**, *70*, 29.
12. Grimes, R. W.; Gale, J. D. *J. Chem. Soc., Chem. Commun.* **1992**, 1222.
13. Grimes, R. W.; Gale, J. D. *J. Phys. Chem.* **1993**, *97*, 4616.
14. Pauling, L. *Proc. Nat. Acad. Sci.* **1992**, *89*, 8175.
15. Ceulemans, A.; Fowler, P. W. *J.Chem. Soc., Faraday Trans.* **1992**, *88*, 2797.
16. Rantala, T. T.; Jelski, D. A.; Bowser, J. R.; Xia, X.; George, T. F. *Z. Phys. D* **1993**, *26*, 255.
17. Chen, H.; Feyereisen, M.; Long, X. P.; Fitzgerald, G. *Phys. Rev. Lett.* **1993**, *71*, 1732.
18. Rohmer, M.-M.; Benard, M.; Bo, C.; Poblet, J.-M. *J. Am. Chem. Soc.* **1995**, *117*, 508.
19. Rohmer, M.-M.; Benard, M.; Henriet, C.; Bo, C.; Poblet, J.-M. *J. Chem. Soc., Chem. Commun.* **1993**, 1182.
20. Wachters, A. J. H. *J. Chem. Phys.* **1970**, *52*, 1033.
21. Hay, P. J. *J. Chem. Phys.* **1977**, *66*, 4377.
22. Hood, D. M.; Pitzer, R. M.; Schaefer, H. F. *J. Chem. Phys.* **1979**, *71*, 1705.
23. Huzinaga, S. *J. Chem. Phys.* **1965**, *42*, 1293.
24. Dunning, T. H. *J. Chem. Phys.* **1970**, *53*, 2823.
25. MOLPRO is a package of ab initio computer packages written by H.-J. Werner and P.J. Knowles with contributions from J. Almlöf, R. Amos, S. Elbert, W. Meyer, E. A. Reinsch, R. Pitzer, and A. Stone.
26. Frisch, M. J.; Trucks, G. W.; Schlegel, H. B.; Gill, P. M. W.; Johnson, B. G.; Robb, M. A.; Cheeseman, J. R.; Keith, T.; Petersson, G. A.; Montgomery, J. A.; Raghavachari, K.; Al-Laham, M. A.; Zakrzewski, V. G.; Ortiz, J. V.; Foresman, J. B.; Cioslowski, J.; Stefanov, B. B.; Nanayakkara, A.; Challacombe, M.; Peng, C. Y.; Ayala, P. Y.; Chen, W.; Wong, M. W.; Andres, J. L.; Replogle, E. S.; Gomperts, R.; Martin, R. L.; Fox, D. J.; Binkley, J. S.; Defrees, D. J.; Baker, J.; Stewart, J. P.; Head-Gordon, M.; Gonzalez, C.; Pople, J. A. *Gaussian 94*; Gaussian, Inc.: Pittsburgh, PA, 1995.
27. Frisch, M. J.; Trucks, G. W.; Schlegel, H. B.; Gill, P. M. W.; Johnson, B. G.; Robb, M. A.; Cheeseman, J. R.; Keith, T.; Petersson, G. A.; Montgomery, J. A.; Raghavachari, K.; Al-Laham, M. A.; Zakrzewski, V. G.; Ortiz, J. V.; Foresman, J. B.; Cioslowski, J.; Stefanov, B. B.; Nanayakkara, A.; Challacombe, M.; Peng, C. Y.; Ayala, P. Y.; Chen, W.; Wong, M. W.; Andres, J. L.; Replogle, E. S.; Gomperts, R.; Martin, R. L.; Fox, D. J.; Binkley, J. S.; Defrees, D. J.; Baker, J.; Stewart, J. P.; Head-Gordon, M.; Gonzalez, C.; Pople, J. A. *Gaussian 95, Development Version (Revision B.2)*; Gaussian, Inc.: Pittsburgh, PA, 1995.
28. Janssen, C. L.; Seidl, E. T.; Scuseria, G. E.; Hamilton, T. P.; Yamaguchi, Y.; Remington, R. B.; Xie, Y.; Vacek, G.; Sherrill, C. D.; Crawford, T. D.; Fermann, J. T.; Allen, W. D.; Brooks, B. R.; Fitzgerald, G. B.; Fox, D. J.; Gaw, J. F.; Handy, N. C.; Laidig, W. D.; Lee, T. J.; Pitzer, R. M.; Rice, J. E.; Saxe, P.; Scheiner, A. C.; Schaefer, H. F. *PSI 2.0*; PSITECH Inc.: Watkinsville, GA, 1995.
29. Becke, A. D. *J. Chem. Phys.* **1993**, *98*, 5648.
30. Lee, C.; Yang, W.; Parr, R. G. *Phys. Rev. B* **1988**, *37*, 785.
31. Becke, A. D. *Phys. Rev. A* **1988**, *38*, 3098.
32. Perdew, J. P.; Wang, Y. *Phys. Rev. B* **1992**, *45*, 13244.
33. Scheiner, A. C.; Scuseria, G. E.; Lee, T. J.; Rice, J. E.; Schaefer, H. F. *J. Chem. Phys.* **1987**, *87*, 5391.
34. Scuseria, G.E. *Chem. Phys. Lett.* **1991**, *176*, 27.
35. Hack, M.D.; Maclagan, R.G.A.R.; Scuseria, G.E.; Gordon, M.S. *J. Chem. Phys.* **1996**, *104*, 6628.
36. Maclagan, R. G. A. R.; Scuseria, G. E. *Chem. Phys. Lett.* **1996**, *262*, 87.
37. Maclagan, R. G. A. R.; Scuseria, G. E. *J. Chem. Phys.* **1996**, *106*, 1491.

38. Benard, M.; Rohmer, M.-M.; Poblet, J.-M.; Bo, C. *J. Phys. Chem.* **1995**, *99*, 16913.
39. Dunning, T. H.; Hay, P. J. In *Modern Theoretical Chemistry*; H. F. Schaefer, Ed.; Plenum: New York, 1976, p. 1.
40. Hay, P. J.; Wadt, W. R. *J. Chem. Phys.* **1988**, *82*, 270.
41. Bauschlicher, C. W.; Partridge, H.; Langhoff, S. R.; Rossi, M. *J. Chem. Phys.* **1991**, *95*, 1057.

STRUCTURE AND MECHANICAL FAILURE IN NANOPHASE SILICON NITRIDE:

LARGE-SCALE MOLECULAR DYNAMICS SIMULATIONS ON PARALLEL COMPUTERS

Andrey Omeltchenko, Aiichiro Nakano,
Kenji Tsuruta, Rajiv K. Kalia, and Priya Vashishta

Advances in Metal and Semiconductor Clusters
Volume 4, pages 263–298

ISBN: 0-7623-0058-2

I. INTRODUCTION

In recent years, there has been a great deal of interest in nanocrystalline materials containing ultrafine microstructures on the order of few nanometers.[1–14] Most of the work relates to the synthesis process, the determination of structure by different experimental probes, and measurements of various physical properties.

Nanophase materials are usually synthesized by the gas-condensation method[5–7,10,12,15–23] shown schematically in Figure 1. This method allows the synthesis of novel nanophase materials with unique electrical, optical, magnetic, and mechanical properties. The apparatus for this synthesis technique consists of an ultrahigh vacuum (UHV) system with evaporation sources of the starting precursor materials, a liquid nitrogen-cooled hollow tube on which clusters condense, and a scraper assembly to remove and compaction devices to consolidate ultrafine powders. Elemental nanophase materials are produced with the UHV system filled with a high-purity inert gas; oxide and nitride materials can be formed with oxygen and nitrogen gases. When the precursor sources are evaporated, the atoms condense in the supersaturated region near the source and are transported to the surface of the tube by the convective flow of the gas. The clusters are removed

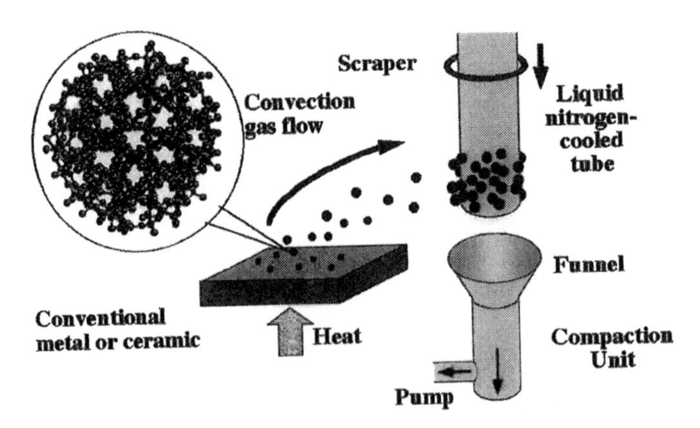

Figure 1. Schematic of the gas-phase condensation method for the synthesis of nanophase materials.

from the tube surface and consolidated into the desired shape and size. The gas-condensation method is quite flexible—it allows the synthesis of a wide variety of nanophase metals, ceramics, and intermetallic compounds. (Other methods can also be used to synthesize nanophase materials.[24–35])

A number of experimental techniques have been used to gather information about the structure of nanophase materials: electron microscopy;[36–40] X-ray diffraction;[41] extended X-ray absorption fine structure (EXAFS);[12,42,43] small-angle neutron scattering (SANS);[44–46] and Mössbauer,[7,20] Raman,[47–49] and positron annihilation spectroscopies (PAS).[50–52] All experimental probes reveal that nanophase materials are structurally different from crystalline or glassy materials of the same chemical composition. The interpretation of many structural measurements was based on the viewpoint that atoms in grain boundaries of nanophase solids were randomly located and had a much wider distribution of interatomic separations than glasses or crystalline materials of the same chemical composition.

Nanophase materials are highly stable against grain growth.[52] Transmission electron microscopy on nanophase metals[53] and ceramics[52] reveals only slight changes in the average grain growth with annealing below half the melting temperature. Grain growth may develop rapidly at elevated temperatures when the average atomic diffusion length becomes comparable to the mean grain size.

Nanophase materials have much better physical properties than their coarse-grained counterparts. For example, nanophase solids sinter at lower temperatures than coarse-grained materials.[15,52,54–56] Consider the case of TiO_2: at ambient pressures it sinters at 400 to 600 °C lower than conventional TiO_2;[52] small-angle neutron scattering[44] experiments indicate that porosity in nanophase TiO_2 is considerably reduced above 500 °C; and fracture characteristics of nanophase TiO_2 are found to be better than those of conventional TiO_2.[55,57,58]

Nanostructured ceramics exhibit ductility at temperatures where conventional ceramics show brittleness.[1,59] Measurements on nanophase TiO_2 and ZnO[56,60] show a significant increase in the strain-rate sensitivity with a decrease in the average grain size. Experiments find strain-rate sensitivities indicative of ductile behavior. For example, Karch and Birringer[61] have observed that nanophase TiO_2 can be readily formed into the desired shape just below 900 °C; Hahn and coworkers have shown that TiO_2 can undergo significant deformation without fracture at elevated temperatures.[62–64]

Self-diffusion and impurity-diffusion measurements in as-consolidated nanophase ceramics and metals reveal orders-of-magnitude higher atomic diffusivities than in coarse-grained materials.[55,65–70] Sintering suppresses these large values of atomic diffusion,[55] which indicates that they are due to porosity in nanophase materials.

Vibrational spectra of nanophase materials are also different from those of crystalline and glassy solids with the same composition. New, low-frequency phonon modes associated with ultrafine microstructures may appear and affect thermodynamic properties such as specific heat. Increased values of specific heat and entropy have been reported in nanophase materials.[71,72]

In this review we present results of large-scale molecular dynamics simulations for structural and mechanical properties including crack propagation and fracture in nanophase Si_3N_4. The outline of this review is as follows: Section II contains the description of the multiresolution MD approach, its implementation on parallel machines, and the interaction potential used in the work reported here; in Section III we summarize the experimental information about the structure and mechanical and thermal properties of crystalline and amorphous Si_3N_4; Section IV deals with the sintering of nanocrystals and amorphous nanoclusters of Si_3N_4 and the preparation of nanophase Si_3N_4 on computers; the simulation results for structural properties, elastic moduli, and crack propagation in nanophase Si_3N_4 are presented in Section V.

II. PARALLEL MOLECULAR DYNAMICS SIMULATIONS

A. Molecular Dynamics Approach

Molecular dynamics (MD) is a computer-simulation technique in which classical equations of motion are solved for a set of atoms or molecules. Consider a system of N atoms with coordinates $\{\mathbf{r}_i\}_{i=1,...,N}$ and momenta $\{p_i\}_{i=1,...,N}$. The atoms are treated as point-like particles described by a classical Hamiltonian,

$$H = \sum_i \frac{\mathbf{p}_i^2}{2m_i} + V \tag{1}$$

where m_i are atomic masses and V is the potential energy:

$$V = V(\mathbf{r}_1, \mathbf{r}_2, \ldots, \mathbf{r}_N) \tag{2}$$

The potential function Eq. 2 is often approximated by a simple pair-wise interaction,

$$V = \frac{1}{2}\sum_{i,j} \nu_{ij}(r_{ij}) \tag{3}$$

where the pair potential, ν_{ij}, depends on the atomic species (C, H, Si, etc.) and the interatomic distance, $r_{ij} = |\mathbf{r}_i - \mathbf{r}_j|$. The Hamiltonian equations of motion,

$$\dot{\mathbf{r}}_i = \frac{\partial H}{\partial \mathbf{p}_i}, \quad \dot{\mathbf{p}}_i = -\frac{\partial H}{\partial \mathbf{r}_i}. \tag{4}$$

for Hamiltonian (1) reduce to Newton's second law,

$$m_i \ddot{\mathbf{r}}_i = \mathbf{f}_i, \tag{5}$$

where $\mathbf{f}_i$ is the force on the i-th particle,

$$\mathbf{f}_i = -\frac{\partial V(\mathbf{r}_1, \mathbf{r}_2, \ldots, \mathbf{r}_N)}{\partial \mathbf{r}_i} \tag{6}$$

The only input information in MD simulations is the expression for the potential energy (2). Given an initial configuration $\{\mathbf{r}_i(0), \mathbf{p}_i(0)\}$, Eqs. 5 are integrated numerically to yield phase-space trajectories:

$$\mathbf{r}_i = \mathbf{r}_i(t),$$

$$\mathbf{p}_i = \mathbf{p}_i(t). \tag{7}$$

By taking averages over phase-space trajectories (Eq. 7), it is possible to compute equilibrium properties of the system (thermodynamic quantities, structural and dynamical correlations). MD simulations may be also used to study nonequilibrium processes, such as microstructural evolution, thermal transport, fracture, etc.

In a MD simulation the atoms are usually placed in a "box" and periodic boundary conditions are applied at the box boundaries. Whenever a particle leaves the box on one side, it immediately enters from the opposite side. Furthermore, atoms near the boundaries of the MD box interact with particles in the appropriate periodic images of the box. (In the case of short-range interactions, only the closest image particles are taken into account; this is known as the "minimum image convention".) Periodic boundary conditions allow us to simulate an infinite system and thus eliminate surface effects in a well-defined manner.

B. Interatomic Potentials

The usefulness of MD simulations depends on how well the model interatomic potential describes the real system. To study material-specific properties, one needs realistic interatomic potentials. Two-body potentials are adequate for monatomic systems with close-packed structures. However, it is often necessary to go beyond the simple two-body approximation, particularly in the presence of strong covalent bonds.

An arbitrary potential may be expanded into one-body, two-body, three-body, etc.:

$$V = \sum_i \nu^{(1)}(\mathbf{r}_i) + \frac{1}{2!}\sum_{i,j} \nu^{(2)}(\mathbf{r}_i, \mathbf{r}_j) + \frac{1}{3!}\sum_{i,j,k} \nu^{(3)}(\mathbf{r}_i, \mathbf{r}_j, \mathbf{r}_k) + \ldots. \tag{8}$$

The next step, beyond the two-body form (Eq. 3), is to keep the three-body term in Eq. 8. (Four-body and higher terms are rarely used.) Assuming translational and rotational invariance, such a potential reduces to the following form,

$$V = \frac{1}{2} \sum_{i,j} V_{ij}^{(2)}(r_{ij}) + \frac{1}{2} \sum_{i,j,k} V_{jik}^{(3)}(r_{ij}, r_{ik}, \cos\theta_{jik}) \tag{9}$$

where θ_{jik} stands for the angle formed by $\boldsymbol{r}_{ij}$ and,

$$\cos\theta_{jik} = \frac{\mathbf{r}_{ij} \cdot \mathbf{r}_{ik}}{r_{ij} r_{ik}} \tag{10}$$

and the three-body terms in Eqs. 8 and 9 are related as:

$$\nu^{(3)}(\mathbf{r}_i, \mathbf{r}_j, \mathbf{r}_k) = V^{(3)}(r_{ij}, r_{ik}, \cos\theta_{jik}) + V^{(3)}(r_{ji}, r_{jk}, \cos\theta_{ijk}) + V^{(3)}(r_{ki}, r_{kj}, \cos\theta_{ikj}) \tag{11}$$

Adding the three-body part introduces an explicit dependence on bond angles θ_{jik}, which is crucial for an accurate description of covalent materials. Stillinger–Weber potential [73] is one of the most widely used models of this type. The three-body part of the Stillinger–Weber potential has the following form,

$$V_{jik}^{(3)}(r_{ij}, r_{ik}, \cos\theta_{jik}) = B_{jik} f_{ij}(r_{ij}) f_{ik}(r_{ik}) [\cos\theta_{jik} - \cos\bar{\theta}_{jik}]^2 \tag{12}$$

where B_{jik} is the strength of the interaction and $\bar{\theta}_{jik}$ is a constant. The function $f_{ij}(r)$ is given by,

$$f_{ij}(r) = \begin{cases} \exp[l/(r - r_o)], \text{ for } r < r_o \\ 0, \text{ for } r \geq r_o \end{cases} \tag{13}$$

where r_o is the cutoff distance and l is a parameter. The function $f_{ij}(r)$ vanishes unless atoms i and j are within the cutoff distance, r_o, which is chosen to be slightly larger than the length of the covalent bond. The factor $[\cos\theta_{jik} - \cos\bar{\theta}_{jik}]^2$ in Eq. 12 represents the energetics of bond bending: it increases the potential energy whenever the bond angle, θ_{jik}, deviates from its optimal value, $\bar{\theta}_{jik}$.

One of the major deficiencies of the three-body potentials (Eq. 9) is that the strength of the covalent bond is assumed to be independent of the coordination number, which limits the ability of the potential to describe atoms in different bonding environments. There has been some progress in describing these effects using the concept of bond order[74,75] and dynamic charge transfer.[76]

MD simulations of Si_3N_4 presented here are based on an interatomic potential developed by Vashishta et al.[77–79] The two-body part of the potential has the following form:

$$V_{ij}^{(2)}(r_{ij}) = \frac{H_{ij}}{r_{ij}^{\eta_{ij}}} + \frac{Z_i Z_j}{r_{ij}} e^{-r_{ij}/r_{1s}} - \frac{\frac{1}{2}(\alpha_i Z_j^2 + \alpha_j Z_i^2)}{r_{ij}^4} e^{-r_{ij}/r_{4s}} \tag{14}$$

The first term in Eq. 14 represents the steric repulsion, the second term is the Coulomb interaction due to charge transfer, and the last term corresponds to the charge–dipole interaction due to large polarizability of negative ions. The parameters of the potential are the strength (H_{ij}) and exponent (η_{ij}) of the steric repulsion, along with the effective atomic charges (Z_i) and polarizabilities (α_i) of the ions. (The subscripts i and j stand for different atomic species: Si or N.) Charge–charge and charge–dipole interactions are screened by exponentially decaying factors with the respective decay lengths of r_{1s} and r_{4s}. This allows the potential to be cut off at a finite distance, $r_c = 5.5$ Å. In another version of the potential, the Coulomb term is not screened and is treated with the fast multipole method.[80]

The three-body part is of the Stillinger–Weber form (Eq. 12) which includes the effects of bond bending and stretching. The three-body term $V_{jik}^{(3)}$ is introduced only if the triplet (j-i-k) forms either Si–N–Si or N–Si–N bond angle. Other possible bond angles (e.g. Si–Si–N) are not realized under any reasonable conditions because of strong Coulomb repulsion between atoms of the same species. Due to its special form, the three-body potential (Eq. 12) may be evaluated by performing summation over pairs instead of triplets of particles.

As discussed in Section III, the results of MD simulations using these interatomic potentials[77–79,81–83] are in good agreement with experiments for elastic constants, phonon density-of-states, and specific heat of α-crystalline Si_3N_4. The calculated static structure factor for amorphous silicon nitride also agrees well with the neutron-scattering results.[84]

C. Multiresolution Molecular-Dynamics Scheme

In MD simulations, the $3N$ coupled equations of motions (Eq. 5) are integrated for a certain interval of time. The time interval is discretized by dividing it into time steps, Δt. At each time step, it is necessary to calculate the forces on all the particles and update the positions using an appropriate finite-difference scheme. The time step should be sufficiently small so that the time derivatives are well approximated by the finite-difference expressions. An important test of the integration algorithm is the energy conservation. For a meaningful simulation in the microcanonical ensemble, the Hamiltonian (Eq. 1) should be conserved to the desired accuracy (typically, 10^{-4}–10^{-5}) over the course of the simulation. An acceptable value of the time step is usually a fraction of the typical time scale of an atomic oscillation. Typical MD simulations involve time scales between $10^3\ \Delta t$ and $10^6\ \Delta t$ with $\Delta t \sim 10^{-15}$ s.

The most compute-intensive part of MD simulations is the calculation of the interatomic interaction at each time step. In a naive implementation, the calculation of two-body forces requires $O(N^2)$ CPU time. This can be improved dramatically by using multiresolution techniques to efficiently manage multiple length and time scales. Finite-range potentials and forces may be computed in $O(N)$ operations using the linked-cell list technique.[85] (For the long-range Coulomb interaction, $O(N)$ algorithms have been developed using divide-and-conquer schemes, such as the fast multipole method (FMM).[80])

The efficiency of the multiresolution algorithm may be further improved by using a multiple time-step (MTS) technique.[86,87] In this approach, the force experienced by a particle is separated into a rapidly varying primary component and a slowly varying secondary component. Usually short-range forces are included in the primary component, while the long-range contributions form the secondary part. The primary interaction is computed every time step, while the secondary component is calculated at intervals of a few time steps. This scheme may be extended to include several different time scales.

In recent years, MD simulations involving millions of atoms have become possible with the emergence of powerful parallel architectures. Large-scale MD simulations are naturally suited for parallelization by domain decomposition. In this approach the system is decomposed into subdomains which are assigned to different processors. The inter-processor communication is small, since it only involves atoms near the boundaries of the subdomains. Therefore, the problem may be efficiently parallelized. One of the main difficulties with the parallel MD approach arises from the bookkeeping of intra- and inter-processor contributions to forces, potential energy, and other quantities. In our implementation we use an efficient parallelization algorithm which treats the intra- and inter-processor contributions on the same footing, thereby making it relatively easy to incorporate new interatomic potentials into the code.

Multiresolution in Space: Linked-Cell and Neighbor Lists

In the linked-cell list technique for short-range potentials, the MD box is divided into cells. The atoms are sorted into cells and the information is stored in a linked list. The cell size in real space, d, is chosen to be,

$$d = r_c + \delta \tag{15}$$

where r_c is the potential cutoff, and δ is the "skin". To calculate the force on a particle i we only need to add the contributions due to particles in the neighboring cells (see Figure 2a). The computation therefore scales as MN, where M is the average number

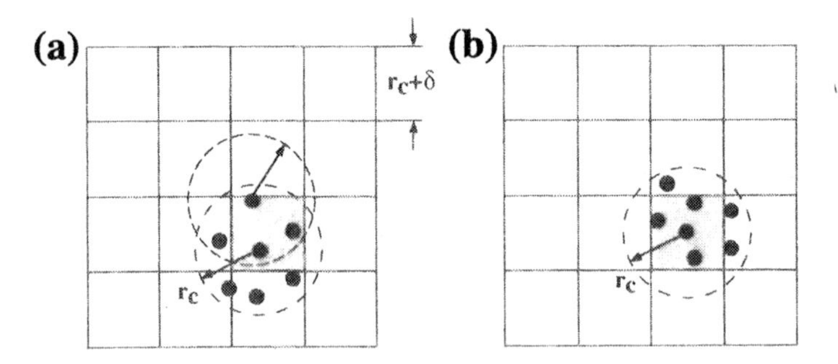

Figure 2. Force calculation using the linked-cell list illustrated in two dimensions.

of atoms involved in calculating the force on atom i. If ρ is the average number density, then:

$$M = \rho(3d)^3 = 27\rho(r_c + \delta)^3 \tag{16}$$

Thus the linked-cell method leads to an O(N) algorithm. The computation may be further reduced by a factor of two, if we exploit Newton's third law and perform the summation over half the number of neighboring cells, as illustrated in Figure 2b. Adding a "skin", δ, in Eq. 15 allows us to skip the calculation of the linked-cell list for several time steps, since particles in this time interval are not likely to move by more than $\delta/2$.

The linked-cell list approach may be further improved by using a cell size,

$$d = (r_c + \delta)/k \tag{17}$$

where k is an integer representing the number of cells per cutoff length. In this case the number of neighbors, M, becomes:

$$M = \rho((2k+1)d)^3 = \left(2 + \frac{1}{k}\right)^3 \rho(r_c + \delta)^3 \tag{18}$$

Choosing the cell size to be half the cutoff length reduces the computation by a factor of 1.728. The cell size is limited from below by requiring that each cell contain several particles, or otherwise additional computation is required to loop over empty cells.

Instead of directly using the linked-cell list to perform the computation of forces, one can first use it to construct a list of neighbors for each particle i. The list indexes the particles lying within $(r_c + \delta)$ of the particle i. The neighbor list needs to be updated only when particles move by more than $\delta/2$. Once the neighbor list is constructed, the average number of particles involved in the calculation of the force on a particle is further decreased to:

$$M = \frac{4}{3}\pi\rho(r_c + \delta)^3 \approx 4.2\rho(r_c + \delta)^3 \tag{19}$$

This number may be again reduced by a factor of two using Newton's third law. A major problem with the neighbor list arises due to large storage requirement (~MN), which may turn out to be excessive for intermediate-range potentials.

Parallelization by Domain Decomposition

Molecular-dynamics simulations are implemented on the nodes of a parallel machine using the domain-decomposition approach. The MD box is divided into subdomains and the data for particles in a given subdomain reside in the memory of a particular node. Each node is divided into cells and a linked-cell list is constructed. The internode communication involved in parallel MD simulations is illustrated in Figure 3. Whenever an atom crosses a node boundary, it has to be reassigned to the new node and all the data pertaining to this atom has to be transferred to the corresponding processor. To calculate forces on particles near the node boundary, one has to obtain the appropriate information for the particles in the boundary cells on the neighboring nodes. When taking advantage of the Newton's third law, additional communication is necessary to collect the contributions to the forces computed on other nodes.

In our implementation, the reassignment of atoms to other nodes is combined with periodic boundary conditions. This procedure is applied whenever the linked-cell list has to be updated, which happens at intervals of several time steps. First, the particles in the cells adjacent to the node boundaries are scanned to determine if any of them has moved outside the boundaries. Then we apply periodic boundary conditions and determine which atoms have to be transferred to other nodes. The coordinates, velocities, and other attributes of those atoms are packed into messages and sent to appropriate processors. The array used to store atoms on a node now has empty slots corresponding to the atoms that have left. These empty slots are first filled with the particles received from other nodes. If more atoms are received than the number of available vacant slots, the extra particles are appended to the end of the array. In the opposite case, the remaining vacancies are filled with the particles taken successively from the end of the array. This procedure requires a minimum rearrangement of particles, while still maintaining a simple array structure. However, there is one side effect—the atoms in the array are being reordered over the course of the simulation. To keep track of particle's identities, we assign tags which may be used to trace the motion of each particle.

A cell on a node is described by its position in X-, Y-, and Z-directions (see Figure 4a),

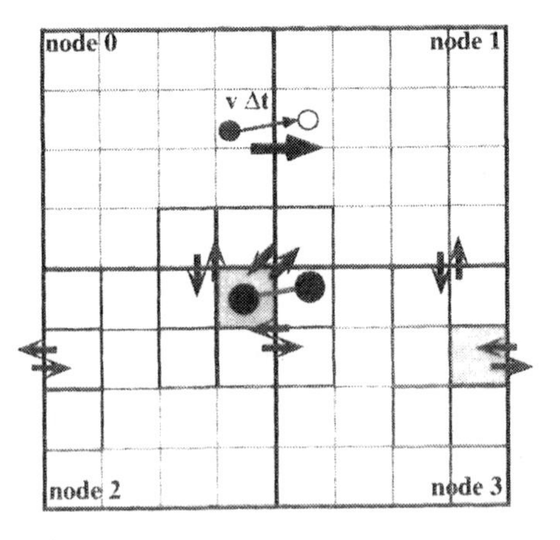

Figure 3. Implementation of parallel MD using domain decomposition.

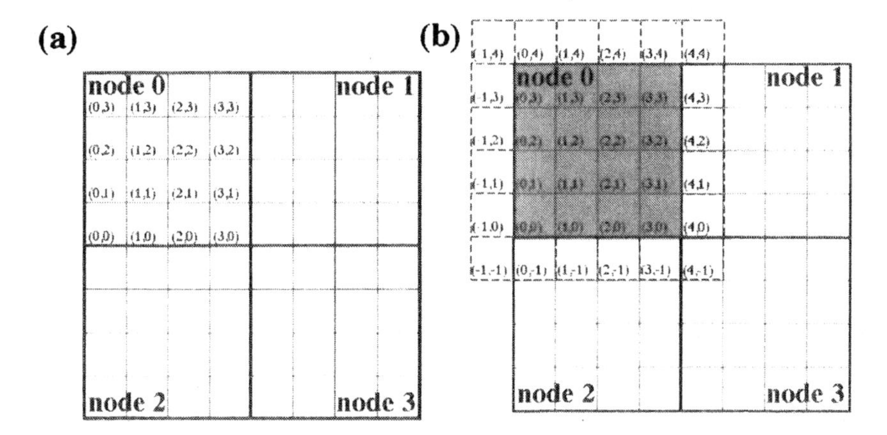

Figure 4. (**a**) Numbering of cells on a node. (**b**) Extended node.

$$
\begin{aligned}
&(i_x, i_y, i_z)\\
&i_x = 0, \ldots, n_x - 1\\
&i_y = 0, \ldots, n_y - 1\\
&i_z = 0, \ldots, n_z - 1
\end{aligned}
\tag{20}
$$

where n_x, n_y, n_z represent the number of cells on the node in X-, Y-, and Z-directions. The bookkeeping associated with interprocessor communication may be significantly simplified by defining an "extended node", as shown in Figure 4b. An extended node represents a regular node surrounded by layers of "skin cells" corresponding to other nodes. The cells on an extended node are numbered as:

$$
\begin{aligned}
&(i_x, i_y, i_z);\\
&i_x = -k, \ldots, n_x + k - 1\\
&i_y = -k, \ldots, n_y + k - 1\\
&i_z = -k, \ldots, n_z + k - 1
\end{aligned}
\tag{21}
$$

Whenever the forces are to be calculated, each node sends the positions of the particles in the cells adjacent to its boundary to other processors. The atoms received from neighboring nodes for the purpose of force computation (referred to as "image" particles as opposed to "real" particles) are appended to the local array of particles and assigned to appropriate skin cells. In addition, it turns out to be convenient to apply the same procedure to the interaction across the periodic boundaries. In the latter case we just create an image of a real particle and place it in a skin cell. As a result, we obtain (1) a single array including real and image particles, and (2) a linked-cell list including normal and skin cells. The intra- and internode contributions to the forces can be therefore treated in the same fashion. For each real particle i, we go over particles j in the adjacent cells (including the skin cells) and add the appropriate contribution to the forces on particles i and j. When using Newton's third law, the image particles may also pick up force contributions. The forces on image particles should be returned to appropriate nodes and added to forces on the corresponding real particles.

This scheme is further modified to improve data locality. Instead of directly using the link-cell list to loop over atoms, an "easy-reference list" is created, which contains particle coordinates and other information ordered in a convenient way. The atoms in the same cell occupy a contiguous portion of an array, and particles close in real space are stored in relatively close locations in the list. Easy-reference list requires additional O(N) storage and O(N) computation. These overheads are far outweighed by the computation speedup due to improved data locality.

Multiresolution in Time

A wide variety of numerical integration algorithms are available to solve the equations of motion (Eq. 5). Since the calculation of forces at each time step is computationally expensive, it is desirable to use integration algorithms which permit large time steps while maintaining sufficient accuracy and stability. The key property of an integration scheme is its long-time behavior. It is well-known that two phase-space trajectories with an arbitrarily small difference in the initial conditions diverge exponentially after sufficiently long time. Therefore, no integration algorithm will provide an exact solution over a long time interval. However, due to the statistical nature of the problem, such an exact solution is not necessary. A good integration algorithm should provide: (1) an accurate approximation of the differential equations on short time scales, and (2) long-time conservation of the constants of motion such as the total energy and momentum of the system. Sophisticated high-order finite-difference algorithms may be used to improve the short-time accuracy. However, higher order schemes do not necessarily improve the long-time behavior. For example, the simple velocity–Verlet algorithm is much more stable over long time scales than Gear's predictor-corrector algorithm.[85]

Recently, there has been much interest in symplectic (or canonical) integration schemes,[88–90] which preserve certain invariants of Hamiltonian systems. An arbitrary integration algorithm may be viewed as a transformation of variables applied at each time step,

$$(x, p) \rightarrow (X, P) \tag{22}$$

where x and p represent positions and momenta of all the particles:

$$\begin{aligned} x &= (\mathbf{r}_1, \mathbf{r}_2, \ldots, \mathbf{r}_N) \\ p &= (\mathbf{p}_1, \mathbf{p}_2, \ldots, \mathbf{p}_N) \end{aligned} \tag{23}$$

The transformation (Eq. 22) is symplectic if it defines a canonical transformation of variables, i.e.,

$$\begin{bmatrix} \frac{\partial X}{\partial x} & \frac{\partial X}{\partial p} \\ \frac{\partial P}{\partial x} & \frac{\partial P}{\partial p} \end{bmatrix}^T \begin{bmatrix} 0 & \mathbf{I} \\ -\mathbf{I} & 0 \end{bmatrix} \begin{bmatrix} \frac{\partial X}{\partial x} & \frac{\partial X}{\partial p} \\ \frac{\partial P}{\partial x} & \frac{\partial P}{\partial p} \end{bmatrix} = \begin{bmatrix} 0 & \mathbf{I} \\ -\mathbf{I} & 0 \end{bmatrix} \tag{24}$$

where $\mathbf{I}$ is a unit $N \times N$ matrix. Such symplectic transformations conserve the phase-space volume, which is crucial for the long-time performance of the integration algorithm. The velocity–Verlet algorithm turns out to be symplectic, while Gear's predictor–corrector scheme does not have this property.

In our MD simulations the equations of motion are integrated with the velocity–Verlet algorithm.[85,91,92] This algorithm is symplectic and time-reversible, which ensures its long-time stability. The velocity–Verlet algorithm works as follows. Suppose we are given atomic positions ($\mathbf{r}_i(t)$), velocities ($\mathbf{v}_i(t)$), and accelerations ($\mathbf{a}_i(t)$) at a certain moment t. First we find the velocities at the midpoint:

$$\mathbf{v}_i(t + \Delta t/2) = \mathbf{v}_i(t) + \mathbf{a}_i(t)\Delta t/2 \tag{25}$$

The atomic positions at time $t + \Delta t$ are calculated from:

$$\mathbf{r}_i(t + \Delta t) = \mathbf{r}_i(t) + \mathbf{v}_i(t + \Delta t/2)\Delta t \tag{26}$$

Subsequently, new values of forces, $\mathbf{f}_i(t + \Delta t)$, and accelerations, $\mathbf{a}_i(t + \Delta t)$, are computed using new atomic positions. Finally, the velocities at time $t + \Delta t$ are found from:

$$\mathbf{v}_i(t + \Delta t) = \mathbf{v}_i(t + \Delta t/2) + \mathbf{a}_i(t + \Delta t)\Delta t/2 \tag{27}$$

As a result, we have new values of positions, velocities, and accelerations. The phase-space trajectories are calculated by applying this procedure recursively.

In our MD simulations, we find that the velocity–Verlet algorithm permits a time step of 2–3 fs depending on the temperature. These values are a factor of 2 larger than the time steps permitted by Gear's predictor–corrector method (for the same level of energy conservation).

Recently Tuckerman et al.[87] have developed an elegant time-reversible integrator using the Trotter factorization of the Liouville operator. This formulation has been used to derive the reversible reference system propagator algorithm (RESPA). The RESPA method offers several advantages over the original MTS approach. The RESPA integrator can be made time-reversible and symplectic, which dramatically improves its stability. Furthermore, RESPA offers a framework to treat various time-scale problems, such as stiff oscillators in a soft fluid, disparate masses, as well as the separation of long- and short-range forces.

For a Hamiltonian system, the Liouville operator, L, is defined as,

$$iL = \{..., H\} \tag{28}$$

where H is the Hamiltonian function and $\{..., ...\}$ denotes the Poisson bracket. For the Hamiltonian (Eq. 1), the Liouville operator is given by:

$$iL = \sum_i \left[\mathbf{v}_i \frac{\partial}{\partial \mathbf{r}_i} + \mathbf{f}_i \frac{\partial}{\partial \mathbf{p}_i} \right] \tag{29}$$

A formal solution for the Hamiltonian equations of motion (Eq. 4) may be written in terms of the Liouville operator,

$$\Gamma(t) = e^{iLt}\Gamma(0) = U(t)\Gamma(0) \tag{30}$$

where $\Gamma(t) = \{\mathbf{r}_i, \mathbf{p}_i\}$ represents a state of the system, and $U(t) = e^{iLt}$ is the classical propagator. If the Liouville operator is decomposed into two parts,

$$iL = iL_1 + iL_2 \tag{31}$$

the Trotter factorization may be applied:

$$U(\Delta t) = e^{iL\Delta t} = e^{iL_1\Delta t/2}e^{iL_2\Delta t}e^{iL_1\Delta t/2} + O(\Delta t^3) \tag{32}$$

The time interval $[0,t]$ may be decomposed into a number of small intervals Δt and the state of the system at time t is obtained by successive application of the propagator $U(\Delta t)$. As an example, let us assume the following decomposition of the Liouville operator (Eq. 29):

$$iL_1 = \sum_i \mathbf{r}_i \frac{\partial}{\partial \mathbf{r}_i},\ iL_2 = \sum_i \mathbf{f}_i \frac{\partial}{\partial p_i} \tag{33}$$

In this case, the Trotter formula yields:

$$U(\Delta t) = U_1(\Delta t/2)U_2(\Delta t)U_1(\Delta t/2) = \\ e^{(\Delta t/2)\sum_i \mathbf{v}_i \frac{\partial}{\partial \mathbf{r}_i}} e^{\Delta t \sum_i \mathbf{f}_i \frac{\partial}{\partial \mathbf{p}_i}} e^{(\Delta t/2)\sum_i \mathbf{v}_i \frac{\partial}{\partial \mathbf{r}_i}} \tag{34}$$

Using the identity,

$$e^{t\frac{\partial}{\partial x}} f(x) = f(x+t) \tag{35}$$

we can identify the operators $U_1(\Delta t/2)$, $U_2(\Delta t)$ with translations of positions and velocities:

$$U_1(\Delta t/2)\text{: } \mathbf{r}_i \rightarrow \mathbf{r}_i + \mathbf{v}_i \Delta t/2 \\ U_2(\Delta t)\text{: } \mathbf{v}_i \rightarrow \mathbf{v}_i + \frac{\mathbf{f}_i}{m_i} \Delta t/2 \tag{36}$$

It may be seen from Eqs. 34 and 36 that the operator $U(\Delta t)$ is equivalent to one step of the velocity–Verlet algorithm. Using different decompositions for the Liouville operator, it is possible to design time-reversible integrators with desired properties.

Multiple time-step algorithms may be derived by using the decomposition (Eq. 31) to separate rapidly varying primary forces ($\mathbf{f}_i^p$) and slowly varying secondary forces ($\mathbf{f}_i^s$):

$$
\begin{aligned}
iL &= iL_p + iL_s \\
iL_p &= \sum_i \dot{\mathbf{r}}_i \frac{\partial}{\partial \mathbf{r}_i} + \mathbf{f}_i^p \frac{\partial}{\partial \mathbf{p}_i} \\
iL_s &= \sum_i \mathbf{f}_i^s \frac{\partial}{\partial \mathbf{p}_i}
\end{aligned}
\tag{37}
$$

The Trotter expansion yields:

$$U(\Delta t) = U_s(\Delta t/2)U_p(\Delta t)U_s(\Delta t/2) \tag{38}$$

The propagator $U_s(\Delta t/2)$ increments the velocities due to secondary forces:

$$U_s(\Delta t)\colon \mathbf{v}_i \rightarrow \mathbf{v}_i + \frac{\mathbf{f}_i^s}{m_i}\Delta t/2 \tag{39}$$

By dividing the time step Δt into n smaller time steps δt, the primary part, $U_p(\Delta t)$, may be further factorized into $U_p(\delta t)$, and each of the elementary propagators $U_p(\delta t)$ is approximated by the velocity–Verlet integrator. The resulting expression for $U_p(\Delta t)$ is,

$$U_p(\Delta t) = [U_{p1}(\delta t/2)U_{p2}(\delta t)U_{p1}(\delta t/2)]^n \tag{40}$$

where $U_{p1}(\delta t/2)$, $U_{p2}(\delta t)$ are given by equations similar to Eq. 36:

$$
\begin{aligned}
U_{p1}(\Delta t/2)&\colon \mathbf{r}_i \rightarrow \mathbf{r}_i + \mathbf{v}_i\Delta t/2 \\
U_{p2}(\Delta t)&\colon \mathbf{v}_i \rightarrow \mathbf{v}_i + \frac{\mathbf{f}_i^p}{m_i}\Delta t/2
\end{aligned}
\tag{41}
$$

This resulting MTS integrator works as follows. First, the secondary forces are computed at time t, and the velocities are incremented using Eq. 39. Subsequently, the velocities and positions are updated by running the usual velocity–Verlet algorithm for n small time steps δt, excluding the secondary forces. Finally, the velocities are again incremented using the secondary forces computed with the new atomic positions.

We have implemented this scheme for the silicon nitride system. The two-body potential (Eq. 14) is divided into short- and long-range parts,

$$V^{(2)}(r) = V^{(2)}_{sh}(r) + V^{(2)}_{lo}(r) \tag{42}$$

where:

$$\begin{aligned} V^{(2)}_{sh}(r) &= f(r)V^{(2)}(r) \\ V^{(2)}_{lo}(r) &= [1 - f(r)]V^{(2)}(r) \end{aligned} \tag{43}$$

Following Tuckerman et al.,[87] the switching function, $f(r)$, is taken to be:

$$f(r) = \begin{cases} 1, & r < r_{sh} - \Delta \\ 1 + \dfrac{(r - r_{sh} + \Delta)^2}{2(r - r_{sh})\Delta - \Delta^2} & r_{sh} - \Delta < r < r_{sh} \\ 0, & r > r_{sh} \end{cases} \tag{44}$$

The values of the cutoff parameters ($r_{sh} = 2.3$Å and $\Delta = 0.2$Å) are chosen so that the short-range potential involves only nearest-neighbor interactions. The primary forces in Eq. 37 are given by the short-range part of the two-body interaction; the secondary interaction consists of the long-range two-body potential and the three-body contributions (Eq. 12):

$$\begin{aligned} V_p &= V^{(2)}_{sh} \\ V_s &= V^{(2)}_{lo} + V^{(3)} \end{aligned} \tag{45}$$

The small time step, δt, in Eq. 40 is chosen to be the same as the time step we would use without the MTS scheme ($\delta t = 2$ fs); the large time step is $\Delta t = 2\delta t$. Compared with the regular velocity–Verlet algorithm, the reversible MTS algorithm reduces the CPU time by 40% without sacrificing the energy conservation.

III. STRUCTURE AND PHYSICAL PROPERTIES OF CRYSTALLINE AND AMORPHOUS Si_3N_4

Silicon nitride is a very promising ceramic material for high-temperature applications.[93] It is widely used in turbine engines, ball bearings, pressure sensors, and microelectronic devices. Silicon nitride has two crystalline structures, α and β. In both cases, each Si atom is bonded to four N atoms in a tetrahedral configuration and each N atom is bonded to three Si atoms in approximately trigonal planar configuration. These structures can also be regarded as interleaved sheets of 8- and 12-membered Si and N rings. In β-Si_3N_4 the layers of Si–N are stacked parallel to the hexagonal basal plane in an alternate sequence of ABABAB.... In the α phase, the sequence is ABCDABCD... where CD layers are related to the AB layers by a c-glide plane. This causes the c-axis of the α phase to be approximately twice as long as that of the β phase. The unit cell of the α phase contains 12 Si and 16 N atoms and the unit cell of the β phase has 6 Si and 8 N atoms. The two phases tend to coexist, since the difference in their enthalpy is only 30 kJ/mol. Also it is hard to avoid small amounts of impurities, especially oxygen. Evaporation of the α phase allows α–β transformation. The reverse transformation, from the β to the α phase, does not seem to occur. The single-crystal densities of α and β phases are 3.185 g/cc and 3.196 g/cc, respectively.

The measured values of Young's moduli in α-Si_3N_4 are 419, 375, and 386 GPa for $\theta = 5°$, 64.2°, and 83.1°, respectively (the orientation angle, θ, is measured with respect to the *c*-axis). The system has a low thermal expansion coefficient, ~ 3 MK^{-1} between 0 and 1300 K. Silicon nitride does not melt under normal pressure; instead it sublimates at 2155 K. Recently Loong et al. measured the one phonon density-of-states (DOS) of α-Si_3N_4 by inelastic neutron scattering. The DOS extends to approximately 170 meV, and it has well-defined peaks around 50 and 112 meV. The specific heat of α-Si_3N_4 has also been measured over a wide range of temperatures.

Silicon nitride also exists in the amorphous state, which is grown by the chemical vapor deposition (CVD) technique. The structure of the amorphous system at a density of 2.8 g/cc was determined by Misawa et al.[84] using the neutron-scattering technique. Their results reveal that the amorphous system also contains tetrahedrally bonded Si and threefold coordinated N atoms.

Using the interatomic potential described in the previous section, MD simulations were performed to determine structural correlations, elastic moduli, phonon densities-of-states, and specific heats of crystalline and amorphous silicon nitride.[77–78,83] These results are in good agreement with experimental measurements mentioned above.

IV. CONSOLIDATION OF NANOPHASE SILICON NITRIDE

One of the most outstanding features of nanophase ceramics is their enhanced ductility. Although it is generally attributed to the structure of intergranular regions,[2,3] there is little quantitative understanding of the relation between the interfacial structure and the physical properties of nanophase materials. The issue of structure–property relationship can be addressed through large-scale atomistic simulations (at least 10^2 clusters, each containing 10^3–10^4 atoms) and long processing times (~10^5–10^6 time steps). Because of large system size and long processing time requirements, only few calculations of nanophase materials have been attempted thus far.[94,95]

We have performed million-atom MD simulations to investigate the sintering process in nanophase Si_3N_4, the structure and mechanical properties as a function of the grain size and porosity, and crack propagation in the consolidated nanophase Si_3N_4. These results are discussed in this section.

A. Sintering of Si_3N_4 Nanoclusters

The sintering process is of great importance in fabrication technology. Despite a great deal of experimental work, there is very little understanding of sintering of ultrafine particles of ceramic materials. In the past, several mechanisms were invoked to interpret sintering experiments on macroscopic grains (diameter ~ 1 μm–1 mm). These included models based on various transport mechanisms (evaporation/condensation, surface diffusion, volume diffusion, viscous flow of matter, etc.).[96–99] Attempts have also been made to include the effect of surface-energy anisotropy in sintering of crystalline particles.[100] Also a new scaling approach has been proposed recently to model sintering of aerogels[101] (this approach does not address dynamical aspects of sintering).

Recent experiments on the synthesis of nanophase materials[1,2] have questioned the validity of macroscopic models of sintering. The experiment of Bonevich and Marks[102] on aluminum oxide reveals that sintering of nanoscale particles is quite different from the sintering of macroscopic particles. Recently the time evolution of the neck growth between faceted nanoclusters of magnesium oxide has also been studied experimentally,[103] and Zhu and Averback[104] have used computer simulation to study the sintering of copper nanoclusters.

A couple of years ago we started the investigation of sintering of Si_3N_4 nanoclusters using MD simulations.[105] Faceted α-Si_3N_4 nanocrystals of size 80 Å (20,335 atoms per cluster) were obtained by Wulff's construction. The clusters were heated to 2500 K over a time period of 150 ps; at several intermediate temperatures the clusters were thermalized for 20 to 50 ps. From 2500 K the clusters were cooled down to 2000 K and again thermalized for 50 ps each.

Figure 5 shows snapshots of the two nanocrystals at 2000 K. Initially the nanocrystals were aligned so that their c-axes were coincident and the (0001) facets were parallel to each other (Figure 5a) at a separation of 3.5 Å. (The initial total angular momentum was set equal to zero.) Subsequently the MD simulation of nanocrystals were allowed to evolve at 2000 K for 220 ps. In the beginning, the two nanocrystals simply rotated relative to each other. After 40 ps a couple of Si atoms on one of the nanocrystals formed bonds with N atoms on the other nanocrystal (see Figure 5b). Between 60 and 100 ps, a highly asymmetric neck formed between the rough small facets of the two nanocrystals (Figure 5c). Between 100 and 220 ps, the neck size remained stable while the relative motion of the clusters began to die. The analysis of the neck region revealed that 55% of Si atoms were fourfold coordinated and the remaining Si atoms had threefold coordination.

Figure 6 shows the time variation of the mean-square displacements of Si and N atoms in the neck region and inside the two nanocrystals after 200 ps. The atoms in the neck region diffuse much more rapidly than those inside the nanocrystals. From the mean-square displacements, we estimate the self-diffusion coefficients of Si and N in the neck region to be approximately 10^{-6} cm^2/s.

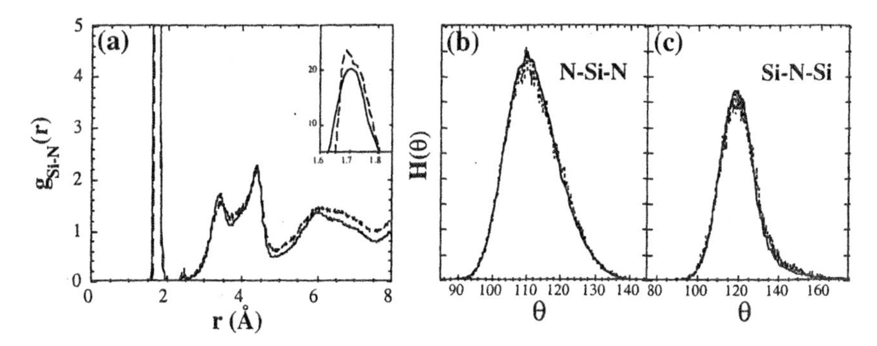

Figure 10. Structural correlations in bulk amorphous Si_3N_4 at a mass density of 2.0 g/cc (*solid lines*) and the intercluster regions in the nanophase system (*dashed lines*): (**a**) Si–N pair-distribution functions; (**b**) and (**c**) N–Si–N and Si–N–Si bond-angle distributions.

vibrational density-of-states is dramatically enhanced at lower densities due to contributions from floppy modes.

In Figure 10 the structural properties of intercluster regions are compared with those of bulk amorphous Si_3N_4.[83] We have chosen an amorphous silicon nitride system whose density is the same as that of interfacial regions in nanophase Si_3N_4. The average coordination (3.58) of silicon atoms in this amorphous system is also close to that in the interfacial regions of nanophase Si_3N_4. Figure 10 shows that the Si–N pair-correlation function and the bond-angle distributions in the interfacial regions are very close to those in the bulk amorphous system. This indicates that the intercluster regions in nanophase Si_3N_4 are structurally very similar to bulk amorphous Si_3N_4. Intergranular glassy layers have in fact been observed experimentally in Si_3N_4 ceramics sintered at high temperature.[111,112] Experiments indicate that the residual amorphous layers cannot be completely recrystallized, and their thickness (~1 nm) varies systematically with the chemical composition of the sample.[112] Keblinski et al.[113] have observed that disordered grain boundaries in their MD simulations of silicon are similar to those in the bulk amorphous silicon system.

V. MECHANICAL PROPERTIES OF NANOPHASE Si_3N_4

We have also investigated the effect of consolidation on elastic moduli of nanophase Si_3N_4.[107] Figure 11 shows the porosity dependence of the bulk modulus, K, and the shear modulus, G. The open circles are the MD results for the bulk modulus of the three nanophase systems with 60 Å clusters; open triangles and squares are the bulk and shear moduli, respectively, of systems with 45 Å clusters. The dependence of elastic moduli on porosity and cluster size can be understood in terms of a

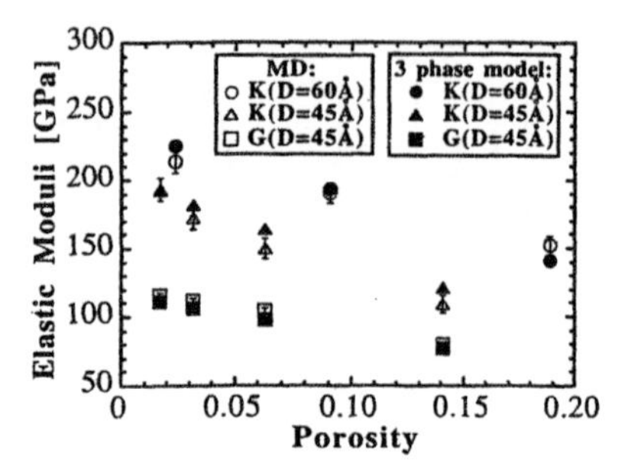

Figure 11. Bulk (K) and shear (G) moduli as a function of porosity. Open circles—MD results for bulk moduli (cluster size D = 60 Å); open triangles and squares—MD results for bulk and shear moduli, respectively, of nanophase systems with 45 Å clusters. Solid circles and triangles denote bulk moduli and solid squares are shear moduli calculated from the multiphase model.

multicomponent model for heterogeneous materials.[114] According to this model, the bulk and shear moduli of a heterogeneous material are given by,

$$\sum_{i=1}^{n} \frac{c_i}{1 - \alpha(1 - K_i/_K)} = \sum_{i=1}^{n} \frac{c_i}{1 - \beta(1 - G_i/_G)} = 1 \tag{46}$$

where n is the number of phases and c_i, K_i, and G_i are their concentrations and bulk and shear moduli, respectively. The quantities α and β are related to the Poisson's ratio, ν:

$$\alpha = \frac{1 + \nu}{3(1 - \nu)}; \; \beta = \frac{2(4 - 5\nu)}{15(1 - \nu)}; \; \nu = \frac{3K - 4G}{6K + 2G} \tag{47}$$

In the nanophase systems we have pores, interior crystalline regions of nanoclusters, and amorphous interfacial regions. The pore concentration, c_1, is the ratio of the pore volume to the total volume of the nanophase system; the concentration, c_2, is obtained from the effective volume of the crystalline part of nanoclusters, and the concentration of amorphous interfacial regions is determined from the condition, $c_1 + c_2 + c_3 = 1$. The bulk and shear moduli of individual phases are obtained from MD calculations for the α-crystal and the amorphous Si_3N_4 system (note, $K_1 = G_1 = 0$): For the α-crystal[81,115] $K_2 = 289$ GPa and $G_2 = 145$ GPa (this is an average over different directions), and for the amorphous system $K_3 = 181$ GPa and $G_3 = 109$ GPa. Using these values of elastic moduli[1] we solve Eqs. 46 and 47 for K and G. These results are also shown in Figure 11. Clearly the three-phase model can successfully explain the MD results for the porosity dependence of elastic moduli.

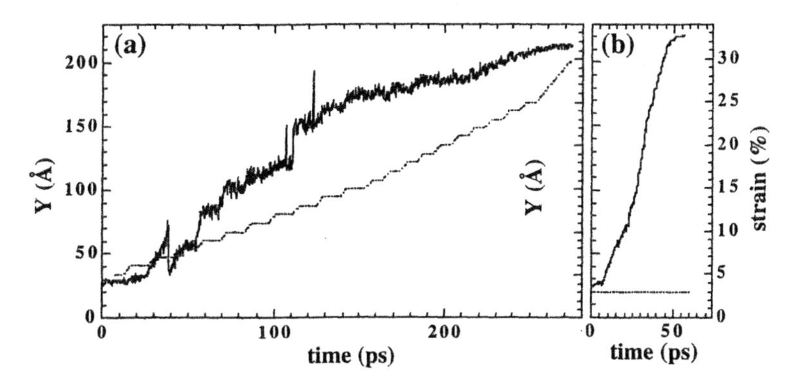

Figure 16. Average crack-tip position for (**a**) nanophase and (**b**) crystalline Si_3N_4 samples. The dotted line shows the time variation of the applied strain.

We have also estimated the elastic energy dissipated per unit area during fracture. For the nanophase system the fracture energy is 24 J/m^2, whereas for the crystal its value is 4 J/m^2. In nanophase Si_3N_4, crack branching and meandering and plastic deformation provide efficient mechanisms for energy dissipation. This makes the nanophase system much tougher than crystalline Si_3N_4.

B. Morphology of Fracture Surfaces

Nature provides many examples of surfaces and interfaces which are irregular and cannot be described by smooth functions. These so-called rough surfaces occur on various length scales as a result of surface growth, fracture, erosion, diffusion, and other processes. Fractal description has proven to be extremely useful in studying the formation, morphology, and properties of such surfaces. While the term "fractal" implies invariance with respect to isotropic dilation (self-similarity), fractal surfaces are usually self-affine, i.e. invariant with respect to anisotropic dilation,

$$(x, y, h) \rightarrow (\lambda x, \lambda y, \lambda^{\alpha} h) \tag{48}$$

where the surface (embedded in 3D space) is defined by its height ($z = h(x, y)$); λ is the scaling parameter and α is the roughness exponent. Equation 48 can be easily generalized to the case of 2D embedding space (in which case the surface is replaced by a curve) or to higher dimensions. The characteristic width of a self-affine surface, observed over length L, scales as,

$$w(L) \sim L^{\alpha} \tag{49}$$

where $\alpha < 1$. The roughness exponent, α, is related to the local fractal dimension, D_f, of the surface,

$$D_f = D - \alpha \tag{50}$$

where D is the dimensionality of the embedding space.

Mandelbrot et al.[118] were the first to apply this fractal analysis to fracture surfaces. Their experiments involved transgranular fracture in steel studied with Charpy impact tests. Mandelbrot et al. found the fracture surfaces to be self-affine objects with a roughness exponent in the range of 0.72–0.90. They concluded that higher roughness exponents corresponded to higher fracture toughness. Subsequent experimental studies failed to confirm a definite correlation between roughness exponents and fracture toughness of materials. In fact, the opposite correlation has been observed by Mecholsky et al.[119] in ceramics. Bouchaud et al.[120] studied ductile fracture in four samples of the same material (an aluminum alloy) subjected to different heat treatments. Different fracture modes and fracture toughness values were observed, but the roughness exponent turned out to be the same (≈ 0.8) for all four samples. This observation lead them to suggest that the roughness exponents of 0.8 in 3D and 0.7 in 2D are universal, i.e. independent of the material or the fracture mode. In many subsequent experimental studies the roughness exponents were found to have the universal values. Måløy et al.[121] performed an extensive study of fracture surfaces in six different brittle materials and reported they all had the same roughness exponent of 0.87 ± 0.07. However, smaller roughness exponents (in the range 0.4–0.6) have also been reported, mostly in STM measurements on the nanometer scale.[122,123] Based on these results, Milman et al.[122,123] argued against the universality of the roughness exponent. Bouchaud et al.[124] investigated fracture surfaces in Ti_3Al alloy over a wide range of length scales (10 nm–1 mm) and for different crack velocities. They found that fracture profiles are characterized by two different roughness exponents: $\alpha \sim 0.45$ below a certain crossover length scale and $\alpha \sim 0.84$ above the crossover length. It has also been determined that the crossover length shifts to smaller length scales as the crack velocity increases. Such a crossover between two different roughness exponents (0.82 and 0.44), corresponding to two different fracture regimes, has been observed in MD simulations of fracture in an amorphous silicon nitride film.[81] The simulations reveal that the smaller roughness exponent corresponds to slow continuous crack growth, while fast crack propagation by coalescence with pores opening in front of the crack tip yields larger roughness exponents. These experimental and numerical results indicate that the universal value of the roughness exponent (~0.8) is relevant to fast crack propagation or above a certain crossover length. On smaller length scales or for sufficiently slow cracks, the fracture process crosses over to a different regime which results in smaller roughness exponents.

Hansen et al.[125] tested the universality of the roughness exponent numerically for a 2D fuse model. The same roughness exponent of 0.7 (within 10% accuracy) was found for different distributions of the fuse strength. Furthermore, it has been argued that brittle fracture is related to the problem of directed polymers in a random medium which gives a roughness exponent of 2/3 in two dimensions. Careful experimental studies of fracture in 2D systems,[126,127] such as paper and various kinds of wood, yield $\alpha = 0.68 \pm 0.04$, which is consistent with the theoretical value of 2/3. However, Måløy et al.[121] have shown that this explanation fails for the case of 3D fracture. The 3D generalization of the directed polymer problem is the problem of finding the minimum energy surface for a random distribution of energy in space. The roughness exponent in this case has been estimated to be 0.50 ± 0.08,[128] which is in sharp disagreement with the typical value of $\alpha \sim 0.8$ for fracture in 3D materials. Furthermore, Bouchaud et al.[129] have argued that the minimum energy surface may not be found dynamically.

To investigate the nature of self-affine fracture surfaces in the nanophase Si_3N_4, we calculate the height–height correlation functions both in and out of the fracture plane *y-z*.[108] Figure 17a shows that the results for the out-of-plane

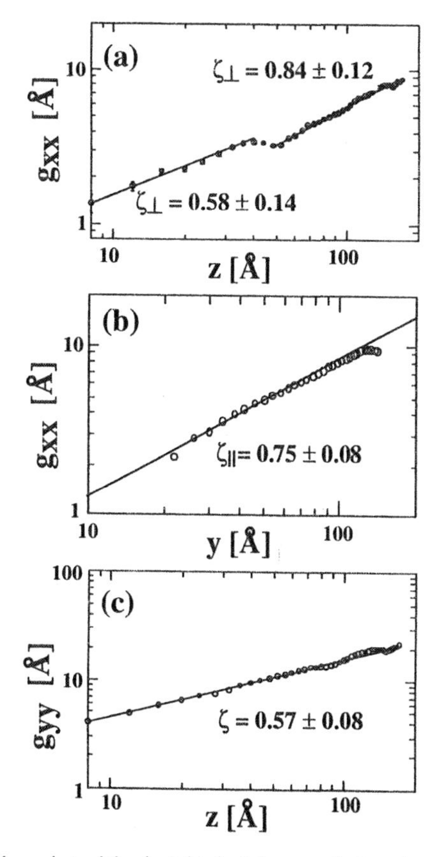

Figure 17. (**a**) Log–log plot of the height–height correlation function, $g_{xx}(z)$, versus z for the out-of-plane fracture profile $x(z)$; (**b**) the results for the out-of-plane profile $x(y)$; and (**c**) $g_{yy}(z)$ for the in-plane fracture profile $y(z)$.

height–height correlation function $g_{xx}(z)$ $(= <(x(z+z_0)-x(z_0))^2>^{1/2})$ can be fit to two roughness exponents: $\zeta_\perp = 0.84 \pm 0.12$ above a certain length scale (64 Å) and $\zeta_\perp = 0.58 \pm 0.14$ otherwise. Figure 17b shows the MD results for the other out-of-plane height–height correlation function $g_{xx}(y)$. (These results are plotted beyond the length of the initial notch in the system.) In this case the best fit to the results gives a roughness exponent $\zeta_\parallel = 0.75 \pm 0.08$. The MD results for $\zeta_\perp$ and $\zeta_\parallel$ are very close to experimental values.[130–132] We have also determined the roughness exponent for the in-plane fracture profile $y(z)$. (Recently Schmittbuhl et al.[131] pointed out that the self-affine correlation length scales with the distance to the notch as $\xi \sim y^{1/1.2}$; as a result, the roughness exponent, $\zeta_\parallel$, for the out-of-plane fracture profile, $x(y)$, should be $\zeta_\perp/1.2$[130].) Figure 17c shows that the best fit to the corresponding height–height correlation function gives $\zeta = 0.57 \pm 0.08$. This is very close to the measured values of the in-plane roughness exponents.[130]

VII. CONCLUSIONS

We have developed a multiresolution approach to implement large-scale MD simulations on parallel machines. Using this approach we have performed a series of MD simulations, involving ~10^6 atoms, for nanophase silicon nitride. The simulations are based on realistic interatomic potentials, which have been validated by comparison with various experiments.

The nanophase systems are consolidated by applying an external pressure at an elevated temperature (2000 K). Under a pressure of 15 GPa the nanophase system is consolidated to 92% of the density of α-Si_3N_4 crystal. The MD calculations for partial pair-correlation functions and bond-angle distributions indicate: (1) interior regions of nanoclusters are crystalline, (2) intercluster regions are amorphous, and (3) voids are present in the system. The structure of intercluster regions is very similar to that of amorphous Si_3N_4 at the corresponding density.

We have also determined the effects of porosity and grain size on the bulk and shear moduli of nanophase Si_3N_4. The results of MD calculations are in excellent agreement with a multiphase model, proposed by Budiansky, for heterogeneous materials.

Finally, we have also examined the fracture process in nanophase silicon nitride. We find that nanostructures in the system tend to induce crack branching and meandering and significant relative motion of nanoclusters takes place due to plastic deformation in disordered interfacial regions. These mechanisms allow the nanophase system to dissipate the elastic energy much more efficiently than α-Si_3N_4 crystal. As a result, it takes a strain of 30% to completely fracture the nanophase system, while crystalline Si_3N_4 cleaves at a strain of only 3%. Fracture surfaces in the nanophase system are characterized by distinct out-of-plane and in-plane roughness exponents. The MD results for roughness exponents are in good agreement with experimental measurements on various other materials. This gives credence to the conjecture that roughness exponents may be "universal," i.e.

independent of material characteristics, length scales of observations, and the mode of fracture.

ACKNOWLEDGMENTS

This work was supported by DOE (Grant No. DE-FG05-92ER45477), NSF (Grant No. DMR-9412965), AFOSR (Grant No. F 49620-94-1-0444), USC-LSU Multidisciplinary University Research Initiative (Grant No. F 49620-95-1-0452), Army Research Office (Grant No. DAAH04-96-1-0393), and Louisiana Education Quality Support Fund (LEQSF) (Grant No. LEQSF(96-99)-RD-A-10). Simulations were performed on parallel machines in the Concurrent Computing Laboratory for Materials Simulations (CCLMS) at Louisiana State University. These facilities were acquired with equipment enhancement grants awarded by LEQSF.

NOTES

1. In the model calculation ν varies between 0.23 and 0.27. From MD calculations we obtain ν = 0.27 for the α-crystal and the amorphous system at 3.2 g/cc. The experimental value for the α-crystal is 0.22–0.28.

REFERENCES

1. Karch, J.; Birringer, R.; Gleiter, H. *Nature* **1987**, *330*, 556.
2. Siegel, R. W. In *Materials Interfaces: Atomic-Level Structure and Properties*; D.Wolf, S. Yip, Eds.; Chapman Hall: London, 1992, p. 431.
3. Stern, E. A.; Siegel, R. W.; Newville, M.; Sanders, P. G.; Haskel, D. *Phys. Rev. Lett.* **1995**, *75*, 3874.
4. Beardsley, T. *Scientific American* **1992**, *114*.
5. Birringer, R.; Herr, U.; Gleiter, H. *Suppl. Trans. Jpn. Inst. Met.* **1986**, *27*, 43.
6. Siegel, R. W.; Hahn, H. In *Current Trends in the Physics of Materials*; M. Yussouff, Ed.; World Scientific: Singapore, 1987, p. 403.
7. Birringer, R.; Gleiter, H. In *Encyclopedia of Material Science and Engineering*; Cahn, R. W., Ed.; Pergamon Press: Oxford, 1988, p. 339.
8. Schaeffer, H. E.; Würschum, R.; Birringer, R. *J. Less-Common Metals* **1988**, *14*, 161.
9. Birringer, R. *Mat. Sci. Eng.* **1989**, *A117*, 33.
10. Andres, R. P. et al. *J. Mater. Res.* **1989**, *4*, 704.
11. Suryanarayana, C.; Froes, F. H. In *Physical Chemistry of Powder Metal Production and Processing*; Murray, W.; Robertson, D. G. C., Eds.; TMS Publ.: Warrendale, PA, 1989, p. 269.
12. Gleiter, H. *Prog. Mat. Sci.* **1990**, *33*, 223.
13. Gleiter, H. *Nanostruct. Mater.* **1992**, *1*, 1.
14. Pechenik, A.; Piermarini, G. J.; Danforth, S. C. *J. Am. Ceram. Soc.* **1992**, *75*, 3283.
15. Kear, B. H. et al. *Research Opportunities for Materials with Ultrafine Microstructures*; National Academy: Washington, DC, 1989, Vol. NMAB-454.
16. Gleiter, H. In *Deformation of Polycrystals: Mechanisms and Microstructures*, N. Hansen et al., Eds.; Risø National Laboratory: Roskilde, 1981, p. 15.
17. Kimoto, K. et al. *Jpn. J. Appl. Phys.* **1963**, 2, 702.
18. Granqvist, C. G.; Buhrman, R. A. *J. Appl. Phys.* **1976**, *47*, 2200.
19. Thölén, A. R. *Acta Metall.* **1979**, *27*, 1765.

20. Birringer, R.; Gleiter, H.; Klein, H.-P.; Marquardt, P. *Phys. Lett.* **1984**, *102A*, 365.
21. Hahn, H.; Eastman, J. A.; Siegel, R. W. In *Ceramic Transactions, Ceramic Powder Science*; Messing, G. L. et al., Eds.; American Ceramic Society: Westerville, OH, 1988, Vol. 1B, p. 1115.
22. Eastman, J. A.; Liao, Y. X.; Narayanaswamy, A.; Siegel, R. W. In *Mater. Res. Soc. Symp. Proc.*, 1989, Vol. 155, p. 255.
23. Siegel, R. W.; Eastman, J. A. In *Mater. Res. Soc. Symp. Proc.*, 1989, Vol. 132, p. 3.
24. Berkowitz, A. E.; Walter, J. L. *J. Mater. Res.* **1987**, *2*, 277.
25. Hellstern, E.; Fecht, H. J.; Fu, Z.; Johnson, W. L. *J. Appl. Phys.* **1989**, *65*, 305.
26. *Processing Science of Advanced Ceramics*; I. A. Aksay, G. L. McVay, D. R. Ulrich, Eds.; *Mater. Res. Soc. Symp. Proc.*, 1991, Vol. 155.
27. *Clusters and Cluster-Assembled Materials*; R. S. Averback, D. L. Nelson, J. Bernholc, Eds.; *J. Mater. Res. Soc. Symp. Proc.*, 1991, Vol. 206.
28. Bowles, R. S.; Kolstad, J. J.; Calo, J. M.; Andres, R. P. *Surf. Sci.* **1981**, *106*, 117.
29. Oya, H.; Ichihashi, T.; Wada, N. *Jpn. J. Appl. Phys.* **1982**, *21*, 554.
30. Hahn, H.; Averback, R. S. *J. Appl. Phys.* **1990**, *67*, 1113.
31. Iwama, S.; Hayakawa, K.; Arizumi, T. *J. Cryst. Growth* **1984**, *66*, 189.
32. Iwama, S.; Hayakawa, K. *Surf. Sci.* **1985**, *156*, 85.
33. Baba, K.; Shohata, N.; Yonezawa, M. *Appl. Phys. Lett.* **1989**, *54*, 2309.
34. Matsunawa, A.; Katayama, S. In *Laser Welding, Machining and Materials Processing*; C. Albright, Ed.; IFS Publ.: Bedford, UK, 1985, p. 206.
35. McCandlish, L. E.; Kear, B. H.; Kim, B. K. *Mat. Sci.* **1990**, *6*, 953.
36. Thomas, G. J.; Siegel, R. W.; Eastman, J. A. *Scripta Metall. et Mater.* **1990**, *24*, 201.
37. Thomas, G. J.; Siegel, R. W.; Eastman, J. A. In *Mater. Res. Soc. Symp. Proc.*; 1989, Vol. 153, p. 13.
38. Siegel, R. W.; Thomas, G. J. *Ultramicros* **1992**, *40*, 376.
39. Ganapathy, S. K.; Rigney, D. A. *Scipta Metall. et Mater.* **1990**, *24*, 1675.
40. Trudeau, M. L.; Neste, A. V.; Schultz, R. In *Mater. Res. Soc. Symp. Proc.*, 1991, p. 206.
41. Zhu, X., et al. *Phys. Rev.* **1987**, *B35*, 9085.
42. Haubold, T., et al. *J. Less-Common Metals* **1988**, *145*, 557.
43. Haubold, T., et al. *Phys. Lett.* **1989**, *A135*, 461.
44. Epperson, J. E., et al. In *Mater. Res. Soc. Symp. Proc.*, 1989, Vol. 132, p. 15.
45. Epperson, J. E., et al. In *Mater. Res. Soc. Symp. Proc.*, 1990, Vol. 166, p. 87.
46. Jorra, E., et al. *Phil. Mag.* **1989**, *B60*, 159.
47. Melendres, C. A., et al. *J. Mater. Res.* **1989**, *4*, 1246.
48. Parker, J. C.; Siegel, R. W. *J. Mater. Res.* **1990**, *5*, 1246.
49. Parker, J. C.; Siegel, R. W. *Appl. Phys. Lett.* **1990**, *57*, 943.
50. Schaefer, H. E., et al. *Mater. Sci. Forum* **1987**, *15–18*, 955.
51. Schaefer, H. E., et al. *Phys. Rev.* **1988**, *B38*, 9545.
52. Siegel, R. W., et al. *J. Mater. Res.* **1988**, *3*, 1367.
53. Hort, E. Diploma Thesis, Universität des Saarlandes, Saarbrücken, 1986.
54. Hahn, H.; Logas, J.; Averback, R. S. *J. Mater. Res.* **1990**, *5*, 609.
55. Averback, R. S., et al. In *Mater. Res. Soc. Symp. Proc.*, 1989, Vol. 153, p. 3.
56. Mayo, M. J., et al. *J. Mater. Res.* **1990**, *5*, 1073.
57. Höfler, H. J.; Averback, R. S. *Scripta Metall. et Mater.* **1990**, *24*, 2401.
58. Li, Z., et al. *Mater. Lett.* **1988**, *6*, 195.
59. Birringer, R.; Hahn, H.; Hofler, H.; Karch, J.; Gleiter, H. *Defect Diffusion Forum* **1988**, *59*, 17.
60. Mayo, M. J.; Siegel, R. W.; Liao, Y. X.; Nix, W. D. *J. Mater. Res.* **1992**, *39*, 67.
61. Karch, J.; Birringer, R. *Ceramics Int.* **1990**, *16*, 291.
62. Hahn, H.; Logas, J.; Höfler, H. J.; Kurath, P.; Averback, R. S. In *Mater. Res. Soc. Symp. Proc.*, 1990, Vol. 196, p. 71.
63. Hahn, H.; Averback, R. S. *J. Am. Ceram. Soc.* **1991**, *74*, 2918.
64. Hahn, H.; Averback, R. S. *Nanostruct. Mater.* **1992**, *1*, 95.

65. Horváth, J.; Birringer, R.; Gleiter, H. *Solid State Commun.* **1987**, *62*, 319.
66. Horváth, J. *Defect Diffusion Forum* **1989**, *66–69*, 207.
67. Hahn, H.; Höfler, H.; Averback, R. S. *Defect Diffusion Forum* **1989**, *66–69*, 549.
68. Schumacher, S.; Birringer, R.; Straub, R.; Gleiter, H. *Acta Metall.* **1989**, *37*, 2485.
69. Mütschele, T.; Kirchheim, R. *Scripta Metall.* **1987**, *21*, 135.
70. Mütschele, T.; Kirchheim, R. *Scripta Metall.* **1987**, *21*, 1101.
71. Rupp, J.; Birringer, R. *Phys. Rev. B* **1987**, *36*, 7888.
72. Korn, D., et al. *J. de Phys. C5* **1988**, *49*, 769.
73. Stillinger, F. H.; Weber, T. A. *Phys. Rev. B* **1985**, *31*, 5262.
74. Abell, G. C. *Phys. Rev. B* **1985**, *31*, 6184–6190.
75. Tersoff, J. *Phys. Rev. B* **1988**, *37*, 6991–7000.
76. Rappé, A. K.; Goddard, W. A. *J. Phys. Chem.* **1991**, *95*, 3358.
77. Vashishta, P.; Kalia, R. K.; Ebbsjö, I. *Phys. Rev. Lett.* **1995**, *75*, 858.
78. Loong, C.-K.; Vashishta, P.; Kalia, R. K.; Ebbsjö, I. *Europhys. Lett.* **1995**, *31*, 201–206.
79. Vashishta, P.; Kalia, R. K.; Nakano, A.; Li, W.; Ebbsjö, I. In *Amorphous Insulators and Semiconductors;* M. F. Thorpe, M. I. Mitkova, Eds.; NATO ASI, 1996.
80. Greengard, L.; Rokhlin, V. *J. Comput. Phys.* **1987**, *73*, 523.
81. Nakano, A.; Kalia, R. K.; Vashishta, P. *Phys. Rev. Lett.* **1995**, *75*, 3138–3141.
82. Vashishta, P.; Nakano, A.; Kalia, R. K.; Ebbsjö, I. *J. Non-Crystal. Sol.* **1995**, *182*, 59–67.
83. Omeltchenko, A.; Nakano, A.; Kalia, R. K.; Vashishta, P. *Europhys. Lett.* **1996**, *33*, 667.
84. Misawa, M.; Fukunaga, T.; Niihara, K.; Hirai, T.; Suzuki, K. *J. Non-Cryst. Sol.* **1979**, *34*, 314.
85. Allen, M. P.; Tildesley, D. J. *Computer Simulation of Liquids*; Clarendon Press: Oxford, 1987.
86. Streett, W. B.; Tildesley, D. J.; Saville, G. *Mol. Phys.* **1978**, *35*, 639.
87. Tuckerman, M.; Berne, B. J. *J. Chem. Phys.* **1992**, *97*, 1990–2001.
88. Okunbor, D. I.; Skeel, R. D. *J. Comp. Chem.* **1994**, *15*, 72–79.
89. Zhang, G.; Schlick, T. *Mol. Phys.* **1995**, *84*, 1077.
90. Skeel, R. D.; Zhang, G.; Schlick, T. *SIAM J. Sci. Comput.* **1997**. In press.
91. Verlet, L. *Phys. Rev.* **1967**, *159*, 98.
92. Swope, W. C.; Andersen, H. C.; Berens, P. H.; Wilson, K. R. *J. Chem. Phys.* **1982**, *76*, 637.
93. Somiya, S.; Mitomo, M.; Yoshimura, M. *Silicon Nitride*; Elsevier: Essex, 1990, Vol. 1.
94. Liu, C.-L.; Adams, J. B.; Siegel, R. W. *Nanostruct. Mater.* **1994**, *4*, 265–274.
95. Zhu, H.; Averback, R. S. *Mat. Sci. Eng.* **1995**, *A204*, 96.
96. Kuczynski, G. C.; Hooton, N. A.; Gibbon, C. F. In *Sintering and Related Phenomena*; Kuczynski, G.C.; Hooton, N.A.; Gibbon, C.F., Eds.; Gordon and Breach Science Publishers: University of Notre Dame, Notre Dame, IN, 1965.
97. Somiya, S.; Moriyoshi, Y., Eds.; *Silicon Nitride*; Elsevier Applied Science: London, 1990.
98. Ashby, M. F. *Acta Metall.* **1974**, *22*, 275.
99. Brinker, C. J.; Sherer, G. W. *Sol-Gel Science: The Physics and Chemistry of Sol-Gel Processing*; Academic Press: London, 1990.
100. Searcy, A. W. *J. Am. Ceram. Soc.* **1985**, *68*, C-267.
101. Sempéré, R.; Bourret, D.; Woignier, T.; Phalippou, J.; Jullien, R. *Phys. Rev. Lett.* **1993**, *71*, 3307.
102. Bonevich, J. E.; Marks, L. D. In *Nanophase and Nanocomposite Materials*; S. Komarneni, J. C. Parker, G. J. Thomas, Eds.; MRS: Pittsburgh, 1993, Vol. 286.
103. Rankin, J.; Boatner, L. A. *J. Am. Ceram. Soc.* **1994**, *77*, 1987.
104. Zhu, H.; Averback, R. S. *Phil. Mag. Lett.* **1996**, *73*, 27.
105. Tsuruta, K.; Omeltchenko, A.; Kalia, R. K.; Vashishta, P. *Europhys. Lett.* **1996**, *33*, 441–446.
106. Kuczynski, G. C. *Sintering Processes*; Plenum Press: New York, 1980, Vol. 13.
107. Kalia, R. K.; Nakano, A.; Tsuruta, K.; Vashishta, P. *Phys. Rev. Lett.* **1997**, *78*, 689.
108. Kalia, R. K.; Nakano, A.; Omeltchenko, A.; Tsuruta, K.; Vashishta, P. *Phys. Rev. Lett.* **1997**, *78*, 2144.

109. Press, W. H.; Teukolsky, S. A.; Vetterling, W. T.; Flannery, B. P. *Numerical Recipes in FORTRAN*; Cambridge University Press: Cambridge, 1992.
110. Parrinello, M.; Rahman, A. *J. Chem. Phys.* **1982**, *76*, 2662.
111. Clarke, D. R. In *Surfaces and Interfaces in Ceramic Materials*; L. C. Dufour, C. Monty, G. Petot-Ervas, Eds.; Kluwer Academic Publishers: Boston, MA, 1989, p. 57.
112. Kleebe, H.-J.; Cinibulk, M. K.; Cannon, R. M.; Rühle, M. *J. Am. Ceram. Soc.* **1993**, *76*, 1969.
113. Keblinski, P.; Phillpot, S. R.; Wolf, D.; Gleiter, H. *Phys. Rev. Lett.* **1996**, *77*, 2965.
114. Budiansky, B. *J. Mech. Phys. Solids* **1965** , *13*, 223–227.
115. Cartz, L.; Jorgensen, J. D. *J. Applied Phys.* **1981**, *52*, 236–244.
116. Marder, M.; Fineberg, J. *Physics Today* **1996** , *49*, 24
117. Marder, M.; Gross, S. *J. Mech. Phys. Solids* **1995**, *43*, 1–48.
118. Mandelbrot, B. B.; Passoja, D. E.; Paullay, A. J. *Nature* **1984**, *308*, 721–722.
119. Mecholsky, J.; Passoja, D. E.; Feinberg-Ringel, K. S. *J. Amer. Ceram. Soc.* **1989**, *72*, 60.
120. Bouchaud, E.; Lapasset, G.; Planès, J. *Europhys. Lett.* **1990**, *13*, 73.
121. Måløy, K. J.; Hansen, A.; Hinrichsen, E. L.; Roux, S. *Phys. Rev. Lett.* **1992**, *68*, 213–215.
122. Milman, V. Y.; Blumenfeld, R.; Stelmashenko, N. A.; Ball, R. C. *Phys. Rev. Lett.* **1993**, *71*, 204.
123. Milman, V. Y.; Stelmashenko, N. A.; Blumenfeld, R. *Prog. Mat. Sci.* **1994**, *38*, 425–474.
124. Bouchaud, E.; Navéos, S. *J. Phys. I France* **1995**, *5*, 547–554.
125. Hansen, A.; Hinrichsen, E. L.; Roux, S. *Phys. Rev. Lett.* **1991**, *66*, 2476.
126. Engoy, T.; Måløy, K. J.; Hansen, A.; Roux, S. *Phys. Rev. Lett.* **1994**, *73*, 834.
127. Kertész, J.; Horváth, V.; Weber, F. *Fractals* **1993**, *1*, 67.
128. Kardar, M.; Zhang, Y.-C. *Europhys. Lett.* **1989**, *8*, 233.
129. Bouchaud, E.; Lapasset, G.; Planès, J.; Navéos, S. *Phys. Rev.* **1993**, *B48*, 2917.
130. Daguier, P.; Bouchaud, E.; Lapasset, G. *Europhys. Lett.* **1995**, *31*, 367–372.
131. Schmittbuhl, J.; Roux, S.; Berthaud, Y. *Europhys. Lett.* **1994**, *28*, 585.
132. Bouchaud, E.; Daguier, P.; Lapasset, G. *ASM International Conference on Metallography*, Colmar, France, 1995.

PROBING THE ELECTRONIC STRUCTURE OF TRANSITION METAL CLUSTERS FROM MOLECULAR TO BULK-LIKE USING PHOTOELECTRON SPECTROSCOPY

Lai-Sheng Wang and Hongbin Wu

Advances in Metal and Semiconductor Clusters
Volume 4, pages 299–343

ISBN: 0-7623-0058-2

I. INTRODUCTION

Metal clusters are aggregates of atoms which bridge the properties of isolated atoms on one hand and condensed matter on the other. Clusters provide a unique means to view how bulk properties of matter evolve with increasing size in an atom-by-atom manner. Discrete metal clusters may be ideal models of metal surfaces in understanding the processes of chemisorption and catalysis, where local interactions are important. Hence, studies of atomic clusters not only offer opportunities for understanding the origins of bulk properties, but also may yield new insight into the physics and chemistry on surfaces and catalysis.

The electronic structure of clusters and its evolution with cluster sizes are some of the most important questions in the cluster research. However, our understanding of the electronic structure of clusters is rather poor and there is not yet a unified description of the electronic structure of metal clusters and its evolution with cluster size, except perhaps that of the simple alkali clusters.[1] This is particularly true for the transition metal clusters because of their enormous complexity. A variety of techniques have been applied to the investigation of clusters.[2,3] One of the most powerful techniques to elucidate the electronic structure of metal clusters is photoelectron spectroscopy (PES) in which the individual electron binding energies of a cluster are measured, yielding directly the electronic energy levels of a cluster system.[4–6] Such questions can be addressed directly by PES: How does the electronic structure of clusters evolve with cluster size? How and when do electrons forming bonds and molecular orbitals (MOs) in small clusters transform to bands in bulk condensed phase? The electronic structure plays one of the most important roles, along with the geometrical structure, in determining the chemical and physical properties of a cluster. Thus, understanding the electronic structure is essential to understand the chemical reactivity, magnetism, and other chemical and physical properties of metal clusters.

However, transition metal clusters have posted considerable challenges both theoretically and experimentally due to the open d-shell. Very little was known about the electronic structure of the transition metal clusters beyond the dimers.[7] Over the past several years, we have engaged in an extensive effort to elucidate the electronic structures of the first-row transition metal clusters and their evolution as a function of cluster size using size-selected anion PES.[8–16] This is made possible due to some significant advancements of the PES technique in our laboratory. In this chapter, we will provide an overview of the PES technique as applied to metal clusters and a summary of our progress in this endeavor including both our experimental details and recent results.

II. PHOTOELECTRON SPECTROSCOPY: AN IDEAL PROBE OF THE ELECTRONIC STRUCTURE OF METAL CLUSTERS

A. Photoelectron Spectroscopy

In PES, one measures the kinetic energy distribution of photoemitted electrons of a underlying system (atoms, molecules, clusters, surfaces, etc.) at a fixed photon energy.[4–6] From energy conservation, this spectrum represents the binding energy distribution of the electrons in the system. PES is one of the most powerful techniques to study the electronic structure of matter. It has contributed much to our understanding of the electronic structure of atoms, bulk solids, surfaces, and surface-adsorbate systems. When applied to molecules, it has provided unique evidence for the molecular orbital theory,[4,5] and at high resolution it also provides much spectroscopic information about molecular ions.

In the application of PES to metal clusters, it is advantageous to study the negative ions for size-selectivity because clusters are always produced with a distribution of sizes. PES involves single-photon nonresonant processes and itself does not have size-selectivity. There are two other advantages to study the negative clusters. First, anions usually have low electron binding energies, allowing the widely available visible and ultraviolet lasers to be used for photodetachment, while neutral metal clusters except the alkali clusters have high binding energies and require higher photon energies for ionization. More importantly, when applied to negative clusters, PES yields electronic and spectroscopic information about the neutral clusters which are of most interest.

In anion PES, the detachment transitions occur from the ground state of the anion clusters to the ground and excited states of the neutral clusters. Therefore, the obtained spectroscopic information—i.e. electronic energy levels, vibrational frequencies, and electron affinities—is about the neutral clusters. Figure 1 shows schematically how the electronic structure evolve from the atoms to the bulk through clusters as cluster size increases. The atoms have discrete energy levels; so do the small clusters. As the cluster size increases from the left to the right, the density of states (DOS) increases from discrete energy levels to the bulk band structure at the infinitive cluster size limit. The top of Figure 1 displays schematically the resulting PES spectra, which mirror the underlying electronic energy levels of the systems and provide direct electronic structure information about the clusters. In principle, some of the energy level information can be obtained from optical spectroscopy, such as resonant two-photon ionization (R2PI). However, it is well known that the R2PI technique is only applicable to transition metal dimers due to the high density of states of the larger clusters. Therefore, PES is a unique experimental technique that is capable of providing electronic structure information for a wide range of cluster sizes.

B. Photoelectron Spectroscopy of Metal Clusters

It was long realized that the PES technique would be important in the study of the electronic structure of clusters. A variety of experimental methods have been developed. In this section, we intend to give a brief overview of the current technologies, emphasizing the advantages and deficiencies in each method. In the next section we will show what might constitute an ideal PES apparatus for studying metal clusters. In the "Experimental Setup" section we will describe the details of our magnetic-bottle time-of-flight PES (MTOF-PES) apparatus, which is nearly the ideal apparatus due to its high mass and energy resolution.

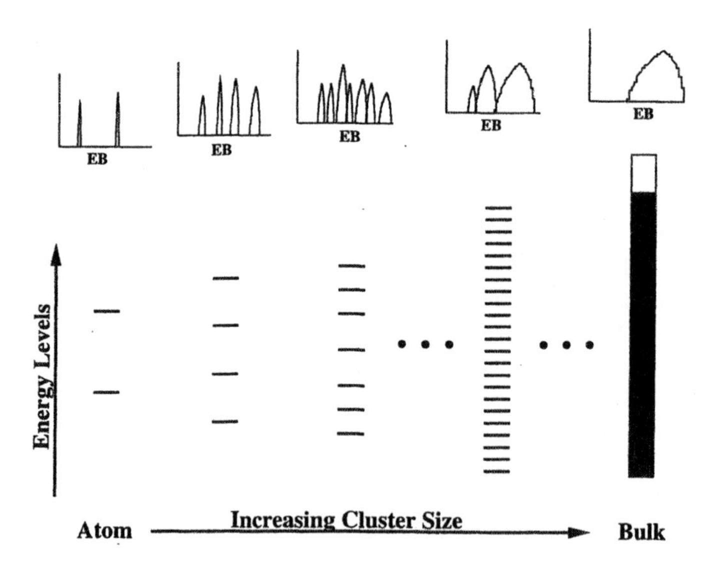

Figure 1. Schematics showing the evolution of the electronic structure from a single atom to clusters and the bulk, and the corresponding photoelectron spectra.

In the early 1980s, one of the authors (LSW) tried to extend a unique high-resolution molecular beam PES apparatus (HeI ionization, 21.2 eV),[17] which was equipped with a high-resolution hemispherical electrostatic electron analyzer and a supersonic molecular beam, to neutral metal clusters.[18] The hemispherical analyzer required an intense, CW neutral cluster beam.[19] Due to the lack of size-selectivity and low collecting efficiency, this apparatus was limited to study only a few systems,[20–22] where clusters could be easily produced, even though the high-temperature molecular beam source was able to produce temperatures up to 2000 K.[18]

The photoelectron–photoion coincidence technique could circumvent the size-selectivity problem of neutral clusters. Dehmer et al. first performed such measurements on neutral Xe clusters.[23] Rademann et al. have coupled this technique with a high-temperature oven cluster source and obtained interesting results for clusters of low melting point elements.[24–27] This method does require a very intense cluster beam and it has been difficult to extent this technique to study less volatile materials. In addition, the coincidence measurement is very time consuming and is difficult to obtain high-resolution spectra.

Lineberger's group first applied the high resolution capability of the hemispherical electrostatic electron analyzer to size-selected negative clusters with an intracavity Ar^+ ion laser and a CW discharge cluster source.[28–30] Bowen's group uses a similar technique but with a high-temperature oven cluster source.[31,32] Leopold's group improved the technique to even higher resolution.[33,34] Interesting vibrationally resolved PES spectra of small clusters have been obtained with this technique. However, there are two intrinsic limits of this technique when applied to metal clusters: (1) the low collecting efficiency requires an intense, CW cluster beam, limiting it to the study of smaller and abundant clusters; (2) the low detachment energies limited to the Ar^+ ion lasers only allow electrons with relatively low binding energies to be detached and it is difficult for it to probe more tightly bound electrons in clusters or clusters with high electron affinities.

Smalley's group[35] combined the laser vaporization cluster beam technique with a MTOF-PES machine which had nearly 100% collecting efficiency.[36] This apparatus had the advantages of having high sensitivity and allowing high energy photons to be used.[37–40] It had the potential as an electronic structure probe for a wide range of clusters and with high photon energies. Meiwes–Broer et al. also independently designed a similar spectrometer with a pulsed arc cluster ion source and applied it to several metal cluster systems.[41–43] The disadvantage of the MTOF technique as it was implemented was the poor energy resolution which often obscured the otherwise important electronic structure information contained in a PES spectrum as mentioned above.

Neumark's group[44,45] has applied the extremely high resolution capability of the ZEKE (zero electron kinetic energy) technique[46] to negative cluster ions with a laser vaporization cluster source and has obtained interesting vibrationally-resolved spectra for several small semiconductor clusters. While this is unquestionably a high resolution technique, it is nonetheless a laser spectroscopy technique rather than a photoelectron spectroscopy technique. The information obtained is mostly about the ground state and perhaps a few low-lying excited states and it is difficult for it to probe deep into the inner electronic energy levels of a cluster. In addition, ZEKE of negative ions depends on the Wigner's threshold law[47] which limits its usefulness only to those clusters which have a *p*-type LUMO and probably prevents its application to transition metal clusters where the LUMO of the neutral clusters is expected to be mostly of s and d characters.

C. An Ideal PES Apparatus for Studying Transition Metal Clusters

In order to obtain well-resolved PES spectra for transition metal clusters over a wide cluster size range, an ideal PES apparatus should at least have the following attributes: (1) a versatile cluster source to generate a variety of cold clusters, (2) size-selection capability with reasonably high mass resolution for complicated mixed clusters, (3) high sensitivity, (4) high photon energy to probe more tightly bound electrons, and (5) high electron energy resolution to obtain detailed energy level information.

While it is difficult to achieve all these conditions at the same time, clearly the MTOF-PES technique with the laser vaporization cluster source is most promising because of its high collecting efficiency and the versatile cluster source, making it applicable to clusters of any elements in the periodic table. It also has the advantage of allowing high-energy photons to be used because its unique magnetic field geometry discriminates against background signals resulting from scattered laser light. This can be a serious problem for photon energies above about 4.5 eV (that is, above the work function of the vacuum chamber materials).

Its resolution, however, must be improved before its full potential for studying metal clusters can be realized. The original apparatuses had poor energy resolution (~100 meV).[35,41] PES spectra obtained from these apparatuses often yielded just the electron affinities of the clusters. The detailed electronic structure information unique to PES was lost. As indicated in Figure 1, for small clusters there should be sharp and discrete spectral features in the PES spectra. These spectral features represent the energy levels of the cluster and can be used to compare with quantum calculations to understand the electronic structure and chemical bonding properties of the clusters. As the cluster size increases, it is expected that the spectral features will broaden and eventually the molecular orbitals in the small clusters will transform into energy bands in large clusters (Figure 1). High energy resolution is essential to discern if a spectral broadening is intrinsic to the cluster or it is due to the instrumental resolution. Achieving higher resolution should make the MTOF-PES a much more powerful and versatile technique for detailed electronic structure studies of transition metal clusters.

Over the past few years, we have significantly improved the magnetic-bottle PES technique and developed a state-of-the-art apparatus based on the original Rice design.[9] In particular, we have combined high-mass and high-electron energy resolution which make this apparatus one of the most powerful to study a wide range of clusters from pure to mixed clusters and at various detachment photon energies from 1060 nm (1.169 eV) to 193 nm (6.424 eV).[8–16,48–63]

Handschuh et al. have also improved the MTOF-PES technique based on the original Meiwes–Broer design.[64] They have achieved impressive resolution for very low energy electrons by fully decelerating the cluster anions, demonstrating the potential of the MTOF-PES technique. Currently, there are several other MTOF-PES apparatuses around the world for studying clusters, all with moderate resolution, including one in Bloomfield's group dedicated to salt clusters,[65] one in Cheshnovsky's group focused on solvated clusters,[66] one in Kondow's group which is described in the current volume, one in Kaya's group,[67] and one in Fuke's group.[68] The details of our apparatus and its performance have been well documented in our publications.[8–16,48–63] A summary is provided in the following section.

III. EXPERIMENTAL SETUP

Figure 2 shows a schematic of our experimental apparatus, which consists of a pulsed laser vaporization cluster source, a modified Wiley–McLaren TOF mass

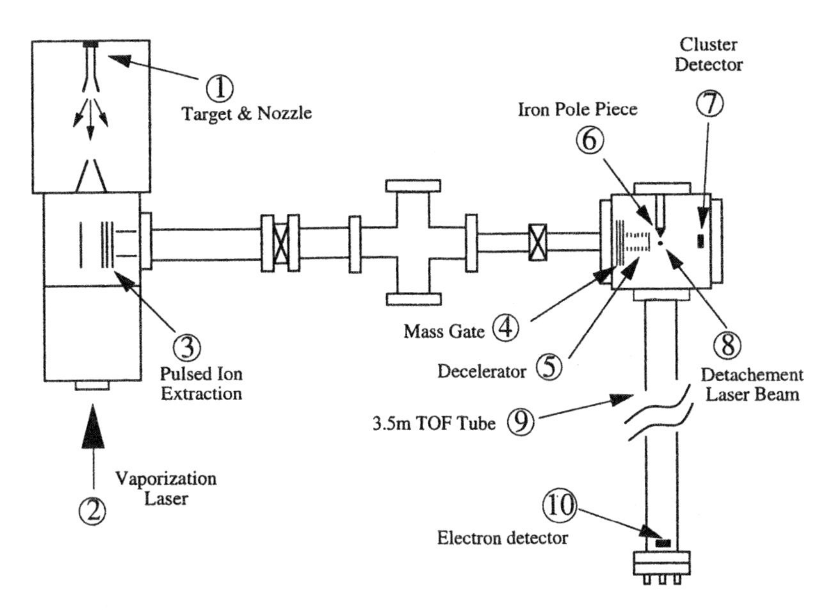

Figure 2. Schematic view of the magnetic-bottle time-of-flight photoelectron apparatus coupled with a laser vaporization supersonic cluster source and a time-of-flight mass spectrometer.

spectrometer,[69,70] a mass gate, a momentum decelerator, and a magnetic-bottle TOF photoelectron analyzer. We describe each part in the following.

A. Pulsed Laser Vaporization Cluster Source

A cold and intense negative cluster beam is essential for high-resolution PES experiments. Negative cluster ions carry an extra electron attachment energy, compared to neutral clusters, and are more difficult to cool. We use a very intense supersonic expansion to maximize cluster cooling. As shown in Figure 2, the vaporization laser beam is directed collinearly with the cluster beam and hits the target disk after passing the 2 mm nozzle orifice. The laser beam, typically 10–20 mJ from the second harmonic of a Nd:YAG laser, is focused down to a 1 mm diameter spot onto the target. Two pulsed molecular beam valves (R. M. Jordan Co., CA), symmetrically mounted, are used to deliver a short and intense helium carrier gas pulse. The stagnation pressure is 10 atm at each valve. This intense carrier gas pulse mixes with and cools the laser-induced plasma, producing clusters in both neutral and charged states. The cluster/He mixture undergoes a supersonic expansion and is skimmed to form a collimated cluster beam into the ion extraction chamber. We use two types of nozzles, as shown in Figure 3. The large waiting room nozzle (Figure 3a) provides a clustering chamber and a well-defined supersonic expansion orifice. It produces slightly colder clusters and favors large cluster formation. On the other hand, the long conical nozzle (Figure 3b) is preferred to produce very small clusters and atomic anions. A liquid-nitrogen-cooled nozzle can be used to supplement the supersonic cooling. However, we found that it has limited effect due to the strong supersonic expansion afforded by the dual-pulsed valve design. In practice, we found that the timing delay between the vaporization laser

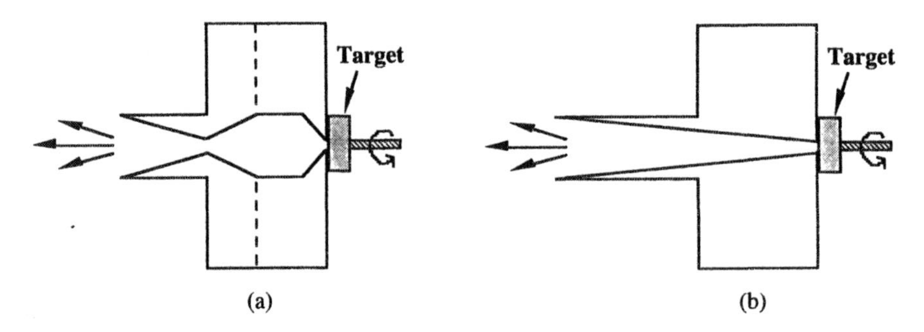

Figure 3. Two types of cluster nozzles used for generating cluster anions. (**a**) Large waiting room nozzle; (**b**) long conical nozzle.

pulse and the He carrier gas pulse has a stronger effect on the cluster temperatures. Colder clusters are produced if the laser is fired at the peak of the carrier gas pulse. However, this condition is often not favorable for producing small cluster anions, such as dimers and trimers, because their formation does not require a high He pressure. These small clusters are often produced more abundantly at the leading edge of the He pulse and consequently cannot achieve optimal supersonic cooling in our source.

Figure 4 shows two Cr_n^- cluster mass distributions at two size ranges using the long conical nozzle. Due to the perpendicular ion extraction, a deflector (not shown in Figure 2) has to be used to compensate for the transverse beam velocity. At a given deflector voltage, only a certain range of clusters can be directed onto the detector. If the cluster abundance is smooth, this leads to a bell-shaped cluster distribution as shown in Figure 4b for the larger Cr_n^- clusters. Transition metal clusters usually do not exhibit magic size distributions and tend to give smooth size distributions such as that shown in Figure 4b. However, in the small size range, Cr_2^- and Cr_4^- exhibit unusually low abundance, suggesting that either the neutral

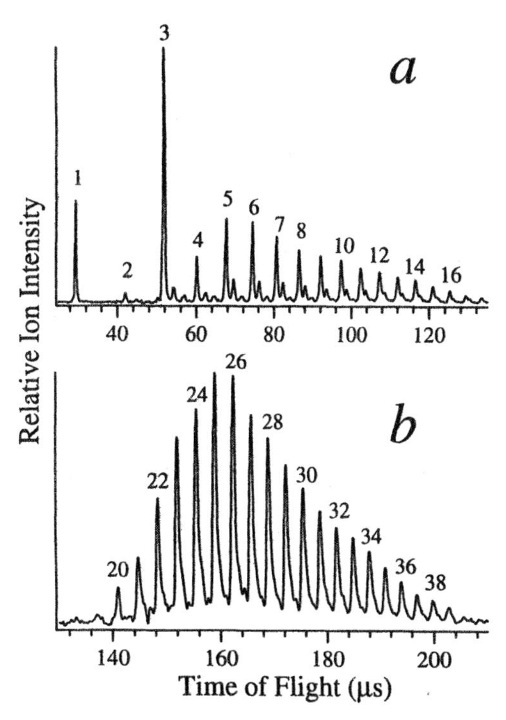

Figure 4. Mass spectra of Cr_n^- cluster distributions at two cluster size ranges. Note the weak abundance of the Cr_2^- and Cr_4^- anions. The weak signals in between the bare clusters are due to oxide contamination.

clusters have low electron affinities (EAs) or low stability. The weak peaks in between the bare clusters are due to oxygen contamination which is very serious for transition metal clusters. The oxide contamination can be minimized by using ultrahigh purity He carrier gas and baking the He gas tubing and the pulsed valve assembly. We usually observe more severe oxide contamination for a fresh sample due to the surface oxide layer and the contamination decreases as the target surface gets ablated several times.

B. Mass Selection and Momentum Deceleration

The negative clusters are extracted perpendicularly from the beam by a 1 kV high-voltage pulse (EUROTEK, HTS50-21) and are subjected to a TOF mass analysis (Figure 4). Our mass spectrometer is a modified Wiley–McLaren type for large volume ion extraction and simultaneous high mass resolution.[70] The major modification is an addition of a short free-flight zone in between the two acceleration stages of the original Wiley–McLaren design.[69] This modification allows us to achieve a mass resolution ($M/\Delta M$) of more than 500 at low masses with a large extraction volume. The resolution deteriorates slightly at higher masses, mainly limited by the fringe field effect due to the ion steering optics to compensate for the transverse velocity of the clusters.

During PES experiments, only clusters of interest are allowed to enter the interaction zone. The clusters are selected by a mass gate and decelerated by a momentum decelerator.[9,71] A three-grid mass gate is used for mass selection as shown in Figure 5. The first and third grids are grounded, and the middle grid is at a negative high voltage (–1 kV) so that no negative clusters are able to pass. Once the desired cluster arrives at the first grid, the high voltage is pulsed to ground for a short period allowing the cluster to pass unaffected. A fast transistor switch (EUROTEK, HTS-30GM) is used to deliver sharp and variable width pulses for the mass gate.

Following the mass gate, the selected cluster beam enters a momentum decelerator as shown in Figure 5. Once the cluster packet passes the third grid of the mass gate, a positive square high-voltage pulse (EUROTEK, HTS-30GM) is applied to this grid for the momentum deceleration. The high voltage is pulsed to ground before the ion packet leaves the deceleration stack, which consists of 10 guarded rings to ensure an uniform deceleration field. Both the pulse amplitude and the pulse width can be varied to achieve the best deceleration effect. All ions experience the same decelerating force within the same time period, thus will be decelerated by the same amount of linear momentum. In this deceleration method, ions are decelerated in the momentum space and the initial ion energy spread is decreased after the deceleration, in contrast to the conventional retarding field deceleration in which the initial ion energy spread remains the same after the deceleration. The deceleration step is crucial to improve the MTOF electron energy resolution due to the minimization of the Doppler-broadening caused by the anion beam velocity.

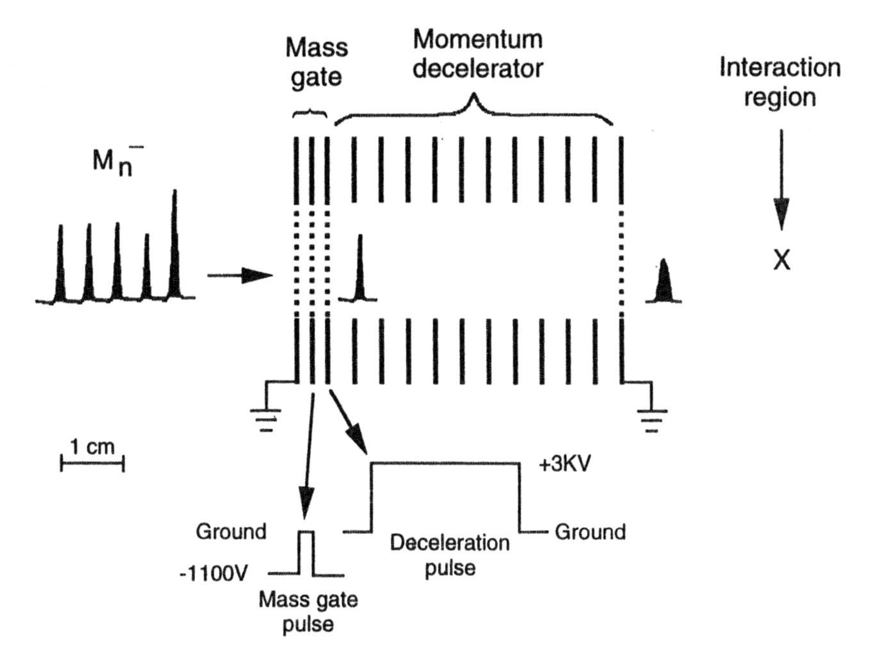

Figure 5. The mass gate and the momentum decelerator.

The momentum decelerator allows us to decelerate a given cluster ion packet down to such low kinetic energies that the Doppler-broadening is no longer a factor.

Figure 6 shows the operation of the mass selection and deceleration. The top panel displays the two isotopes of the Cu atom at a mass resolution of about 550. The middle panel shows the mass-selected peak of the ^{65}Cu isotope. The bottom panel is the same ^{65}Cu peak after a deceleration by using a 2 kV and 550 ns high-voltage pulse. The ion energy is reduced to about 30 eV from the original 1000 eV. Further deceleration to even lower energy is routinely used in PES experiments as will be shown below.

C. Magnetic-Bottle Time-of-Flight Photoelectron Analyzer

The magnetic-bottle type TOF photoelectron spectrometer first described by Kruit and Read is ideal for the study of clusters due to its high collecting efficiency (2π solid angle),[36] which compensates for the low cluster number density typically found with the laser vaporization cluster source. This design has been implemented for negative cluster studies and configured to collect all 4π solid angles as mentioned above.[35,41] Our apparatus is a modified 4π version of the MTOF.

As shown in Figure 2, the MTOF is located at the end of the TOF mass spectrometer. The high field is generated by a permanent magnet which is mounted

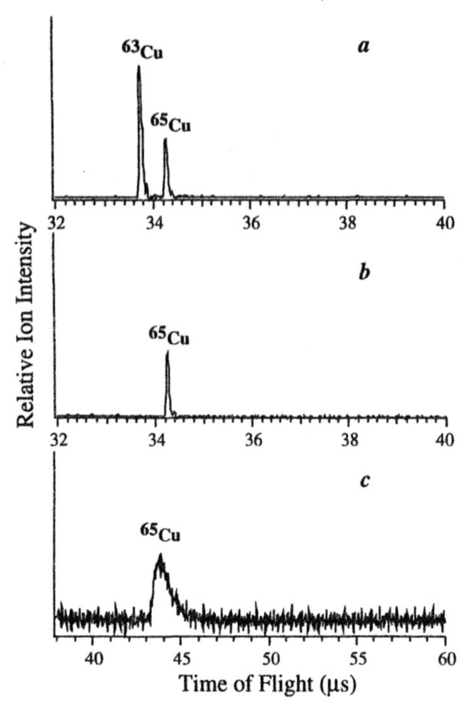

Figure 6. (**a**) Mass spectrum of Cu^-, showing the two well-resolved isotopes of Cu (^{63}Cu, 69%, and ^{65}Cu, 31%). The mass resolution here is $M/\Delta M = 550$. (**b**) Mass-selected $^{65}Cu^-$ isotope peak. (**c**) Decelerated $^{65}Cu^-$ isotope peak with a high voltage pulse (2 kV pulse height, 550 ns pulse width).

outside of the UHV interaction chamber. A soft iron cone is used as a pole piece to focus the magnetic field to a spot. The tip of the soft iron pole is 8 mm away from the center of the detachment zone, which is also 2.5 cm down stream from the momentum decelerator. A set of in-line micro-channel plates, about 12 cm down stream from the detachment zone, are used as the cluster detector. The MTOF tube is 3.5 m long, at the end of which is located a fast Z-stack electron detector (R. M. Jordan Co., CA). Not shown in the figure are the low field solenoid and the double μ-metal shielding along the 3.5 m MTOF tube.

Our detachment lasers include the four harmonics of a Nd:YAG laser [1064 nm (1.16 eV), 532 nm (2.33 eV), 355 nm (3.49 eV), 266 nm (4.66 eV)], and an ArF Excimer laser [193 nm (6.42 eV)]. Significant background electrons can arise at high photon energies due to photoemission from surfaces near the interaction zone. At 3.49 eV or lower photon energies which do not produce background electrons, photoelectron TOF spectra are measured for the selected clusters at 10 Hz repetition rate. At 4.66 and 6.42 eV, spectra are taken at 20 Hz with the cluster beam on and off at alternating laser shot for background subtraction. For relatively weak signals, the 4.66 and 6.42 eV spectra can still exhibit significant noise at low electron energies due to imperfect background subtraction. The electron TOF spectra are converted to kinetic energy distributions, calibrated with known spectra and smoothed with a 5 or 10 meV square window function. The binding energy spectra are obtained by subtracting the kinetic energy spectra from the corresponding photon energies.

D. Performance: PES of Cu^- at 3.49, 4.66, and 6.42 eV

Figure 7 shows the effect of parent ion deceleration on the electron energy resolution for the Cu^- PES spectrum at 3.49 eV photon energy. Figure 7c shows the best resolution achieved with our current apparatus. The peak widths (FWHM) for the $^2S_{1/2}$, $^2D_{5/2}$, and $^2D_{3/2}$ states are 47, 22, and 19 meV, respectively, indicating the energy resolution dependence on the electron kinetic energies. We can decelerate the Cu^- ion beams down to such low energies that the Doppler-broadening is

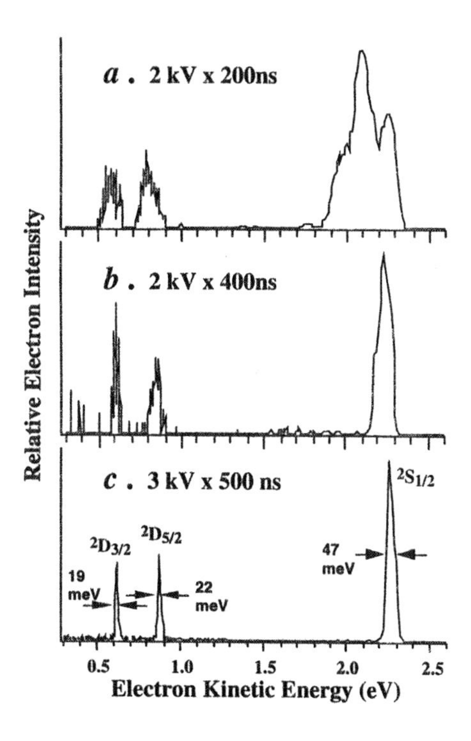

Figure 7. Photoelectron spectra of Cu^- at 3.49 eV photon energy, showing the effect of ion deceleration on the photoelectron energy resolution. The deceleration pulses and the best resolution are indicated.

no longer the dominating effect. For the spectrum in Figure 7c, the Cu^- anion beam was decelerated to <10 eV mean kinetic energy.

The PES spectrum of Cu^- is used routinely for the spectrometer calibration in the TOF-to-kinetic energy conversions. Figure 8 shows the Cu^- PES spectra in electron binding energies at three photon energies: 3.49, 4.66, and 6.42 eV. The increasing spectral widths with photon energies reflect the resolution dependence on the electron kinetic energies. The band width of the excimer laser (~20 meV) also contributes to the broadening of the 193 nm spectrum. From Figure 8, it can be seen clearly that the noise begins to show up at the high BE side of the 266 nm spectrum above ~3.5 eV BE. The 193 nm spectrum indicates that the noise becomes significant above 4.6 eV BE and often low photon flux has to be used at the higher photon energies to reduce the noise problem.

We also observe two other new features in Figure 8: (1) the photon energy dependence of detachment cross sections for different detachment transitions, and (2) a new peak near 5 eV BE at 193 nm. Since our spectrometer collects nearly 100% of the photoelectrons the relative peak intensities represent the relative total detachment cross sections. We observe that the cross section for the ^2D states increases with photon energies relative to that for the ^2S state. This is consistent with the general observation in PES that cross sections for emitting electrons from higher angular momentum states increase with photon energies.[6]

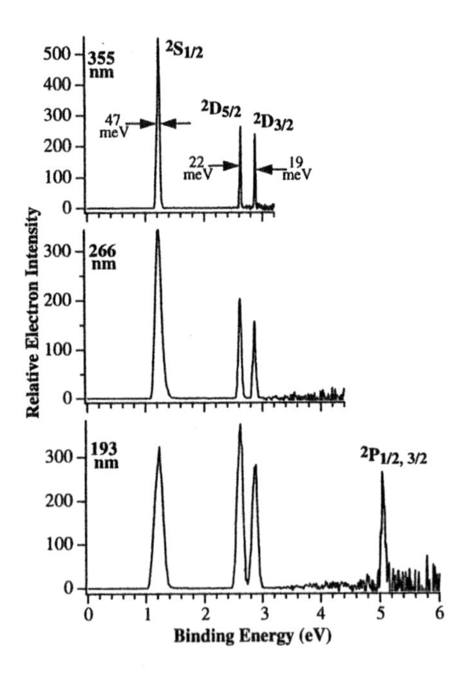

Figure 8. Photoelectron spectra of Cu^- at 355, 266, and 193 nm.

The new peak observed at 5 eV BE is interesting since it is due to a two-electron transition. This peak can be assigned from the Cu atomic energy levels.[72] It is due to the ^2P excited state of a Cu atom with an $3d^{10}4p^1$ electron configuration. There is a small spin-orbit splitting (30 meV), that is not resolved in the 193 nm spectrum. Cu^- has a $3d^{10}4s^2$ configuration. Thus, the ^2P state is resulted from detaching a 4s electron and at the same time exciting another 4s electron to the 4p orbital. Such two-electron transitions are common in PES, which is the chief experimental technique to study electron correlation effects in atoms and molecules.[73] These transitions are usually called satellites and exhibit very weak intensities. It is surprising that the intensity of the ^2P state observed here are almost comparable to the main transitions. This suggests very strong electron correlation effects between the two 4s electrons in Cu^-.

Figure 8 shows the advantages of performing PES experiments at various photon energies using the MTOF technique. Low photon energies allow optimum spectral resolution and are noise-free while high photon energies can reveal more highly excited states. Additionally, photon energy dependence of detachment cross sections can provide useful information in spectral assignments concerning the initial states of the photoemitted electrons.

IV. RESULTS AND DISCUSSION

Our focus has been on the first row transition metal clusters including, Ti_n^- (n = 1–65),[12] V_n^- (n = 1–65),[14] Cr_n^- (n = 1–70),[13,16] Fe_n^- (n = 1–33),[8–10] Co_n^- (n = 1–23),[11] and Ni_n^- (n = 1–50).[15] Except the Co system, we have obtained PES spectra of these clusters at several detachment photon energies ranging from the second harmonic (532 nm) to fourth harmonic (266 nm) of the YAG laser and 193 nm from an ArF excimer laser. As mentioned above, the photon energy dependence studies are highly valuable. The lower photon energies give better resolved spectra, revealing the discrete energy levels of the neutral clusters. On the other hand, the high photon energies allow us to probe deeper into the valence band of the clusters and are indispensable in order to follow the evolution of the electronic structure from the molecular to bulk-like behavior. We will summarize and discuss each system in this section.

A. Titanium Clusters, Ti_n^- (n = 1–65): Onset and Evolution of the 3d Band

Figure 9 shows the PES spectra of Ti_n^- for n = 1, 3–26 at 355 nm and Figure 10 displays the PES spectra of Ti_n^- for n = 3–65 at 266 nm. The 355 nm spectra are slightly better resolved as can be seen clearly for the smaller clusters. Discrete spectral features are observed for Ti_{3-8}^- . The spectra in this size range vary from one cluster to the other, reflecting their molecular nature. Starting from Ti_8^-, one prominent feature appears near the detachment threshold and the spectra begin to show similarities. No sharp features are observed beyond Ti_8^- and the discrete

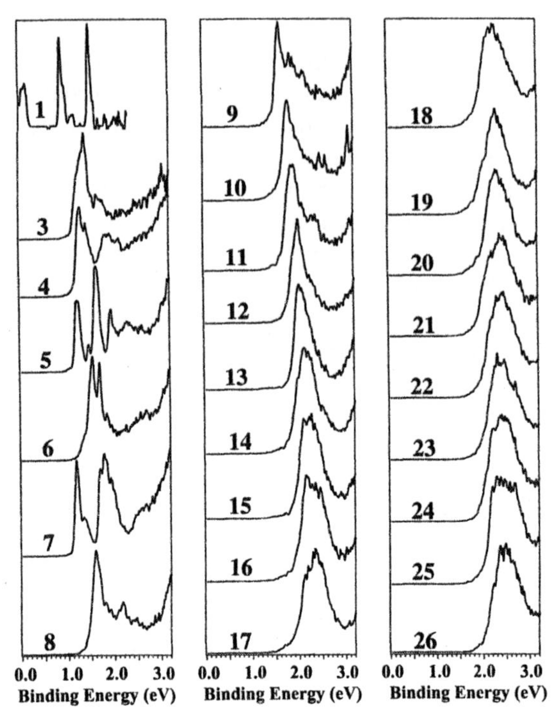

Figure 9. Photoelectron spectra of Ti_n^- (n = 1, 3–26) at 355 nm. Note that the spectra become essentially featureless above Ti_8^-.

features resolved for the smaller clusters seem to congest into the broad feature, whose width increases with cluster size. Interestingly, this broad feature becomes somewhat narrow for Ti_{55}^- and broadens again for the higher clusters. The spikes at the high binding energy side of the 266 nm spectra are due to statistical noises, part of which is caused by the imperfect background subtraction.

The detachment threshold yields the EA of the corresponding neutral cluster. Due to the lack of vibrational structure, the EA is obtained by drawing a straight line alone the leading edge of a PES spectrum and taking the intercept with the BE axis plus a constant which takes into account the instrumental resolution and a finite thermal broadening. When a sharp threshold feature is observed, the EA can be determined within 0.05 eV. The obtained EAs for the Ti_n (n = 3–65) clusters are plotted in Figure 11. The uncertainty of the EA values is ±0.08 eV for broad features, but the relative changes for the different cluster sizes are seen clearly in Figures 9 and 10. The Ti_7 cluster has a particularly low EA. The EAs for Ti_8 and above show roughly a monotonic increase and should approach the bulk work function (4.33 eV) at the infinitive size limit. For metallic clusters, the threshold level will evolve

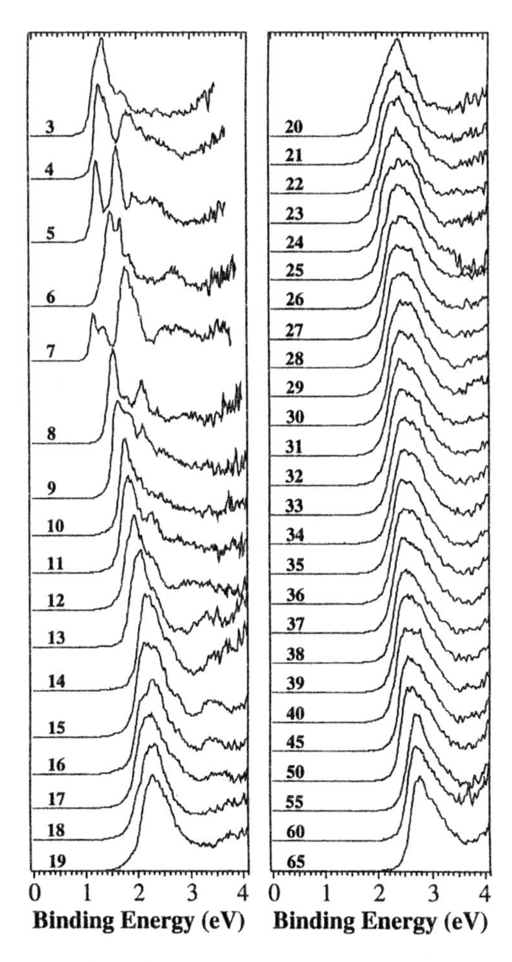

Figure 10. Photoelectron spectra of Ti_n^- (n = 3–65) at 266 nm.

into the Fermi level of the bulk crystal, and the PES spectra should eventually resemble the bulk valence photoemission spectrum of titanium.

Our most striking observation is the spectral change from Ti_7^- to Ti_8^- and the appearance that after Ti_8^- all the spectral features seem to be congested into a broad band. This broad feature is already quite similar to the bulk titanium valence photoemission spectrum, which shows a single broad feature near the Fermi level with a width (FWHM) of about 2 eV due to the 3d band.[74,75] The cluster spectral feature broadens as a function of the cluster size and attains a width of about 1 eV in the range between Ti_{20} and Ti_{50}. Ti_{55} exhibits a particularly narrow band and it broadens again at larger cluster sizes. It is anticipated that this feature will broaden and evolve into the bulk valence photoemission feature as the cluster size is

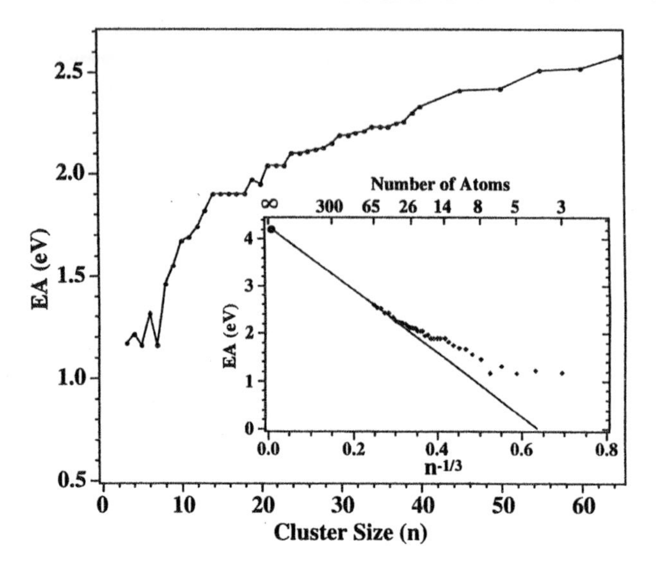

Figure 11. Electron affinity (EA) of the Ti_n clusters as a function of cluster size. Inset: EA vs. $n^{-1/3}$.

continuously increased. Therefore, it was concluded that the bulk 3d band already emerges near Ti_8 and evolves towards bulk as the cluster size increases.

Besides the diatomics,[7] the only previous experimental study on the titanium clusters is a collision-induced dissociation experiment on the positive clusters, Ti_n^+ (n = 2–22), by Lian et al.[76] With only two d electrons, it is expected that the electronic structure of the small titanium clusters would be the simplest among the transition metal clusters and sharp spectral features should be resolved at larger cluster sizes. The onset of the bulk 3d band at relatively small cluster size is quite surprising. This is likely due to the strong bonding nature of the 3d orbitals of titanium. The size of the 3d orbitals in the first row transition metals decreases from the left to the right as a result of the increased nuclear charge. The 3d orbitals of titanium are completely valence orbitals that participate in chemical bonding in all the titanium chemical compounds. The d orbitals of titanium are expected to be substantially delocalized and overlap with neighboring atoms in the clusters. Since there are fewer d electrons in titanium, there will be no antibonding d orbitals filled, facilitating a more close-packed and more tightly bound clusters. Indeed, the dissociation energies of the small Ti_n^+ clusters are among the highest in the first-row transition metal clusters, and highly close-packed icosahedral structures have been suggested for certain titanium clusters.[76] The observation of a narrower PES feature (Figure 10) for Ti_{55}^- is also consistent with a highly symmetric icosahedral structure for this cluster. Therefore, the rapid convergence of the titanium cluster electronic structure toward that of the bulk is attributed to the strong delocalization of the 3d orbitals and the anticipated close-packed cluster structures. Recent theoretical calculations by Izquierdo et al. support the experimental observations and conclusions.[77] In particular, the DOS of Ti_8 was found in the calculation to be similar to that of bulk titanium as suggested by the PES spectra.

B. Vanadium Clusters, V_n^- ($n = 1$–65): Molecular to Bulk-Like Transition

Bulk vanadium is paramagnetic with a body-centered cubic (BCC) crystal structure.[78] Many studies have been focused on the magnetic properties of the vanadium clusters to see if these clusters would exhibit novel magnetism and how it would evolve towards the bulk paramagnetism.[15,79–85] Central to these questions are the electronic structures of these clusters. However, the electronic structures of these clusters have posted considerable challenges theoretically, and many of the theoretical works on the vanadium clusters are model calculations and there are few fully optimized studies.

We investigated the PES of V_n^- clusters ($n = 1$–70) at three photon energies: 3.49, 4.66, and 6.42 eV, as shown in Figures 12–14, respectively. Extensive discrete and

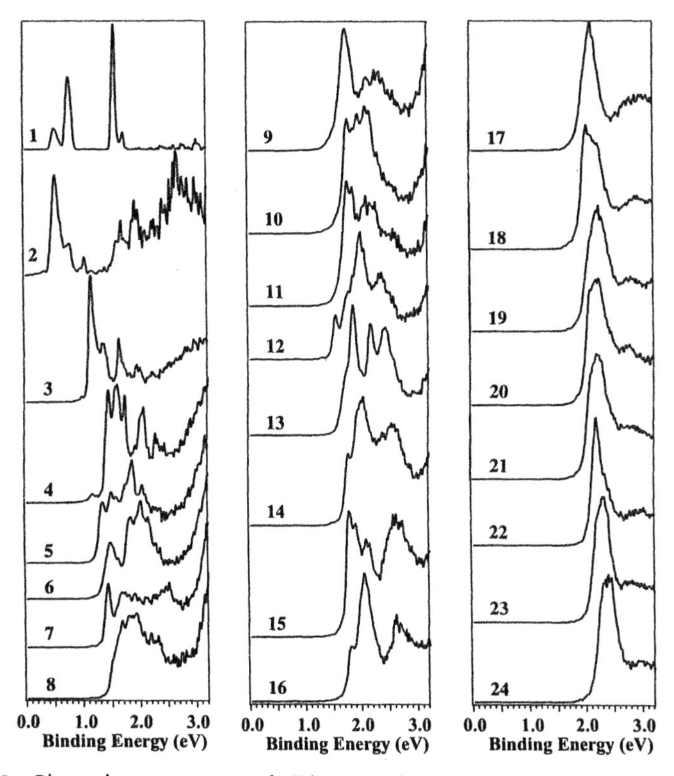

Figure 12. Photoelectron spectra of V_n^- ($n = 1$–24) at 355 nm. Note that sharp spectral features can be no longer resolved above V_{16}^-.

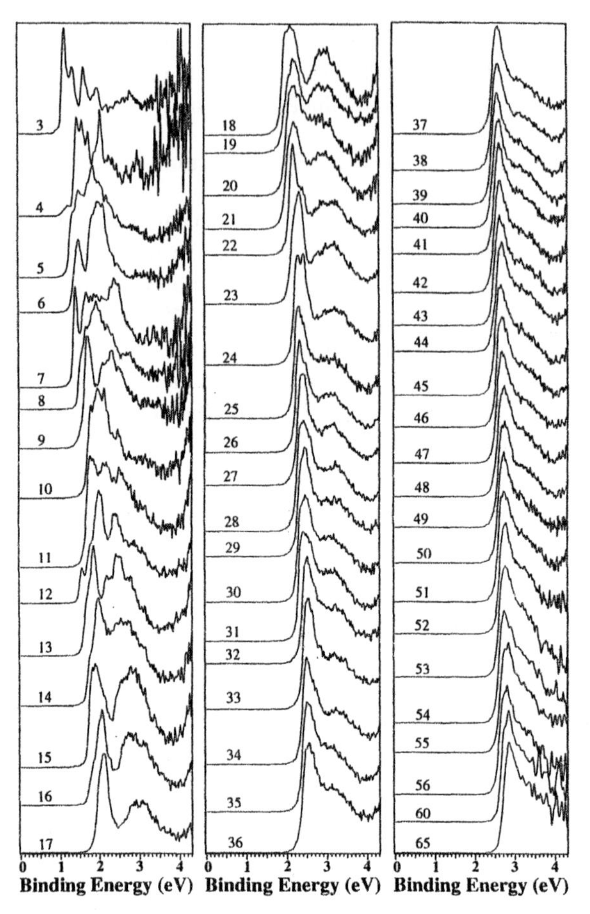

Figure 13. Photoelectron spectra of V_n^- (n = 3–65) at 4.66 eV photon energy. Note the appearance of only two spectral features at larger sizes and their convergence as cluster size increases.

sharp features are resolved for clusters up to V_{16}. The 3.49 eV spectra are best resolved. In particular, the shoulders at the low BE side of V_{13}, V_{14}, and V_{16} in the 4.66 eV spectra are well resolved as distinct peaks and the first band of V_{15} is resolved into several features. However, above V_{16}, such sharp features disappear abruptly and the 4.66 eV spectra basically consist of two bands: a narrower band near the threshold, and a second broader band at a BE of about 3.2 eV (Figure 13). Interestingly, the BE of the second band does not seem to change much with cluster size and the two bands eventually merge into one broader band near V_{60}.

The two-band pattern in the 4.66 eV spectra for the larger clusters between n = 17–60 can already be seen to emerge at V_{13}, although each of the bands for V_{13} is

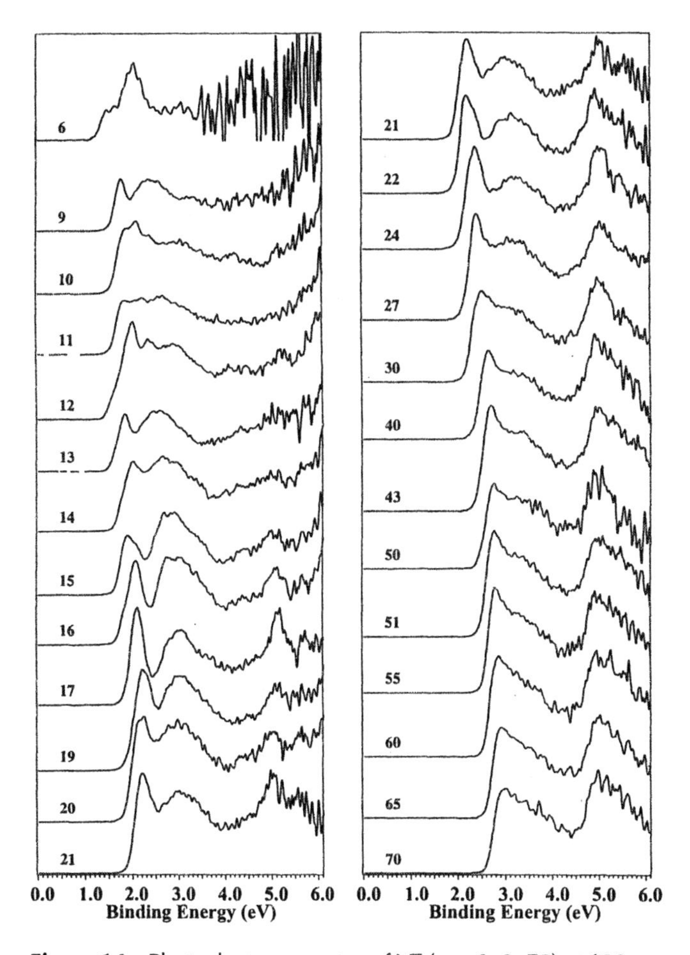

Figure 14. Photoelectron spectra of V_n^- (n = 6, 9–70) at 193 nm.

still composed of several features which are resolved better in the 3.49 eV spectrum. The clusters from V_{13} to V_{16} seem to represent the transition size range from the discrete spectra to the two-band spectra starting at V_{17}: the BE of the second band from V_{13} to V_{16} is systematically shifted to higher values until V_{17} where from the BE of the second feature changes very little with cluster size.

The 6.42 eV spectra basically reproduce the two-band pattern as revealed in the 4.66 eV spectra. However, the higher photon energy allows a third feature to be observed at a BE of about 5.2 eV. This feature is observed more clearly starting at V_{17} and seems to emerge from V_{13}. Similar to the second band, the BE of the third band also does not seem to change with cluster size. It is important to note that the discrete features in the small cluster size range are all smoothed out in the lower resolution spectra at the 6.42 eV detachment energy.

The spectra shown in Figures 12–14 represent the most comprehensive data on the electronic structure of the V_n clusters and reveal the most detailed electronic structure evolution with size. These PES spectra can be roughly divided into four regions of evolution: (1) from $n = 3–12$, distinctly molecular behavior with extensive discrete features and dramatic size variations; (2) from $n = 13–16$, transition region from the discrete features to the three-band features at $n = 17$ and above; (3) from $n = 17$ to about 60, gradual convergence of the first two bands into a single broad feature; and (4) above $n = 60$, two broad spectral bands with little change as cluster size increases further.

The invariance of the PES spectra with cluster size above V_{60} suggests the onset of bulk features. Hence, it is useful to compare the cluster spectra to bulk PES spectra of vanadium. Angle-resolved photoemission of V(100) surface has been studied and shows two broad features at normal emission angle,[86] one near the Fermi level and one about 2 eV below the Fermi level. These have been compared to band structure calculations and they are understood to be due to the d bands of bulk vanadium.[87] The bulk spectrum is compared to the cluster spectra in Figure 15. Amazingly, the V_{65} cluster spectrum already resembles the bulk feature to a remarkable degree. It is clear that the first two cluster PES features in the size range of V_{17}–V_{60} converge to the first bulk feature and the third cluster feature in the same size range evolves into the second bulk feature. Thus, the onset of bulk features can be traced to V_{17}. An obvious question arises: what is the nature of the first sharper peak that emerges from V_{13} and gradually merges with the second broad peak between V_{17} and V_{60}?

The most significant observation is the appearance of both bulk-like features at V_{17} near 3.2 and 5.2 eV (Figure 14). The BEs of these features are seen to change little except becoming broader with increasing cluster size. These bulk-like features are attributed to be due to the interiors of the clusters, suggesting that the clusters may already possess structures similar to the bulk crystal lattice. Thus, one reasonable interpretation of the first sharp feature is that it is due to the cluster surfaces. This interpretation suggests that starting from V_{13} there is already a surface and an interior for the cluster and starting from V_{17} the interior of the clusters may already possess some similarity to the bulk. This is not surprising for V_{13} since any three-dimensional arrangements of a 13-atom cluster will produce an interior. There would be at least one interior atom for either an icosahedral (I_h) V_{13} or BCC V_{13} although the cluster structure is not exactly known. One recent tight-binding calculation suggests that V_{13} may have an I_h structure[88] while most calculations on V clusters have assumed BCC bulk structures.[79–83]

The photon energy dependence of the relative intensities between the first sharp feature and the 3.2 eV feature provides another important clue regarding the nature of these features. From Figures 12–14, it is seen that the relative intensity of the 3.2 eV feature is increased at the higher photon energy. A similar trend is also observed by comparing the 4.66 eV to the 3.49 eV spectra. This suggests that the 3.2 eV feature is most likely due to d-like states and the first sharp feature is due to

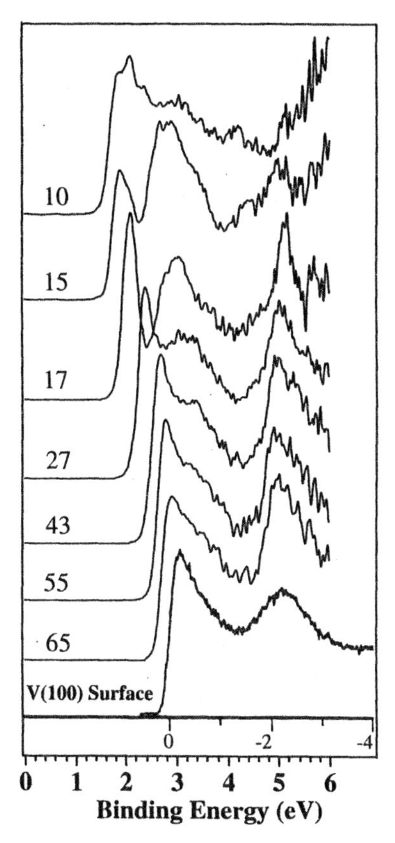

Figure 15. Photoelectron spectra of V_{10}^-, V_{15}^-, V_{17}^-, V_{27}^-, V_{43}^-, V_{55}^-, and V_{65}^- at 6.42 eV photon energy, compared to the bulk photoelectron spectrum of V(100) surface at 21.21 eV photon energy. It shows the appearance of bulk features at V_{17} and how the cluster spectral features evolve toward the bulk. The bulk spectrum (Fermi level at origin) is taken from Ref. 86 and plotted in the same energy scale as the cluster spectra.

s-like states.[6] This is consistent with the assignment that the 3.2 eV feature corresponds to the first bulk band. It is then expected that the first sharp feature should merge to the d-like band as the clusters approach the bulk since in the bulk the density-of-states near the Fermi level is mostly due to the d band.[87]

Theoretical studies by Lee and Callaway, who calculated the DOS for a BCC V_{15} at the bulk lattice constant,[81,82] lend more support to the surface-interior interpretation of the cluster PES spectra. They found two features for the DOS near the Fermi level. Their local DOS curves suggested that the first feature closer to the Fermi level was mainly due to the outer shell while the second feature was mainly due to the second shell. Even though these calculations were not for fully optimized clusters, this agreement is still informative. More systematic and optimized calculations are needed to confirm our assignment and interpretation.

The EA as a function of cluster size provides additional insight into the evolution of the V_n clusters and these are shown in Figure 16. The EAs for smaller clusters between V_3 and V_{16} exhibit significant size-dependence with distinct local minima at V_5 and V_{12}, and maxima at V_4 and V_{10}. They abruptly become more smooth starting from V_{17}. The metallic drop model, widely used to understand the behavior of metal cluster properties,[89–91] predicts that for a metallic sphere the EA is linearly proportional to 1/R: $EA = WF - \beta(e^2/R)$, where R is the effective radius of a cluster, WF is the bulk workfunction, and β is a constant. The inset of Figure 16 shows such a plot for the cluster EAs vs. $n^{-1/3}$ (proportional to 1/R). Starting from V_{17}, the EAs are clearly seen to fall on a straight line that extrapolates to 4.2 eV at infinitive cluster size. This value agrees well with the bulk workfunction of vanadium (4.3 eV).

It is expected that the PES spectra of the higher clusters will change little beyond V_{70} and the EAs will slowly approach the bulk workfunction according to the metallic drop model. The PES spectra thus provide a rather complete view of the evolution of the electronic structure from molecular to bulk-like as a function of cluster size.

Most of the previous theoretical studies have focused on the magnetic properties of the V_n clusters and suggested that the cluster magnetic properties depend on the

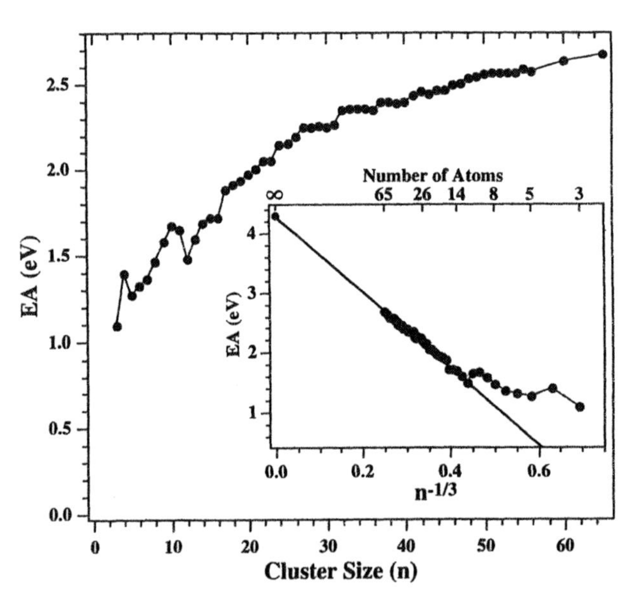

Figure 16. EA of the V_n clusters as a function of cluster size. Inset: EA vs. $n^{-1/3}$.

atomic distances.[79–83] Assuming bulk BCC structure and bulk bond distances, several works have predicted that at V_{15} the cluster abruptly becomes paramagnetic similar to the bulk while smaller clusters show ferromagnetic and antiferromagnetic couplings. The PES spectra shown in Figures 1 and 2 are likely to have major implications for the theoretical understanding of the magnetic properties of these clusters. The electronic structures obtained from the PES spectra suggest the appearance of bulk-like features between V_{13} to V_{17}. This seems to be consistent with the previous predictions that the bulk magnetic property turns on near V_{15}, although our results really should be compared to fully optimized calculations.

More importantly, the PES results also have significant implications for the geometrical structures of the V_n clusters. The appearance of bulk-like PES features in the clusters and their convergence to the bulk photoemission from a BCC crystal suggest that the clusters may grow according to the BCC structures or possess interiors similar to BCC structures. Alternatively, it may be that the electronic structure of the clusters as probed by PES is insensitive to the details of the cluster geometrical structures. The latter is quite unlikely and needs to be settled by accurate electronic structure calculations. BCC types of cluster structures have been suggested for clusters of other bulk BCC transition metals, e.g. Cr and Fe clusters.[13,92] Therefore, it is plausible for the V_n clusters to grow like BCC structures as well. This would considerably simplify future efforts to obtain fully optimized structures for these clusters.

C. Chromium Clusters, Cr_n^- (n = 1–70): Dimer Growth and Even–Odd Alternations

Chromium clusters have attracted broad interests. In particular, the Cr_2 dimer has drawn considerable attention as a prototype system to understand d–d bonding.[7,34,93–99] Several calculations have been performed on selected Cr clusters by assuming the bulk body-centered cubic (BCC) lattice to gain insight into the structure–property relationships and the cluster–bulk analogies.[79,81,82,92,100] Additionally, a few experimental studies have been focused on small chromium clusters. These include dissociation studies of the Cr_n^+ (n = 2–21) clusters to probe their structure and bonding and a Stern–Gerlach experiment on the magnetic properties of Cr_n (n = 9–31) clusters.[101,102]

Chromium is unique among the transition metals. Its $3d^5 4s^1$ half-filled electronic configuration results in strong d–d bonding in the dimer with an exceptionally short bondlength (1.68 Å),[7] while bulk chromium is antiferromagnetic with a BCC structure (nearest Cr–Cr distance: 2.50 Å).[78] Two issues arise in understanding chromium clusters: (1) the evolution from a tightly bonded dimer towards the bulk structure, (2) the cluster magnetism. We have performed both experimental and theoretical studies to address these issues.[13,16]

PES spectra for Cr_n^- (n = 1–70) have been obtained at three photon energies: 3.49, 4.66, and 6.42 eV. Figure 17 displays the PES spectra of Cr_n^- (n = 1–25) at

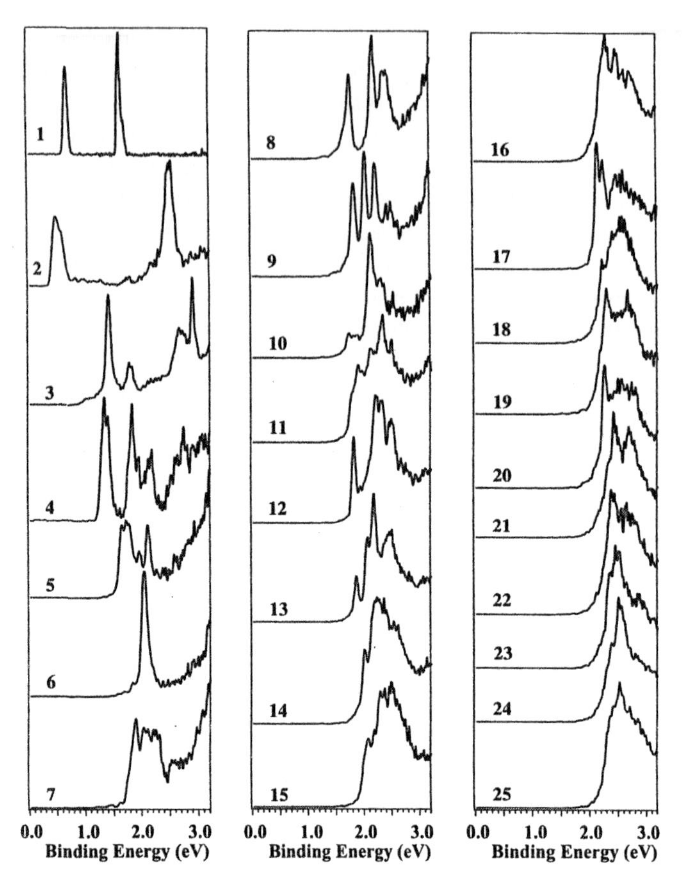

Figure 17. Photoelectron spectra of Cr_n^- (n = 1–25) at 355 nm.

3.49 eV. Extensive discrete spectral features are observed for the smaller clusters. The spectra become more congested as cluster size increases. Interestingly, the spectra of the smaller clusters seem to exhibit an even–odd alternation: the even ones (n = 2–10) give simpler spectra with fewer features near the threshold and a distinct gap; the odd ones (n = 3–11) show more complicated spectra with congested features near the threshold, particularly for n = 5, 7, and 11. Although the spectrum of Cr_9^- is well resolved with five sharp peaks, it exhibits apparent similarity to the spectra of the other odd clusters. This even–odd alternation in the PES spectra disappears above Cr_{12}^-, where a sharp threshold feature is observed. This sharp feature evolves toward high BE and overlaps with higher BE features as cluster size increases. The even–odd alternation is seen more clearly in the EAs of the clusters, as shown in Figure 18: except for Cr_6, all the even clusters from n = 2–10 have lower EAs than their neighboring odd ones. The PES spectra of Cr_n^- were also

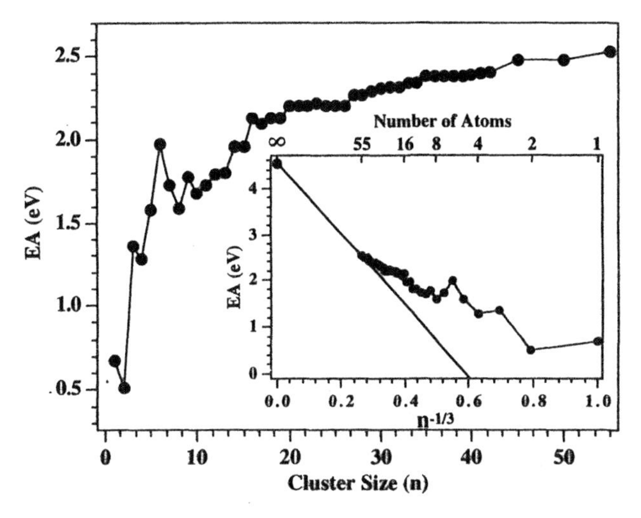

Figure 18. EA of the Cr_n clusters as a function of cluster size. Inset: EA vs. $n^{-1/3}$. Note the even–odd alternations between n = 1–11 (except n = 6).

measured at 4.66 eV for n up to 55. The spectra of the larger clusters at 4.66 eV are similar to the Cr_{25}^- spectrum shown in Figure 17, all with one broad band. The 6.42 eV spectra are similar to that of the V_n^- clusters with two broad bands for the larger clusters up to Cr_{70}^-.

The observation of the apparent even–odd alternations in the small Cr_n clusters is quite unusual among transition metal clusters and must be indicative of the unique electronic structure of the Cr clusters. To understand these effects and the detailed electronic structure of the small Cr_n clusters, we carried out an extensive density functional theory (DFT) study for n up to 15.[13]

The uniqueness of the Cr clusters among the transition metals originates from the special atomic ground state configuration of Cr, $3d^5 4s^1$. This half-filled configuration is ready to form bonds without the need of a 4s→3d promotion as required for other transition metals. This results in a very high bond order of six in Cr_2 dimer in which all the electrons are paired ($\sigma_{3d}^2 \pi_{3d}^4 \delta_{3d}^4 \sigma_{4s}^2 \sigma_{4s}^*$, ${}^1\Sigma_g^+$) with an extremely short bondlength (1.68 Å).[34] The unusual bonding in Cr_2 has been the subject of extensive research and is now well understood. However, one key question that remains unanswered is how larger clusters evolve from the dimer to the bulk where the Cr–Cr distance is much longer. Through an extensive DFT theoretical study, it was found surprisingly that the dimer maintains its identity in the larger clusters up to Cr_{11}.[13] The structures for n up to 15 are shown in Figure 19. These clusters essentially follow a dimer growth route, in which the even clusters are composed of dimers and the odd clusters are formed by attaching a single atom to the preceding even clusters. Above Cr_{11}, the clusters begin to assume more

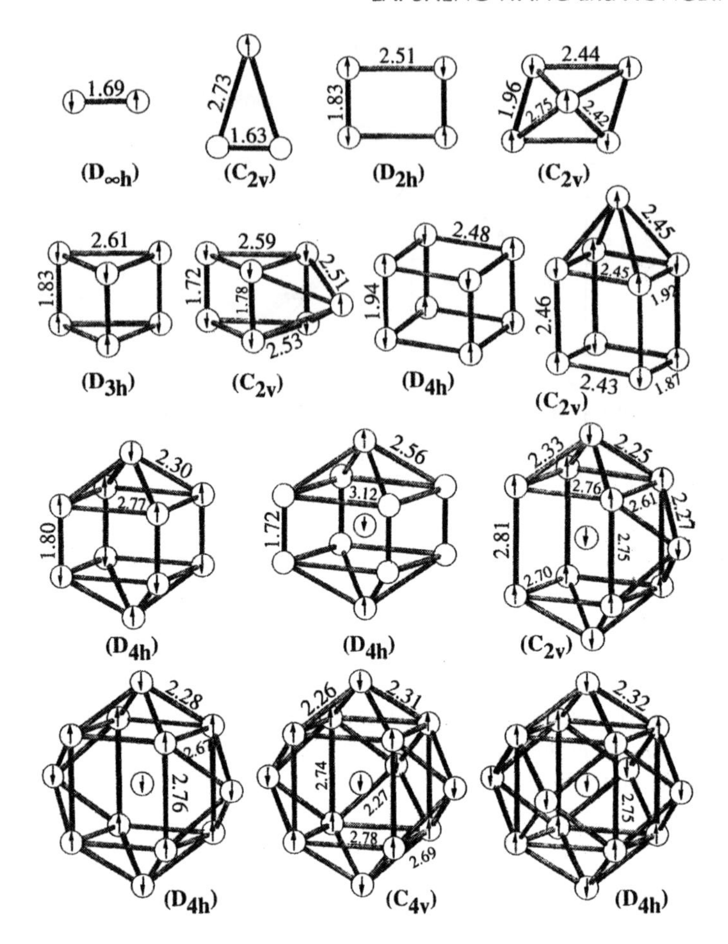

Figure 19. Fully optimized Cr cluster structures using density functional theory calculations. The bondlengths are in Å and the arrows indicate the directions of the local atomic spins.

bulk-like BCC structures. The dimer growth in the small Cr clusters is essentially the cause of the even–odd alternation observed in the PES spectra and the EAs. The details of the DFT calculations and the cluster structures were discussed previously.[13] We will show in the following that the PES spectra and the even–odd alternations can be well understood based on the electronic structure information obtained from the DFT calculations.

The closed-shell electronic configuration of Cr_2 results in a large energy gap between the highest occupied molecular orbital (HOMO), σ_{4s}, and the lowest unoccupied MO (LUMO), σ^*_{4s}, as indicated by the gap between the X and A states of the Cr_2^- $(...\sigma_{4s}^2\sigma_{4s}^{*1}, {}^2\Sigma_g^+)$ spectrum, as shown in Figure 20. A vibrationally resolved

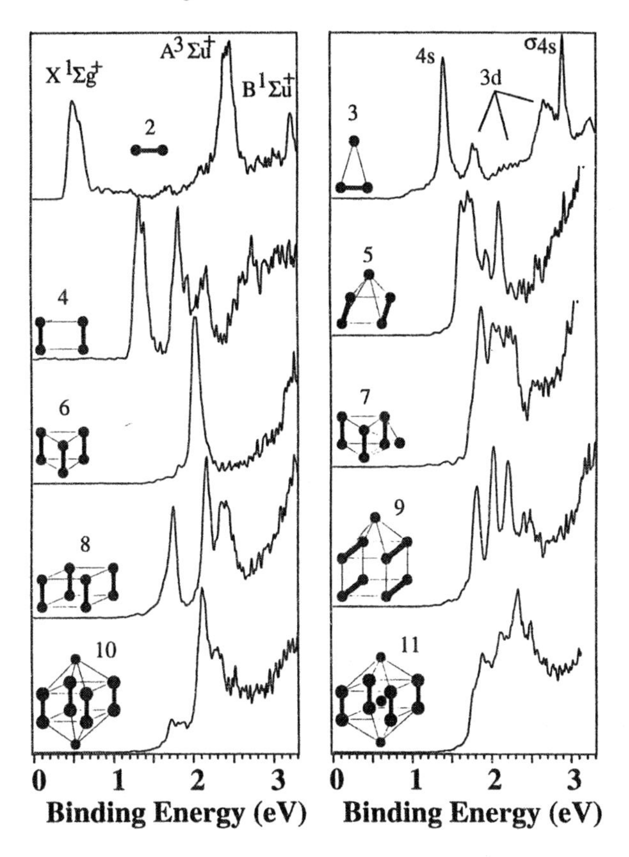

Figure 20. Photoelectron spectra of Cr_n^- (n = 2–11) at 355 nm. Note the even–odd spectral difference and the interpretation on the basis of the dimer growth structures.

PES spectrum of Cr_2^- has been obtained previously for the X and A states.[34] The current spectrum taken at 3.49 eV allows the observation of the B state, which is produced by removing a spin-up electron from the σ_{4s} orbital, similar to the A state formed by removing a spin-down σ_{4s} electron.

The Cr_3 trimer is basically composed of a dimer plus a loosely bonded atom with an overall C_{2v} symmetry. The electronic configuration of Cr_3 obtained from the DFT calculations can be approximately described as $(\sigma_{3d}^2\pi_{3d}^4\delta_{3d}^4\sigma_{4s}^2)3d^5 4s^1$ with the third atom maintaining all its six unpaired electrons. The orbitals in the parentheses are from the dimer. Of course, all the orbital degeneracy will be lifted under the C_{2v} symmetry and, in particular, the 3d orbitals of the odd atom are split into five nondegenerate orbitals. The energy gaps between the adjacent electronic states are small, reflecting the weak interactions between the dimer and the odd atom. This is clearly shown in the PES spectrum of Cr_3^- (Figure 20). The 4s orbital of the odd atom has some mixing with the σ_{4s} orbital of the dimer and becomes the HOMO of Cr_3. The extra electron in the anion resides in the 4s orbital of the odd atom. The first feature of the Cr_3^- PES spectrum at 1.4 eV is mainly due to the removal of a 4s electron from the odd atom. The sharp peak near 2.9 eV is actually due to the removal of a σ_{4s} electron and is of the same orbital origin as the A state in the dimer spectrum. This state is shifted slightly to a higher BE value in Cr_3 due to the interaction between the σ_{4s} orbital on the dimer and the 4s orbital on the third atom. All the features in between the first peak and the 2.9 eV peak are due to the 3d electrons of the odd atom. *It is important to point out that all the states arising from the odd atom in Cr_3 are in fact falling in between the HOMO–LUMO gap of the Cr_2 dimer*. This is essential to understanding the larger odd clusters where similar electronic characters arise, leading to the more complicated features in their PES spectra.

The PES spectra of Cr_4^- and Cr_8^- suggest that the neutrals are closed-shell, each with a HOMO–LUMO gap of about 0.48 and 0.41 eV, respectively. The spectrum of Cr_6^- is unusual with an intense threshold peak and an abnormally high EA (Figures 17 and 18). These can all be accounted for by the DFT calculations, which show that both Cr_4 (D_{4h}) and Cr_8 (D_{4h}) are indeed closed-shell. However, Cr_6 (D_{3h}) is found to be open-shell with two electrons in a doubly degenerate HOMO. The d electrons in Cr_6 are all paired in the three dimers and the six 4s orbitals combine to form four MOs: two nondegenerate orbitals with lower energies and two doublet degenerate orbitals with higher energies. The two nondegenerate MOs are filled with two electrons each. The remaining two electrons fill one of the degenerate MOs, resulting in the open-shell configuration for Cr_6 with a triplet ground state. The extra electron in the anion enters into the partially filled HOMO, in contrast to Cr_4^- and Cr_8^- where the extra electron occupies the empty LUMO of the neutral clusters. This results in a very high EA for Cr_6, compared to Cr_4 and Cr_8, in excellent agreement with the experimental observation.

For the odd clusters from Cr_5 to Cr_9, their electronic configuration can be described as due to the coupling between the electronic structure of the even cluster and that of the odd atom. As expected, the energy levels of the odd atom are all congested near the HOMO and in between the HOMO–LUMO gap of the corresponding even cluster, similar to the situation in Cr_3. These lead to the complicated features in the PES spectra and higher EAs for the odd clusters compared to their even neighbors (except Cr_6) since the extra electron enters into the open-shell HOMO in the odd clusters. The interactions between the odd atom and the corresponding even cluster increase from Cr_5 to Cr_9, resulting in a systematic shift of some 3d levels of the odd atom to higher BE. This is evident in the well-resolved spectrum of Cr_9^-, where the 3d level shifts lead to less DOS near the HOMO, compared to Cr_5 and Cr_7.

Cr_{10} and Cr_{11} deviate slightly from the expected dimer growth structures,[13] each with only four dimers instead of five. In Cr_{10} the four dimers form a cubic framework with two capping atoms giving an overall D_{4h} symmetry. Cr_{11} (D_{4h}) is similar to Cr_{10}, except that the eleventh atom sits in the center of the cube, forming a structure resembling a BCC fragment. The interpretations of their PES spectra are less straightforward compared to the smaller ones. Nevertheless, they each seem to resemble the smaller ones, fitting nicely in the even–odd pattern both in their EAs (Figure 18) and their PES spectral patterns (Figure 20).

From Cr_{11} to Cr_{12}, the DFT results suggest that a significant structural change takes place and the larger clusters begin to resemble the bulk BCC structure (Figure 19). Experimentally, we observe that the even–odd alternations in the PES spectra and the EAs disappear abruptly at the same size range. The PES spectra of the larger clusters begin to exhibit similarities for Cr_{12}^- and up, all with a sharp feature near the threshold. This sharp feature slowly merges with the higher BE features from Cr_{12} to Cr_{15} as seen in Figure 17. The well-resolved features at the higher BEs as observed in Cr_{12} and Cr_{13} become hard to resolve in the larger clusters due to increasing electronic density of states. Discrete spectral features still can be resolved up to Cr_{24}, beyond which only one broad band is observed near the threshold at both the 3.49 and 4.66 eV photon energies. The spectra in this size range resemble the first bulk feature to a significant degree where a sharp feature near the Fermi level is followed by a broad surface feature.[103–105] At 6.42 eV, a second band is further revealed at a much higher BE (>5 eV) and the overall spectral features begin to resemble the full bulk valence photoemission spectrum.

It is noteworthy that even–odd alternations have been observed for alkali and coinage metal clusters where they are explained by the ellipsoidal jellium model due to the itinerary nature of the valence s electrons of these elements.[1] However, we show that the even–odd alternation found in Cr clusters is completely different in nature. Here it is due to the special dimer growth of the small Cr clusters and the even–odd alternations disappear as the cluster structure changes from dimer growth to bulk-like BCC structures.

D. Iron Clusters, Fe_n^- (n = 1–33): Electronic Structure and Chemical Reactivity

One of the most extensively studied transition metal cluster systems experimentally is the iron cluster systems. Research on the chemical reactivity of iron clusters reveals a strong size effect.[106–115] In particular, the reactivity of small iron clusters with H_2 exhibits striking size-dependency with orders of magnitude variations in reaction rate between sizes 15 and 16, 18 and 19, and, to a less extent, between 22 and 23. Because of the unique ferromagnetic properties of bulk iron, studies on the magnetic properties of the small iron clusters have drawn considerable interests. All iron clusters are found to be magnetic with higher moments than those found in the bulk.[116,117] In other experimental aspects, ionization potentials (IPs) and dissociation energies of the positive clusters have been measured.[118–121]

Several possible interpretations have been proposed to account for the dramatic size-dependency of the chemical reactivity of the iron clusters with H_2. Whetten et al. first made the interesting observation between the cluster reactivity and the cluster IPs.[109] Their proposed model for the reactivity involved the cluster-to-hydrogen charge transfer. Subsequently, Siegbahn et al. focused on the detailed electronic structure of the clusters on the reactivity and showed that the correlation between the IP and reactivity is only a secondary consequence of the electronic structure of the clusters.[122] Moreover, Richtsmeier et al. stressed the geometrical structure effect on the cluster reactivity.[108] Recently, Conceicao et al. found that there is a better correlation between the cluster reactivity and the difference of the cluster IP and EA.[123] They proposed that the reactivity is determined by an entrance channel barrier, with a magnitude proportional to the (IP–EA) difference. Theoretically, several studies have attempted to understand the electronic structure of the iron clusters.[92,124–130] However, the complicated nature of these systems only allows a few selected high-symmetry and very small systems to be studied.

We have measured the PES spectra of Fe_n^- clusters up to $n = 33$ at three photon energies: 3.49, 4.66, and 6.42 eV. Figure 21 displays the spectra for Fe_n^- ($n = 3$–33) at 3.49 eV. The only previous PES work on any iron clusters is by Leopold and Lineberger, who obtained a vibrationally resolved PES spectrum of Fe_2^- at 2.54 eV photon energy.[28] The most surprising observation is the extensive sharp spectral features exhibited in the PES spectra, even for clusters as large as Fe_{23}^-. This was thought not to be possible for clusters beyond two or three atoms due to the enormously high DOS expected for these complicated clusters.

Well-resolved spectral features shown for the small clusters are quite different from each other, indicating the molecular nature of the small clusters. The spectra of the larger clusters get increasingly diffuse and they seem to show certain systematic changes. One of the most striking features in the spectra is the sharp peak near the threshold for Fe_{8-15}^-. In particular, this sharp feature in the spectra of Fe_{13}^- to Fe_{15}^- is nearly identical. There is an abrupt spectral change from Fe_{15}^- to Fe_{16}^-, where an extra feature appears at the lower BE side. The spectra from Fe_{16}^- to Fe_{18}^- are similar—the extra feature appearing in the Fe_{16}^- spectrum seems to increasingly gain intensity in Fe_{17}^- and Fe_{18}^-. In Fe_{19}^-, this small feature becomes a well-resolved peak, and at the same time the whole spectrum shifts to lower binding energy. This pattern seems to repeat from Fe_{20}^- to Fe_{23}^-, where a well-resolved threshold peak appears and the spectrum is again shifted to a lower binding energy. Beyond Fe_{23}^-, no sharp spectral features can be resolved even though a threshold shoulder can be discerned up to about Fe_{29}^-. At higher clusters, the spectra change very little except the threshold (EA) steadily shifts to higher BE. The EA as a function of cluster size is plotted in Figure 22.

We have attempted to explain the electronic structures of these clusters using simple extended Hückel MO calculations.[8,9] The goal was to understand the dramatic size-dependence of the cluster chemical reactivity with the H_2 molecule. The chemical reactivity of the clusters is ultimately determined by their electronic structure which is revealed in these PES spectra. In principle, the understanding of the cluster PES spectra should allow an explanation of the cluster reactivity. The

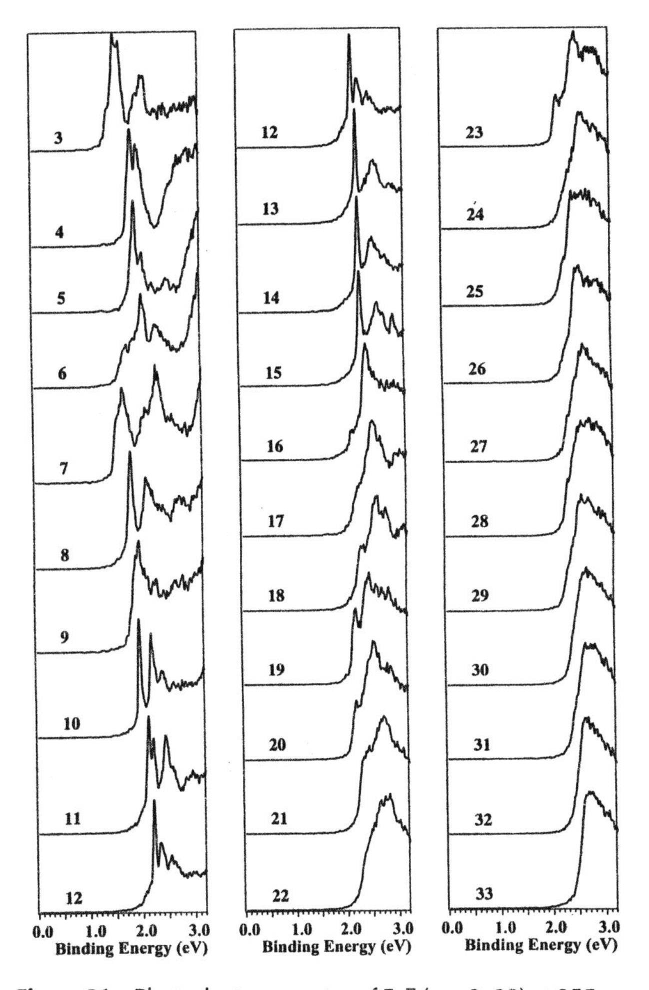

Figure 21. Photoelectron spectra of Fe_n^- (n = 3–33) at 355 nm.

reactivity of Fe_n clusters with H_2 shows a dramatic decrease from Fe_{15} to Fe_{16} by about 2 orders of magnitude. However, a nearly 2 orders of magnitude increase is observed from Fe_{18} to Fe_{19} and then again a significant increase from Fe_{22} to Fe_{23}. Interestingly, these spectral pattern changes coincide exactly with the chemical reactivity changes of the iron clusters, indicating the importance of the cluster electronic structure to their chemical reactivity. A frontier orbital picture based on the extended Hückel MO calculations was qualitatively used to provide a detailed interpretation of the size-dependence of the reactivity. However, more accurate calculations would be much desirable, which unfortunately have been limited only to very small clusters.

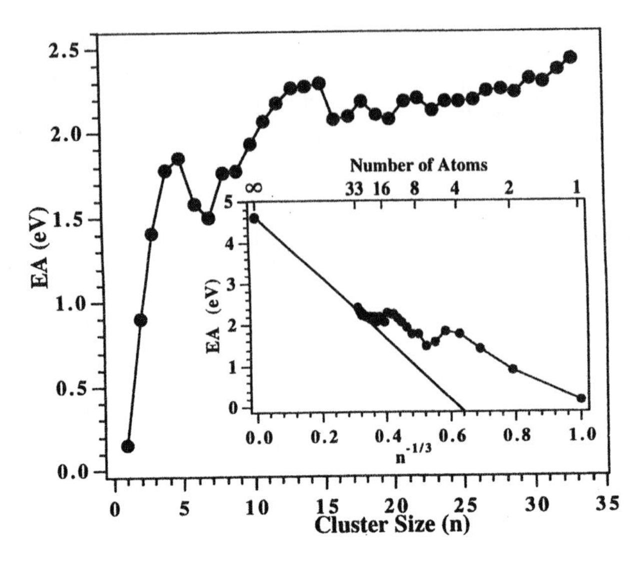

Figure 22. EA of the Fe_n clusters as a function of cluster size. Inset: EA vs. $n^{-1/3}$.

E. Cobalt Clusters, Co_n^- (n = 1–23): Electronic and Geometrical Structure

Cobalt clusters have received a fair amount of attention as other late transition metals because of their interesting magnetic properties. Magnetic moments have been measured experimentally for larger Co_n clusters.[116,131,132] Several theoretical calculations on the magnetism of cobalt clusters have also been carried out.[7,133] The ionization potentials have been measured for a wide range of cobalt cluster sizes.[118,119] The bond energies and dissociation mechanisms of positive Co_n^+ clusters were determined by using collision-induced dissociation experiments.[134] Unlike other first row transition metal clusters for which there are no systematic PES studies other than our own, one previous PES study has appeared for Co_n^- (n = 3–70) along with one previous vibrationally resolved PES study on Co_2^- dimer.[28,135] Yoshida et al. conducted a similar PES study on Co_n ($3 \leq n \leq 70$) at 3.49 eV. But only eight PES spectra were presented for n = 3–7, 13, 20, and 30. As will be shown below, the full range of cluster PES spectra are required to give a complete picture about the electronic structure evolution as a function of cluster size.

We have only measured the PES spectra of Co_n^- clusters for n = 1–23 at 3.49 eV photon energy, which are shown in Figure 23. The Co_2^- was taken at lower resolution with poor counting rate due to the very weak mass abundance from our cluster source which favors formation of larger anion clusters. The spectra from Co_3^- to Co_7^- agree with the previous measurement, but at slightly higher resolution. One interesting observation is that the Co_n^- clusters seem to exhibit much less discrete spectral features than the corresponding Fe_n^- clusters. Sharp and discrete

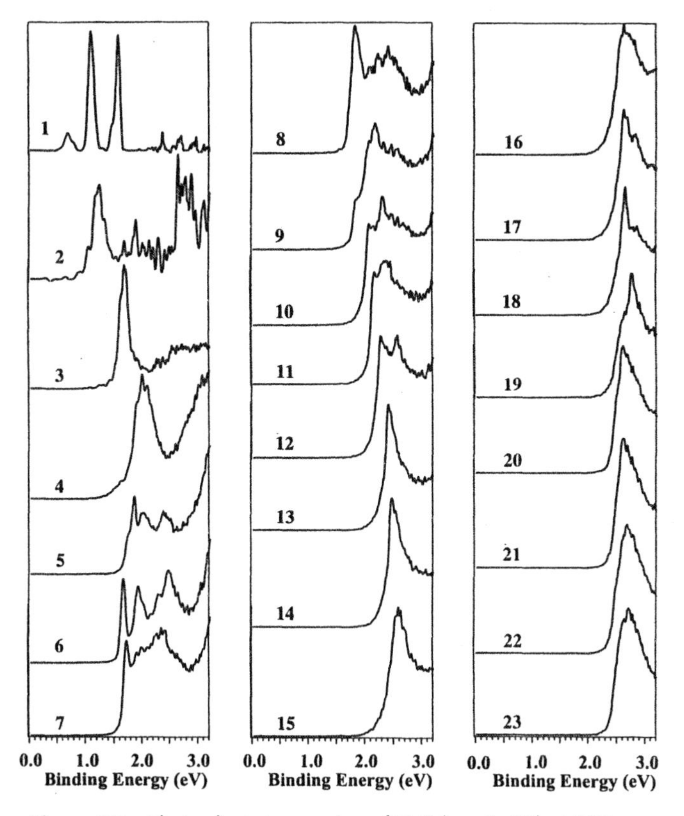

Figure 23. Photoelectron spectra of Co_n^- (n = 1–23) at 355 nm.

spectral features are observed only to Co_{12}^-. The most surprising observation is the abrupt spectral change from Co_{12}^- to Co_{13}^-, where only a single sharp feature is observed. This spectral pattern seems to persist to Co_{16}^- with gradually increasing spectral width. From Co_{17}^- to Co_{19}^-, discrete spectral features are again resolvable. Beyond Co_{20}^-, only a broad feature is observed at the 3.49 eV photon energy and the spectra become quite similar to each other. The EAs as a function of cluster size is plotted in Figure 24.

The unusually narrow PES spectrum observed for Co_{13}^- suggests that it may have a highly symmetric geometrical structure. A highly symmetric cluster would result in high electronic degeneracy, leading to the narrow and sharp PES spectrum. This is consistent with the high-symmetry I_h structure proposed for the Co_{13} clusters.[133] It would be interesting to study the Co_{13}^- cluster at higher photon energies to see how the overall spectral pattern changes with photon energies and if the sharp spectral feature remains. In any case, the electronic and geometric structures of the clusters are intimately related. An unusual PES spectrum must imply a special cluster structure.

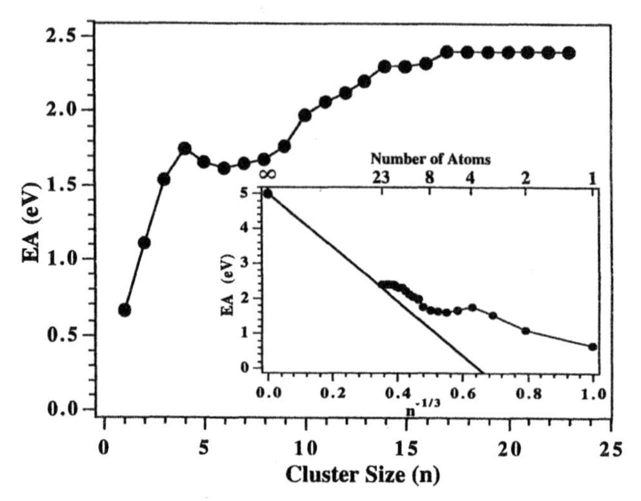

Figure 24. EA of the Co_n clusters as a function of cluster size. Inset: EA vs. $n^{-1/3}$.

F. Nickel Clusters, Ni_n^- ($n = 1–50$): Electronic Structure and Cluster Magnetism

Nickel clusters perhaps represent some of the most theoretically investigated transition metal cluster systems because the 3d orbitals are expected to be localized and may be treated with effective potentials similar to that in Cu.[136–142] Indeed, PES spectra of Ni dimer, trimer, and Ni–coinage mixed dimer and trimer exhibit similar electronic features as in the corresponding coinage clusters.[30,143,144] A recent PES work has found this similarity between Ni_n and Cu_n clusters up to $n = 6$ and used the 3d localization in the small Ni clusters as a basis to explain the high magnetic moments exhibited by the Ni clusters.[145]

We have obtained the PES spectra of Ni_n^- clusters at three photon energies: 3.49, 4.66, and 6.42 eV. The spectra for the Ni_n^- clusters at 3.49 eV are shown in Figure 25 for $n = 1–50$. The spectra of Ni^- to Ni_3^- have been well studied before and that of Ni_2^- to Ni_9^- have been reported recently. Our spectra are consistent with the previous works except that of Ni_7^-. A sharp peak at the detachment threshold is resolved in the current work for Ni_7^- while only a broad feature was observed in the previous work.

Figure 25 represents the most extensive PES data set for the Ni_n clusters. Compared to the Co_n^- clusters, the Ni_n^- clusters seem to show even less discrete spectral features, which are only observed up to Ni_9^-. Several other observations are also made immediately from the spectra: (1) from Ni^- to Ni_5^- there seems to be an even–odd alternation, where the even clusters have lower electron affinities and show a distinct sharp feature near the threshold when compared to the odd

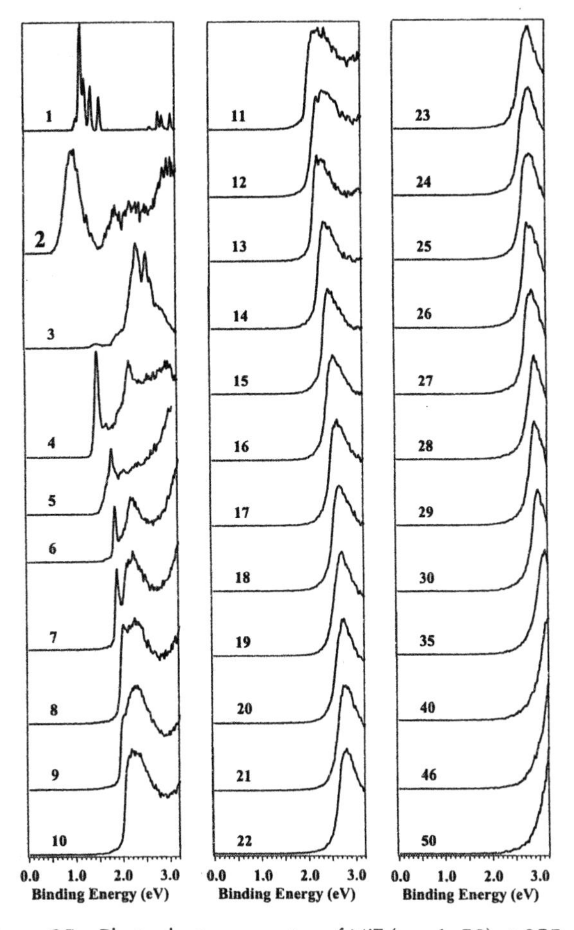

Figure 25. Photoelectron spectra of Ni_n^- (n = 1–50) at 355 nm.

neighbors; (2) a sharp peak is resolved near the threshold for Ni_6^- and Ni_7^-, which seem to have similar PES spectra—this feature exists in the larger clusters (Ni_8^-, Ni_9^-), but appears to move to higher BE and becomes overlapped with the higher BE features; (3) from Ni_{10}^- to Ni_{12}^-, only one broad feature is observed and this feature becomes broader with increasing size. However, this broad feature becomes narrower suddenly at Ni_{13}^-; and (4) the spectra beyond Ni_{14}^- show little change as cluster size increases except that the BE systematically shifts to higher values and the spectral feature becomes increasingly broadened. The BE for the large clusters becomes so high that only an onset is observed at the 3.49 eV detachment energy. The apparent sharpness of the spectral feature for the larger clusters is due to the low photon energy used. The full spectral width for the larger clusters is better revealed at higher photon energies, as shown in Figure 26, where the PES spectra of Ni_4^-, Ni_7^-, Ni_{13}^-, and Ni_{30}^- at three photon energies are compared.

Even–odd effect has been seen in the PES spectra of Cu_n^- clusters where it is caused by the $4s^1$ electronic configuration of the Cu atom.[29] In the small Ni clusters, the Ni atom assumes a $3d^9 4s^1$ configuration. Since the 3d orbitals are highly localized, the bonding in the small Ni clusters is mostly due to the 4s electrons, similar to that in Cu clusters, leading to the even–odd alternation in the PES spectra in the Ni clusters as well. The similarity between the small Ni_n and Cu_n clusters is well discussed in the previous works.

The sharp threshold feature observed in the larger clusters ($n > 5$) appears rather similar to that in the Ni_4^- spectrum. Therefore, they are expected to be due to the 4s orbitals as in the smaller clusters. This is confirmed from photon energy dependent studies. As can be seen clearly in Figure 26, the relative intensity of the sharp feature

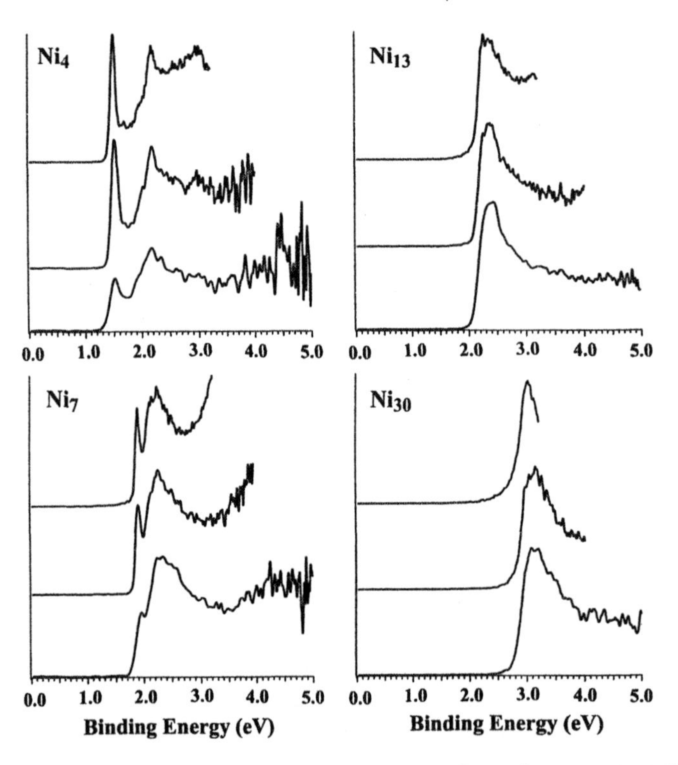

Figure 26. Comparison of photoelectron spectra of Ni_n^- (n = 4,7,13,30) at three photon energies. Top: 3.49 eV; middle: 4.66 eV; bottom: 6.42 eV. Note: (1) the decrease of relative intensity of the sharp threshold feature in Ni_4^- and Ni_7^- at the higher photon energies; (2) the narrow spectral feature in Ni_{13}^-; (3) the real spectral width of Ni_{30}^-, which is only revealed at the higher photon energies.

in the Ni_4^- and Ni_7^- spectra decreases at higher photon energies, consistent with the s-like character of this feature. The merging of this feature with the d-like features as the cluster size increases suggests the increasing degree of s–d hybridization as cluster size increases. The s-like feature is still discernible in the spectra of Ni_8^- and Ni_9^-. Beyond Ni_{10}^-, this feature is completely overlapped with the d-like features.

The observation of a sudden narrowing of the spectrum from Ni_{12}^- to Ni_{13}^- is interesting. The narrow spectral width of the Ni_{13}^- cluster is shown clearly in Figure 26, regardless of the photon energies used. Several calculations as well as chemisorption studies have predicted that the 13-atom cluster possesses a high-symmetry I_h structure.[136,140,141,146] The magnetic moment of the Ni_{13} cluster is also observed to be anomalously small and is interpreted based on the I_h structure.[147] Our observation of a narrow PES spectrum for Ni_{13}^- is consistent with a high-symmetry structure. A high-symmetry structure will lead to high orbital degeneracy, giving rise to the narrower PES feature observed. This observation is similar to that observed in the Co_n^- clusters, suggesting that these two cluster systems may possess similar electronic and geometrical structures.

As the cluster size increases, the PES spectra change very little except that the EAs shift to higher values. The EAs vs. cluster size is displayed in Figure 27. The width of the feature also increases as cluster size increases, although the PES spectra at 3.49 eV show rather sharp features. This is due to the cutoff and reduced collecting efficiency for very low energy electrons at the low photon energy. Figure 26 shows the PES spectra of Ni_{30}^- at the three photon energies used. The real width of the feature is revealed at the higher photon energies. The PES spectra of the larger

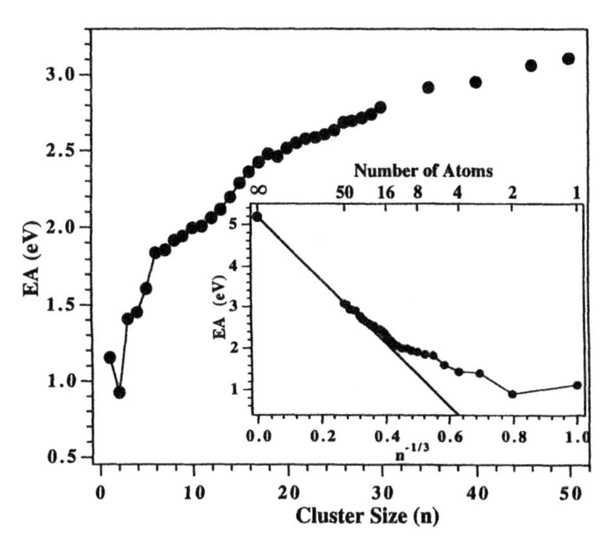

Figure 27. EA of the Ni_n clusters as a function of cluster size. Inset: EA vs. $n^{-1/3}$.

clusters are already quite similar to that of the bulk, which shows a d-band feature near the Fermi level with a width of about 1 eV.[148] It is anticipated that the cluster PES spectra will gradually converge to that of the bulk as the cluster size continues to increase. Magnetic moment measurements on the Ni_n clusters show significant size variations up to $n = 70$, above which the cluster moment slowly converges to the bulk value at about $n > 700$.[116,147] Therefore, it appears that the cluster magnetic moment is more sensitive to the cluster size than the electronic structure observed from the PES spectra. This is not surprising, since the enhancement of the magnetic moments in the clusters are mainly due to the surfaces of the clusters.[147] On the other hand, the electronic density of states in the clusters increases so fast with cluster size such that once the cluster attains a certain size the influence of one atom on the overall density of states becomes rather small, thus leading to the very small change observed in the PES spectra.

G. Discussion: Electronic Structure and EA vs. Cluster Size

As discussed above, the PES spectra provide the detailed electronic structures of the individual clusters. The evolution of the cluster electronic structure with increasing size is viewed atom-by-atom from these PES spectra. As shown in Figure 1, the electronic structure of the clusters should eventually converge to that of the bulk as cluster size increases. This is clearly born out from the results presented here. For all the systems, the PES spectra of the small clusters exhibit strong size-dependence, reflecting the molecular nature of the small clusters. The transition from molecular to bulk-like is distinctly observed. The size range that this transition occurs seems to vary from Ti to Ni. The onset of bulk-like behavior is observed to occur at a surprisingly small size for the Ti_n clusters at a size of 8.[12] For V_n clusters, although bulk-like PES spectra are only observed for $n > 60$, the onset of the bulk-like features can be clearly identified starting at V_{13}.[14] The Ti_n and V_n clusters provide the most distinct behavior of a molecular to bulk-like transition, compared to the late transition metal clusters. For Cr_n clusters, although the DFT calculations suggest that the cluster structures begin to take on the bulk-like BCC structures, the PES spectra begin to show similarity only for $n > 25$.[13,16] Most surprising is the Fe_n cluster system, for which discrete spectral features can even be observed up to $n = 23$. For Fe_n to Ni_n, it becomes harder to resolve discrete spectral features.[9,15] For Ni_n, no discrete spectral features can be observed beyond a relatively small size of 8.

The EA as a function of cluster size provides additional insight into the evolution of the cluster electronic structure from molecular to bulk. The EAs as a function of cluster size are plotted in Figures 11, 16, 18, 22, 24, and 27 for the six cluster systems, respectively. The EAs for smaller clusters in each system exhibit significant size-dependence, again reflecting the molecular nature of these small species. As the cluster size increases, the EA variation with size becomes steady and the EA should eventually approach the bulk workfunction at infinitive cluster size. The metallic drop model, widely used to understand the behavior of metal cluster properties,[1,89–91] predicts that for a metallic sphere the EA is linearly proportional to 1/R: $EA = WF - \beta(e^2/R)$, where R is the effective radius of a cluster, WF is the bulk workfunction, and β is a constant. The insets of Figures 11, 16, 18, 22, 24, and 27 show plots of the EA vs. $n^{-1/3}$ (proportional to 1/R) with a straight line extrapolating to the corresponding bulk workfunctions. For the Ti_n clusters, the EA begins to follow the metallic drop model for $n > 30$, although the onset of the 3d band occurs at $n = 8$. For the V_n clusters, the EA are clearly seen to fall on the straight line precisely at $n = 12$, which coincides with the onset of the 3d band. Thus the Ti_n and V_n clusters are seen to become metallic at relatively small sizes. However, for the late transition metal clusters, particularly Cr_n and Fe_n, the EAs are off the straight line up to much larger cluster sizes. This may be related to the enhanced magnetic moments of the small clusters since the EA, which measures the stability of the neutral clusters to bind an extra electron, may be sensitive to the enhanced magnetic moments found for the clusters of the late transition metals.

V. CONCLUSIONS AND PERSPECTIVE

We have extensively studied the electronic structure evolution of the first-row transition metal clusters as a function of cluster size. We have observed that the electronic structures of the clusters as probed by the PES appear to approach the bulk limit at relatively small sizes. The improved resolution of the magnetic-bottle type photoelectron spectrometer has proved to be critical in these studies. The higher resolution allows detailed electronic structure information of individual clusters to be obtained, fulfilling the promise of the PES technique as a powerful probe of the electronic structures of atomic clusters.

Our work so far has focused on the evolution of the electronic structure from molecular to bulk. The individual PES spectra of the small clusters, which are signatures of the electronic and atomic structures of the clusters and their molecular nature, have not been analyzed in detail, except perhaps for the small Cr_n clusters, for which the PES spectra have been interpreted qualitatively on the basis of DFT calculations.[16] Further theoretical studies and detailed interpretation of the well-resolved and discrete electronic energy levels, so characteristic of the PES technique, should allow a detailed understanding of the structure and bonding of these clusters. This will be a significant challenge to electronic structure theories, as well as to theoreticians. The obtained EAs and electronic energy level information should be of considerable value to validate further theoretical calculations.

ACKNOWLEDGMENTS

We are greatly indebted to Dr. Hansong Cheng for invaluable discussions and collaboration on the transition metal clusters. We also like to acknowledge experimental contributions by Dr. J. Fan. Support for this research by the National Science Foundation (CHE-9404428) are gratefully acknowledged. The apparatus used for this research was funded by Pacific Northwest National Laboratory and the U.S. Department of Energy, Office of Basic Energy Sciences, Chemical Sciences Division. The work is performed at Pacific Northwest National Laboratory, operated for the U.S. Department of Energy by Battelle under contract DE-AC06-76RLO 1830.

REFERENCES AND NOTES

1. de Heer, W. A. *Rev. Mod. Phys.* **1993**, *65*, 611.
2. *Advances in Metal and Semiconductor Clusters*; Duncan, M. A., Ed.; JAI Press: Greenwich, CT, 1992, Vols. 1–4.
3. *Physics and Chemistry of Finite Systems: From Clusters to Crystals*; Jena, P.; Khanna, S. N.; Rao, B. K., Eds.; Kluwer Academic Publishers: Boston, 1992.
4. Rabalais, J. W. *Principles of Ultraviolet Photoelectron Spectroscopy*; Wiley: New York, 1977.
5. Turner, D. W.; Baker, C.; Baker, A. D.; Bundle, C. R. *Molecular Photoelectron Spectroscopy*; Wiley: New York, 1970.
6. Hufner, S. *Photoelectron Spectroscopy - Principles and Applications*; Springer-Verlag: New York, 1995.
7. Morse, M. D. *Chem. Rev.* **1986**, *86*, 1049.
8. Wang, L. S.; Cheng, H. S.; Fan, J. *Chem. Phys. Lett.* **1995**, *236*, 57.
9. Wang, L. S.; Cheng, H. S.; Fan, J. *J. Chem. Phys.* **1995**, *102*, 9480.
10. Wang, L. S.; Fan, J.; Lou, L. *Surf. Rev. Lett.* **1996**, *3*, 695.
11. Wang, L. S.; Wu, H. In *Proc. Int. Symp. of the Sci. and Tech. of Atomically Engineered Materials*; Jena, P.; Khanna, S. N.; Rao, B. K., Eds.; World Scientific: NJ; Richmond, VA, 1995, pp. 245.
12. Wu, H.; Desai, S. R.; Wang, L. S. *Phys. Rev. Lett.* **1996**, *76*, 212.
13. Cheng, H. S.; Wang, L. S. *Phys. Rev. Lett.* **1996**, *77*, 51.
14. Wu, H.; Desai, S. R.; Wang, L. S. *Phys. Rev. Lett.* **1996**, *77*, 2436.
15. Wang, L. S.; Wu, H. *Z. Phys. Chem.* **1998**, *203*, 45.
16. Wang, L. S.; Wu, H.; Cheng, H. *Phys. Rev. B* **1997**, *55*, 12884.
17. Pollard, J. E.; Trevor, D. J.; Lee, Y. T.; Shirley, D. A. *Rev. Sci. Instrum.* **1981**, *52*, 1837.
18. Wang, L. S. Ph.D. Thesis, University of California, Berkeley, 1990.
19. Wang, L. S.; Reutt-Robey, J. E.; Niu, B.; Lee, Y. T.; Shirley, D. A. *J. Electron Spectrosc. Relat. Phenom.* **1990**, *51*, 513.
20. Wang, L. S.; Niu, B.; Lee, Y. T.; Shirley, D. A.; Balasubramanian, K. *J. Chem. Phys.* **1990**, *92*, 899.
21. Wang, L. S.; Lee, Y. T.; Shirley, D. A.; Balasubramanian, K.; Feng, P. *J. Chem. Phys.* **1990**, *93*, 6310.
22. Wang, L. S.; Niu, B.; Lee, Y. T.; Shirley, D. A.; Ghelichkhani, E.; Grant, E. R. *J. Chem. Phys.* **1990**, *93*, 6327.
23. Poliakoff, E. D.; Dehmer, P. M.; Dehmer, J. L.; Stockbauer, R. *J. Chem. Phys.* **1982**, *76*, 5214.
24. Rademann, K. *Ber. Bunsenges. Phys. Chem.* **1989**, *93*, 653.
25. Rademann, K.; Kaiser, B.; Even, U.; Hensel, F. *Phys. Rev. Lett.* **1987**, *59*, 2319.
26. Kaiser, B.; Rademann, K. *Phys. Rev. Lett.* **1992**, *69*, 3204.
27. Ruppel, M.; Rademann, K. *Chem. Phys. Lett.* **1992**, *197*, 280.
28. Leopold, D. G.; Lineberger, W. C. *J. Chem. Phys.* **1986**, *81*, 51.
29. Leopold, D. G.; Ho, J.; Lineberger, W. C. *J. Chem. Phys.* **1987**, *86*, 1715.
30. Ervin, K. M.; Ho, J.; Lineberger, W. C. *J. Chem. Phys.* **1988**, *89*, 4514.
31. McHugh, K. M.; Eaton, J. G.; Lee, G. H.; Sarkas, H. W.; Kidder, L. H.; Snodgrass, J. T.; Manaa, M. R.; Bowen, K. H. *J. Chem. Phys.* **1989**, *91*, 3792.

32. Eaton, J. G.; Sarkas, H. W.; Arnold, S. T.; McHugh, K. M.; Bowen, K. H. *Chem. Phys. Lett.* **1992,** *193,* 141.
33. Casey, S. M.; Villalta, P. W.; Bengali, A. A.; Cheng, C. L.; Dick, J. P.; Fenn, P. T.; Leopol, D. G. *J. Am. Chem. Soc.* **1989,** *113,* 6688.
34. Casey, S. M.; Leopold, D. G. *J. Phys. Chem.* **1993,** *97,* 816.
35. Cheshnovsky, O.; Yang, S. H.; Pettiette, C. L.; Craycraft, M. J.; Smalley, R. E. *Rev. Sci. Instrum.* **1987,** *58,* 2131.
36. Kruit, P.; Read, F. H. *J. Phys. E: Sci. Instrum.* **1983,** *16,* 313.
37. Yang, S. H.; Pettiette, C. L.; Conceicao, J.; Cheshnovsky, O.; Smalley, R. E. *Chem. Phys. Lett.* **1987,** *139,* 233.
38. Cheshnovsky, O.; Yang, S. H.; Pettiette, C. L.; Craycraft, M. J.; Liu, Y.; Smalley, R. E. *Chem. Phys. Lett.* **1987,** *138,* 119.
39. Cheshnovsky, O.; Taylor, K. J.; Conceicao, J.; Smalley, R. E. *Phys. Rev. Lett.* **1990,** *64,* 1785.
40. Taylor, K. J.; Pettiette-Hall, C. L.; Cheshnovsky, O.; Smalley, R. E. *J. Chem. Phys.* **1992,** *96,* 3319.
41. Gantefor, G.; Meiwes-Broer, K. H.; Lutz, H. O. *Phys. Rev. A* **1988,** *37,* 2716.
42. Gausa, M.; Gantefor, G.; Lutz, H. O.; Meiwes-Broer, K. H. *Int. J. Mass Spectrosc. Ion Proc.* **1990,** *102,* 227.
43. Gantefor, G.; Gausa, M.; Meiwes-Broer, K. H.; Lutz, H. O. *Z. Phys. D - Atoms, Molecules, and Clusters* **1989,** *12,* 405.
44. Kitsopoulos, T. N.; Chick, C. J.; Zhao, Y.; Neumark, D. M. *J. Chem. Phys.* **1991,** *95,* 1441.
45. Arnold, C. C.; Zhao, Y.; Kitsopoulos, T. N.; Neumark, D. M. *J. Chem. Phys.* **1992,** *97,* 6121.
46. Sander, M.; Chewter, L. A.; Muller-Dethlefs, K.; Schlag, E. W. *Phys. Rev. A* **1987,** *36,* 4543.
47. Wigner, E. P. *Phys. Rev.* **1948,** *73,* 1002.
48. Fan, J.; Wang, L. S. *J. Phys. Chem.* **1994,** *98,* 11814.
49. Fan, J.; Lou, L.; Wang, L. S. *J. Chem. Phys.* **1995,** *102,* 2701.
50. Fan, J.; Nicholas, J. B.; Price, J. M.; Colson, S. D.; Wang, L. S. *J. Am. Chem. Soc.* **1995,** *117,* 5417.
51. Nicholas, J. B.; Fan, J.; Wu, H.; Colson, S. D.; Wang, L. S. *J. Chem. Phys.* **1995,** *102,* 8277.
52. Fan, J.; Wang, L. S. *J. Chem. Phys.* **1995,** *102,* 8714.
53. Wu, H.; Desai, S. R.; Wang, L. S. *J. Chem. Phys.* **1995,** *103,* 4363.
54. Wang, L. S. *Surf. Rev. Lett.* **1996,** *3,* 423.
55. Wang, L. S.; Wu, H.; Desai, S. R.; Lou, L. *Phys. Rev. B* **1996,** *53,* 8028.
56. Wang, L. S.; Wu, H.; Desai, S. R.; Fan, J.; Colson, S. D. *J. Phys. Chem.* **1996,** *100,* 8697.
57. Wu, H.; Desai, S. R.; Wang, L. S. *J. Am. Chem. Soc.* **1996,** *118,* 5296.
58. Wang, L. S.; Wu, H.; Desai, S. R. *Phys. Rev. Lett.* **1996,** *76,* 4853.
59. Wang, L. S.; Li, S.; Wu, H. *J. Phys. Chem.* **1996,** *100,* 19211.
60. Desai, S. R.; Wu, H.; Wang, L. S. *Int. J. Mass Spectrom. Ion Processes* **1996,** *159,* 75.
61. Desai, S. R.; Wu, H.; Rohfling, C.; Wang, L. S. *J. Chem. Phys.* **1997,** *106,* 1309.
62. Wang, L. S.; Desai, S. R.; Wu, H.; Nicholas, J. B. *Z. Phys. D - Atoms, Molecules, and Clusters* **1997,** *40,* 36.
63. Wu, H.; Desai, S. R.; Wang, L. S. *J. Phys. Chem. A* **1997,** *101,* 2103.
64. Handschuh, H.; Gantefor, G.; Eberhardt, W. *Rev. Sci. Instrum.* **1995,** *66,* 3838.
65. Xia, P.; Bloomfield, L. A. *Phys. Rev. Lett.* **1994,** *72,* 2577.
66. Markovich, G.; Pollack, S.; Giniger, R.; Cheshnovsky, O. *J. Chem. Phys.* **1994,** *101,* 9344.
67. Zhang, N.; Kawamata, H.; Nakajima, A.; Kaya, K. *J. Chem. Phys.* **1996,** *104,* 36.
68. Misaizu, F.; Tsukamoto, K.; Sanekata, M.; Fuke, K. *Laser Chem.* **1995,** *15,* 195.
69. Wiley, W. C.; McLaren, I. H. *Rev. Sci. Instrum.* **1955,** *26,* 1150.
70. de Heer, W. A.; Milani, P. *Rev. Sci. Instrum.* **1991,** *62,* 670.
71. Markovich, G.; Giniger, R.; Levin, M.; Cheshnovsky, O. *J. Chem. Phys.* **1991,** *95,* 9416.
72. Moore, C. E. *Atomic Energy Levels*; Natl. Bur. Stand. Circ.: Washington DC, 1971, Vol. II.
73. Suzer, S.; Lee, S. T.; Shirley, D. A. *Phys. Rev. A* **1976,** *13,* 1842.
74. Hanson, D. N.; Stockbauer, R.; Madey, T. E. *Phys. Rev. B* **1981,** *24,* 5513.

75. Feibelman, P. J.; Himpsel, F. J. *Phys. Rev. B* **1980**, *21*, 1394.
76. Lian, L.; Su, C.-X.; Armentrout, P. B. *J. Chem. Phys.* **1992**, *97*, 4084.
77. Izquierdo, J.; Torres, M. B.; Vega, A.; Rubio, A.; Balbas, L. C. In *ISSPIC 8*; Lindelof, P. E., Ed.; Univ. of Copenhagen: Copenhagen, Denmark, 1996.
78. Kittel, C. *Introduction to Solid State Physics*; Wiley: New York, 1966.
79. Salahub, D. R.; Messmer, R. P. *Surf. Sci.* **1981**, *106*, 415.
80. Liu, F.; Khanna, S. N.; Jena, P. *Phys. Rev. B* **1991**, *43*, 8179.
81. Lee, K.; Callaway, J. *Phys. Rev. B* **1993**, *48*, 15358.
82. Lee, K.; Callaway, J. *Phys. Rev. B* **1994**, *49*, 13906.
83. Dorantes-Davila, J.; Dreysse, H. *Phys. Rev. B* **1993**, *47*, 3857.
84. Alvarado, P.; Dorantes-Davila, J.; Dreysse, H. *Nanostruct. Mater.* **1993**, *3*, 331.
85. Dreye, H.; Dorantes-Davila, J.; Vega, A.; Balbas, L. C.; Bouarab, S.; Nait-Laziz, H.; Demangeat, C. *J. Appl. Phys.* **1993**, *73*, 6207.
86. Pervan, P.; Valla, T.; Milun, M. *Solid State Comm.* **1994**, *89*, 917.
87. Moruzzi, V. L.; Janak, J. F.; Williams, A. R. *Calculated Electronic Properties of Metals*; Pergamon: New York, 1978.
88. Zhao, J.; Chen, X.; Sun, Q.; Liu, F.; Wang, G.; Lain, K. D. *Physica B* **1995**, *215*, 377.
89. Wood, D. M. *Phys. Rev. Lett.* **1981**, *45*, 749.
90. Makov, G.; Nitzan, A.; Brus, L. E. *J. Chem. Phys.* **1988**, *88*, 5076.
91. Seidl, M.; Meiwes-Broer, K.-H.; Brack, M. *J. Chem. Phys.* **1991**, *95*, 1295.
92. Pastor, G. M.; Dorantes-Davila, J.; Bennemann, K. H. *Phys. Rev. B* **1989**, *40*, 7642.
93. Salahub, D. R. *Ab Initio Methods in Quantum Chemistry*; ACS: New York, 1987, Vol. 2.
94. Goodgame, M. M.; Goddard, W. A., III. *Phys. Rev. Lett.* **1985**, *54*, 661.
95. Andersson, K.; Roos, B. O.; Widmark, P.-O. *Chem. Phys. Lett.* **1994**, *230*, 391.
96. Andersson, K. *Chem. Phys. Lett.* **1995**, *237*, 212.
97. Roos, B. O.; Andersson, K. *Chem. Phys. Lett.* **1995**, *245*, 215.
98. Edgecombe, K. E.; Becke, A. D. *Chem. Phys. Lett.* **1995**, *244*, 427.
99. Bauschlicher Jr., C. W.; Partridge, H. *Chem. Phys. Lett.* **1994**, *231*, 277.
100. Reddy, B. V.; Khanna, S. N. *Phys. Rev. B* **1992**, *45*, 10103.
101. Su, C. X.; Armentrout, P. B. *J. Chem. Phys.* **1993**, *99*, 6506.
102. Douglass, D. C.; Bucher, J. P.; Bloomfield, L. A. *Phys. Rev. B* **1992**, *45*, 6341.
103. Klebanoff, L. E.; Robey, S. W.; Liu, G.; Shirley, D. A. *Phys. Rev. B* **1984**, *30*, 1048.
104. Klebanoff, L. E.; Victora, R. H.; Falicov, L. M.; Shirley, D. A. *Phys. Rev. B* **1985**, *32*, 1997.
105. Klebanoff, L. E.; Robey, S. W.; Liu, G.; Shirley, D. A. *J. Magn. Magn. Mater.* **1986**, *54*, 728.
106. Geusic, M. E.; Morse, M. D.; Smalley, R. E. *J. Chem. Phys.* **1985**, *82*, 590.
107. Morse, M. D.; Geusic, M. E.; Heath, J. R.; Smalley, R. E. *J. Chem. Phys.* **1985**, *83*, 2293.
108. Richtsmeier, S. C.; Parks, E. K.; Liu, K.; Pobo, L. G.; Riley, S. J. *J. Chem. Phys.* **1985**, *82*, 3659.
109. Whetten, R. L.; Cox, D. M.; Trevor, D. J.; Kaldor, A. *Phys. Rev. Lett.* **1985**, *54*, 1494.
110. Parks, E. K.; Liu, K.; Richtsmeier, S. C.; Pobo, L. G.; Riley, S. J. *J. Chem. Phys.* **1985**, *82*, 5470.
111. Brucat, P. J.; Pettiette, C. L.; Yang, S.; Zheng, L.-S.; Craycraft, M. J.; Smalley, R. E. *J. Chem. Phys.* **1986**, *85*, 4747.
112. Parks, E. K.; Weiller, B. H.; Bechthold, P. S.; Hoffman, W. F.; Nieman, G. C.; Pobo, L. G.; Riley, S. J. *J. Chem. Phys.* **1988**, *88*, 1622.
113. Cox, D. M.; Reichmann, K. C.; Trevor, D. J.; Kaldor, A. *J. Chem. Phys.* **1988**, *88*, 111.
114. Parks, E. K.; Nieman, G. C.; Pobo, L. G.; Riley, S. J. *J. Chem. Phys.* **1988**, *88*, 6260.
115. Conceicao, J.; Loh, S. K.; Lian, L.; Armentrout, P. B. *J. Chem. Phys.* **1996**, *104*, 3976.
116. Billas, I. M. L.; Chatelain, A.; de Heer, W. A. *Science* **1994**, *265*, 1682.
117. Cox, D. M.; Trevor, D. J.; Whetten, R. L.; Rohlfing, E. A.; Kaldor, A. *Phys. Rev. B* **1985**, *32*, 7290.
118. Yang, S.; Knickelbein, M. B. *J. Chem. Phys.* **1990**, *93*, 1533.
119. Parks, E. K.; Klots, T. D.; Riley, S. J. *J. Chem. Phys.* **1990**, *92*, 3813.
120. Rohlfing, E. A.; Cox, D. M.; Kaldor, A.; Johnson, K. H. *J. Chem. Phys.* **1984**, *81*, 3846.

121. Lian, L.; Su, C.-X.; Armentrout, P. B. *J. Chem. Phys.* **1992**, *97*, 4072.
122. Panas, I.; Siegbahn, P.; Wahlgren, U. In *The Challenge of d and f Electrons*; American Chemical Society: Washington DC, 1989, pp. 125.
123. Conceicao, J.; Laaksonen, R. T.; Wang, L. S.; Guo, T.; Nordlander, P.; Smalley, R. E. *Phys. Rev. B* **1995**, *51*, 4668.
124. Yang, C. Y.; Johnson, K. H.; Salahub, D. R.; Kaspar, J.; Messmer, R. P. *Phys. Rev. B* **1981**, *24*, 5673.
125. Lee, K.; Callaway, J.; Dhar, S. *Phys. Rev. B* **1984**, *30*, 1724.
126. Holland, G. F.; Ellis, D. E.; Trogler, W. C. *J. Chem. Phys.* **1985**, *83*, 3507.
127. Pastor, G. M.; Dorantes-Davila, J.; Bennemann, K. H. *Chem. Phys. Lett.* **1988**, *148*, 459.
128. Chen, J. L.; Wang, C. S.; Jackson, K. A.; Pederson, M. R. *Phys. Rev. B* **1991**, *44*, 6558.
129. Castro, M.; Salahub, D. R. *Phys. Rev. B* **1994**, *49*, 11842.
130. Bouarab, S.; Vega, A.; Aloso, J. A.; Iniguez, M. P. *Phys. Rev. B* **1996**, *54*, 3003.
131. Douglas, D. C.; Cox, A. J.; Bucher, J. P.; Bloomfield, L. A. *Phys. Rev. B* **1992**, *47*, 12874.
132. Bucher, J. P.; Douglass, D. C.; Bloomfield, L. A. *Phys. Rev. Lett.* **1991**, *66*, 3052.
133. Miura, K.; Kimura, H.; Imanaga, S. *Phys. Rev. B* **1994**, *50*, 10335.
134. Hales, D. A.; Su, C.-X.; Lian, L.; Armentrout, P. B. *J. Chem. Phys.* **1994**, *100*, 1049.
135. Yoshida, H.; Terasaki, A.; Kobayashi, K.; Tsukada, M.; Kondow, T. *J. Chem. Phys.* **1995**, *102*, 5960.
136. Stave, M. S.; DePristo, A. E. *J. Chem. Phys.* **1992**, *97*, 3386.
137. Gropen, O.; Almlof, J. *Chem. Phys. Lett.* **1992**, *191*, 306.
138. Rosch, N.; Achermann, L.; Pacchioni, G. *Chem. Phys. Lett.* **1992**, *199*, 275.
139. Nygren, M. A.; Siegbahn, P. E. M.; Wahlgren, U.; Ackeby, H. *J. Phys. Chem.* **1994**, *96*, 3633.
140. Reuse, F. A.; Khanna, S. N. *Chem. Phys. Lett.* **1995**, *234*, 77.
141. Lathiotakis, N. N.; Andriotis, A. N.; Menon, M.; Connolly, J. *J. Chem. Phys.* **1996**, *104*, 992.
142. Nayak, S. K.; Khanna, S. N.; Rao, B. K.; Jena, P. *J. Phys. Chem.* **1997**, *101*, 1072.
143. Ho, J.; Polak, M. L.; Ervin, K. M.; Lineberger, W. C. *J. Chem. Phys.* **1993**, *99*, 8542.
144. Dixon-Warren, S. J.; Gunion, R. F.; Lineberger, W. C. *J. Chem. Phys.* **1996**, *104*, 4902.
145. Gantefor, G.; Eberhardt, W. *Phys. Rev. Lett.* **1996**, *76*, 4975.
146. Parks, E. K.; Zhu, L.; Ho, J.; Riley, S. J. *J. Chem. Phys.* **1994**, *100*, 7206.
147. Apsel, S. E.; Emmert, J. W.; Deng, J.; Bloomfield, L. A. *Phys. Rev. Lett.* **1996**, *76*, 1441.
148. Eberhardt, W.; Plummer, E. W. *Phys. Rev. B* **1980**, *21*, 3245.

OPTICAL INVESTIGATIONS OF SURFACES AND INTERFACES OF METAL CLUSTERS

U. Kreibig, M. Gartz, A. Hilger, and H. Hövel

Advances in Metal and Semiconductor Clusters
Volume 4, pages 345–393

ISBN: 0-7623-0058-2

I. INTRODUCTION

In cluster science, clusters/nanoparticles/dots usually are classified as being *molecular* or *solid state*. The former, consisting of only few atoms of the order of 10^1, are characterized by the facts that their structure, in general, changes drastically when one or few more atoms are added, and that they are too small for developing a well-defined surface. The latter, the larger ones, usually consist of an inner core with more or less regular atomic order (sometimes different from the bulk-like lattice structure) which, usually, is only slightly changed when further atoms are condensing, and this core can be distinguished from the cluster surface region (which, yet, may contain more atoms than the core). We will restrict our review to solid-state clusters/nanoparticles.

Clusters and nanoparticles may be characterized by their confinement, i.e. as condensed matter which is small in all three dimensions. The resulting *size effects* (for compilations, see e.g. refs. 1, 2, 7) have kept physicists busy for half a century. The alternative way of characterizing clusters is to stress the importance of their *surface* which researchers, mainly chemists, have also carefully examined, as suggested by the example of the heterogeneous catalysis.

Free clusters produced in UHV are best suited for basic research. In nature and for technical and practical applications, however, particles commonly are produced in, and stabilized by, materials surrounding them. This is supported by Table 1: all examples given there belong to what is called *cluster-matter*, i.e. composites of many clusters and materials surrounding them. These surrounding materials may be inorganic or organic, adsorbates and ligands, deposition substrates, or liquid and

Table 1. Examples of Nanoparticles in Nature and Technology

Geo-Colloids (minerals, hydrosols, aerosols)	Quantum dots
Heterogeneous catalysts	Island films
Photographic materials	Sol-gel systems
Color filters	Ferro-fluids
Glass coloration (ruby, yellow etc.)	Nanocrystalline material
Solar absorbers	Sintered nano-ceramics
Cermets, Varistors	Cytochemical markers (medicine)
Recording tapes	Autoradiographic markers (medicine)

solid embedding media. In all cases, the *free cluster surfaces* are changed into *interface regions* due to the coating with foreign substances.

Surfaces are drastically more complex when transformed into interfaces. Chemical interactions between the nanoparticle and the material outside may induce variations, both of the properties of the cluster and of the closeby regions of the surrounding material. In particular, *static* and—in the case of metallic clusters—*dynamic charge transfer processes* in this interface region are often of larger influence than the better known cluster size effects. These former effects will be treated in Section V.

Due to the small size of the clusters, the real interface should not be approximated by a two-dimensional area; it is, in fact, a three-dimensional shell with non-homogeneous properties that extend into both materials. In the case of small size clusters, the interface volume may even surpass that of the cluster core. The natural unit then is no longer the cluster but cluster *plus* the surrounding interface region, i.e. the *dressed cluster.*

Such interface effects are not restricted to clusters, but are also important for arbitrary nanostructured material and mesoscopic systems. However, clusters appear to be the most simple and clear model systems for their investigation and interpretation. Hence, this article, devoted to interface properties, is restricted to the most simple model of realistic cluster matter: (almost) spherical particles of (almost) equal size, embedded in statistical disorder, and with a low filling factor in some foreign material which is assumed to be homogeneous far away from the interface region.

Metal clusters exhibit unique optical properties due to the excitation of *Mie plasmon resonances* (i.e. cluster surface plasmon polaritons) which have been phenomenologically explained by Mie[3] and will be treated in Section III. It is well known for decades that measured resonances of metal clusters which are embedded in some liquid or solid surrounding medium are *not* described quantitatively by Mie's theory (Section IV). Recently[4] we traced back several of these discrepancies to the complex structure and behavior of the interface between the cluster and its surroundings region and ascribed them to chemical and physical interface effects. They proved in some cases to be drastically more important than cluster size effects. They will be presented and discussed in Section VI and complemented by various recent, and, in part, not yet published experimental results of our cluster-matter research group.

It will be shown that these findings open a new field of surface/interface research where the *deviations* of measured Mie resonances from the predictions of Mie's theory are used as sensitive sensors for the investigation of chemical interface effects. By combining optical experiments on free clusters in UHV and on the same clusters after embedding, this method can be calibrated to separate matrix effects from other cluster effects (refs. 4–6).

The recent experimental results, presented in Section VI, cover a broad field of composites. However, in order to allow quantitative comparisons among these results, the same kind of clusters was used throughout all experiments: silver clusters with about 2 to 3 nm mean size produced thermally and characterized in a supersonic beam before they were embedded in various surroundings. We refer to recent books (refs. 2 and 7) for the broad field of optical investigations on clusters of other materials.

There is an important reason why just silver clusters were selected here: these clusters exhibit the most selective and intense Mie resonance of all known metals, and hence, any changes in the electronic system are indicated most sensitively. The observed matrix effects are discussed by modeling qualitatively the static and dynamic charge transfer processes at the interfaces in view of their implications on the excitation of the cluster plasmon polaritons and their observed ultrafast relaxation times. The investigations presented here are limited to linear optical response.

II. CLUSTER INTERFACES

Provided that cluster sizes are large enough to enable the development of a well-defined cluster surface, all kinds of surfaces and interfaces, well known from surfaces of planar, spatially extended geometry, can, in principle, be obtained (Figure 1). However, there are essential differences between surfaces of the two kinds of symmetries. First, the amount of surface atoms may easily surpass the number of atoms forming the inner portion of the cluster (Figure 2b). The resulting atomic topology of the clusters is determined by a struggle of the ordering tendencies of these two groups: usually the surface atoms try to minimize the free surface energy while the core atoms favor the formation of a lattice-like structure. It is easily imagined that the change of the free surface into an interface will influence the involved energetics and, hence, embedding of (very small) clusters into some foreign matrix may induce modifications of the atomic structure of the clusters as a whole.

Second, the long-range ordering of well-defined planar crystalline surfaces is absent in clusters. Even, if the surface is formed by "crystal planes", the numbers of atoms at edges and corners, which are a consequence of the closed topology of the cluster surface, cannot be neglected. Even worse is when the surface atoms create *curved surfaces* to minimize the surface area. In any case, cluster surfaces, in general, exhibit broad varieties of topologically different atomic sites at the surface with different *coordination numbers* and, hence, different energetics. As an example (Figure 2a), in the densely packed tetrahedron the coordination numbers take the values 3, 6, and 9. In an icosahedron, values of 5, 7, 8, and 9 occur.[8] So, covering a cluster surface by some foreign material will induce a broad spectrum of more or less strong chemical bonds, physical adsorption sites, and electronic surface states.

Third, as mentioned in the Introduction, the extension of the surface region—and even more pronounced—the interface region in three dimensions (and hence, the volume of the surface/interface region) cannot be neglected against the core volume. Three-dimensional non-homogeneous shells are to be regarded rather than two-di-

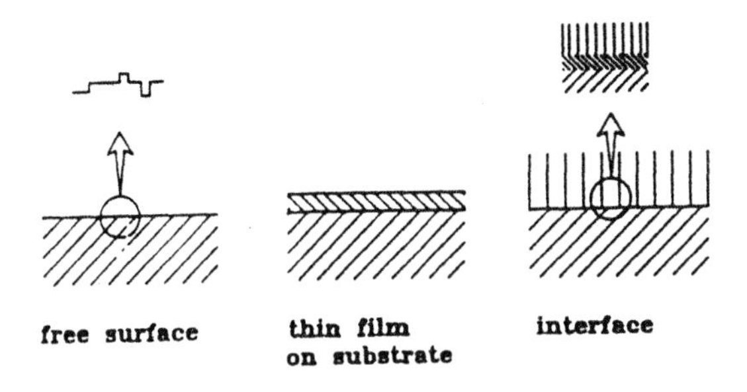

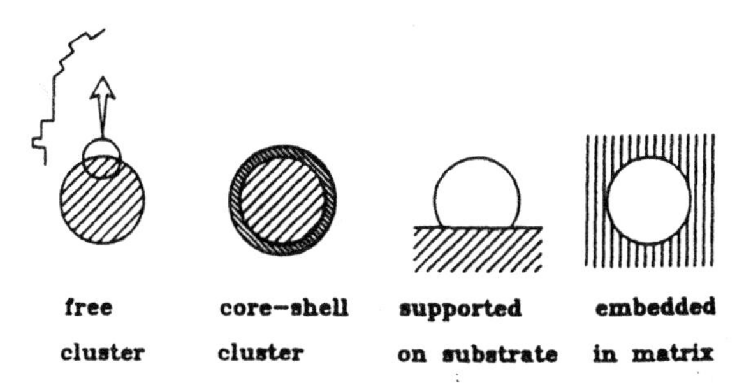

Figure 1. Topologies of surfaces and interfaces of planar and of spherical symmetry.

mensional areas, and complex combinations of electronic volume and surface states from cluster and outer material are expected to occur. As a consequence, covered or embedded clusters have to be treated as *dressed* clusters and not metal clusters alone.

As will be shown in Section V, there are other important features that render the cluster interface to be clearly distinct from the planar one. Several cluster surface and interface properties are compiled in Table 2 to demonstrate the influences of foreign surrounding material by changing the free surfaces into interfaces.

Probably, as a consequence of its complex nature, the cluster interface region has not yet gained the attention it deserves. In theoretical simulations of realistic systems, even box models with step-like two-dimensional potential walls are introduced. A next and essentially more realistic step of approximation is the treatment of the cluster in the *spherical jellium model* which has successfully been done in the *local density approximation* and the *time-dependent local density approximation*.[9,10]

Figure 3 demonstrates the resulting free surface region of the jellium cluster. Charge densities are plotted versus the radial coordinate (r = 0 indicates the

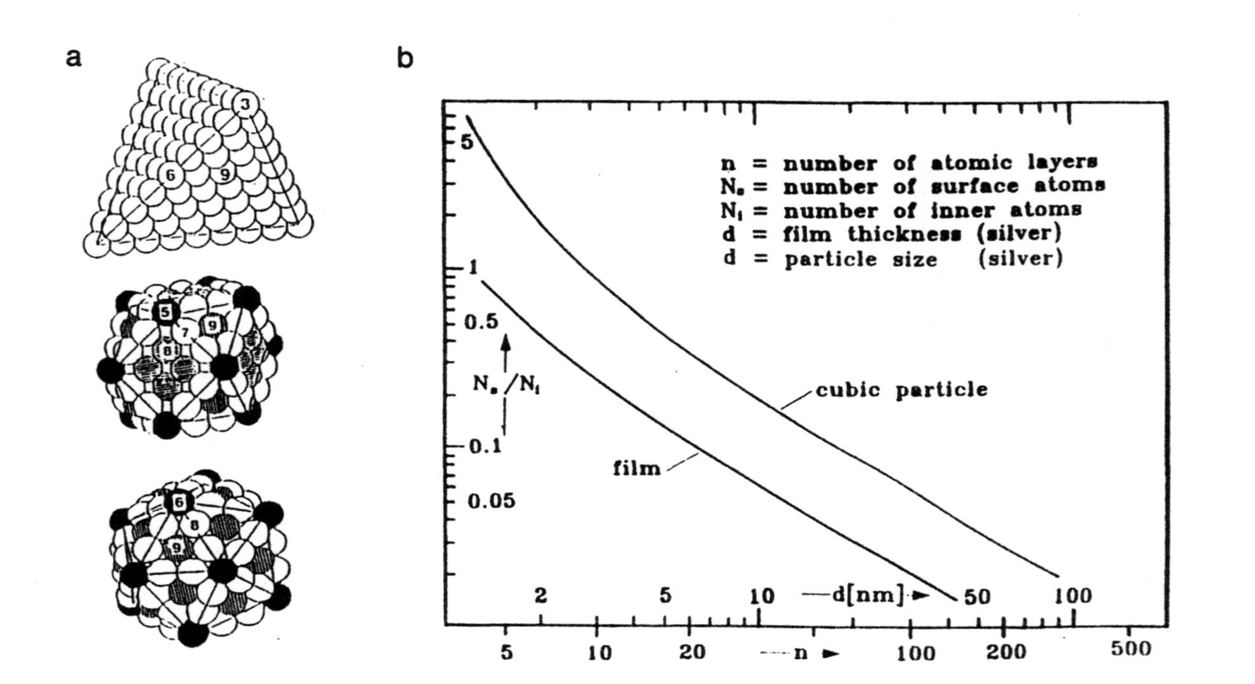

Figure 2. (**a**) Coordination numbers of surface atoms of a tetrahedral cluster (3,6,9), an icosahedral cluster (5,7,8,9) and a cuboctahedron (6,8,9). After Fritsche et al., ref. 8. (**b**) Ratio of numbers of surface and inner atoms in a planar film compared to a cubic cluster.

Table 2. Properties of Cluster Surfaces and Interfaces

Free Cluster Surface	*Cluster Interface*
• Surface atomic structure	
• Reconstruction	> • Induced surface structure changes
• Electronic surface states	> • Changes of surface states
• Electronic surface resonances	> • Changes of surface resonances
• Electronic "spill out"	> • Changes of the "spill out"
	• Electric charging of the cluster (*Static charge transfer*)
	• Electric double layers (e.g. in colloidal systems)
	• Surface pinned metal electrons
	• Chemical interface reactions at different sites
	• Electron transfer through interface Tunneling into/from adatom states (*Dynamic charge transfer*)
	• *Chemical Interface Damping* of Mie plasmon polaritons

Figure 3 demonstrates the resulting free surface region of the jellium cluster. Charge densities are plotted versus the radial coordinate ($r = 0$ indicates the *geometrical* surface given by the step function of the positive background charge). Filling electrons into the cluster until overall electrical neutrality is reached, and letting the electrons find their stationary ground states, results in the plotted *smooth* negative charge density profile of the three-dimensional *spill-out effect.* This indicates clearly that even uncovered metallic clusters always exhibit naturally an electrical double layer at the surface, the spatial extension of which is determined by the de-Broglie wavelength.

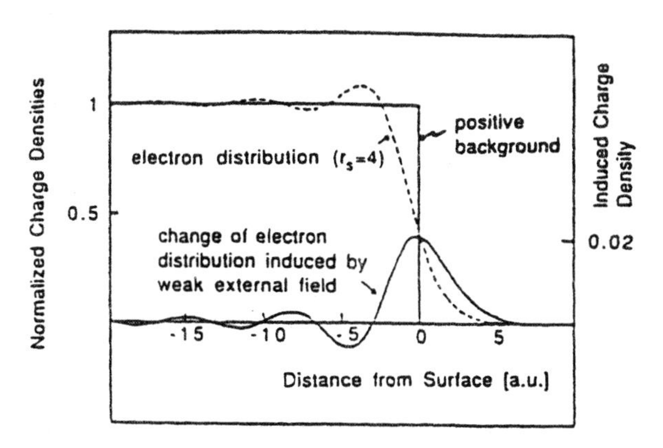

Figure 3. Electron charge density distribution $n(r)$ at the surface of a jellium cluster (parameters of Na) and its changes δn by an external electric field normal to the surface. After Liebsch, ref. 10.

III. OPTICAL EXCITATIONS IN METAL CLUSTERS: MIE'S THEORY

Linear optical properties of the single spherical cluster of arbitrary size and material, embedded in a homogeneous arbitrary matrix, have been derived from classical electrodynamics by Mie.[3] In Mie's work the matrix is assumed to be nonabsorbing. His theory has been extended later into various directions, such as absorbing matrix materials.[2,38] The unique feature of this famous theory (the only theoretical ansatz that yields numerical data to be compared quantitatively with experimental results from clusters of an *arbitrary* material) is the strict separation of the optical response into the response of the sphere onto the incident electromagnetic wave and the optical cluster material properties. The first is treated in Mie's theory by applying Maxwell's equations and developing the formally exact solution by a partial wave expansion technique. In that sense, Mie's theory is the spherical symmetry analogon to Fresnel's formulae, and exact to the same formal extent. The second part, however—the optical/electronic properties of the cluster material—remains disregarded in this phenomenological theory. Frequently, it is assumed that they might be represented by the dielectric functions of the according bulk material which have to be taken from the literature and are treated merely as a set of *free parameters*. In fact, the dielectric properties have to be treated in the frame of solid-state theories. Already, Mie recognized that the above assumption has its limits towards small cluster sizes. Up to now, the question of the proper choice of these parameters and the modeling of their differences to the bulk dielectric function received only less satisfying and fragmentary answers and is topic of intense research. In Section IV this problem will be treated in more detail. First we present a brief introduction to the dielectric function of the bulk material.

A. Optical Properties of Bulk Materials

Macroscopically, the dielectric constant is defined via the dielectric displacement vector. This definition is usually extended to the optical frequency region, and the dielectric function becomes frequency-dependent and complex-valued. The real part expresses the polarization, the imaginary part the energy dissipation (absorption):

$$\text{Dielectric displacement: } D(\omega) = \varepsilon_0 \varepsilon(\omega, E) E(\omega) \tag{1}$$

$$\text{Linear response: } \varepsilon(\omega) = \varepsilon_1(\omega) + i \cdot \varepsilon_2(\omega)$$

$$= 1 + \chi^{\text{lattice}}(\omega) + \chi^{\text{free}}(\omega) + \chi^{\text{inter}}(\omega) \tag{2}$$

Here, the corresponding susceptibility contributions due to lattice-, conduction (free) electron- and interband-transition excitations are written down separately, assuming (as usually done) that they are additive. A basic expression for metals (disregarding χ^{lattice}) was given by Bassani et al.,[11]

$$\varepsilon(\omega) = \varepsilon_1(\omega) + i \cdot \varepsilon_2(\omega) = 1 - \underbrace{\frac{(ne^2/\varepsilon_0 m_{\text{eff}})}{\omega^2 + i\gamma\,\omega}}_{\text{conduction electrons}} + \underbrace{\chi^{\text{inter}}}_{\text{interband transitions}}$$

with,

$$\chi^{\text{inter}} = \frac{8\hbar^3 \pi e^2}{m_{\text{eff}}^2} \sum_{\text{i,f}} \int_{\text{BZ}} \frac{2\,d\mathbf{k}}{(2\pi)^3} \left|\mathbf{e}\cdot M_{\text{if}}(\mathbf{k})\right|^2 \cdot$$

$$\left\{\frac{1}{[E_{\text{f}}(\mathbf{k}) - E_{\text{i}}(\mathbf{k})][(E_{\text{f}}(\mathbf{k}) - E_{\text{i}}(\mathbf{k}))^2 - \hbar^2\omega^2]} + \text{i}\frac{\pi}{\hbar^2\omega^2}\,\delta[E_{\text{f}}(\mathbf{k}) - E_{\text{i}}(\mathbf{k}) - \hbar\omega]\right\} \tag{3}$$

where n = conduction electron density, m_{eff} = effective mass, γ = relaxation frequency, M_{if} = transition matrix elements, E_{i},E_{f} = initial and final energy band states. BZ: Brillouin zone; **k**: wave factor.

Mie's theory treats $\varepsilon(\omega)$, as was stated above, as a free parameter which is taken from the literature for the according bulk material. Hence, no information about its origin and magnitude enters the theory or can, vice versa, be directly drawn therefrom.

B. The Mie Theory

The mathematical treatment of this theory,[3] being quite voluminous, will only be sketched as follows:

- Assumption: One spherical cluster/nanoparticle of arbitrary material and size, embedded in a homogeneous (nonabsorbing) surrounding matrix.
- Formulation of the electromagnetic potentials in spherical coordinates r, ϑ, φ. Maxwellian ("sharp") boundary conditions for the electric and magnetic fields, E, M, at the cluster/matrix interface, e.g.:

$$E_\vartheta^{\text{outside}} = E_\vartheta^{\text{inside}}\,;\; E_\varphi^{\text{outside}} = E_\varphi^{\text{inside}}\,;\; \varepsilon^{\text{outside}} E_r^{\text{outside}} = \varepsilon^{\text{inside}} E_r^{\text{inside}} \quad \text{at } r = \text{R} \tag{4}$$

  with the spectra of dielectric functions to be taken from the bulk material.
- Solution of the Helmholtz wave equation for the potentials by product-ansatz.
- Multipole expansion of the fields: Solutions are obtained for the incident, interior and scattered waves in terms of infinite series of Bessel and Neumann functions and Legendre polynomials, of different polar order.

- Application of Poynting's law to obtain incident, absorbed, and scattered intensities.
- Derivation of the spectra of absorption, scattering, and extinction cross sections or the spectra of absorption, scattering, and extinction constants.

Careful analyses by comparing quantum mechanical solid-state theory with the phenomenological results of Mie's theory have, meanwhile, given a clear picture about the physics behind these spectra and their peculiarities. Specializing to metal particles, the electrical wave modes contain collective excitations of the conduction electrons, i.e. *spherical surface plasmon polaritons*, and the formal multipole expansion of their fields is attributed to the *infinite* series of plasmons of different polar order (dipole, quadrupole symmetry, etc.). In addition, the magnetic components of the electromagnetic fields are effective, too; however not via magnetic interactions but by exciting an infinite series of *eddy current modes* of the electrons (which do not exhibit resonances like the electric excitations). In ionic clusters the analog phenomenon can occur for atomic vibrations (*Mie cluster phonons*).

For practical purposes, the general property of Mie's solutions is important that, if clusters are very small compared to the electromagnetic wavelength, among all possible plasmon modes only the dipolar one has sufficient oscillator strength to be observable and, in addition, this electrodynamic mode renders independent of cluster size. It then collects almost the total oscillator strength of the conduction electron ensemble in the cluster. In the special case of silver clusters, the limitation to the size-independent dipolar mode requires sizes below 15 nm. This has a very important consequence for the experimentalist: if, in this size region, the Mie resonance develops some size-dependence, its origin cannot be found in the electrodynamics but exclusively in size-dependent cluster material properties. Such size effects are often labeled *intrinsic* to distinguish from the *extrinsic* ones caused by electrodynamic effects of the electromagnetic wave, like retardation, excitation of higher order multipoles, etc.

In the following we shall focus upon the electronic Mie resonances. Both, spectral positions $\hbar\omega_s$ and resonance band widths Γ strongly vary with the cluster material. In many metals, like Pt, the resonances are fully suppressed due to strong energy dissipation. As we will see below, this statement, however, deserves restrictions with regard to the influence of surrounding materials.

Mie resonances, in general, depend drastically on the cluster size. In the limit of very small sizes compared to the wavelengths of the light, they lose this dependence as mentioned above, whereas towards larger sizes, retardation effects usually cause strong red shifts and drastic broadening. Besides, higher order excitation modes become important, e.g. the quadrupolar plasmon is distinct in silver clusters from, say, 30 nm upward. These cluster effects following from Mie's theory are due to the electrodynamics only (since the bulk ε is included). Hence, we call them *electromagnetic effects* or *dielectric effects* or *extrinsic effects* to distinguish them from those cluster effects which are due to the material properties of the clusters themselves and their deviations from the bulk (*intrinsic effects*). These latter effects will be the topics of the subsequent sections.

In realistic materials, "lattice" excitations and individual interband transitions (electron-hole excitations) contribute to the dielectric function of Eq. 2, and hence, acting by their complex polarizability, influence the collective Mie excitations both by additional polarization effects and resonance shifts and by opening additional dissipation channels compared to the free electron cluster.

A simple heuristic picture of the Mie resonances, which will be helpful to comprehend the role of cluster interfaces for the optical properties, treated in Sections IV and V, is to identify the cluster with a spherical electromagnetic *nano-antenna* which behaves simultaneously as receiver and transmitter. The electric vector of the incident electromagnetic wave $E(t)$ causes the (freely moving) conduction electrons of the cluster to be collectively displaced against the fixed positive ion charges. Of course, this concerns only the tiny amount of *drift motion* superimposed to the statistically isotropic electron motions with velocities following from Fermi statistics. The important feature is that the drift motion is a *collective*, *coherent* one and all of the cluster conduction electrons are involved *in same phase* which is forced upon the electrons by the incident light field. Hence, in k space, the whole Fermi sphere is oscillating coherently with the external frequency, and the *drift wavevector* amounts to:

$$k_{\text{drift}} = (\hbar)^{-1} m v_{\text{drift}};\ \hbar v_{\text{drift}}(t) = e\, m^{-1} \int E(t)\, dt \tag{5}$$

Thus, periodically alternating electric charges are induced at the cluster surface which give rise to restoring forces for the electron motion. In full contrast to the bulk case, where the Drude–Lorentz–Sommerfeld conduction electrons behave like *relaxators* (with eigenfrequency zero), the collective of conduction electrons in the cluster therefore performs *oscillations* with the Mie resonances as eigenstates. As stated above, these resonances of complex spatial symmetries can be formally decomposed into an *infinite* series of multipolar eigenmodes (plasmon polaritons), of dipolar, quadrupolar, etc. symmetry, and the analogous eddy current modes. By quantization of these oscillations, the total quantum energy $\hbar\omega_s$ is stored and distributed among all cooperating electrons of the whole cluster. The additional excitation energy *per* conduction electron thus remains small, though it increases with decreasing cluster size (and, hence, decreasing number of collaborating electrons). We can extend this picture: the plasmon life-times being typically of the order of 10 fs (as derived from the Mie band widths Γ), we can compare with the average time *between* subsequent excitation acts. Assuming highly focused light of some source of, say, 1 watt of suitable frequencies, we find that, due to the small absorption cross section of each cluster, the times between excitations exceed the plasmon lifetimes for various orders of magnitude. As a consequence, the possibility of nonlinear multiplasmon excitations is small under normal irradiation conditions.

In Figure 3 the resulting charge distribution changes δn in the jellium cluster, brought about by a homogeneous static external electric field, are visualized: they are concentrated close to the surface region, and it is obvious that every change of the structure and the composition of the surface/interface region will have strong impact upon them. Vice versa, the high frequency resonances are expected to act as sensitive monitors of such changes.

In fact, Mie's theory, including *all* optical excitations contained in the dielectric function formally correctly, calculates the behavior of the *inner electric field* in the cluster which can be resonance-like surmounted, compared to the outer field, irrespective of its origin. This Mie resonance can be of electronic and of ionic origin. It should be noted that Mie's theory also describes correctly the *near-field* configurations and their continuous transformation into the far-field which are important concerning electromagnetic interactions among neighboring clusters, and, at present, have gained interest in view of the scanning near-field optical microscope.

Narrow Mie resonances occur in many cluster–matrix systems. As stated above, the ones in silver cluster systems are the most pronounced. Free sodium clusters are the next best. In general, not the cluster material but the combination of cluster and matrix are decisive for the spectral selectivity and strength. Figure 4 demonstrates the outstanding properties of silver clusters by comparison with gold and copper. The resonance profile on the quantum energy scale is close to Lorentzian in the case of silver and can be well characterized by the *peak position* $\hbar\omega_s$ and the

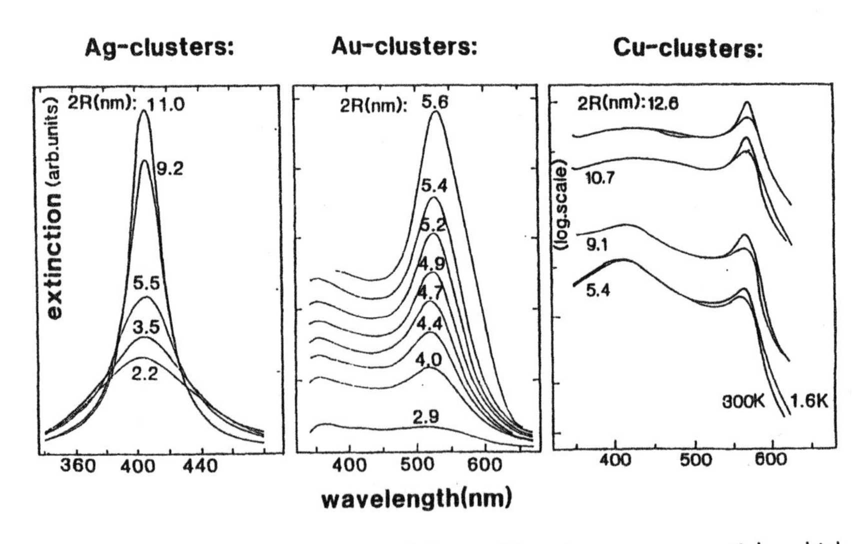

Figure 4. Measured extinction spectra of silver, gold, and copper nanoparticles which are prepared and embedded in a glass matrix (see ref. 2). Sample temperatures: Ag, Au 300 K; Cu 1.6 and 300 K.

halfwidth Γ. The marked dependence of the line profile on the cluster size in Figure 4 is exclusively due to *intrinsic*, i.e. material effects. In Section V it will be shown that the magnitudes of $\hbar\omega_s$ and Γ are sensitive indicators of chemical effects at cluster surfaces and interfaces. In contrast to what might be expected from simple oscillator models, these two quantities are not exclusively connected to dispersion and absorption, respectively. Instead, the band width results from, both, the real and the imaginary part of the dielectric function of the cluster material.

In the approximation of very small silver clusters these quantities can be analytically derived from Mie's theory. For the frequency position of the plasmon polariton peak we have then:

$$\omega_{resonance} \approx (n\, e^2/\varepsilon_0\, m_{eff})^{1/2}\,(2\,\varepsilon_{matrix} + 1 + \chi_{1,\,interband})^{-1/2} \tag{6}$$

It is obvious that the resonance position is not only determined by the conduction electron parameters n and m_{eff} (though their influence is strong), but also by the interband transition part, i.e. the electron-hole excitations which, though due to single-electron excitations contribute by their background polarization to the coherent collective excitation.

Another important feature of Eq. 6 is the dependence on the dielectric constant of the matrix material: changes of ε_{matrix} strongly shift the position. So, if some clusters are embedded in a matrix material, the peak will already be shifted due to this dielectric effect without considering any substantial changes of atomic or electronic structures at the surface/interface. The reason for this shift is clearly electromagnetic: the embedding material alters the Maxwellian boundary conditions (Eq. 4) for the fields.

The Mie absorption band halfwidth follows to be:[2]

$$\Gamma \approx 2\varepsilon_2(\omega)\,[(d\varepsilon_1/d\omega)^2 + (d\varepsilon_2/d\omega)^2]^{-1/2}\,(1+\beta) \tag{7}$$

with $\beta\,(\omega_s, \varepsilon_2, d\varepsilon_1/d\omega, d\varepsilon_2/d\omega) \leq 10^{-2}$ for the case of very small silver clusters. The derivatives are taken at the resonance frequency. In contrast to Eq. 6, this equation does not include any explicit dependence on ε_{matrix}. Since in our case, $d\varepsilon_1/d\omega >> d\varepsilon_2/d\omega$, the influence of the real part of ε on Γ of Eq. 7 is obvious. It is this contribution that renders the resonance of silver clusters to be markedly narrower than that of a corresponding free Drude electron cluster. This effect is elucidated in Figure 5.

In Figure 5, four different Mie absorption spectra are shown which were calculated with the dielectric functions plotted below each of the spectra. These dielectric functions are fully hypothetical, arbitrarily chosen for demonstration purposes only. The spectra (A) and (B) exhibit drastically different band widths, but these are *not* due to different imaginary parts of ε, thus are not due to electronic dissipation or relaxation processes but rather to different spectra of its real part. This, again indicates the complex interplay of polarization and dissipation in the Mie reso-

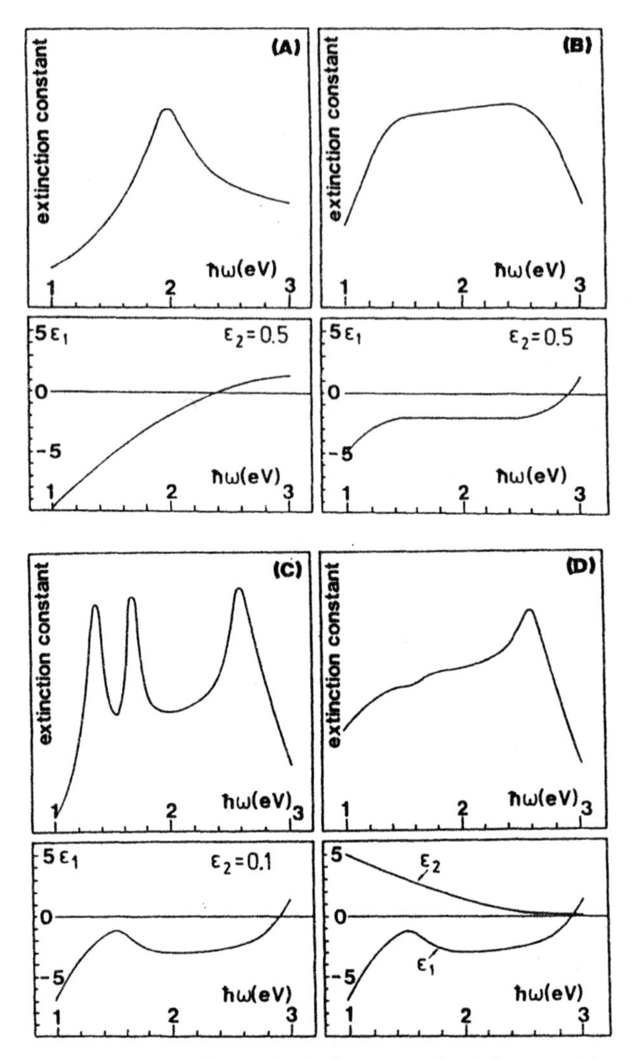

Figure 5. Mie resonances of spherical clusters with fictitious dielectric material functions, the spectra of which are plotted in the lower parts of the figures.[7] Widths, peak positions and number of Mie peaks are mainly determined by ε_1 (ω).

nances. The spectra (C) and (D) are, in contrast, based upon the same real part of ε, but now the imaginary part is different in both cases. In (C), the ε_1-spectrum was chosen such that *three* Mie resonances occur instead of one, while in (D), two of them are damped away again by increased ε_2 in the left-hand part of the spectrum.

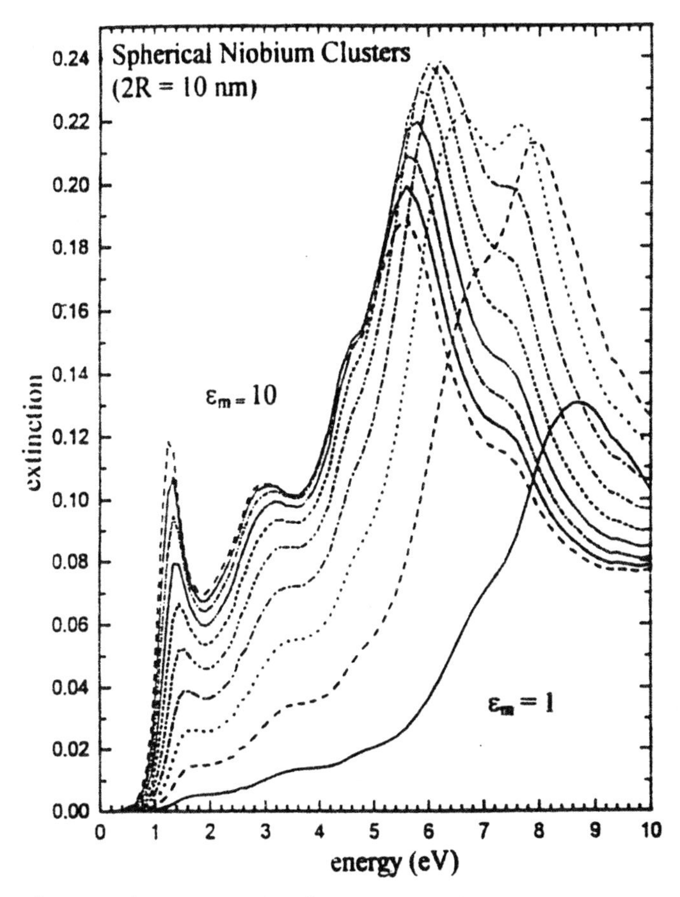

Figure 6. The optical extinction of well separated Niobium clusters (2R = 10 nm) as calculated from Mie's theory with dielectric material function of the bulk material. The parameter is the dielectric constant ε_m of the embedding material varying in steps of 1 from 1 to 10. The differences of these spectra represent the *dielectric effect.*

Concluding this section we present Mie absorption spectra for the case of niobium clusters (Figure 6), a material which was arbitrarily chosen to demonstrate the influence of the *dielectric effect*, where increasing the magnitude of ε_{matrix} changes the spectra dramatically. By this *immersion effect* existing peaks are changed both in position and height but also new peaks appear (described in some detail in ref. 2).

This section has covered the original Mie theory based upon the application of the dielectric function of bulk material. In the next section, the assumptions of this theory will be reconsidered in detail, and we will see that Mie's theory, though being formally correct, does not meet reality in certain important aspects.

IV. BEYOND MIE'S THEORY

Mie's theory is a formally correct solution to Maxwell's theory under certain assumptions and restrictions. Real cluster systems, however, are not in agreement with all of these assumptions as will be shown in the following and, hence, their optical behavior cannot be expected to agree quantitatively with the predictions of Mie's theory. Vice versa, it will be shown that the differences between experimental and theoretical resonance spectra allow a deeper insight into the properties of realistic cluster systems.

The following compilation is not intended to be complete; its purpose is merely to demonstrate where Mie's theory does not hold, namely if:

(1) Cluster shapes differ from the sphere.
(2) Cluster-dependent dielectric functions differ from the functions of the bulk.
(3) The (discontinuous) Maxwellian boundary conditions (Eq. 4) at the cluster–matrix interface are violated.
(4) Dielectric functions of cluster and/or matrix are nonlinear.
(5) Surrounding/embedding material is absorbing.
(6) Cluster structures are heterogeneous (core-shell structures, multigrain clusters, etc.).
(7) Clusters are forming close-packed cluster-matter instead of being well-separated single clusters.

These effects have been previously reviewed (ref. 2). In accordance with the objective of this chapter we will concentrate in what follows on items (2) and (3) above.

A. The Parameter Called the Dielectric Function of the Cluster Material

The *macroscopic* definition of ε in Eq. 1 appears to be inappropriate for volume elements much smaller than the wavelength, which constitute our clusters. Yet, the larger problem is that clusters are non-homogeneous on the nanometer scale due to the cluster surface/interface region and nonlocal effects. The latter are strong in metal clusters since their sizes are below the bulk mean free paths of the conduction electrons. It is, on the other hand, advantageous that clusters are too small to exhibit the anomalous skin effect.

The alternative way of a *microscopical* definition of ε can be used in view of application in Mie's theory, since, in fact, only the *volume-averaged complex cluster polarization* is required. It is in the tradition of Maxwell's theory to use, instead, the formally related dielectric function. The total cluster polarization is

$$P = \Sigma p_i = \Sigma \alpha_i E_i^{loc} = \alpha E^{average} \tag{8}$$

with α_i = local polarizabilities in the cluster volume and of the cluster surface/interface region. The resulting (total) polarizability α can formally be related to ε by:

$$\alpha(\omega,...) = \varepsilon_0(\varepsilon(\omega,...) - 1) \tag{9}$$

It is this, in general complex-valued, $\varepsilon(\omega)$, that enters the theory of Mie.

In planar geometries, dielectric functions are determined by measuring (at least) two independent optical properties and applying Fresnel's formulae or similar relations. (As an alternative, one experiment can, in principle, be replaced by a Kramers–Kronig analysis). In the case of spherical geometry, the basically same procedure can be followed: measuring two independent optical properties and applying Mie's and Gans–Happel's formulae (see e.g. ref. 2). Also, a Kramers–Kronig analysis can replace one experiment (see ref. 2).

What is obtained by the above is the *cluster averaged* polarizability $\alpha(\omega,R..)$ or dielectric function $\varepsilon(\omega,R..)$, the average being extended over, both, the cluster core and cluster surface region. Problems with this quantity are obvious, especially for the *dressed* clusters of Section II, since the inhomogeneous interface shell also includes some parts of the matrix which contribute to the (complex) polarization of the cluster differently than considered in the electrodynamic boundary conditions where the discontinuous jump from ε to ε_{matrix} in Eq. 4 is assumed.

Taking the thus limited precision of the quantity $\varepsilon(\omega,R..)$ into account, a definition of the free parameter entering the electromagnetic theory under the label of *dielectric function of the cluster material* is to state that, performing Mie calculations with varied $\varepsilon(\omega)$, the proper spectrum of ε has been found when the Mie theory reproduces quantitatively the according measured optical absorption (extinction) spectrum. The start values for $\varepsilon(\omega)$ may be bulk data of the cluster composition, taken from the literature. By this means, Mie's ansatz of introducing *one* spatially averaged quantity for the cluster material is maintained. An important basic property is its convergence toward the dielectric function of the bulk for vanishing size and surface influences. Convergence towards the molecular polarizability at small sizes is, however, not expected. The advantage of this procedure is evident: this kind of theory is the only one which allows the computation of cluster absorption or scattering spectra of nickel as well as of thorium or sodium, to give some arbitrary examples and, thus, goes beyond jellium and related model substances.

The next step towards a deeper understanding must be to introduce models on the basis of solid-state theory, to take the real non-homogeneous structure of the cluster into account and, thus, to express deviations from the dielectric function of the bulk material quantitatively. By this way, the deviations from bulk behavior are described by correction terms introduced into the dielectric function of the bulk. This strategy is followed throughout this review.

Such correction models were developed by focusing upon effects determined by the confinement (*size effects*) and effects concerning the surface (*surface/interface effects*). Of course, these two kinds of effects are not independent; however, separate models can be and have been developed. Such models, tested by comparison of Mie calculations with experiments, render the free parameter of Mie's theory into a quantity which is helpful to better understand atomic and electronic structures and properties of real clusters in cluster-matter. Vice versa, the cluster-ε derived by *inverse* application of Mie's theory on *measured* absorption, scattering or extinction spectra, contains—by its deviations from the bulk-ε—information about size and surface/interface effects of the clusters under investigation which may be extracted by proper quantum mechanical microscopic models.

B. Cluster Size Effects

A wealth of cluster-size corrections to the bulk-ε has been developed since half a century for metallic and semiconductor clusters. We refer to reviews (e.g. refs. 2,7) for compilations. They can be classified according to their theoretical basis:

- Classical conductivity theory (electronic mean free path limitation (*free path effects*)); quasi-continuous electronic band structure).
- Quantum box models (discrete electronic ground states in the otherwise empty box; vertical potential walls (*quantum size effects*)).
- Linear response theories (inclusion of ions; discrete electron energy levels (*quantum size effects*)).
- Jellium models (uniform positive background; selfconsistent *smooth* potential; inclusion of exchange and correlation effects (*quantum size effects*)).

Both, the case of quasicontinuous bandstructure and of discrete, *quantized* electronic level structure can occur. This depends on the relation between the level spacings ΔE and level widths δE including lifetime effects: $\delta E > \Delta E$ means that the size quantization structure is smeared out and electrons may be continuously accelerated as presumed for classical conductivity, while for $\delta E < \Delta E$ separated levels exist.[12] While level spacings were explained on various routes (roughly estimated, we have $\Delta E \approx E_{\text{Fermi}}/N$; N = number of involved electrons in the cluster), the level widths are still controversely discussed.

In the case of metallic clusters, only few experimental hints have been published which point to a discrete level structure in clusters consisting of more than, say some 10 atoms. The recent investigation of Sinzig et al.[13] on ligand-stabilized monosize Pd clusters gives clear indications. Since magnetic and thermal properties were examined, only levels very close to E_{Fermi} were involved with, probably, longer lifetimes, in contrast to the *hot metal electrons* usually excited by optical means which, up to now, did not point clearly to level discretization in any experiment.

An interesting feature of the many cluster size-effect theories is the high degree of correspondence concerning the resulting correction terms for the dielectric function. While only few theories deal with the real part of the dielectric function, and mostly only small corrections of the bulk ε were predicted (see, however ref.

9), drastic changes of its imaginary part were found which, despite the different bases of the models, conformably are described by the famous *1/R* law:

$$\varepsilon_2(\omega, R) \approx \varepsilon_2^{\text{bulk}}(\omega) + \Delta\varepsilon_2(\omega, R) \quad \text{with} \quad \Delta\varepsilon_2(\omega, R) \propto \mathbf{A}_{\text{size}}/R \tag{10a}$$

We have introduced here the **A** *parameter* which, as we will show below, plays a key role for the following by determining the quantitative amount of size dependences. In the frame of the quantum size effect, the **A**/*R* term is derived from the size dependence of the level spacing, while in the free path effect, this term is attributed to a reduction of the conduction electron mean free path due to collisions at the cluster surface.

These collisions contribute with an additional relaxation frequency,

$$1/\tau = 1/\tau_{\text{bulk}} + 1/\tau_{\text{surface}} \quad \text{with} \quad 1/\tau_{\text{surface}} = \mathbf{A}_{\text{size}}\, v_{\text{Fermi}}/R \tag{11}$$

where v_{Fermi} is the Fermi velocity of the electrons which are most effective for the electron relaxation processes.

For this latter case the following explicit expression of the dielectric function was obtained (e.g. ref. 2) which holds, among other cluster materials, for silver clusters around the visible spectral region,

$$\varepsilon(\omega, R) \approx 1 + \chi_{\text{interband}} - \omega_p^2/\omega^2 + i\,\omega_p^2/\omega^3\,(1/\tau_{\text{bulk}} + \mathbf{A}_{\text{size}}\, v_{\text{Fermi}}/R) \tag{10b}$$

where ω_p is the Drude plasma frequency. In this approximation, a small size dependence of the real part of ε was neglected and the main size dependence remains with the imaginary Drude part. In fact, not only the optical response of the Drude electrons is influenced by the confinement, but also are the interband contributions $\chi_{\text{interband}}$. As has been shown,[13] however, these size dependences only become relevant at essentially smaller sizes due to the closer localization of electron-hole pair excitations.

So, the $\mathbf{A}_{\text{size}}$ parameter measures the magnitude of the predicted size effects of the conduction electrons. Numerous different theories yield values of $\mathbf{A}_{\text{size}}$ which lie between 0.1 and ~1 (see ref. 2). For instance, the quantum size theory of Zaremba et al.[14] for free electron clusters resulted in $\mathbf{A}_{\text{size}} = 0.25$, while the free path effect theory[12] gives values between 0.75 and ~1, depending, in case of realistic clusters, also on their internal multiple grain structure.[15]

Introducing the ansatz (Eqs. 11, 10b) into Mie's theory, we obtain the approximative size dependence of the dipolar resonance band width of Eq. 7:

$$\Gamma(\mathbf{A},R) \approx \Gamma_0 + (2\omega_p^2/\omega^3)v_{\text{Fermi}}\,[(\partial\varepsilon_1/\partial\omega)^2 + (\partial\varepsilon_2/\partial\omega)^2]^{-1/2}\,\mathbf{A}_{\text{size}}/R \tag{12}$$

Here, Γ_0 is the width which would result from the bulk dielectric function. This equation, based upon the quasistatic dipole approximation of Mie's theory, only holds, in the case of silver, for clusters smaller than 15 nm in diameter. Precise numerical values for arbitrary cluster materials and sizes are easily obtained by

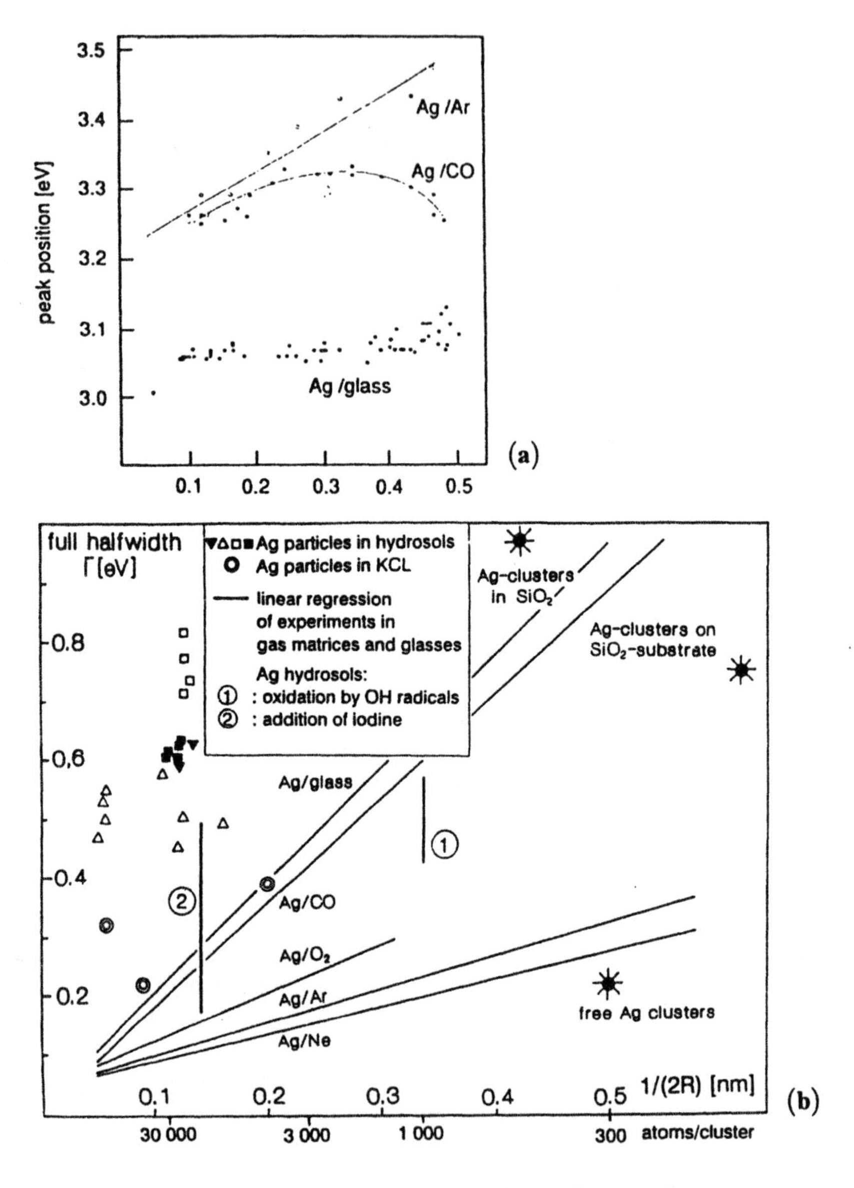

Figure 7. Influences of various surrounding media on the Mie plasmon resonance position (**a**) and width (**b**) of spherical silver clusters of various sizes. In these plots, a linear dependence reflects the 1/R law. Obviously, the slopes in (**b**) differ strongly for different embedding media. In colloidal systems this dependence is absent. Figure 7(**a**) from ref. 16; Figure 7(**b**) from ref. 4.

introducing the size-dependent cluster-ε of Eq. 12 immediately into the complete Mie theory and evaluating the resulting spectra numerically.

A general comment appears to be apt here: From the above models of increased damping effects in small particles, **A** follows to be always positive (**A** > 0) and, hence, the Mie resonance always is broadened (see refs. 1,2). However, if $\varepsilon_1 = \varepsilon_1(R)$, the influence of $\partial\varepsilon_1/\partial\omega$ in Eqs. 7 and 10b, visualized in Figure 5, can also cause the band to be narrowed with decreasing cluster size. Such an effect occurs e.g. when the interband transition edge is extending into the Mie band region and is flattened at small sizes. This size effect was experimentally observed, both from gold (see ref. 2) and from silver clusters (see refs. 6, 26).

Experimental confirmations and quantitative determinations of the size effects in the Mie resonance of very small clusters are less numerous than the according theories, but are compelling. Several of them, performed with silver clusters, are compiled in Figure 7.

It is obvious that markedly different Ag/matrix systems develop a clear $1/R$ dependence of the Mie resonance width. It holds for glass matrices as well as for the various solid rare-gas matrices investigated in the pioneering work of the Schulze group.[16] Other systems, however, such as Ag hydrosols or Ag clusters in KCl crystals, exhibit exorbitant widths where, as it appears, strong additional effects conceal a $1/R$ dependence. The slopes of the linear dependences in Figure 7b directly reflect the magnitude of the **A** parameter and, surprisingly, this parameter exhibits drastical variations between different cluster–matter systems.

In concluding this section, we point out that the additional dependence of **A** on the nature of the matrix surrounding the clusters is a consequence of differently formed interface regions and interface effects. It is mainly the *chemical interface effect* which is the subject of the next section.

V. CHEMICAL INTERFACE EFFECTS

The first indications of the previously underestimated role of interfaces on the optical properties of metal clusters came from experiments like those shown in Figure 6. The Schulze group[16] clearly formulated the effect from the extended experimental work on silver clusters, matrix-isolated in different solid gases. Possibly it is because of the complex structure of the created interfaces that still up-to-date satisfying theoretical treatments are not in sight.

Attempts in the frame of extensions of Mie's theory demonstrated that a *dielectric smoothing* of the electrodynamic boundary conditions yields corrections into the right direction of a Mie band broadening, yet it is probable this model does not reflect the real situation comprehensively. In these attempts, Mie's theory was extended to a heterogeneous multishell structure (onion structure) with concentric shells of different local Drude contributions to $\varepsilon(\omega)$.[2,17] Hence, a smooth change of $\varepsilon(\omega)$ at the surface was simulated by a multistep function. Figure 8 gives an example.

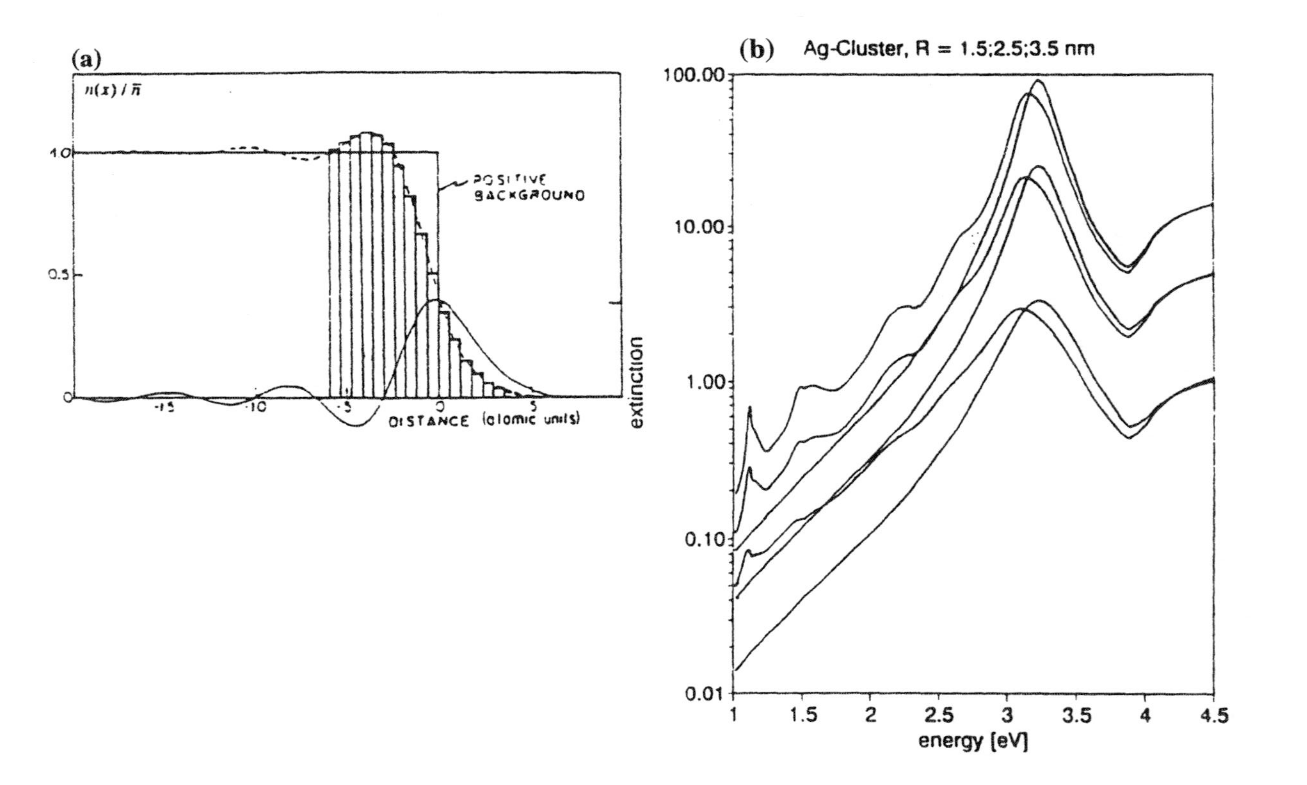

Figure 8. Multi-shell cluster simulation of the cluster surface smoothened by the *spill-out effect.* (**a**) The surface (see Figure 3) was discretizised into 17 shells according to the shown conduction electron densities which were introduced in the Drude term of the size-dependent dielectric function. The parameters were chosen similar to silver. (**b**) The optical absorption spectra of these multishell (*onion*) particles were calculated by an extension of Mie's theory[17] for sizes $2R = 3, 5, 7$ nm (from bottom to top). For comparison, normal Mie spectra are added. Their Γ's are smaller.

The drawback of this method is that the local $\varepsilon(\omega,R)$ was applied for extremely thin layers and the nonlocality of the conduction electrons and their collective excitations are disregarded. Hence, effects of charge shifts and charge transfer through the onion skins were not taken properly into account.

A promising alternative approach is to model, in detail, the formation of adsorption at the cluster surface. Figure 9 gives a scheme of the chemisorption of adatoms at the surface of a metal (after ref. 18): During the deposition at a metal surface the energy levels ε_a of the atoms are shifted and broadened, thus forming a broad distribution of density of states, which is also representative for a multitude of atoms deposited at different surface sites. Usually, the adsorbed atoms are separated from the metal by a tunnel barrier.

A. Static Charge Transfer in the Interface

Depending on the energetic overlap of the adsorbate states with the Fermi level in the metal and on their occupation, electrons may tunnel through the barrier to or from the adsorbate. In the case of the electrically isolated metal cluster the electron density in the cluster is changed thereby. This may give rise to drastic changes of the electric double layer, existing due to the spill-out effect (Figure 3), and Coulomb barriers at this modified interface may be of remarkable magnitude. This Coulomb barrier energy Φ is known to follow a $1/R$ dependence (e.g. ref. 19):

$$\Phi = e^2/(4\pi\varepsilon_0\varepsilon_{matrix}R) \tag{13}$$

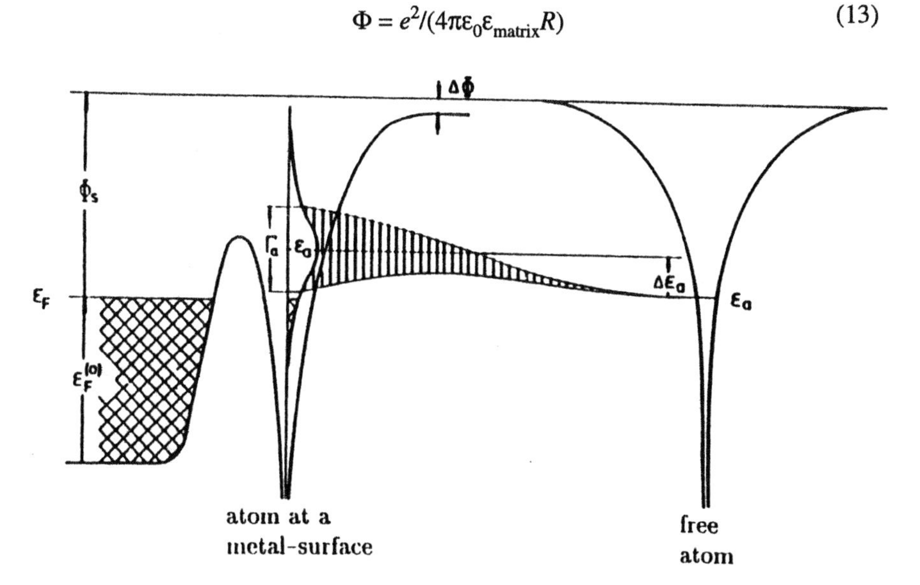

Figure 9. Electron energy scheme for adsorption at a metal surface. (After ref. 18). The free atom (*right*) approaches the metal surface (*left*). Its energy levels are thereby shifted and broadened. They are separated from the metal by a tunnel barrier.

In our clusters the full removal ($+ \rightarrow \infty$) of one electron gives Φ of, roughly, 0.1 eV. It is smaller, of course, for charge shifts *within* the interface region, which change the electric double layer only.

In addition, new combination wavefunctions are created including the adsorbate states. As shown in Figure 10 the tendency toward closed electronic subshells in the adsorbed atom may, e.g., cause Cl and Li atoms deposited on an Al cluster to inject an electron into or to eject an electron out of the metal cluster, respectively. This is a *static* (i.e. permanent) charge transfer.

The Coulomb energy involved in the creation or change of an electric double layer on the atomic scale is essentially smaller than of the total charge removal of Eq. 13. Hence, in the case of a dense surface adsorption layer or compact embedding medium, a multitude of charges can be shifted through the interface. It is not yet clear, however, to which extent this kind of charge transfer can take place in a given realistic cluster/matrix combination and how the upper limit is determined. In the next section an experimental example will be given which points to the static charge transfer of about 10^2 electrons from a 500 atom silver cluster into the surrounding matrix of fullerite! This opens the possibility of electron density manipulations in metals, usually assumed to be the exclusive domain of semiconductors.

Such charge transfer will change the Mie resonance since plasmon energies depend on the electron density n. From Eq. 6 follows the root dependence of the peak shift for a change $n_1 \rightarrow n_2$:

$$\Delta\omega_{\text{resonance}} \approx \{(n_1)^{1/2} - (n_2)^{1/2}\}\ (e^2/\varepsilon_0\ m_{\text{eff}})^{1/2}\ (2\varepsilon_{\text{matrix}} + 1 + \chi_{1,\ \text{interband}})^{-1/2} \quad (14)$$

This relation is based upon the implicit assumption that the charge transfer leads to a uniform change of the *mean* sp electron density n in the whole cluster. This assumption appears not to be trivial: alternatively, gradients of the electron density, particularly close to the interface, might occur and charges may be localized in the surface region of the cluster which no longer contribute by a Drude behavior.

From Eq. 7 follows the relation for the according change of the Mie-band width $\Delta\Gamma \propto (n_1 - n_2)$.

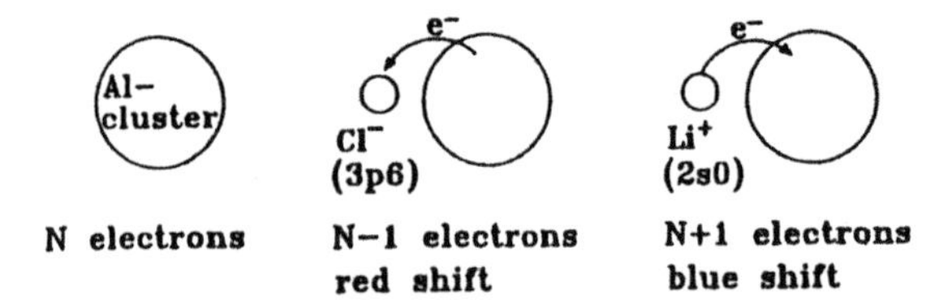

Figure 10. Static charge transfer from/to a *N* electron cluster to/from some adsorbed atom, in order to obtain closed subshells. As a consequence, the Mie resonance is red or blue shifted.

Strictly, Eq. 14 only holds under the assumption that all other terms in Eq. 6 remain unchanged. It is interesting to see that this may not be the case: In fact, a change of the conduction electron density will influence the Fermi energy and, hence, the conduction electron effective masses and the low-frequency interband transition edge which, in the case of silver, is due to 4d $\rightarrow$ 5sp (E_{Fermi}) transitions, i.e. the Fermi level is involved. Moreover, the charging of the cluster may not only change the occupation of band levels but the bandstructure itself: a typical self-consistence problem which awaits a solution. Size-dependent changes of interband excitations have supplementary implications on the Mie resonances: a reduction of the interband transitions in the edge region is followed by a blue shift of the Mie peak in the case of silver.[2,6]

As pointed out above, the experiments of embedding clusters into some matrix material include, additionally, dielectric shifts due to changes of ε_m in Eq. 14. To separate these two shift effects in measured absorption spectra, additivity is assumed since the changes usually are small:

$$\Delta\omega_{\mathrm{experimental}} \approx \Delta\omega_{\mathrm{dielectric}} + \Delta\omega_{\mathrm{interface}} \tag{15}$$

B. Dynamic Charge Transfer in the Interface

At first glance it may be surprising that the broadening of Mie resonances in clusters, observed after their embedding into some matrix or after covering the surface by some adsorbate, can be traced back as well to charge transfer processes involving adsorbed atom states in the interface. These are *dynamic charge transfer* processes due to tunneling.

If, after equilibration, the cluster Fermi level cuts some of the broadened adsorbate levels which are not fully occupied, cluster electrons with energies close to E_{Fermi} or higher (*hot electrons*) can tunnel through the interface barrier into the adsorbate level. After some statistical *residence time* there, the electrons will return to the cluster and the play begins anew. This dynamic charge transfer induces fluctuations of the electron density in the cluster. It may be limited by the involved Coulomb forces. This charge transfer may become important if it takes place *during* the excitation of a Mie plasmon in the cluster. As pointed out above, the main feature of this excitation is the coherence of the collective drift motion of all the individual electrons, i.e. the *phase coherence* of the conduction electrons. Due to the statistically disordered tunnel processes back and forth and residence times between them, the electron returning to the metal cluster is, in general *out-of-phase*, i.e. it no longer contributes to the collective *drift* motion.

In the case of a dense embedding medium, there are many of these tunnel channels. These *dephasing* relaxation processes of the single electrons are then frequent and the collective Mie excitation in the cluster fades away. Of course, the relaxation velocity due to this process depends on chemical details of the regarded cluster/matrix system (mainly the position and width of adsorbate states and the tunnel barrier). As will be shown in Section VI, our embedding experiments point to ultrashort phase relaxation times for many different embedding substances. They are of the order of 10 fs and hence shorter by various orders of magnitude than the according *energy* relaxation times in Ag clusters, which have been recently measured successfully by femtosecond spectroscopy.[20]

In order to describe this merely qualitatively introduced effect, which we call *chemical interface damping*,[4] the size-dependent band width of Eq. 12 is *formally* extended by an additional interface term. This procedure finds some justification by the theory of Persson which will be outlined below. Now we assume the **A** parameter to consist of two additive contributions:

$$\mathbf{A} = \mathbf{A}_{\text{size}} + \mathbf{A}_{\text{interface}} \tag{16}$$

This **A** parameter expresses both the strengths of the size effects and the chemical interface damping, i.e. the effect of adsorbate-induced tunneling. For one fixed cluster size R, the magnitude of **A** directly characterizes the probability of temporal charge transfer, the number of contributing adsorbate atoms, the involved adsorbate states, and their density of states, hence, important properties of the adsorption.

In Section VI we shall interpret our embedding experiments by numerically determining **A** parameters from the measured Mie resonance spectra (see Table 3 of Section VIII). Of course, this is only a first and rough step towards better understanding of the cluster–matrix interaction within the interface region.

C. The Theory of Persson

In a series of papers, Persson et al.[21,22] developed a quantitative model of the magnitude of the **A** parameter for a given energetic position and density of electronic adsorbate states. Here, as before, adsorbates also include liquid and solid embedding media.

The model system is a jellium sphere with foreign atoms at the surface (adsorbates, chemisorbates, substrates, liquid or solid embedding media). A value of $\mathbf{A}_{\text{size}} = 0.25$, as calculated earlier[14] from the quantum size effect, was presumed. Then, the (complex) polarization of the cluster as a whole due to the excitation of a Mie plasmon is written down as,

$$\alpha(\omega) = \frac{1-\varepsilon_{\text{matrix}}}{1+2\varepsilon_{\text{matrix}}} R^3 + \frac{3\varepsilon_{\text{matrix}}}{1+2\varepsilon_{\text{matrix}}} R^3 \frac{\Omega^2}{\Omega^2 - \omega^2 - \Sigma(\omega)},$$

$$\text{with } \Omega = \frac{\omega_p}{\sqrt{2\varepsilon_{\text{matrix}} + 1}} \tag{17}$$

where $\Sigma(\omega)$ is the *self-energy*, the real part and the imaginary part of which determine peak shifts and additional bandwidths, respectively. The bandwidths, in turn, include the **A** parameter:

$$\text{Metal electron relaxation} \quad \gamma(R) = \gamma_{\text{bulk}} + \mathbf{A} v_{\text{Fermi}}/R$$
$$\text{Mie resonance width} \quad \Gamma(R) = \Gamma_0 + \mathbf{A}^* \, v_{\text{Fermi}}/R \tag{18}$$

where $\mathbf{A}\, v_{\text{Fermi}}/R \approx \text{Im}\Sigma(\Omega)/\Omega$ and $\mathbf{A}^* \approx \mathbf{A}\,(2\,\omega_p^2/\omega_s^3)(d\varepsilon_1/d\omega)^{-1}$. The adsorbate density of states is modeled as shown in Figure 11.

The electric (near field) field vector *in* the cluster is decomposed into components *parallel* and *normal* to the surface tangential plane, and the same holds for the according current densities. In the first step, the power absorption of the parallel current component by adatoms is calculated via a friction force $F = M\eta v$ with M = adsorbate mass, η = the friction coefficient, and v the electron drift velocity parallel to the surface. The real part $\text{Re}\{\eta(\omega)\}$ proved to be proportional to $\sigma^{\text{diff}}(\omega)$, the diffuse scattering cross section of one adatom which was calculated earlier.[23] From the power absorption of N_A adatoms, the change of the absorption band width is then determined with Eq. 17. As a result,

$$\Gamma_{\text{parallel}}(R) - \Gamma_0 = \text{const} \cdot v_{\text{Fermi}}\,(1/R)\,(n_A/n_e)\,J_{\text{parallel}}(\rho_A, E_{\text{Fermi}}, \hbar\omega) \tag{19}$$

where n_A, n_e are densities of surface adsorbates and electrons, respectively. The function J_{parallel} is plotted in Figure 12a. The definition of the adatom density of states ρ_A follows from Figure 11.

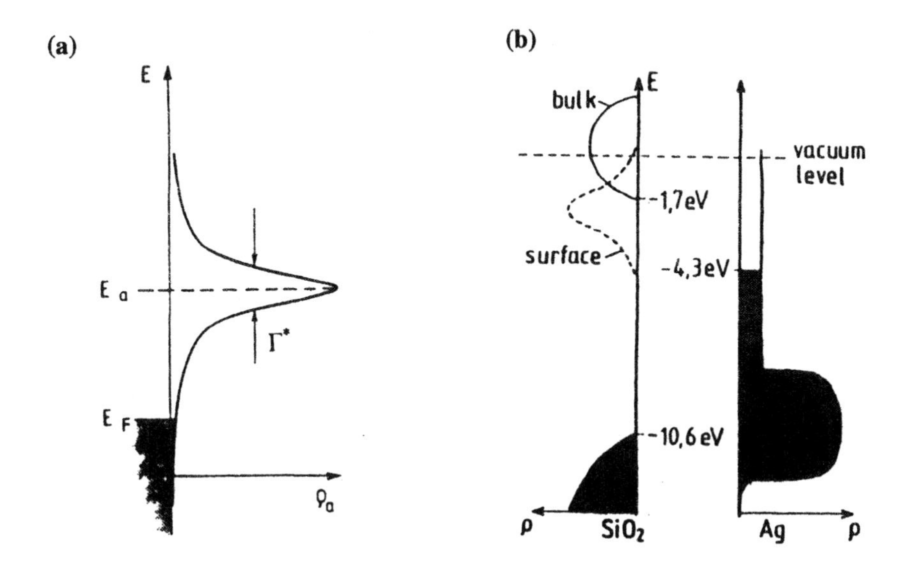

Figure 11. (**a**) Adsorbate induced resonance state centered at the energy $E = E_a$ and with the width Γ^* (After ref. 21). (**b**) Left : the density of states of compact SiO_2 as embedding medium of an Ag cluster with the density of states shown right. Black: occupied states. (After ref. 21).

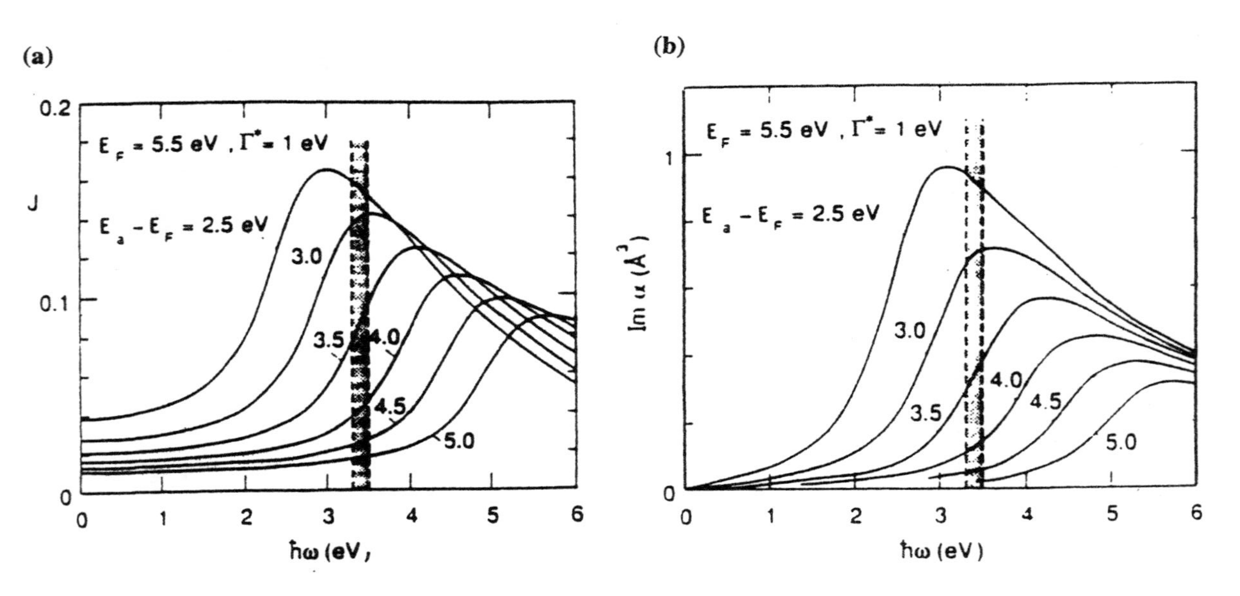

Figure 12. (**a**) and (**b**): The functions $J_{parallel}(\omega)$ and $Im\{\alpha_A(\omega)\}$, respectively. The parameters are given as insets. (After ref. 21).

In the second step, the corresponding contributions of the field and current components *normal* to the surface tangential plane were determined. A similar expression was obtained,

$$\Gamma_{normal}(R) - \Gamma_0 = \text{const} \cdot (1 + 2\varepsilon_{matrix})^{-1} (n_A/R)\, d^2 \cdot J_{normal}(\rho_A, E_{Fermi}\, \hbar\omega) \quad (20)$$

where $J_{normal}(\omega) \propto \text{Im}\{\alpha_A(\omega)\}$, $\alpha_A(\omega)$ being the polarizability of the adsorbate atom[22] and $\text{Im}\{\alpha_A(\omega)\}$ describing tunneling of cluster electrons into or out of adsorbate states, i.e. the *dynamic charge transfer*. d is the distance between the center of the adsorbate orbital and the image plane.[23] $\text{Im}\{\alpha_A(\omega)\}$ is given in Figure 12b.

Combining both calculated bandwidth contributions which are present simultaneously due to the spherical cluster shape, the total band width is obtained and therefrom the total **A** parameter:

$$\Gamma_{total}(R) \approx \Gamma_{parallel}(R) + \Gamma_{normal}(R) + \Gamma_0 + \Gamma_{size}(R) \quad (21)$$

$$\mathbf{A} \approx \mathbf{A}_{parallel} + \mathbf{A}_{normal} + \mathbf{A}_{size} \quad (22)$$

In principle, the according determination of $\text{Re}\{\Sigma(\omega)\}$ would give information about corresponding *peak shifts*, yet this was not performed by the authors.

Persson evaluated numerical values of the **A** parameter for several matrix materials, experimental results of which are shown in Figure 7. They are compiled in Table 3 at the end of this chapter, together with experimental results for the same matrix materials but with clusters of silver replacing the jellium. The correspondence is impressive.

VI. EXPERIMENTAL INVESTIGATIONS OF INTERFACE EFFECTS

A. Cluster–Matter Production

In contrast to the experiments on size effects noted in Figure 7, the investigation of interface effects requires clusters with sizes that can be kept constant for all embedding experiments in order to allow quantitative comparisons. Our experiments, some results of which will be presented in the following, were thus performed with a cluster source generating clusters of constant size with narrow size distribution. Due to the limited possibility of *reproducing* identical or even similar cluster–matter systems, an additional condition was posed that clusters should be produced with clean, *uncontaminated* surfaces, and only after a thorough characterization are they embedded in a matrix material. This insures that *changes* due to interface effects can be separated and calibrated. The requirement of reference experiments with the *same* clusters but with free, uncovered surfaces, excluded all chemical sample preparation methods; instead the production and manipulation of clusters in vacuum had to be chosen.

Experimental conditions to be fulfilled, were:

1. Production of a free beam of silver clusters in UHV of fixed mean size and moderate size distribution.
2. Number density of the clusters in the beam high enough to allow optical absorption spectroscopy of the free, uncovered clusters.
3. In-situ embedding of the clusters in various matrix materials to produce cluster–matter samples.
4. In-situ spectroscopy of the optical absorption of these cluster–matter samples.
5. Determination of sizes and size distributions by TEM.

Figure 13 gives a schematic overview of the cluster machine developed for these purposes.[4] The right hand part is the cluster source. Metal vapor pressures of about 0.5 bar are obtained thermally. Silver is a high melting material, so high temperatures of about 2300 K are required. Ar gas of up to 10 bar stagnation pressure acts as a seeding gas. The technical source parameters were chosen with the help of Hagena's criteria.[24]

The beam is produced by ultrasonic expansion. It is cooled by the carrier gas. This process is supported by a beam collimator, parts of which operating at 16 and 40 K, respectively, to remove the Ar gas before the cluster beam enters the experiment chamber. The beam then passes through small apertures into the UHV experiment chamber (left hand part of Figure 13). The cluster temperature is not known, but definitely is below the cluster melting temperature; we estimate that cluster temperatures are around room temperature. Beam densities in the experiment chamber correspond to an equivalent evaporation rate of 50 to 230 nm/s at a distance of 0.3 m from the nozzle. This exorbitantly high value is a prerequisite for the optical experiments in the free beam. The gas dynamic velocity was estimated to be $\approx 1.5\ 10^3$ m/s. After passing through the optical spectroscopy system, consisting of an optical multireflection arrangement and a parallel recording diode array spectrometer of high sensitivity, the beam can be deposited upon some substrate. We usually chose quartz glass slides.

We carefully analyzed the cluster shapes after deposition on this substrate by grazing incidence polarized light.[25] The results clearly demonstrate that only slight plastic deformations of the cluster occur during the collision with the substrate: The resulting axial ratio of the emerging spheriodal (pancake) shapes is about 0.8. This proved also that the relatively high beam velocity does *not* impede the embedding procedure by cluster damage.

The cluster topologies were, as a routine, controlled and characterized from TEM samples produced in situ on carbon foils. Figure 14 gives an example. The cluster size is about $2R \approx 2$ to 2.5 nm, i.e. the silver clusters consist of about 2.5 to 5×10^2

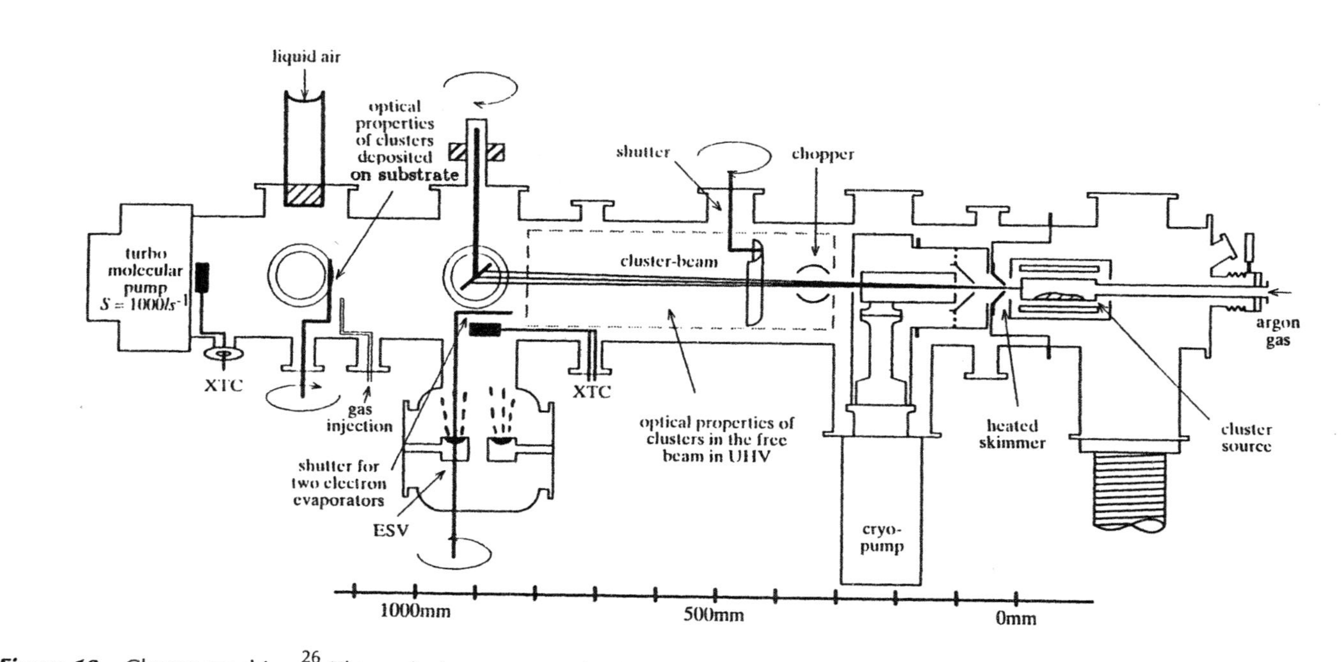

Figure 13. Cluster machine:[26] Thermal cluster source (*right*) producing a supersonic, gas seeded cluster beam. In the experiment chamber (*left*) the clusters are deposited on substrates or embedded, by coevaporation of some matrix materials. The optical extinction spectra of the free beam, of the deposited and the embedded clusters are measured in situ.

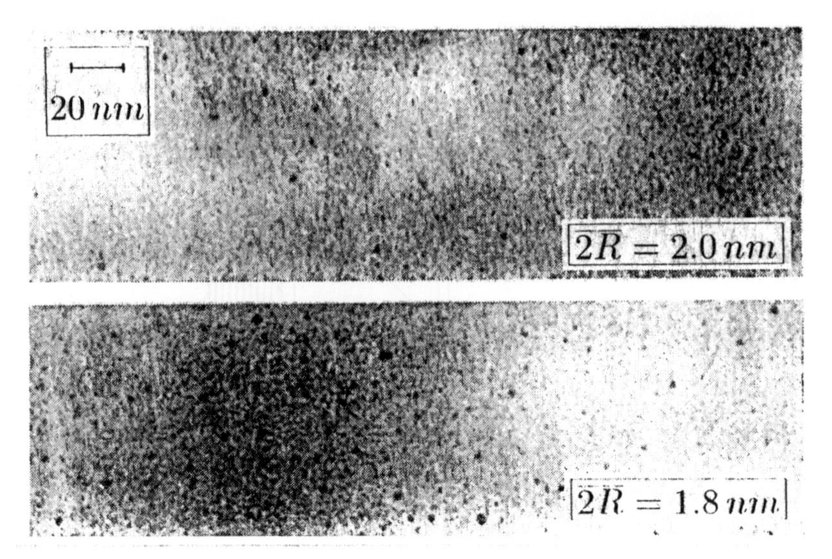

Figure 14. Transmission electron micrographs of silver clusters deposited on carbon foil. (Philips 400 T; 80 keV).

atoms (and 2.5 to 5×10^2 conduction electrons) each. The reasons why just silver was selected as cluster material were given above. Attempts are presently being made to change to other cluster metals.

To embed the clusters, various matrix materials were chosen to be codeposited simultaneously onto the substrate by electron beam evaporation. A broad variety of embedding materials was investigated. The following examples and the corresponding discussions give an overview and also demonstrate extreme cases.

B. Optical Spectra of Free, Supported, and Embedded Silver Clusters

In the following sections various experimental spectra are compiled. The first ones, with SiO_2 as substrate and as embedding material will be described in more detail than the following cluster–matter systems. In a first step and in accordance with the theoretical considerations of Section V we mainly concentrate upon the discussion of the observed energetic positions and widths of the dipolar Mie plasmon.

SiO_2 Substrate and Matrix

Figure 15 shows recorded absorption spectra of (a) the free silver clusters, (b) the clusters after deposition on a SiO_2 glass slide, and (c) after embedding in compact SiO_x (with x close to 2).[4,27] The *free* particles exhibit a surprisingly sharp Mie resonance, the width of which is described by the **A** parameter value of $\mathbf{A}_{size} = 0.25$, i.e. exactly the value calculated by Zaremba and Persson for the quantum size effect

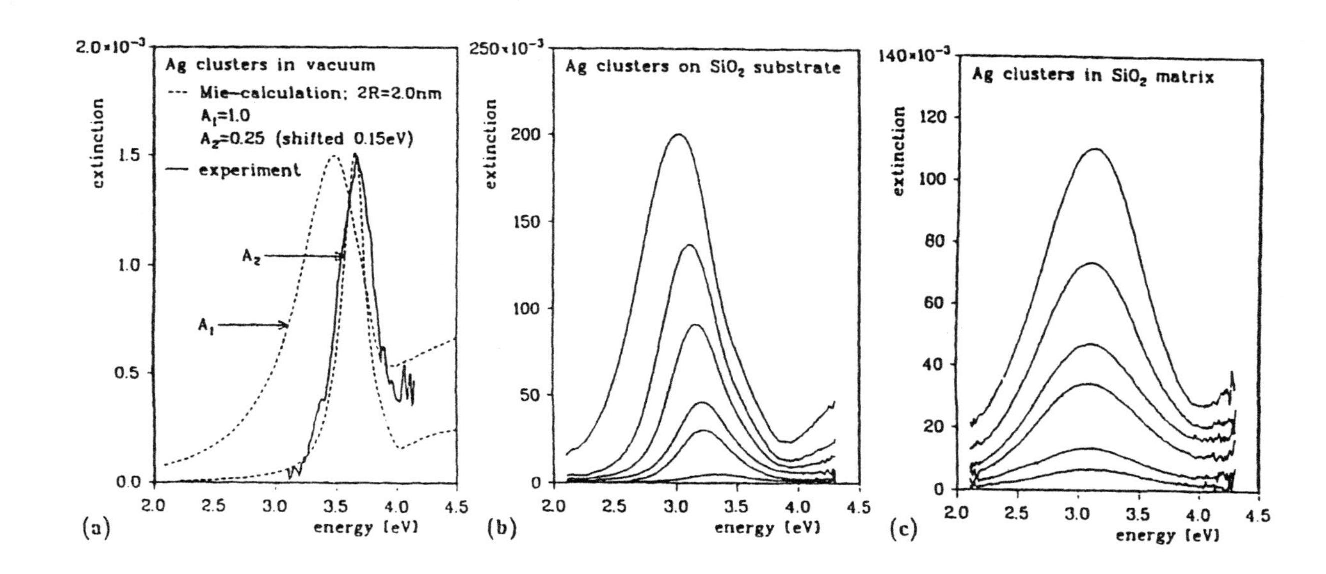

Figure 15. Absorption spectra of silver clusters ($2R \approx 2$ nm), (**a**) in the free beam, (**b**) deposited upon quartz glass substrate, and (**c**) embedded in SiO_x ($x \approx 2$). In (**a**) Mie calculations are added; the parameters are given as insets. In (**b**) and (**c**) from bottom to top: spectra for increasing coverage and sample thickness, respectively.

mentioned before. Figure 15a shows, in addition, some simulated Mie curves. (Their evaluation indicate interband excitations which clearly differ from the bulk behavior. Thus, previous results were confirmed which point to size-dependent interband transitions and the breakdown of the band structure in this size region. We will not treat this size effect here and only point to the literature.[2,6])

The spectra are drastically shifted and broadened after *deposition* on the SiO_2 slab (Figure 15b). In fact, now an essential part of the cluster surface is covered by foreign material. The shift of –0.4 eV is mainly due to the *dielectric* or *electromagnetic* effect described above. Because of the strong anisotropy of the cluster surroundings in this case, the shift can only be roughly estimated.[25]

Figure 15b contains spectra of different cluster packing densities which demonstrate that already at coverages larger than 5 % (!) cluster–cluster interactions lead to additional red shifts. The resulting **A** parameter now contains an additional interface contribution of $\mathbf{A}_{\text{interface}} \approx 0.5$. Even more drastic are the observed changes when the clusters are fully *embedded* in the SiO_2 matrix (Figure 15c). The red shift has increased to –0.55 eV and the according interface effect contribution to the **A** parameter amounts to $\mathbf{A}_{\text{interface}} \approx 1$. Several spectra for different filling factors and sample thicknesses are shown: due to the matrix high extinction values can now be obtained before cluster–cluster interactions cause the peak to shift.

All spectra were recorded against reference samples of the pure matrix, which were produced simultaneously with the samples at the same substrate but withholding the cluster beam. Since the dielectric functions of the evaporated matrix materials, which are required for the Mie calculations, were directly determined by measuring the transmittance and the reflectance of these reference samples, effects of film topology, stoichiometry, etc. which are the same in sample and reference were taken into account in the evaluations.

An open question is: to which extent is the bandwidth observed from the macroscopic cluster–matter sample really attributed to the *homogeneous linewidth* of the single cluster. One might argue that, e.g., non-homogenities of the matrix in the surroundings of the cluster give rise to dielectric shifts of differing amounts in different clusters and, hence, the measured band as the envelope of the contributions of all clusters is *non-homogeneously* broadened. In fact, an argument *pro* is that during the evaporation of the SiO_2 its stoichiometry is lost by slight diminuation of the O content, thereby changing the peak positions. As an argument *contra*, silver clusters produced in glass melts[12] only yielded moderately smaller **A** values ($\mathbf{A}_{\text{interface}} \approx 0.75$) and almost corresponding peak positions for the same cluster sizes. So, we preliminarily state that essential parts of the observed Mie resonance bandwidth are due to the single cluster and can be explained from microscopical models.

As a consequence we ascribe the $\mathbf{A}_{\text{interface}}$ parameter values, evaluated from the cluster matter spectra, to the limited *phase lifetime* of the plasmon excitation in the individual cluster, as was discussed in the last section. The result is a surprisingly low lifetime of the order of 10^1 fs for the free cluster which is reduced to several fs after embedding in matrix materials. This will be discussed in Section VII.

Lifetimes of this duration in clusters could up to now only be measured once by femtosecond spectroscopy.[20c] Other femtosecond experiments,[20] which will be discussed in Section VII, yielded relaxation times of the order of ps and more. The apparent discrepancy to our results falls by considering that these experiments measured *energy relaxation times*. In the experiments cited above, electromagnetic radiation with frequencies beyond the interband transition edge was applied, hence, mainly electron-hole pair excitations took place, the relaxation of which influenced the Mie resonance. In contrast, we measure, as described in Sections VI and VII, the *phase relaxation time*, which, as also known from semiconductors, is essentially smaller.

A very recent additional investigation[28] concerning peak positions of non-stoichiometric silicon oxide yielded the results presented in Figure 16: here, Si doped with oxygen, Si:O, was produced by evaporating Si in presence of low partial

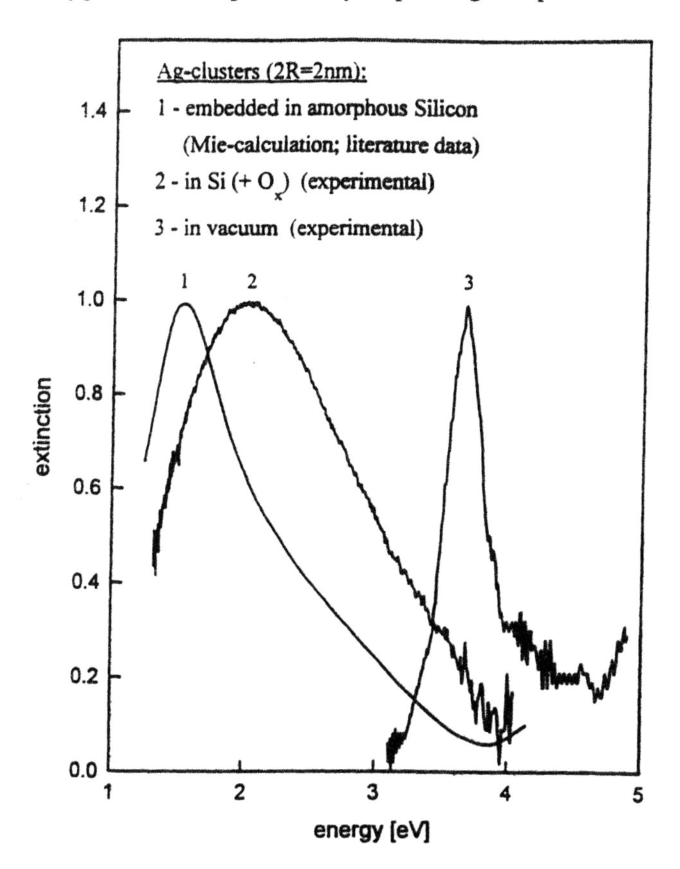

Figure 16. Optical extinction of silver clusters ($2R \approx 2$ nm). The spectra were measured in the free beam (*right*), in Si with small amounts of oxygen (*center*), and are calculated for pure amorphous Si (*left*).

pressure of oxygen in the rest gas of the vacuum system. So, x is now essentially smaller than 2, even than 1.

The figure shows a spectrum of the free silver cluster absorption (right), the calculated Mie spectrum for Ag clusters in pure amorphous Si (left), and the measured spectrum of the weakly oxidized Si matrix.[28] Compared to the free clusters, the measured peak position is lowered for about 1.5 eV; it is the largest shift we ever investigated and should even amount to about 2 eV for the pure Si matrix! The calibration of the oxygen content is now underway. As a consequence, the colors of these samples are shifted all over the visible spectrum.

Other Oxidic Matrix Materials

In the course of an extended research program,[28] the same 2 nm silver clusters were embedded in various dielectric metal oxides. The matrix materials all were evaporated by e-beam and codeposited together with the cluster beam on quartz glass substrates. The optical absorption spectra were recorded in-situ as described above.

In Figure 17 several examples are compiled. For comparison, a measured spectrum of the free, uncontaminated clusters is shown again. Both band position and width of the Mie plasmon prove to be dramatically influenced by the different interfaces, in several cases much stronger than with SiO_2. As a consequence, the colors of these samples are shifted all over the visible spectrum. These changes, shown in Figure 18, are far beyond the predictions of Mie's theory: from classical Mie there should be *no* increase of the bandwidth, i.e. **A** = 0, and the shifts should

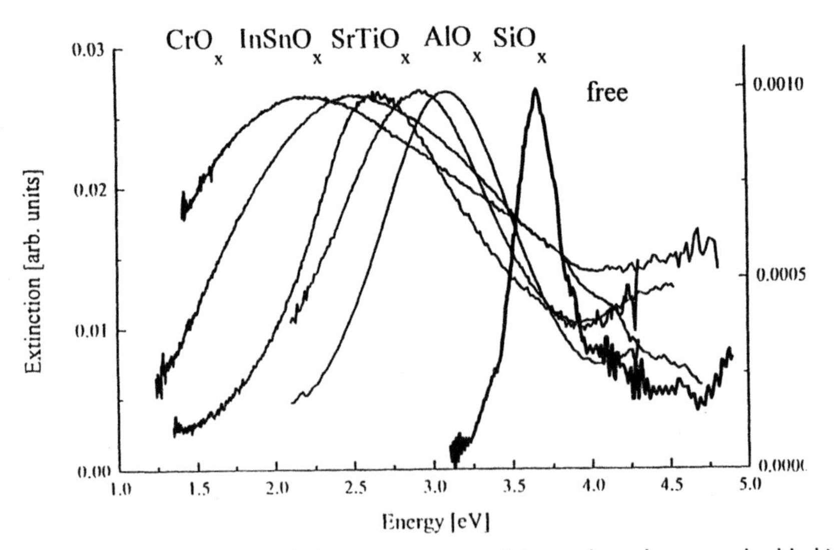

Figure 17. Measured optical absorption spectra of 2 nm silver clusters embedded in various oxidic embedding materials.[28] For comparison, a spectrum of free clusters is added (right-hand scale). (x is close to the compounds SiO_2, Al_2O_3, $SrTiO_3$, CrO_2).

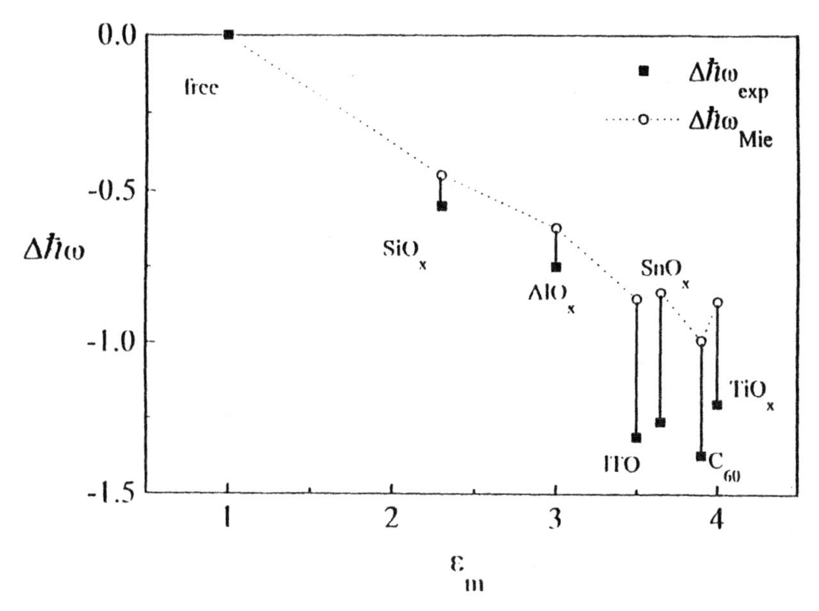

Figure 18. Measured Mie peak shifts of silver clusters embedded in various solid matrix materials. The shifts are plotted against the dielectric constant of the matrix material. Open circles and dashed line indicate the *dielectric shift* following from Mie's theory; the deviations of the experimental data are assumed to be due to *static charge transfer* processes.

be restricted to the *dielectric* ones (which, however, give the essential contribution as shown by the open circles and the dashed line in Figure 18). A summarizing discussion will be published later.

Silver Clusters in Solid Triphenylphosphine

A milestone in the development of metal clusters was the invention by Schmid[29] to precipitate chemically stoichiometric metal-organic compounds consisting of a core of metal atoms (Au, Pd, Pt, Ru, Rh, etc.) surrounded by (few) stabilizing, nonlinear organic molecules. One of these organic components is $P(C_6H_5)_3$. To a large extent, these clusters are monodisperse—their numbers of atoms following the sequence of geometric magic numbers (13, 55, 147, 309, 561...).[2] Later, similar monodisperse Au clusters were produced with linear stabilizing organics.[30]

The advantage of these clusters to be almost monodisperse in macroscopic amounts is accompanied by the drawback of the existence of a stabilizing organic shell with strong chemical bonds towards the clusters, which were expected to influence strongly the electronic and optical properties of the latter. This influence was observed as demonstrated in Figure 19.[28] The figure shows absorption spectra of our silver clusters (a) deposited on quartz glass, (b) deposited on triphenyl-phosphine covered quartz glass, and (c) embedded in triphenyl-phosphine. Obvi-

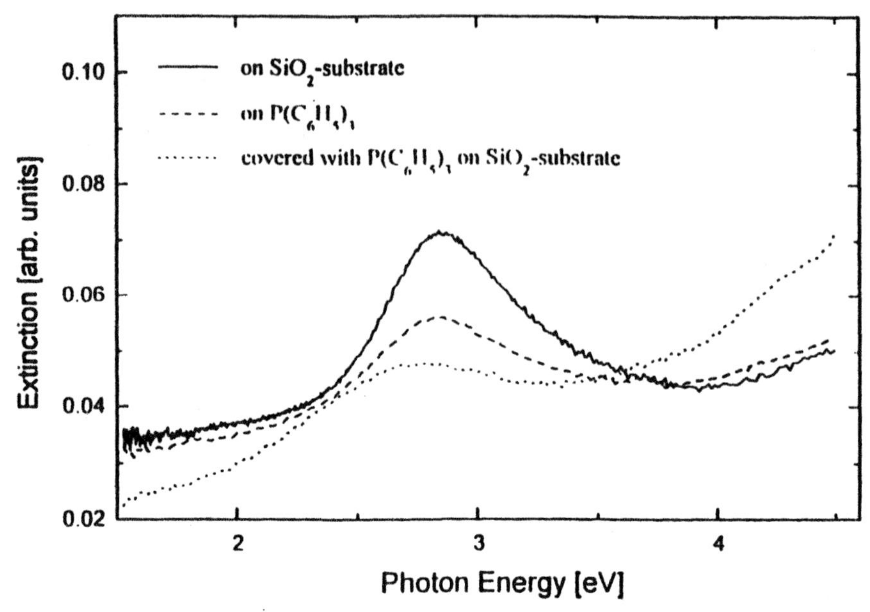

Figure 19. Measured absorption spectra of 2 nm silver clusters. From top to bottom: Clusters deposited on quartz glass slide, clusters on triphenylphosphine-covered slide, clusters embedded in triphenylphosphine.

ously, the phosphine bonds contribute a strong additional broadening which renders the total **A** parameter value to increase from $\mathbf{A} \approx 0.5$ at the clean quartz substrate to $\mathbf{A} \approx 2$. Since we know the $\mathbf{A}_{size}$ contribution to amount to 0.25, there is an interface contribution of $\mathbf{A}_{interface} \approx 1.7$, i.e. drastic chemical interface damping. The corresponding shift amounts to about –0.1 eV.

These results confirm the statement made in Section I that in cluster-matter the appropriate building units are the *dressed* clusters rather than their metallic cores only.

Silver Clusters in ITO

Mixed indium–tin oxide (ITO) is frequently used for practical purposes due to its uncommon combination of high electric conductivity and high transparency in the optical region. It contains mainly In_2O_3, with SnO_2 in the order of 10%. When deposited by e-beam evaporation in a vacuum, the composition is usually violated (mainly, by oxygen losses) changing the ITO into an optically absorbing material with lower electrical conductivity.[31] The desired properties can be regained by annealing the films in a vacuum under low oxygen pressure.

It was expected that metal clusters embedded in ITO would exhibit especially large spill-out effects and we therefore performed an extended investigation of this cluster-matter system. Figure 20 compiles some results which reveal a surprisingly

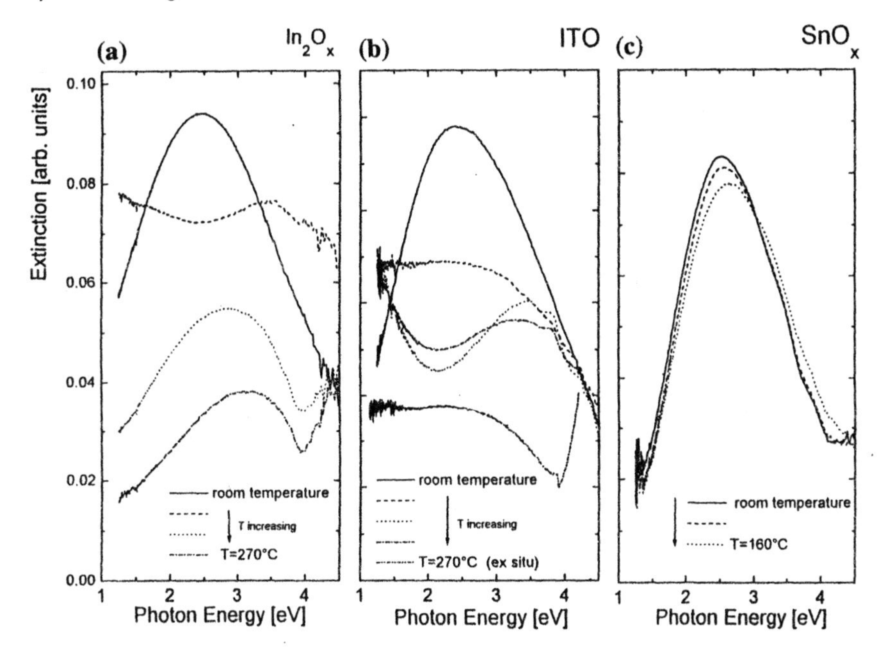

Figure 20. Measured absorption spectra of silver clusters in ITO matrix (**b**), and, for comparisons, in matrices of In oxide (**a**) and Sn oxide (**c**). The different spectra were taken after varying amounts of annealing and measured against cluster-free reference samples.

complex behavior.[32] The figure in the center shows the extremely broad spectrum ($\mathbf{A}_{total} \approx 1.5$) of our silver clusters in as-evaporated ITO and after increasing the amounts of annealing in situ. The as-evaporated Mie band is red shifted for about 0.4 eV against the theoretical spectrum. (All calculations in this section were performed, as described above, with optical dielectric functions for the matrix materials measured for that very purpose from analogous, silver-free specimens.) In addition, the *absorption minimum* at about 4 eV, which is characteristic for silver and indicates the onset of interband transitions, is absent. Hence, the existence of silver clusters is to be doubted, and the formation of some alloy may be regarded. The existence of Ag_2O in the as-deposited spectrum is rather improbable since the volume oxidation of silver usually takes place only above 600 K. In addition, previous experiments on the oxidation at 300 K of our silver clusters deposited on quartz glass substrates resulted in only tiny changes of the Mie band which point to only weak surface oxidation.[27] This will be discussed in the next section.

At the highest annealing temperature (540 K), the spectrum in the central figure no longer exhibits any similarity to the silver plasmon band. Instead, there is a high-frequency peak at about 3.7 eV and another peak below the lower spectral limit of our experimental equipment of 1.2 eV. In order to identify this complex behavior, the analysis was extended by embedding the silver clusters in a pure SnO_x ($x \lesssim 2$) matrix (Figure 20, right) and in a pure InO_x ($x \lesssim 1.5$) matrix (Figure 20, left), respectively. The obtained spectra point to the following interpretation:[32]

1. Annealing Ag clusters in a SnO_x matrix yields only slight changes of the spectra. The interband minimum of silver is not well resolved. The red shift against the theoretical spectra amounts to only 0.27 eV. Hence, the formation of Ag–Sn alloys cannot explain the observed spectra in ITO.
2. The similarity of the spectra of the clusters embedded in InO_x and in ITO point to the formation of Ag_xIn_y alloys. First, as in the silver/ITO system the interband edge of silver is absent in InO_x. Second, the appearance of a two-peak spectrum at elevated temperature is similar in both samples.
3. As a peculiarity of Ag–In alloys, it is documented in the literature[33] that the *inner oxidation*, taking place at 1000 K under oxygen pressure of 0.1 bar, is converted into the *external oxidation* by lowering to 700 K and 10^{-4} mbar. The latter process means that the In content of the alloy is reduced by extricating In_2O_3 through the cluster surface. Hence, the Ag content of the alloyed clusters would be increased. In fact, the lower spectra in (Figure 20, left) exhibit the interband transition edge of silver already after moderate annealing. The alloying of silver clusters in an In_2O_3 matrix appears to be not completely removed by moderate annealing under O_2 pressure of 10^{-4} mbar. Similar behavior of the silver clusters in ITO was observed (see the lower spectra of Figure 20b).

This extremely complex behavior indicates that chemical reactions between silver and ITO are present around room temperature, that they increase towards higher temperature and that the alloying can be reduced by annealing under oxygen. However, we estimate this interpretation only as preliminary since, as a second effect, these samples showed that the optical clearing up of the matrix material during the annealing process is different in the sample and the reference. Hence, the subtraction of the reference spectra may cause contributions in the spectra of Figure 20 which stem from differences of the matrix material rather than from the absorption of the clusters, only.

Silver Clusters in Solid C_{60}

C_{60} of our own production was used as matrix material by sublimating it onto the quartz glass substrate which was kept at room temperature. The optical spectra, obtained with embedded Ag clusters with mean size of 5×10^2 atoms, which are shown in Figure 21, point again to strong static and dynamic charge transfer at the interface.[5,35] Detailed simulations of this experiment, based upon Mie's theory, indicate a static electron transfer from the silver towards the C_{60} of not less than 20% of the conduction electrons in the cluster! The number of shifted electrons

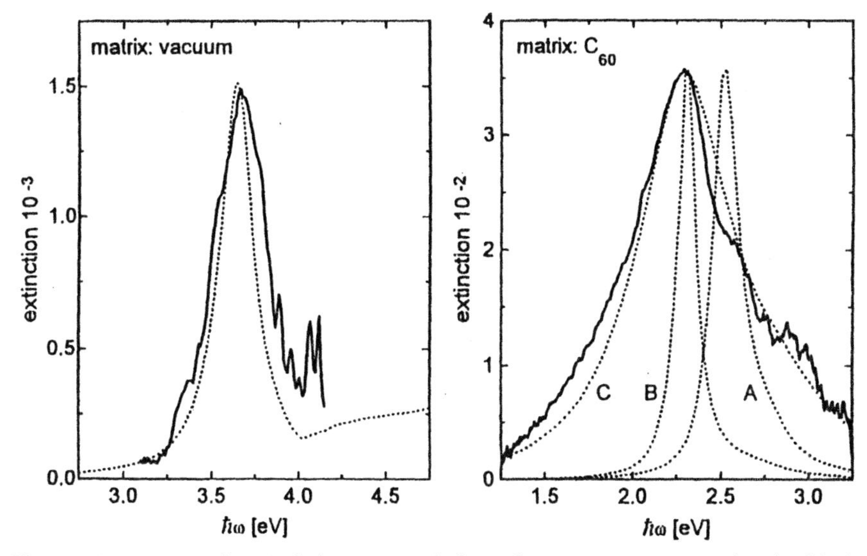

Figure 21. Measured optical absorption of silver clusters in vacuum and embedded in solid C_{60} film. For analysis, Mie-spectra were calculated (**A**) with bulk $\varepsilon(\omega)$, (**B**) with conduction electron density reduced for 20 %, (**C**) as (**B**) but with **A** = 1. Reference: same film without clusters.

corresponds with the number of C_{60} molecules in *direct* contact with the cluster surface, hence, we state the transport of *one* electron per directly touching C_{60} into the LUMO of the molecule. This is in correspondence with recent PES and EELS investigations.[36]

Concerning the dynamical charge transfer, Figure 21 shows that an **A** parameter of $\mathbf{A}_{total} \approx 1$ fits quantitatively the measured spectrum. Applying the proposed interpretation of the optical spectra, several interesting conclusions are induced:

1. A strong electric double layer at the interface, as also known from colloidal systems, is formed, storing a high amount of Coulomb energy.
2. The occupation of the 5sp electron levels in Ag (i.e. the conduction electron band) is essentially reduced, hence the Fermi level is lowered. Thus the effects derived from manipulations of the electron density, which are usually ascribed only to semiconductors, can also be obtained in metals. Since the optical interband transition edge in silver involves the Fermi level (transitions $4d \rightarrow 5sp(E_{Fermi})$ close to X and L in the Brillouin zone), a red shift of this edge is expected and, consequently, also a red shift of the Mie peak.[6,26]
3. The C_{60} molecules directly covering the cluster surface are electrically charged with *one* electron. It is expected that this C_{60}^{-} shell is electrically conductive. In samples of high cluster concentrations, percolation of these C_{60}^{-} *shells* should occur and induce strong increase of electric conductivity of the sample. Further experiments on this phenomenon are in progress.

Recent efforts[34] to perform optical spectroscopy of Ag/C_{60} samples at higher temperatures up to 550 K indicated migration of the whole clusters and strong Ostwald ripening through the C_{60} film which favors the growth of larger silver clusters. The bond strengths between C_{60}^{-} and clusters appear to exceed those between neutral C_{60} molecules, probably due to additional ionic contributions, hence after intense annealing, extended islands of coagulated and coalesced clusters and adhering C_{60}^{-} are separated by regions where the C_{60}^{0} has been reevaporated.

Chemical Reactions at the Free Cluster Surface

Attempts to observe chemical reactions of clusters with gases in the free cluster beam were not convincing up to now. The reason is that at gas pressures above 10^{-3} mbar the free beam is rendered diffuse due to cluster–gas collisions, and, on the other hand, well below this pressure the reaction probabilities of reactive gases with the cluster are too small because of the short lifetime of about 1 ms of the clusters flying at 1 km/s through the experiment chamber.

However, clusters deposited on quartz substrates also exhibit free, uncontaminated surface regions, where the addition of reactive gases can initiate chemical surface reactions. Here, chemical reactions at the cluster surface can take place and can be observed without lifetime limitations.

At the end of this compilation of our recent experimental results, two examples concerning different kinds of chemical reactions at the surface of such deposited silver clusters are depicted in Figure 22. Among many reactive gases, which in the course of our present research project were introduced into the vacuum of the experiment chamber, we chose oxygen and sulfur gas as examples to show optical spectra which were recorded *during* the chemical reactions. These spectra are plotted in Figure 22a and b for the two gases. In both cases, the peak position and the width are influenced increasingly with time. Hence, the Mie resonance proves also to be a highly sensitive sensor of chemical interface reactions.

In the case of oxygen, the effects remain small even at high partial pressure of the gas. This clearly points to a weak reaction. It is well known that at room temperature planar silver surfaces only bind oxygen chemically at imperfections.[37] At elevated temperatures the oxidation is increased. Since cluster surfaces to a large extent exhibit atomic edge and corner positions as pointed out in Section II, oxygen will find binding positions at the surface, hence we attribute the observed changes in the spectra to weak surface oxidation of the silver clusters.

The drastically varying spectra taken during the interaction with free sulphur indicate an essentially different kind of reaction: even at low partial pressure, the Mie band is continuously shifted and broadened until its total disappearance after 100 s. We interpret this behavior as due to the formation of some silver sulfides which, starting at the surface, proceeds through the whole particle and after 100 s no metallic silver is left. Since the newly formed sulfides are dielectric, a Mie resonance is not exhibited, hence the spectra disappear completely.

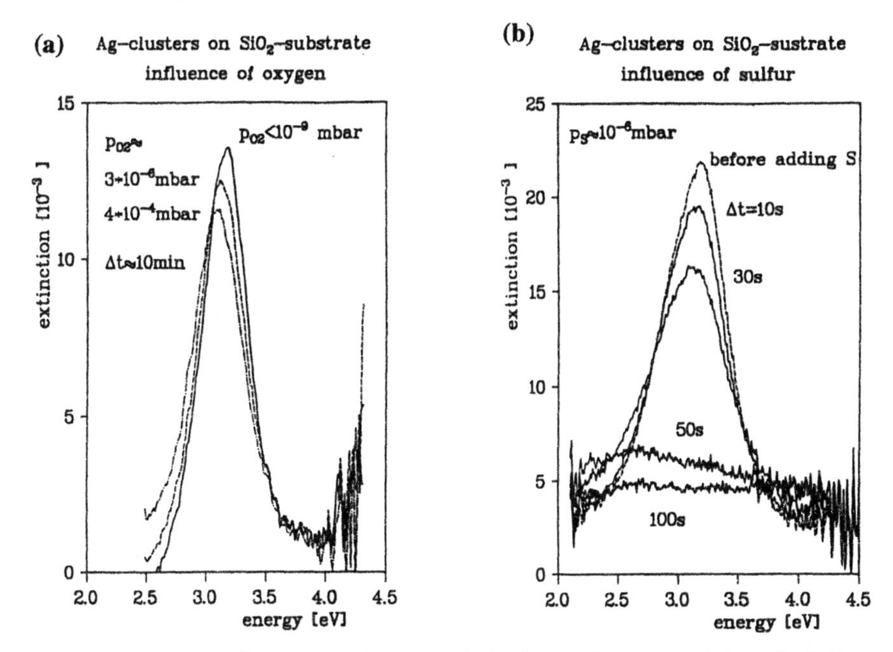

Figure 22. Chemical reactions of gases with the free surface parts of deposited silver clusters, demonstrated by the Mie resonance acting as sensitive sensor. (**a**) Weak surface oxydation, (**b**) strong volume reaction to form silver sulfides.

VII. COMPARISON WITH RELAXATION TIMES DETERMINED BY FEMTOSECOND SPECTROSCOPY

The recent development of femtosecond spectroscopy has given much impact to the experimental investigation of the dynamics of Mie resonances in silver and gold nanoparticles.

There are two groups of experiments which were performed: measurements of phase coherence and of single electron scattering.

A. Phase Coherence

In the direct measurement of the phase coherence of the plasma oscillation[20a,c] the cluster sample is used to create second harmonic radiation. Its efficiency is highly increased during the excitation of the Mie resonance by the involved enhancement of the *local electromagnetic field* in the cluster. The local field enhancement due to coherent electron excitation was described above. The according relaxation time τ_{Mie} is thus obtained by the autocorrelation of the local electromagnetic field E_{loc},[20c]

$$I_{2\omega}(\tau) = \frac{1}{2T}\int_{-T}^{+T} \left| \left| E_{\text{loc}}(t) + E_{\text{loc}}(t-\tau) \right|^2 \right|^2 dt \tag{23}$$

where T is the integration time of the photomultiplier. For evaporated silver island films with, roughly, 20 nm island size on ITO and GaAs substrates,[20a] the *relaxation time* was measured to 40 fs.

With lithographically produced regular arrays of silver islands of triangular (i.e. non-centrosymmetric) shapes and approximately 200 nm in diameter,[20c] the relaxation time of the field enhancement was found to amount to 10 ± 1 fs. In particles of this latter size, radiation contributes to a remarkable extent to the damping. In the according optical extinction spectra, the non-sphericity of the particle shapes and the anisotropy of the surrounding materials both increase the Mie bandwidths.

The order of magnitude of this measured relaxation corresponds with our optical measurements if we deduce lifetimes directly from the halfwidth of the measured Mie resonance band. For instance, from the bandwidth of our free 2 nm silver clusters of 0.25 eV we obtain a lifetime of 7 fs. Regarding the size-effect broadening contribution to the halfwidth, we can extrapolate from these optical results a lifetime of about 20 fs for 15 nm clusters. Particles of this size develop the smallest Mie resonance in the case of silver and, hence, the longest lifetime.

On the other hand, our 2 nm clusters exhibit broadening up to 1 eV halfwidth and larger when embedded in matrix materials. This corresponds to lifetimes of only 2 fs. Hence, the chemical damping effects dominate the dynamics of such cluster systems. As discussed before, non-homogeneous line broadening will, probably, contribute by essential parts, yet, the order of magnitude of τ will remain.

B. Single-Electron Scattering

The second kind of femtosecond spectroscopy is not devoted to the phase coherence but to the single-electron scattering processes involved in the excitation and relaxation of Mie resonances.[20b,d,e] (Their linear contributions in the bulk noble metals were treated much earlier; see e.g. ref. 2.) Typically, pump–probe experiments were performed by irradiating in the spectral region of interband transitions and measuring the irradiation-induced changes of the absorption in the Mie-peak maximum. The nonlinear broadening in 30 nm gold clusters was shown in ref. 20. With each pulse of very high energy, about 4×10^4 hot electrons and the corresponding number of holes were created in each particle by electron-hole excitations 5d - 6sp(E_{Fermi}) at ~ 3 eV. This heats the electrons to 4000 K. The time-dependent changes[20e] begin with a delay of less 1ps after the pump pulse until the nonlinear increase of broadening and decrease of absorption in the Mie maximum are established. This decrease recovers exponentially with two characteristic time constants: ≈4 ps and ≈200 ps.

Three scattering mechanisms are discussed which differ strongly by their time constants:

1. Electron–electron scattering, which equilibrates *electron energies and momenta* in the Fermi gas toward the Fermi distribution at elevated temperature, after some excitation event. This is the most rapid effect with time constants clearly below 200 fs (however, this could not be resolved by the authors). Electron–electron scattering is responsible for the fast rise of the nonlinear effect.
2. Electron–phonon interaction, which dissipates excess energy from the conduction electron system to the cluster atoms until thermal equilibrium of the whole cluster is obtained. This causes a fast component of decreasing nonlinearity, which is of the order of 1 ps.[20b] Electron–phonon interaction proved to depend sensitively on cluster size as shown for ligand-stabilized Au clusters of various sizes:[20d] 7 ps for 15 nm particles, 1 ps for Au_{55} (corresponding with bulk metal values), and 300 ps (dominating contribution) for Au_{13}. The authors interpret these size dependences by the transition from the molecular to the solid state at sizes between Au_{55} and Au_{13}.
3. The dissipation of excitation energy of the cluster into the cluster surroundings. This includes thermal radiation in the case of free particles and thermal conduction through the interface between cluster and surroundings in cluster-matter.

From these present results we can, tentatively, state the following sequence of relaxation time domains in our noble metal clusters:

- Phase relaxation (coherence dissipation), typically of the order of 10^1 fs.
- Electron-electron scattering, typically of the order of 10^2 fs to 1 ps.
- Electron-lattice energy relaxation, of the order of 1 to 10^2 ps.
- Cluster-surroundings energy relaxation, of the order of 10^2 ps.

Our optical experiments are sensitive to the electron–electron and electron–lattice scattering which are enclosed in $1/\tau_{bulk}$ and γ_{bulk} of Eqs. 11 and 18, respectively. As an example, measurements of the Mie absorption at different temperatures gave information about their temperature dependence (see ref. 2). Our optical experiments are, as shown, also sensitive to the phase relaxation and, in particular, to the additional contributions caused by chemical interface effects.

VIII. SUMMARY OF THE EXPERIMENTAL PART

The aim of the experiments and results concerning the optical properties of *dressed* metal clusters, presented here, is to gain detailed and many-sided insights about the complex interface region between clusters and the matrix material surrounding them in cluster matter. This is still a large "white" area and unconquered field on the map of cluster properties, despite its importance for practical purposes as, e.g., the heterogeneous catalysis and, extending to more general sample topologies, the construction of nanostructured devices.

It has been proven by the presented numerous examples that the optical Mie plasmon resonance acts as a highly sensitive sensor for the interface properties of dressed clusters. Data can be and were evaluated from the *deviations* of the measured plasmon band spectra from the predictions of Mie's theory. As a *model* system the cluster matter appears to be suited best, yet this method opens a novel way of analysis of complex interfaces in general. The method is based upon optical extinction/absorption spectroscopy, i.e. a technically quite simple kind of experiment. Yet, the availability of data from the free, *uncontaminated* surfaces of the *same* clusters proved to be an absolute "must" in order to have a reference for quantitative evaluations. This condition increases the technical expenditure to a remarkable extent and renders large-scale experiments as shown in Section VI; it excludes (to large extent) colloidal systems for the present investigations since undressed, free clusters cannot be easily produced by chemical means. (To avoid misinterpretation, of course, the present results are also of high relevance for a better understanding of colloidal systems because of their complex interface structures.)

We have chosen the position and the width of the Mie band as the direct experimental parameters to simplify evaluations. These parameters, however, carry the desired information only indirectly; the development of theoretical paths towards better understanding is in progress. At present, we restrict ourselves to show finally a summarizing list of **A** parameters (Table 3). Concerning the according peak shifts we point to the results given in the respective sections. We hope that it will be possible in near future to extract therefrom important properties of the interface layer.

Our presented results concern the *linear* response of complex cluster–matrix systems. As an important additional relaxation effect due to the interface, the linear dynamic charge transfer will have to be considered for *nonlinear* excitations as in femtosecond spectroscopy as well as its nonlinear extension.

Future aims in the frame of our present project are to obtain, quantitatively, characterizing features of the static and dynamic cluster–matrix interface effects including energetic positions and densities of state of adsorbate atoms/ions, static and dynamic charge transfer processes, and electric double layers in the extended interface regions. In addition, data on the resulting effects of the interface on the volume of the cluster would be desirable, such as effective electron densities, influences of the surroundings upon electronic excitations in the clusters and upon band structure features, upon the molecule–solid transitions or upon the cluster shape and atomic structure, etc. Future developments will determine the realistic capabilities of the outlined novel optical interface analysis of dressed clusters via Mie plasmon excitation.

Table 3. A-parameter Values Describing the Chemical Interface Effect[a]

	Cluster Surroundings	*ΔA Persson Theory*	*A Experiment*	*Reference*
Ag_N in	Solid Ne	~0	0.25	(16)
	Solid Ar	0.1	0.3	
	Solid O_2	0.3	0.5	
	Solid CO	0.8	0.9	
	Na-SiO_2 glass	1.0	1.0	(12)
Au_N in	Na-SiO_2 glass		1.5	(12)
Ag_N in	Ice		0.5	
	$+O_2$		0.5	given in the present paper
	SiO_2 substrate		0.7	
	Solid C_{60}		1.0	
	+S		1.2...1.7	
	SiO_x (~SiO_2)		1.3	
	$InSnO_x$ (ITO)		1.5	
	$SrTiO_x$		1.5	
	AlO_x (~Al_2O_3)		1.7	
	TiO_x (~TiO_2)		1.8	
	Triphenylphosphine		2	
	SbO_x		2	
	Si ($SiO_{x,x<<1}$)		3	
	CrO_x		3	

Note: [a]Compilation of numerical values of the A parameter for silver and gold clusters in contact with various surrounding materials. A(experiment) = A_{total}. The calculated values of Persson[21] are defined as $\Delta A = A_{total} - A_{free} = A_{total} - A_{size}$.

REFERENCES

1. Kreibig, U.; Genzel, L. *Surf. Sci.* **1985,** *156,* 678.
2. Kreibig, U.; Vollmer, M. *Optical Properties of Metal Clusters*; Springer Series in Materials Science; Springer: Berlin, 1995, Vol. 25.
3. Mie, G. *Ann. Phys.* **1908,** *25,* 377.
4. Hövel, H.; Fritz, S.; Hilger, A.; Kreibig, U.; Vollmer, M. *Phys. Rev. B* **1993,** *48,* 18148.
5. Kreibig, U.; Hilger, A.; Hövel, H.; Quinten, M. In *Large Cluster of Atoms and Molecules*; Martin, T., Ed.; Kluwer: Dordrecht, 1996; Kreibig, U.; Gartz, M.; Hilger, A.; Hövel, H. In *Fine Particle Science and Technology*; Pelizzetti, E., Ed.; Kluwer: Dordrecht, 1996; Kreibig, U.; Gartz, M.; Hilger, A. In *Nanoparticles in Solids and Solutions*; Fendler, J.; Dekany, I., Eds.; Kluwer: Dordrecht, 1996; Kreibig, U.; Gartz, M.; Hilger, A.; Hövel, H. In *Science and Technology of Atomically Engineered Materials*; Jena, P.; Khanna, S.; Rao, B., Eds.; World Scientific: Singapore, 1996.
6. Kreibig, U. In *Handbook of Optical Properties*; Hummel, R.; Wißmann, P., Eds.; CRC: Boca Raton, 1997.

7. Haberland, H. *Clusters of Atoms and Molecules*; Springer Series in Chemical Physics, Springer: Berlin, 1993, 1994, Vols. I,II.
8. Fritsche, H.; Benfield, R. *Z. Physik D* **1993**, *26*, 15.
9. Ekardt, W. *Phys. Rev. B* **1984**, *29*, 1558; *Phys. Rev. B* **1985**, *31*, 6360.
10. Liebsch, A. *Phys. Rev. B* **1993**, *48*, 11317.
11. Bassani, F.; Pastori Parravicini, G. *Electronic States and Optical Transitions in Solids*; Pergamon: Oxford, 1975.
12. Kreibig, U. *J. Phys. F: Met. Phys.* **1974**, *4*, 999.
13. Volotkin, Y.; Sinzig, J.; de Jongh, L.; Schmid, G.; Vargaftik, M.; Moiseev, I. *Nature* **1996**, *384*, 621.
14. Zaremba, E.; Persson, B. N. J. *Phys. Rev. B* **1987**, *35*, 596.
15. Kreibig, U. *Z. Physik B* **1978**, *31*, 39.
16. Charlé, K.-P.; Frank, F.; Schulze, W. *Berichte Bunsengesellschaft Phys. Chem.* **1984**, *88*, 350.
17. Sinzig, J.; Radtke, U.; Quinten, M.; Kreibig, U. *Z. Phys. D* **1993**, *26*, 242; Sinzig, J.; Quinten, M.; Kreibig, U. *Appl. Phys. A* **1994**, *58*, 157.
18. Hölzl, J.; Schulte, F.; Wagner, H. In *Solid Surface Physics*; Springer Tracts in Modern Physics; Springer: Berlin, 1979, Vol. 85.
19. Kreibig, U.; Fauth, K.; Granquist, C. G.; Schmid, G. *Z. Physik. Chem.* **1990**, *169*, 11; Kreibig, U.; Hilger, A.; Schmid, G.; Granquist, C. G. In *Science and Technology of Atomically Engineered Materials*; Jena, P.; Khanna, S.; Rao, B., Eds.; World Scientific: Singapore, 1996.
20. (a) Steinmüller-Nethel, D.; Höpfel, R.; Gornik, E.; Leitner, A.; Aussenegg, F. R. *Phys. Rev. Lett.* **1992**, *68*, 389; (b) Bigot, J. Y.; Merle, J.; Cregut, O.; Dannois, A. *Phys. Rev. Lett.* **1995**, *75*, 4702; Halté, V.; Merle, J.; Daunois, A.; Albrecht, M.; Bigot, J. Y. *Phys. Rev. Lett.*, to appear 1997; (c) Lamprecht, B.; Leitner, A.; Aussenegg, F. R. *Appl. Phys. B* **1997**, *64*, 269; (d) Smith, B.; Zhang, J.; Giebel, U.; Schmid, G. *Chem. Phys. Lett.*, accepted 1997; (e) Perner, M.; Bost, P.; Lemmer, U.; von Plessen, G.; Feldmann, J.; Becker, U.; Mennig, M.; Schmitt, M.; Schmidt, H. *Phys. Rev. Lett.*, to appear 1997.
21. Persson, B. N. J. *Surf. Sci.* **1993**, *281*, 153.
22. Persson, B. N. J. *Phys. Rev. B* **1991**, *44*, 3277; Persson, B. N. J.; Dubois, L. H. *Phys. Rev. B* **1989**, *39*, 8220.
23. Feibelman, P. *Progr. Surf. Sci.* **1982**, *12*, 287.
24. Hagena, O. *Z. Physik D* **1991**, *20*, 425; Hagena, O.; Knop, G.; Linker, G. In *Physics and Chemistry of Finite Systems*; Jena, P.; Khanna, S.; Rao, B., Eds.; Kluwer: Dordrecht, 1992.
25. Hövel, H.; Hilger, A.; Nusch, I.; Kreibig, U. *Z. Physik D* **1997**, *42*, 628.
26. Hövel, H. Dissertation, RWTH Aachen, 1995.
27. Hilger, A. Diploma, RWTH Aachen, 1994.
28. Hilger, A.; Cüppers, N.; Maaß, J.; Fröba, C. to be published 1997.
29. Schmid, G.; Pfeil, R.; Böse, R.; Bandermann, F.; Meyer, S.; Calis, G.; van der Velden, J. *Chem. Ber.* **1981**, *114*, 3634; Schmid, G. In *Structure and Bonding*; Springer: Berlin, 1985, Vol. 62, p. 51.
30. Bust, M.; Walker, M.; Bethell, D.; Shiffrin, D.; Whyman, R. *J. Chem. Soc., Chem. Commun.* **1994**, *801*; Whetten, R.; Khoury, J.; Alvarez, M.; Murthy, S.; Vezmar, I.; Wang, Z.; Stephens, P.; Cleveland, C.; Luedke, W.; Landman, U. *Adv. Materials* **1996**, *8*, 428.
31. Hamberg, I. Dissertation, Chalmers Tekniska Högskola, Göteborg, 1984.
32. Maaß, J.; Hilger, A. to be published, 1998.
33. Gmelin, *Handbuch der anorganischen Chemie*; Silber, Syst.; VCH: Weinheim, 1972, No. 61.
34. Fröba, C.; Hilger, A. to be published, 1998.
35. Gartz, M.; Hilger, A.; Nusch, I.; Hövel, H.; Kreibig, U. *Verhandl. DPG (VI)* **1995**, *30*.
36. Wertheim, G. K.; Buchanan, D. N. E. *Phys. Rev. B* **1994**, *50*, 11070; Hunt, M.; Modesti, S.; Rudolf, R.; Palmer, R. *Phys. Rev. B* **1995**, *51*, 10039.

37. Rehren, C.; Muhler, M.; Bao, X.; Schlögel, R.; Ertl, G. *Z. Phys. Chem.* **1991**, *174*, 11; Otto, A.; Mrozek, I.; Grabkern, H.; Akemann, W. *J. Phys. Cond. Mat.* **1992**, *4*, 1143.
38. Quinten, M.; Rostalski, J. *Part. Part. Syst. Charact.* **1996**, *13*, 89.
39. Hollstein, T.; Kreibig, U.; Leis, H. *Physica Stat. Solid B* **1977**, *82*, 545.

INDEX

Printed and bound by CPI Group (UK) Ltd, Croydon, CR0 4YY
08/05/2025
01864830-0005